Process Chemistry of Coal Utilization

Woodhead Publishing Series in Energy

Process Chemistry of Coal Utilization

Impacts of Coal Quality and Operating Conditions

Stephen Niksa, PhD

WP
WOODHEAD PUBLISHING
ELSEVIER
An imprint of Elsevier

Woodhead Publishing is an imprint of Elsevier
The Officers' Mess Business Centre, Royston Road, Duxford, CB22 4QH, United Kingdom
50 Hampshire Street, 5th Floor, Cambridge, MA 02139, United States
The Boulevard, Langford Lane, Kidlington, OX5 1GB, United Kingdom

Library of Congress Cataloging-in-Publication Data
A catalog record for this book is available from the Library of Congress

British Library Cataloguing-in-Publication Data
A catalogue record for this book is available from the British Library

ISBN: 978-0-12-818713-5 (print)
ISBN: 978-0-12-818714-2 (online)

For information on all Woodhead publications visit our website at https://www.elsevier.com/books-and-journals

Publisher: Brian Romer
Acquisition Editor: Maria Convey
Editorial Project Manager: Aleksandra
Production Project Manager: Debasish Ghosh
Cover Designer: Greg Harris

Typeset by SPi Global, India

Transferred to Digital Printing in 2019

Contents

Preface

Coal is used worldwide to produce electricity, process heat, synthetic feedstocks and chemicals, metallurgical cokes for steelmaking, and specialty chemicals. The scale of these industries and the scope of their utilization technologies are among the broadest that humanity has ever devised. Whether you size it in tonnage or kilowatt-hours or monetary value, the coal enterprise is gigantic. Even while the enterprise shrinks in the developed world to contain its environmental impact, it is also expanding at a phenomenal pace across Asia, and coming to life in Africa. For as far into the future as any of us can foresee, coal utilization will keep its place among the essential technologies that can advance societies into the modern world.

During this author's career, how we manage coal utilization technologies and perform R&D has been radically transformed. Coal characterization had been rooted in the mining industry, and pursued by analogy to rocks with optical instruments and a battery of simple, standardized tests in retorts and muffle furnaces. Two OPEC oil shocks during the 1970s prompted a complete reorganization of the coal R&D community in the United States. The commercial focus immediately shifted toward synthetic fuels and feedstocks while the government, for the first time, mobilized basic research scientists to characterize coal with the most advanced diagnostics of the day, and to unravel its conversion mechanisms at the molecular level. Indeed, most of our current understanding of coal constitution was revealed during this period.

As coals' molecular features were exposed, a hope rose that the engineering support for coal technologies would soon be brought to parity with recent advances in petrochemical engineering. The search for reaction mechanisms was on! I vividly remember the largest hotel conference halls in Pittsburgh and Washington, District of Columbia packed to standing room capacity for project review meetings sponsored by the US Department of Energy (DoE). Audiences reacted to the latest test data and modeling approaches with vigorous debates, because the scientific findings were being fast-tracked into technology development programs. Managers at America's largest petrochemical, aerospace, and defense contractors perceived huge payoffs down the road, while they struggled up the steep learning curves for coal handling and processing. But by the middle of the 1980s, petroleum markets had relaxed to an agreeable equilibrium and government funding in the American coal community took a nosedive.

Other coal research communities took their turn on the leading edge. Australians made seminal contributions to char oxidation kinetics throughout the 1970s and 1980s, then introduced and rapidly developed computational fluid dynamics (CFD) simulations for coal-fired furnaces. The physicochemical architecture in these first CFD simulations remains in place today, while CFD grew into a cornerstone of engineering design throughout the world. The German community led from the middle of the 1980s through the beginning of the 1990s with support from the mining industry, making major contributions to pyrolysis and gasification mechanisms, and to testing at elevated pressures. Japanese academic researchers made major contributions in coal liquefaction, and also compiled the databases on pyrolysis, oxidation, and gasification kinetics that clearly revealed the coal quality impacts for the first time. But by the dawn of the Clinton Administration in the United States, the American community had dwindled and largely withdrawn into DoE's National Energy Technology Laboratories.

English, German, Portuguese, and American engineers rapidly expanded applications of CFD furnace simulations through the 1990s by focusing them toward NO_X emissions control with coordinated testing at laboratory and pilot scale. Through the same period, researchers in the United States, United Kingdom, and Australia largely unraveled the scientific basis for ash fouling and slagging problems in utility furnaces, and Northern Europeans described how CFBCs operate. Italians and Scandinavians brought fluidized bed processing into the mainstream. From the mid-1990s through the mid-2000s, Japan re-focused its academic research on support for CFD, to extend these capabilities to gasification and other high-pressure technologies. The Australian community shared this focus on gasification technology development via CFD over most of the same time period, and also pressed forward in the rational basis to manage boiler fouling and slagging problems.

Since then, the South Korean community worked on technology development and CFD, while Australians pioneered the characterization and utilization of brown coals, which are now poised to enter the world coal trade. The Chinese community has become the most active, by far, emphasizing basic research capabilities and incremental advances to established methods for just about every utilization technology covered in earlier decades, including liquefaction, hydropyrolysis, and other long-neglected synfuels schemes. This community is very well positioned to make larger contributions. Japan and, more recently, Germany are committing resources toward coal flame characterization based on large eddy simulations, which may ultimately eclipse conventional CFD. But compared with their former productivity, American, Japanese, English, and most other European research communities have largely receded.

This juncture is an appropriate time to ask, "What was accomplished?" The widespread implementation of modern analytical techniques, especially ^{13}C NMR, FTIR, TGA/MS, GC/MS, and a multitude of new chromatography packages, gave a much clearer picture of coal structure and constitution, albeit in qualitative terms rather than numerical values. Massive volumes of performance data on pyrolysis, combustion, gasification, and solvation revealed the underlying coal quality impacts in the reaction kinetics. As seen throughout this book, the bulk of these observations have already been synthesized into accurate predictive capabilities for devolatilization, and into comprehensive mechanisms for combustion and gasification, even in complex syngas mixtures across a broad pressure range.

The parallel advances in coal utilization technology have been nothing short of spectacular. Several of the synfuels production schemes put forward in the last century are now operating in China at commercial scale. Emissions of NO_X and SO_X from the power sector have fallen dramatically with the near-universal adoption of low-NO_X

burners, deep air staging, SNCR, reburning, SCR, and wet FGD. Both CFBC and AFBC have thoroughly penetrated the power sector. At the same time, efficiencies for coal fired power plants grew with the adoption of ultra-critical steam cycles and entrained-flow gasifiers. And now Hg emissions are being controlled in utility gas cleaning systems with activated carbon injection, new catalyst formulations for SCR, and additives for wet FGD.

Unfortunately, the people who find causality between the renaissance in coal science and the parallel advances in coal utilization technology are misguided. As one who spent the past 20 years at the interface between these two enterprises, I know too well the major barriers to exchange and legitimate collaboration among these tribes. The vast majority of researchers and technology developers have very little contact with each other throughout their careers, no matter where they operate. Their communications are almost always confined to messengers—those select few scientists and consulting engineers who are made aware of the thorniest technological problems and given opportunities to participate directly in their resolution. Even in Japan, which is often touted for large-scale mobilization of human resources in technology development, coal utilization technology advances with relatively little input from academic fuel scientists.

Of course, there are many sensible reasons for this situation. The operating conditions in advanced utilization technologies are usually too dangerous to re-create in academic settings, and the minimum scale of testing that attracts commercial interest is too expensive to sustain. Solid fuel processing is usually too complex to characterize with spectroscopic diagnostics, due to the prominent roles for two-phase mixing and particle dynamics. Often, questions that are important in research settings are actually inconsequential in practical applications, simply because very small test facilities often operate in different hydrodynamic regimes than large vessels. On the other hand, oversimplification of the commercial reaction systems diverts attention away from important practical issues. There are philosophical barriers as well The scientific community must renounce the premise that practical applications demand comprehensive understanding at the molecular scale. Clearly they do not, because the technology continues to develop without thorough scientific underpinnings, as it has for decades.

The deepest and most productive connections between coal researchers and technology developers emerge in quantitative interpretations for design, validation, performance, and troubleshooting of full-scale utilization technologies. Accordingly, this book and the book to follow are written to relate the major advances in coal science to quantitative interpretations of performance data from lab-, pilot-, and commercial scale. They consider a very broad range of quantitative methods, from statistical regressions to rudimentary models to CFD to comprehensive reaction mechanisms, because "optimal" solutions to real-world problems are chosen to satisfy constraints on schedule and budget; usually, deep technical sophistication is something to avoid, if possible. The best solution could be a statistical correlation for carbonaceous deposits within commercial coal burners based on regression variables that are much more powerful than standard coal properties. Or rudimentary model analogs for full-scale coal flames that can connect baseline NO_X levels at one furnace to the emissions from different fuels at the same furnace. Or CFD simulations that accurately describe how switching among different fuels will affect the combustion efficiencies and emissions from a specific coal-fired boiler. Or simulations that combine truly predictive reaction mechanisms for fuel devolatilization and gasification with comprehensive elementary reaction mechanisms for the gaseous species to accurately predict minor and trace components of syngas from different fuels.

The specific focus is on applications where the fuel quality impacts are first-order important. In many of these situations, the aim is to identify bad actors; i.e., the relatively few coals that are problematic among the bulk of fuels from a particular production region. Problems caused by bad actors include burner deposits, the potential for autoignition, excessive unburned carbon emissions in flyash (loss-on-ignition or LOI), and SCR catalyst deactivation by chemical poisons. Fuel switching is another common application that involves closely managing the fuel quality impacts. It can exploit price differentials when a company switches from coal to cheaper opportunity fuels without suffering penalties of furnace de-rating or elevated emissions. It can also be a cost-effective route to compliance with either emissions regulations or new policy imperatives, such as Renewable Portfolio Standards. In general, companies want to know in advance which fuels will likely minimize operational problems, and which will enhance the performance of each specific technology in their generation systems.

Since chemistry largely determines the fuel quality impacts, the process chemistry will be emphasized throughout. Of course, many applications also involve a coupling among transport phenomena and the chemical reaction mechanisms, including bulk mixing patterns, and many situations like this will also be analyzed. So fuel quality impacts determine the scope of the applications of interest, and chemistry is the primary underlying science. The immediate implication is that an analysis should be as close as possible to a molecular description. However, solid fuels like coal cannot be adequately characterized at the molecular level, and the resolution of detail in the process chemistry must always be weighed against the barriers to and costs of laboratory analytical support. CFD practitioners do not want to conduct any laboratory tests before they simulate the performance of a new solid fuel. Note the obvious tension: A thorough chemist would not be satisfied with anything short of a molecular explanation, and revels in monumental analytical support, whereas simulation specialists do not want any connections at all between their computations and a physical laboratory.

The emphasis on the chemistry of fuel quality impacts excludes many important areas of practical consequence, including fuel handling and grinding; momentum-driven phenomena, such as wall impingement of coal flames; and heat transfer considerations. In addition, issues associated with mineral fouling and slagging, such as gas intercooler plugging in IGCC systems, will not be considered. These are certainly among the most important fuel quality impacts, but they are governed by completely different physicochemical principles than the ones used here. Aside from pyrite transformations and superficial references to catalysis of the primary coal conversion channels by metallic elements, analyses where mineral matter transformations play prominent roles will be omitted.

This is the first of a pair of books. It is devoted to reaction mechanisms for coal decomposition and volatiles conversion mostly at the scale of individual fuel particles. Suspension loading effects and the associated impact of mixing and particle dynamics are considered only after all the necessary reaction mechanisms have been developed and validated. Two preliminary chapters survey the most important coal utilization technologies, to define the operating domains for the reaction mechanisms, and give the essential aspects of coal constitution along with the necessary thermophysical properties and transport coefficients. The section on coal processing begins with a chapter on coal drying that focuses on the special difficulties in drying coals of the lowest ranks. Then three chapters on primary devolatilization sequence through the laboratory database, reaction mechanisms from global rate laws through the network depolymerization mechanisms, and validations across the database for FLASHCHAIN®, the author's mechanism. The treatment of volatiles conversion sequences from tar decomposition through volatiles reforming and volatiles combustion, including pollutant formation. A final chapter then extends all these mechanisms for coal hydropyrolysis and hydrogasification at elevated H_2 pressures.

One stage of coal processing is conspicuously absent: The heterogeneous conversion of char and soot by O_2, steam, CO_2, and H_2 is incorporated into the treatment of volatiles conversion at heavy coal loadings, but the char conversion mechanisms are not explicitly developed in this book. The reason is that only one of the many mechanisms developed for char conversion is actually implemented with the reaction mechanisms in this book. And this mechanism—the Carbon Burnout Kinetics (CBK) model—has already been explained and validated in detail in journal publications by Niksa et al. (2003) for char oxidation, and by Liu and Niksa (2004) for char gasification. These articles can be viewed as supplemental chapters to this book, especially since their material is organized in the same way as the chapters here.

Throughout this book the performance of the reaction mechanisms is validated with an enormous database of laboratory measurements. Data from lab-scale experiments is widely regarded as the best foundation for reaction mechanisms and chemical kinetics, because it enables the most accurate recording of the test conditions, and accommodates detailed monitoring of intermediates and products. Conversely, even pilot-scale systems always have gradients in temperature and concentration that are large enough to obscure the reaction kinetics. So our only practical option is to formulate mechanisms and kinetics from the laboratory database, then implement these findings in simulations of the larger systems of commercial interest with a watchful eye on operating conditions beyond the scope of the laboratory database.

The second book will present a host of applications in commercial systems to illustrate how the reaction kinetics for individual particles can be used to interpret fuel quality impacts on conversion efficiencies, emissions, and various operational problems. That discussion progresses from statistical regressions to simple modeling analogs to CFD to comprehensive reaction mechanisms, and emphasizes applications at pilot- and commercial-scale.

It is a pleasure to acknowledge the very thorough review of text by Dr. Brian C. Young, formerly of Envirosafe and CSIRO in Australia. During our first collaboration some 30 years ago, Dr. Young re-defined for me the meaning of a publication in "final form." His extensive edits and scrupulous cross-checking of the text in this book are very much appreciated, and will benefit every reader.

Finally, this book is dedicated to the multitude of program managers, research directors, and government funding agents who sponsored my work over the years. Research sponsorship, like any financial obligation, entails mutual trust among the parties. I am especially grateful for the encouragement and confidence of my following major sponsors:

Dr. George R. Offen (dec.) and Mr. Jeff Stallings, Electric Power Research Institute
Mr. Takeo Yamada, Mr. Naoki Fujiwara, and Dr. Shinji Kambara, Idemitsu Kosan Co., Ltd.
Mr. Ichiro Kajigaya and Dr. Shigehiro Miyamae, Ishikawajima-Harima Heavy Industries (IHI)
Mr. Corey Tyree and Mr. Mark Berry, Southern Company Services, Inc.
Dr. Michiaki Harada, Japan Coal Energy Center (JCOAL)
Mr. Albert Tsang, ConocoPhillips Company
Mr. James D. Hickerson and Mr. Chris Guenther, National Energy Technology Laboratory, US DoE (NETL)
Drs. Ripudaman Malhotra and Donald Eckstrom, SRI International
Dr. Edmundo Vasquez, Alliant Energy Corp.

References

Liu G-S, Niksa S. Coal conversion submodels for design applications at elevated pressures. Part II. Char gasification. Prog Energy Combust Sci 2004;30(6):697–717.

Niksa S, Liu G-S, Hurt RH. Coal conversion submodels for design applications at elevated pressures. Part I. Devolatilization and char oxidation. Prog Energy Combust Sci 2003;29:425–77.

Acronyms

AC/Cl aromatic carbons per unit cluster in a coal macromolecule

AFBC atmospheric fluidized bed combustor

AFR Advanced Fuel Research Inc.

APCS Argonne Premium Coal Samples

ASTM American Society for the Testing of Materials

BTX benzene, toluene, and xylene

C2SM competing two-step chemical reaction model

CBK/E extended version of the Carbon Burnout Kinetics model for char oxidation

CBK/G version of the Carbon Burnout Kinetics model for char gasification by steam, CO_2, and H_2

CCOFA close-coupled overfire air injection

CCSEM computer-controlled scanning electron microscopy

CFBC circulating fluidized bed combustor

CFD computational fluid dynamics

CPD chemical percolation devolatilization mechanism

CPP Curie-Point pyrolyzer

CV calorific value of coal, kJ/kg

DAEM distributed activation energy reaction model

daf dry-ash-free basis for coal composition

dmmf dry-mineral matter-free basis for coal composition

DoE US Department of Energy

EFR entrained flow reactor

EMC equilibrium moisture content of coal, wt.%

FC fixed carbon content of coal

FDA the Flash Distillation Analogy for volatiles escape during coal devolatilization

FFR free-fall reactor

FG functional group model for coal devolatilization

FGD flue gas desulfurization

FG-DVC functional group/devolatilization-vaporization-crosslinking mechanism for coal devolatilization

FTIR Fourier transform infrared spectrometry

GC/MS gas chromatography with mass spectrometry

GEPS General Electric Power Systems

GHC light (C_1-C_4) gaseous hydrocarbons

GPC gel permeation chromatography

HCL hydrocarbon liquids

HPLC high performance liquid chromatography

HRR hot rod reactor

hv high volatile bituminous coal

IEA International Energy Agency

IGCC integrated process for gasification with combined cycle power generation
IRP isothermal reaction period following heatup in a WMR, s
LOI flyash loss on ignition, wt.%
lv low volatile bituminous coal
MALDI matrix-assisted laser desorption mass spectroscopy
M_n number-average molecular weight
MS mass spectrometry
mv medium volatile bituminous coal
MWD molecular weight distribution
NEA Niksa Energy Associates LLC
NMR nuclear magnetic resonance spectroscopy
NO_X oxides of nitrogen
OFA overfire air injection
OPEC organization of petroleum exporting countries
p.f. pulverized fuel size grade with 70 wt.% through 200 mesh
PAH polynuclear aromatic hydrocarbons
PBR packed-bed reactor
PCC pulverized coal combustion
PCD particle collection device
PCX phenol, cresol, and xylenol
PSD particle size distribution
PVM proximate volatile matter content of coal
R&D research and development
RCFR radiant coal flow reactor
RH relative humidity
RTD residence time distribution
SCR selective catalytic NO_X reduction
SFBD steam fluidized bed drying of coal
SFOR single, first-order chemical reaction
SNCR selective noncatalytic NO_X reduction
SOFA separated overfire air injection
SO_X oxides of sulfur
SR stoichiometric ratio evaluated as the ratio of air to fuel flowrates normalized by the ratio for stoichiometric combustion
TFR tubular flow reactor
TGA/MS thermogravimetric analysis with mass spectrometry
UBC unburned carbon in flyash
WMR electrically heated wire mesh reactor
XANES X-ray absorption near-edge structure absorption spectroscopy
XPS X-ray photoemission spectroscopy

Coal utilization technologies

1

About eight billion long tons of coal are produced worldwide each year. According to the International Energy Agency (IEA), just over two-thirds of that coal is used to generate electricity; 11% is associated with steelmaking; 8% with industrial power; 4% with cement; 3% with heating; and the balance with numerous small end-uses. The technical information in this book pertains to all these end-uses except steelmaking. The reason for this exclusion is that nearly all the coal in steelmaking is actually used as coke, and coke is produced in ovens that are too large and too cool to impose the severe thermal processing conditions that are represented in the validation database for the coal conversion mechanisms. Even so, about 90% of world coal consumption is subjected to severe thermal processing to generate electricity, steam, and other forms of heat, and to produce commodity materials for infrastructure.

The technologies that produce electricity from coal impose the broadest domain of operating conditions so, collectively, they determine the conversion conditions that reaction mechanisms for coal conversion must describe. These technologies can be classified by their coal injection scheme, or by the operating pressure range for the furnace or the steam turbine cycle, or by the hydrodynamics within a furnace, or whether the coal is consumed under oxidizing or reducing atmospheres. Table 1.1 introduces the labels for established technologies and for the most important advanced technologies. Thorough descriptions of all these technologies are available in reference books such as "Steam: Its Generation and Use" (Tomei, 2015), or in the extensive catalog of monographs published by IEA Coal Research (www.iea-coal.org). Here the discussion is focused on the terminology and the different operating domains for the most fully commercialized technologies.

The three established technologies are pulverized coal combustion (PCC) with sub-critical steam cycles; wet-bottom, cyclone-fired furnaces; and stoker furnaces. The distinguishing features among these three systems are the size range of the coal and the means of contacting fuel with air streams. PCC furnaces process the finest size grade, by far, and contact the coal with several air streams injected at various furnace elevations through convective, two-phase mixing. Cyclone furnaces inject much coarser coal grinds into flames swirled within barrels by tangential air injectors. The cyclone barrels are hot enough to hold molten slag layers that capture the largest coal particles. Comparable portions of the coal burn in dense suspensions in air, and as inclusions within the molten slag layer. Stoker furnaces spread even larger lumps of coal onto fixed or moving grates within an upward crossflow of air. This configuration has ratings between 10 and 25 MW, and the relatively very small Stoker population is shrinking fast because it is uneconomical to comply with environmental regulations with these furnaces. So they are not considered further.

In Table 1.1, pressures are nominal values and the coal sizes cover a range to the top size. But the maximum gas temperature is not as literal. In PCC furnaces, gas temperatures vary by several hundred degrees from the near-burner region to the furnace exit,

Process Chemistry of Coal Utilization. https://doi.org/10.1016/B978-0-12-818713-5.00001-0

Table 1.1 Technologies for electricity generation from coal.

Technology	Air contacting	Size	T_{MAX} (°C)	*P* (MPa)
Established				
PCC	Convective mixing	40–300 μm	1700	0.1
Cyclone furnace	Convective mixing + slag layer	40 μm–5 mm	2000	0.1
Stoker furnace	Crossflow through a moving bed	1–25 mm	1400	0.1
Advanced				
Supercritical PCC	Convective mixing	40–300 μm	1700	0.1
Oxy-fired PCC	Convective mixing	40–300 μm	1700	0.1
AFBC	Exchange from bubbles into emulsion	Several mm	900	0.1
CFBC	Exchange from bubbles into emulsion	Several mm	950	0.1
PFBC	Exchange from bubbles into emulsion	Several mm	950	1.0
EF gasifier	Convective mixing	40 μm–3 mm	2400	4.0
Transport gasifier	Exchange from bubbles into emulsion	Several mm	1000	2.0
Fixed-bed gasifier	Counterflow through a fixed bed	1–5 cm	1000	3.0

where temperatures are almost always 1000–1050°C. But the profile within a particular furnace is determined by burner design, the partitioning of air among burners and ports in the upper elevations, and coal quality, among other factors. So the hotter maximum in cyclone furnaces in Table 1.1 is a relative indication that cyclone flames are somewhat hotter than the flames in PCC furnaces.

The label "Advanced Technologies" is potentially misleading because supercritical PCCs, CFBCs, and entrained-flow (EF) gasifiers have already been widely implemented in many parts of the world. Among recent applications for new power plants in the United States, these three technologies were selected at more-than-double the rate of subcritical PCC, although the number of coal-fired furnaces of any type that will ever be built in the United States is uncertain. It is safe to say that supercritical PCC, CFBC, and EF and fixed-bed gasifiers will collectively comprise a major portion of the future furnace population. AFBCs have lost out to CFBCs, and PFBCs have not fared well in the international marketplace after a successful demonstration at commercial scale. Oxy-fired PCC is just now being demonstrated at commercial scale in North America, but the transport gasifier demonstration has not fared well.

The distinction between sub- and supercritical PCC technologies only pertains to the steam cycle so, for our purposes, these furnaces are indistinguishable. Oxy-fired PCC recycles flue gas into the coal burners to make CO_2 the largest flue gas component, which facilitates its recovery and sequestration downstream. This scheme raises the possibility that both O_2 and CO_2 participate in coal conversion, and thereby expands the operating domain for reaction mechanisms. The three fluidized combustors use grinds measured in millimeters rather than microns. These combustors operate close to the exit temperatures from PCC furnaces, and none of them sustain coal flames in the conventional sense. AFBCs and PFBCs have no large flame structure at all, because the coal is a very small percentage of the bed fluidization medium, which is often sand and/or coarse ash. The only flames are small ones within some of the bubbles that percolate through the fluidized bed. CFBCs also inject coal into a stationary fluidized bed, although it is relatively small and does not exchange nearly enough air with the bubbles to burn out even the coal volatiles. Most of the combustion occurs above the dense bottom bed, while the fuel species released by the coal are gradually mixed into air pockets from bubbles that rupture at the bed surface. Small flames form at the convoluted mixing interfaces in the flowfield, but there is no large flame structure comparable in size to the combustor riser.

The three gasifiers operate at comparable elevated pressures, but span a very broad temperature range. EF gasifiers are the hottest coal conversion units of all, particularly when they use dense suspensions in O_2 of somewhat finer grinds than used in PCC furnaces. But when EF gasifiers are fed by coal slurries, the grinds must be much coarser to manage the slurry viscosity within tolerable limits. The top size in Table 1.1 is for slurries. The transport gasifier is configured like a CFBC, in that the coal is fed into a dense bottom bed and then moves upward through a splash zone and into a riser. But this system is O_2-deficient, by design, to produce pressurized syngas. The maximum temperatures in fixed-bed gasifiers are comparable to those in transport gasifiers, and can be fired by air in coflow or counterflow with respect to the end of the bed that receives the coal feed.

1.1 Relevant domain of operating conditions

Our goal is to specify the domain of operating conditions that must be described by coal conversion mechanisms, so that the mechanisms can be incorporated into simulations of the utilization technologies. What conditions need to be specified? Reaction rates are evaluated as functions of time at specified temperature, pressure, and the partial pressures of all gaseous reactants. All these conditions must be assigned, along with one or more indices on conversion in the condensed phase, such as weight loss, changes in aromaticity, extent of burnout, etc.

Sample temperature seems like an obvious specification but, in fact, is often the most difficult prerequisite for a kinetic analysis of coal conversion. One reason is that the fuel temperature is hardly ever isothermal, even in lab tests; to the contrary, coal is usually converted while the fuel temperature changes by several hundred to well over a thousand degrees in tests at any scale. Consequently, the temperature requirement is, in practice, a requirement for the thermal history; i.e., fuel temperature as an explicit function of time. Another complication is that the temperature of the fuel particle is often different than the temperatures of surrounding gases and the source of thermal radiation. So even in many lab tests, the particle temperature is calculated from an energy balance among the various heat fluxes onto a fuel suspension. Similarly, the partial pressures of reacting gases are often affected by various mass transfer resistances, so values at and within a reacting fuel particle are calculated, rather than monitored. In these situations, the particle size as a function of time must also be incorporated into the kinetic analysis.

Extremes in the thermal histories of individual coal particles are sketched in Fig. 1.1. The history for PCC conditions is based on the profiles of O_2 concentration and temperatures of the gas and wall through the core of an unstaged wall-fired furnace. It covers the complete burnout history of a 63 μm coal particle, which is shorter than 1.5 s. The thermal history for the AFBC represents a uniform bed temperature of 875°C, and covers only the partial burnout of a 4 mm particle in 25 s. These two thermal histories roughly bracket the thermal histories for all the utilization technologies in Table 1.1.

In the PCC furnace, the particle temperature is initially driven by convective and radiant heat fluxes, first, to the onset temperature for devolatilization and, then, to the ignition temperature for char oxidation. The particle temperature of 1350°C at ignition is reached in 22 ms, so the heating rate for devolatilization is roughly 6×10^4°C/s. The heat released during the ignition stage drives the temperature to its maximum of 1750°C within about 400 ms. Quasi-steady combustion sets in soon after ignition, and maintains the temperature above 1700°C, which is hotter than the local gas temperature and much hotter than the waterwalls. But at some point while the particle is consumed, heat release no longer balances heat losses, and the combustion process is extinguished. The temperature then relaxes to a value that is slightly cooler than the gas temperature, although this stage is not shown in Fig. 1.1.

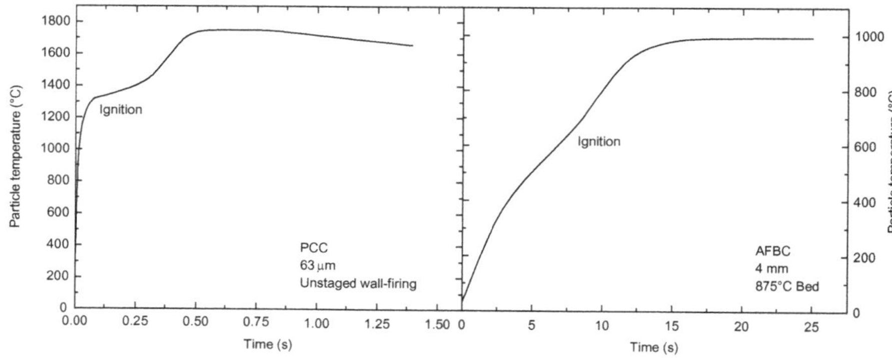

Fig. 1.1 Extremes in the thermal histories for utilization technologies from (left) an unstaged wall-fired PCC furnace and (right) an AFBC.

The thermal history for a much larger particle injected into an AFBC moves through three of the same stages, without extinction. But the heating rate is slower than 100°C/s, and the maximum temperature is only 1000°C. The heatup time is extended beyond 10 s by the much weaker heat fluxes, and the burnout time is longer than a minute, because the burning rate for the lower temperature is much slower.

Even though the thermal histories in Fig. 1.1 have several distinctive features, it is useful to coarsely resolve three nominal characteristics to facilitate comparisons: The heating rate is the ratio of the temperature change at the ignition point to the time to reach the ignition temperature; the reaction temperature is taken as the maximum value; and the reaction time is the elapsed time to a quench cycle or complete burnout, whichever comes first. If there is no quench cycle, then the reaction time would be evaluated as the total transit time of fuel from injection to the system exit. Ranges for these three characteristics, and for pressure, reactant partial pressures, and size determine the operating domain for the reaction mechanisms.

1.2 PCC furnaces

The layout of a typical PCC furnace is shown in Fig. 1.2. Streams of coarse coal are ground into the PCC size grade in pulverizers, then conveyed in primary air streams into a manifold of burners near the base of the furnace walls. Collectively, the

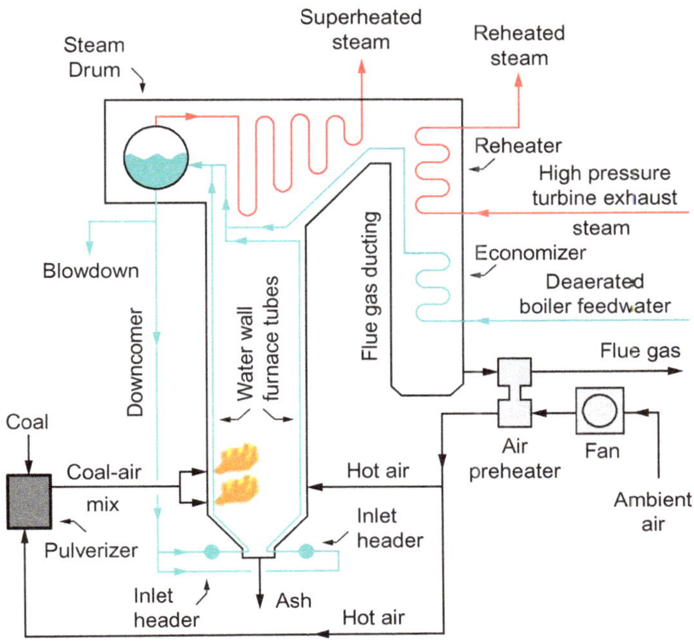

Fig. 1.2 Layout for a typical PCC furnace.

manifold of burners is called the burner belt, and the furnace section that contains it is called the near-burner region. The coal stream ignites upon injection, which stabilizes a large flame across and above the near-burner region. Larger, denser particles of mostly mineral matter fall downward and are recovered as bottom ash in a hopper at the bottom of the furnace. With the most troublesome coals, large mineral deposits dislodged from steam tubes in the upper furnace may also fall into bottom ash hoppers, as may molten mineral slags from a burner belt. Nearly all the coal stream moves upward while the flame mixes with so-called secondary air injected either through annular openings in the burners called registers, or through separate air injection ports. Additional air is usually injected well above the burners through ports called overfire air (OFA) injectors. Beyond the radiant flame zone, the suspension continues upward until it is turned and accelerated through the converging superheater section. Multiple heat exchangers within the superheater section are collectively called the convective passes, based on the primary heat transfer mechanism into the steam tubes in this section. The stream then passes through the furnace exit and turns downward through the reheat section and economizer.

The furnace exit is the interface between the furnace and the gas cleaning system, at the superheater exit. Temperatures of 1000–1050°C into the furnace exit are too cool for appreciable coal conversion. So the gas cleaning system is not relevant to the conversion mechanisms, although it obviously pertains to emissions control and the system thermal efficiency. The notable exceptions are the collection devices that recover flyash containing unburned combustible components. The stream passing through the furnace exit is a suspension of hollow mineral spheres, larger porous mineral particles, and unconverted combustible solids in flue gas. The combustible remnants are called unburned carbon (UBC), and the suspended minerals plus UBC are called flyash. Some flyash is recovered downstream of the economizer, where the stream turns into either a selective catalytic NO_X reduction (SCR) unit or air preheater. But most of the flyash is collected further downstream in a dedicated particle collection device (PCD) such as an electrostatic precipitator (ESP) or fabric (or baghouse) filter (FF). The weight percentage of combustibles in the flyash recovered in the PCD is defined as flyash loss-on-ignition (LOI). LOI is the most important index for the total coal conversion in a furnace. There are unconverted fuel species in the flue gas as well, particularly CO and light hydrocarbons, but these species usually make very small contributions to the total carbon inventory. Similarly, the UBC in bottom ash is another small contribution. Consequently, flyash LOI represents the largest single source of unconverted coal and is widely used as a gauge on the overall combustion efficiency.

Within the generic layout in Fig. 1.2, there are several variations. Firing configuration denotes the type and layout of burners or fuel injectors on the furnace walls, and there are four popular configurations: front wall-fired, opposed wall-fired, tangential-fired, and cyclone-fired. In wall-fired furnaces, either one or two burner belts contain an array of burners staggered in several rows across most of a wall. Furnaces with one belt are front wall-fired, and those with two on opposite walls are opposed wall-fired. In both variations, the streams from all the burners coalesce into one giant flame structure that bends upward (hopefully) before it reaches any of the other walls in the near-burner region. Tangential-fired (or T- or corner-fired) furnaces have fuel injectors near each corner, and several stacks of injector sets at multiple elevations collectively called registers. Multiple air ports alternate with the fuel injectors at different elevations to introduce auxiliary or close-coupled OFA (CCOFA). Additional OFA ports above the fuel registers are called separated OFA (SOFA). The fuel injectors are adjustable in the horizontal and vertical planes, to position a swirling fireball as desired within the radiant section. In T-fired furnaces, the coal streams from different injectors penetrate through much of the swirling fireball, and accumulate in the core of the flowfield while moving upward. But the bulk of gases coalesce into helical streams that remain fairly close to the walls. As noted previously, with cyclone-firing, much coarser coal grinds are injected into swirling flames within barrels. The barrels are mounted within the furnace walls, and some of the coal leaves the barrels within molten slag that falls into the bottom ash collector and the rest leaves in dilute suspension with the primary air streams.

Firing configuration is important for several reasons. In wall- and T-fired furnaces (so-called dry bottom furnaces), about 15%–30% of the coal minerals are recovered as bottom ash, whereas in cyclone furnaces, bottom ash is 60%–70%. This difference makes a substantial difference in the magnitude of LOI for a given UBC level, because the same absolute flow rate of UBC out of the furnace will be weighted by very different flow rates of mineral flyash. Suppose the coal contained 10% mineral matter, and the UBC is 0.5% of the combustible matter. In a wall- or T-fired furnace with 30% bottom ash, these levels would give flyash LOI of 6 wt.% (from the ratio of combustibles in flyash to the sum of combustibles and flyash, which is 90(0.005)/(0.7(10) + 0.45); in a cyclone furnace with 60% bottom ash, flyash LOI would be 10.1 wt.% (from 90(0.005)/(0.4(10) + 0.45). For the same UBC level—which means the same combustion efficiency—LOI levels are greater for progressively greater bottom ash recoveries.

Two other important operating variations pertain to the amount and distribution of the air streams injected into the furnace. Excess air is the amount over and above the air flow needed to completely convert the coal feed into ultimate combustion products. In the United States, the excess air is typically 15%, which gives 3%–4% residual O_2 in the flue gas at the economizer. Elsewhere, excess air levels can be high enough to give 6% economizer O_2. Of course, the stoichiometric air requirement depends on the coal composition. It is usually expressed in terms of the stoichiometric ratio (SR), which is the ratio of the actual air-to-fuel ratio normalized by the value of the ratio for stoichiometric combustion, according to the following expression:

$$\mathrm{SR} = \frac{\dfrac{m_{\mathrm{Oxidizer}}}{m_{\mathrm{Coal}}}}{\left(\dfrac{m_{\mathrm{Oxidizer}}}{m_{\mathrm{Coal}}}\right)_{\mathrm{stoichiometric}}} \tag{1.1}$$

where m_i are mass flowrates of an oxidizer and coal. Excess air is defined as SR minus unity, as a percentage. For typical high volatile (*hv*) bituminous coals, an excess air level of 15% requires a total air flow about ten times the coal feedrate. This multiple diminishes to seven or so for high volatility coals (because of their substantial oxygen contents), and increases slightly for low volatility coals (which contain minimal oxygen).

The other important variation associated with air is the extent of staging, which denotes the partitioning of total air among the various ports on different furnace elevations. The reference condition for staging is an idealized state of complete mixing among all the air streams and the coal feed. This state could be realized if all the air was used as primary air to convey the coal into the burners or injectors. Then the SR-value for the primary air would equal the SR-value for the entire furnace. This situation never arises in practice, so SR values based on portions of the total air flow are used to express the deviation from a premixed state. The SR of the primary air stream, SR_{PR}, incorporates only the primary air flow; that for the near-burner region, SR_{NB}, incorporates primary plus secondary air through burners or primary plus close-coupled OFA in T-firing; and the furnace SR incorporates all air streams into the calculation. This detailed resolution of the various air streams is crucial for aerodynamic NO_X abatement strategies. Values of SR_{NB} are rarely less than 0.80 in the United States, to avoid burner belt corrosion, but are often as low as 0.70 in Japan, where mostly low-sulfur coals are burned. A more cursory index for staging simply assigns the staging level as the percentage of the air injected through OFA ports above burner belts or as SOFA in T-firing. For example, if 30% of total air was injected through OFA ports in a cyclone furnace, then the furnace would be 30% staged.

Both firing configuration and the staging level affect thermal histories through the furnace, and also mean O_2 profiles. The three histories for the T-fired furnace in Fig. 1.3 were based on a CFD simulation of a 550 MW furnace fired with subbituminous coal. Particle tracking data with both massless and inertial particles were used to compile mean temperatures of gases and the radiant surroundings for three primary flowpaths. One path moved most of the reacting fuel particles into a cylindrical furnace core that rose upward along the centerline. Due to their substantial inertia, the fuel particles were able to penetrate the swirling helical flow along the furnace walls, and to entrain small portions of the primary and secondary air streams. The core expanded outward due to the addition of fuel suspension from different fuel injection elevations, and then it entrained SOFA along the upper furnace elevations. As seen in Fig. 1.3, the path along the furnace core features relatively slow gas heating at about 3000°C/s to a maximum temperature of 1600°C, and an abrupt consumption of the available O_2 within the first 250 ms. Since the particle streams entrain only small portions of the injected air, the furnace core remains depleted of O_2 throughout the lower furnace elevation. Oxygen from the SOFA jets eventually mixes into the core flow along the convective passes.

Whereas the furnace core flow moves radially inward and upward along the centerline, the second primary flow path swirls along a helical flow along the furnace walls. These "gas ribbons" contain few fuel particles, but have most of the volatile matter released from the coal during the initial stages of particle heating. Ribbons originate in primary air from the fuel injectors, then rapidly mix with CCOFA and SOFA. As seen in Fig. 1.3, the maximum temperatures are the same as those in the furnace core, but heating rates are slightly faster and the O_2 levels rise much faster

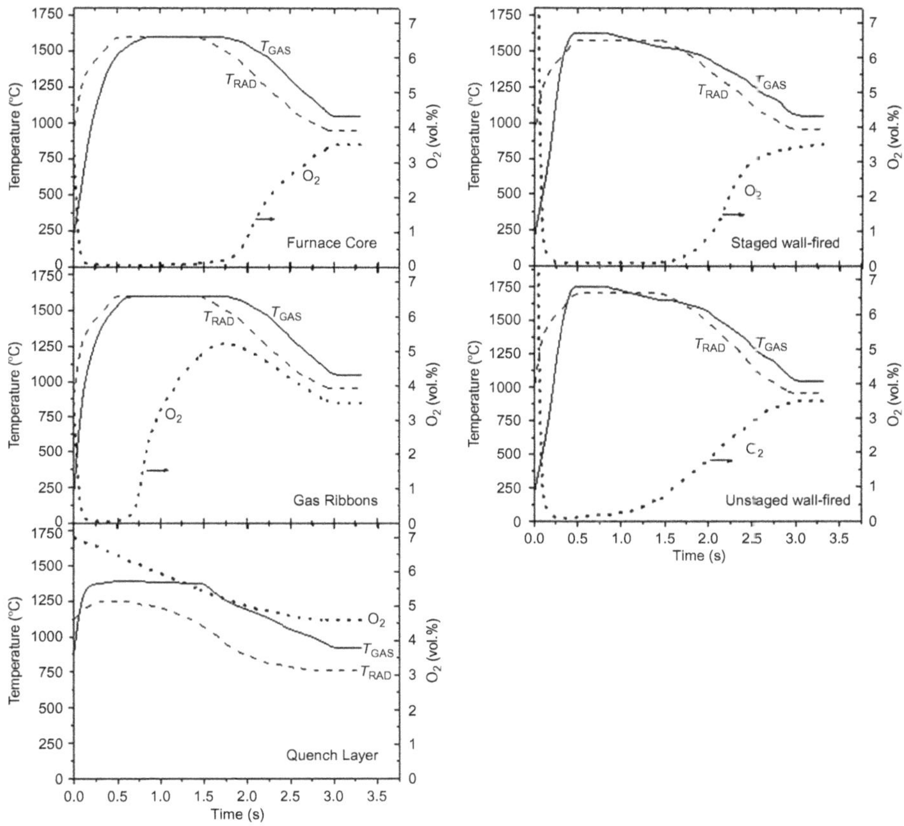

Fig. 1.3 Thermal histories of *(solid curves)* gas, *(dashed)* radiant surroundings, and *(dotted)* O_2 mole fraction for (left panel) paths through a T-fired furnace through the (top) furnace core; (middle) gas ribbons; and (bottom) quench layer; and (right panel) through (top) staged and (bottom) unstaged wall-fired furnaces.

and reach much greater values. The third path remains close to the waterwalls as it moves through the furnace elevations. It is the coolest path, by far, with regard to both gas and radiation temperatures, which prompts its labeling as a "quench layer." But it also sustains much greater O_2 concentrations as well.

The two additional cases in Fig. 1.3 are for staged and unstaged wall-fired furnaces along the centerlines of flows from individual burners. Both cases have faster heating to hotter maximum temperatures than the T-fired cases. And both have O_2 histories that plummet very close to the burners and remain low for at least a second. The O_2 level in the unstaged case never reaches the minimum of the staged case, and gradually rises after about 500 ms, even while the gas temperature is as hot as 1750°C. In the staged case, the minimum O_2 level persists through 1.5 s, and the O_2 level rises faster due to the direct injection of OFA. Whereas the maximum gas temperatures for the staged wall-fired furnace and the core flow through the T-fired furnace are comparable, that for the unstaged furnace is hotter by 150–200°C. Cyclone furnaces usually have thermal histories slightly hotter than unstaged wall-fired furnaces.

During the heatup stage, the thermal histories for particles and gas are very similar. It is a widespread misconception in this field that particle thermal histories in coal flames can be based on the injection of a single particle into a nominally infinite medium at specified gas and radiant temperatures. Such an analysis gives particle heating rates approaching 10^{6}°C/s for PCC grinds. But the coal loading in primary air streams is roughly 0.4 kg-coal/kg-air, and the momentum dominated jets emanating from burners and fuel injectors do not heat nearly as fast as isolated particles. Turbulent mixing of secondary air and radiant and convective heat transfer determine the actual fuel heating rates which, from CFD simulations, are of the order of 10^{4}°C/s. In other words, fuel particles do not heat much faster than their entrainment gas, because the temperatures of both phases are closely coupled into the large sensible enthalpy of the coal feedstream. After ignition, calculated thermal histories must account for the heat release due to char oxidation, and will typically exceed the local gas temperatures by several hundred degrees.

The last specification for PCC furnaces is the particle size. The particle size distribution (PSD) for PCC furnaces is specified as 70 wt.% through a 200 std. mesh sieve [74 μm opening], and under 0.5 wt.% through 50 mesh [297 μm]. The large-size specification, called the top size, is especially important because nearly all LOI comes from the large end of the fuel PSD. For this reason, the conversion mechanisms should target the upper half of the PSD, from a mean size of 50 μm through a top size on 50 mesh [297 μm]. The smaller half will be completely consumed and not contribute to LOI.

Hence, for conventional PCC applications, coal heating rates range from 5×10^{3} to 5×10^{4}°C/s; maximum gas temperatures range from 1500°C to 1800°C; and total transit times are 3–4 s. The pressure is roughly 0.1 MPa, and O_2 partial pressures vary from near-zero to 21 kPa although, immediately after ignition, O_2 levels rarely exceed 5 kPa except in quench layers along the walls. As noted above, coal sizes of interest range from 50 to 300 μm, although only the coarser sizes affect the combustion efficiency.

Oxy-fired PCC retains the basic layout and may be implemented with any of the conventional firing configurations. The only differences are in the composition of the primary and/or secondary air streams. Their compositions are altered toward greater maximum O_2 levels of 27%, and CO_2 concentrations up to 70%. The much larger CO_2 level admits the possibility that CO_2 gasification may come into play across the furnace elevations that have been depleted of O_2.

1.3 Fluidized bed technologies

CFBCs, AFBCs, and PFBCs have well-defined operating conditions, particularly since the gas and radiant source temperatures are very nearly isothermal and all these systems are analyzed in the steady-state. Yet, several of the specifications on reaction mechanisms are more ambiguous than for PCC systems. In PCC furnaces, both the fuel and the oxidizer emanate from fixed ports, which makes it logical to track the combustion process along a time coordinate, albeit from a macroscopic rather than a microscopic perspective. Such Lagrangian tracking accesses the same information as needed to evaluate reaction mechanisms, so coal conversion histories can be evaluated continuously from the Lagrangian trajectories for local operating conditions. Two characteristics of fluidized beds completely undermine this approach. First, beds comprise at least two distinct phases called bubbles and an emulsion and, second, the emulsions are always regarded as completely well-mixed.

Nearly all particles in a fluidized bed, including the coal, remain in the emulsion, except that wakes and clouds attached to bubbles contain appreciable particle loadings in the hydrodynamic regime for AFBCs. All the primary air is dispersed into bubbles at the bed bottom, which expand and coalesce while they percolate upward through the emulsion. The only means to disrupt the inherent segregation of fuel and oxidizer in the bed is called exchange, which is an ambiguous transport process that spontaneously exchanges fluid parcels between the phases. The exchange of gaseous fuels into bubbles can be seen as conversion along a distance coordinate into the bed which, given estimates for bubble gas velocities, directly maps into a time coordinate. But since both particles and gas in the emulsion are regarded as well-mixed, O_2 exchanged from bubbles cannot be associated with any position in the bed, or any increment of the coal feed, or any portion of the residence time distribution (RTD) of bed solids. Consequently, O_2 partial pressures can only be specified from calculations as a single nominal value for the entire emulsion. These calculated values always reflect the large uncertainties in the engineering correlations for the coefficients for the exchange processes, as well as several ambiguities in the solids mixing patterns. Whereas the maximum O_2 level in bubbles is the inlet O_2 level in the air stream, reported O_2 levels in the emulsion vary widely depending on the complexity of the analysis used to estimate them.

Fortunately, this ambiguity does not arise in the analysis of CFBCs, where bubbles travel in the so-called "fast bubble" regime. As their name implies, these bubbles travel so quickly through the bed that only very minor portions of their O_2 is exchanged into the emulsion; in fact, even with anthracites, which release the smallest amounts of volatiles of any coal type, insufficient O_2 is exchanged into emulsions to burn out even the volatile fuel components. So volatiles conversion can be analyzed as a diffusion-limited process (that contains no reaction mechanisms), and char oxidation can be omitted entirely. Moreover, bubble gas travels too fast to reach typical ignition temperatures for hydrocarbon mixtures before they rupture through the bed. So there is minimal fuel conversion of any kind in bubbles.

In some ways the fuel particle size in fluidized systems is even more difficult to estimate than reactant partial pressures. Certainly, the PSD of solid fuel particles within a fluidized bed will bear little resemblance to the PSD of the coal feedstream. The coal PSD may be transformed by swelling and spontaneous fragmentation during the devolatilization stage; impact fragmentation; mechanical attrition; elutriation; and shrinkage during char conversion. Population balances can be formulated to estimate bed PSDs, but these estimates are subject to large uncertainties on the coefficients for spontaneous and impact fragmentation, as well as variations in the attrition rates for chars from different coal types that have not yet been addressed. Whereas the PSDs of char in the bed will not be larger than the coal PSD, reported bed PSDs vary widely depending on the analysis used to estimate them, and on the estimated values of uncertain transport coefficients.

Another complication pertaining to size is counterintuitive, in that smaller char particles from the coal feed make the greatest contributions to flyash LOI. The largest char particles are confined by hydrodynamics to remain in the bed indefinitely. Other than char oxidation and attrition, the only way to eliminate them is through the bottom ash drain. But this extraction is regarded as a representative sampling of the entire bed PSD, so the only contributions that the largest end of the bed PSD makes to incomplete coal conversion are a relatively very modest contribution to UBC in bottom ash, and none whatsoever to flyash LOI. This is opposite to the predominant contributions to flyash LOI by the larger char particles in PCC furnaces.

In fluidized systems, the transit time to the furnace exit and the size of char that meets the threshold for ejection out of the bed are the key parameters, because they determine how much of that char particle will be converted before it exits the system. What remains unconverted contributes to flyash LOI. The elutriation rate is proportional to the bulk density of char, which varies throughout char oxidation at typical bed temperatures and also varies for chars from different coal types; it is also coarsely correlated with the superficial gas velocity in the bed. The transit time is governed by gas velocity above the bed as well as the terminal particle velocity which, in turn, depends on particle size and density and gas viscosity and density. So transit times for ejected particles can be several seconds even when the gas exits in only a few seconds, and the size threshold for ejected particles is inherently ambiguous. The prudent response is to focus on the smaller half of the coal PSD.

As a tangible illustration, consider the flow path through a CFBC in Fig. 1.4. The dense bottom bed occupies only the first meter or so of elevation, at the bottom of the system where fuel, primary air, and circulating ash are introduced. Bed solids are continuously ejected into a chaotic region above the bed called a splash zone. Larger and heavier particles fall back into the bed, while smaller particles are entrained upward along the riser, mix into secondary air, and turn into the cyclone at the top. The author's analysis of the particle dynamics in a commercial-scale CFBC estimated that, for a coal PSD with a mean size of 2.5 mm, the largest char particle that was elutriated into the riser was 650 μm. Of course, this value will vary for different CFBC conditions and coal types. But the case demonstrates that the smaller half of the coal PSD is an appropriate range for the reaction mechanisms. The analysis also showed that the conversion of the smaller char particles and char fines along the riser affected flyash LOI by consuming O_2 that would otherwise be available to accelerate the oxidation of larger entrained char. In the conversion of char suspensions, factors that retard the burning rates of the largest sizes will lead to greater flyash LOI.

To this point, we have seen that fluidized combustors operate with O_2 partial pressures and PSDs for char in the beds that are subject to large uncertainties, although the very long solids RTDs mitigates the uncertain PSDs for larger char particles. The extended RTDs for solids also clarify the conditions needed to analyze coal devolatilization in fluidized beds. Provided that the residence times for coal in the

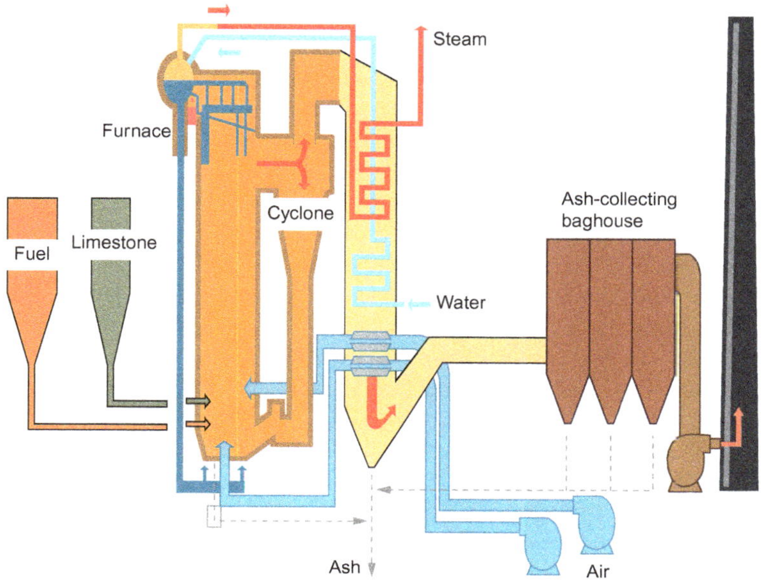

Fig. 1.4 Layout for a typical CFBC.

bed are much longer than the characteristic times for coal devolatilization, it is reasonable to assume that every increment in the coal PSD releases its maximum amount of volatile matter for the heating rate, bed temperature, and pressure under consideration. Whereas the bed temperature and pressure are fixed, the fuel heating rates depend on the particle size increment. An energy balance that accounts for heat transfer via convection with particle-to-particle contact and radiant transfer is the most expedient means to describe this situation. Calculated heating rates for typical mean sizes are roughly 25–75°C/s. Considering that the entire coal PSD is subject to devolatilization, the target range of heating rates for the reaction mechanisms is roughly 1–100°C/s.

For applications in AFBCs, coal heating rates range from 1 to 100°C/s; gas temperatures range from 750°C to 1000°C; and reaction times should extend to at least 30 s. The pressure is 0.1 MPa, and O_2 partial pressures vary from near-zero to several percent in bubbling beds. The same conditions pertain to PFBCs, except that pressures are elevated to 1 MPa. In CFBC risers, O_2 levels rarely exceed 5 kPa except in secondary air layers along the walls. Coal sizes of interest range from fines to about a millimeter.

1.4 Gasification technologies

The three gasification technologies of interest are EF, transport, and fixed-bed gasifiers. EF gasifiers inject coal through the top of a refractory lined pressure vessel, or radially into the midsection from several opposing injectors. The coal is entrained in O_2 or air, and steam is introduced through dedicated injectors with or without slurry water in the coal feedstream. Flames attach to the fuel injectors and rapidly consume the O_2, mostly to burn out the volatile fuel components and finest char particles. Once the local atmosphere turns reducing, char is converted into H_2 and CO, the primary syngas components. Minerals are collected as a molten slag layer that flows along the walls through the bottom of the vessel.

EF gasifier manufacturers do not report detailed operating conditions. However, independent studies have reported gasifier simulations for numerous test programs (Bockelie et al., 2003; Zheng and Furimsky, 2005). Here reported fuel properties, feedstream compositions, exit gas temperatures, and syngas compositions are used to specify operating conditions for the EF gasifiers from Shell and General Electric Power Systems (GEPS). The GEPS gasifier operates at 4.1 MPa, and the Shell gasifier operates from 2.0 to 2.7 MPa. A typical coal PSD for the Shell gasifier has a mean size of about 40 μm, finer than PCC size grades (Zheng and Furimsky, 2005), but the PSD for coal slurries fed to a GEPS gasifier would have a larger mean closer to 200 μm, to manage the slurry viscosity.

The most variable operating conditions for both gasifiers are the O_2/coal and steam/coal ratios, because these ratios are deliberately adjusted to produce essentially uniform syngas compositions from a broad range of fuel quality in the same gasifier design. Reported values for both gasifiers are collected in Table 1.2, along with the C-contents of the coal feeds on a dry, ash-free (daf) basis, and atomic H/C and O/C ratios of the whole feedstreams. These later ratios are evaluated as

$$\frac{H}{C}=\left(\frac{12}{1}\right)\frac{\%H_{dry}+\frac{2}{18}\frac{steam}{coal}}{\%C_{dry}}$$

$$\frac{O}{C}=\left(\frac{12}{16}\right)\frac{\%O_{dry}+\frac{O_2}{coal}+\frac{16}{18}\frac{steam}{coal}}{\%C_{dry}}$$

where the C-, H-, and O-contents are on a dry weight basis; the coal feedrate is for dry coal; and the steam/coal ratio also includes inherent coal moisture. All the hydrogen and oxygen in steam are assumed to participate in syngas reforming, due to the very high operating temperatures.

In Table 1.2, the coals under each gasifier type are arranged in descending order of decreasing volatility. Shell gasifiers are operated with dry coal feeds, and little or no steam is injected downstream. Consequently, the O_2/coal ratio is adjusted to regulate the H/C and O/C ratios for syngas reforming. This ratio is generally increased for coals of progressively lower volatility to maintain H/C ratios from 0.80 to 1.00 and O/C ratios from 0.90 to 1.15. Only the H/C ratio for the very low rank lignite sample (Coal S1) falls out of these ranges. GEPS gasifiers are fed with coal slurry that has 65 wt% coal, so steam/coal ratios fall between 0.50 and 0.70. The O_2/coal ratios are generally increased for coals of progressively lower volatility although, as for Shell gasifiers, the variation in this parameter for coals with C-contents over 77%

Table 1.2 Feed characteristics for EF gasifiers.

Coal	C, daf wt.%	O_2/coal	H_2O/coal	H/C	O/C
Shell					
Coal S1	61.9	0.40	0.01	1.30	1.01
Coal S2	73.2	0.73	0.02	0.81	1.12
Coal S3	78.2	0.84	-	0.83	0.94
Coal S4	79.0	0.90	0.09	1.00	1.14
Coal S5	82.6	0.93	-	0.86	1.09
Coal S6	82.6	0.88	0.07	0.99	1.10
GEPS					
Coal G1	61.9	0.64	0.54	2.75	2.10
Coal G2	73.2	0.84	0.54	1.94	1.82
Coal G3	77.3	0.92	0.51	1.88	1.58
Coal G4	78.8	0.87	0.65	2.03	1.65
Coal G5	79.0	0.97	0.54	1.86	1.64
Coal G6	79.1	0.89	0.68	2.08	1.68
Coal G7	79.4	0.87	0.62	1.98	1.62
Coal G8	78.2	0.90	0.58	1.89	1.54
Coal G9	82.5	0.89	0.59	1.90	1.61
Coal G10	83.9	0.98	0.69	2.00	1.59

Data from Zheng L, Furimsky E. Energy Convers Manag 2005; 46(11/12):1767–79, with permission from Elsevier.

is minimal. For all but the high volatility coals, the H/C varies from 1.85 to 2.10 and O/C only varies from 1.54 to 1.65.

The thermal histories in EF gasifiers are largely unknown, because such a hostile environment provides no access to any form of diagnostics. The thermal histories for the mean sizes in Fig. 1.5 are coarse estimates that relax to calculated exit gas temperatures for various cases in Shell and GEPS gasifiers (Bockelie et al., 2003). Nominal heating rates are 6000°C/s, which is slower than the estimates for PCC furnaces, because the sensible enthalpies of feedstreams at elevated pressure are much greater than at atmospheric pressure, all else the same. The respective maximum temperatures are 2400°C and 2100°C, which are significantly hotter than flame temperatures in PCC furnaces. The same histories can be applied to both gases and the radiant environment. Since the overall elemental compositions into the gasifiers are similar with all coals, the same thermal history can be used for all fuel samples, except that the exit gas temperatures should be adjusted. Nominal residence times are 2.0 and 2.8 s for the Shell and GEPS gasifiers, respectively, and are probably similar for all coals. One strategy to maintain uniform residence times is to adjust the times at the maximum temperature to compensate for the variable times to reach the different exit gas temperatures at an assumed uniform gas quench rate.

The gasification agents in these systems are O_2, H_2O, and CO_2 and the inhibitors are H_2 and CO. Their partial pressures cover broad ranges along the primary flow path because all O_2 is consumed soon after injection, and the feedstream contains no CO_2,

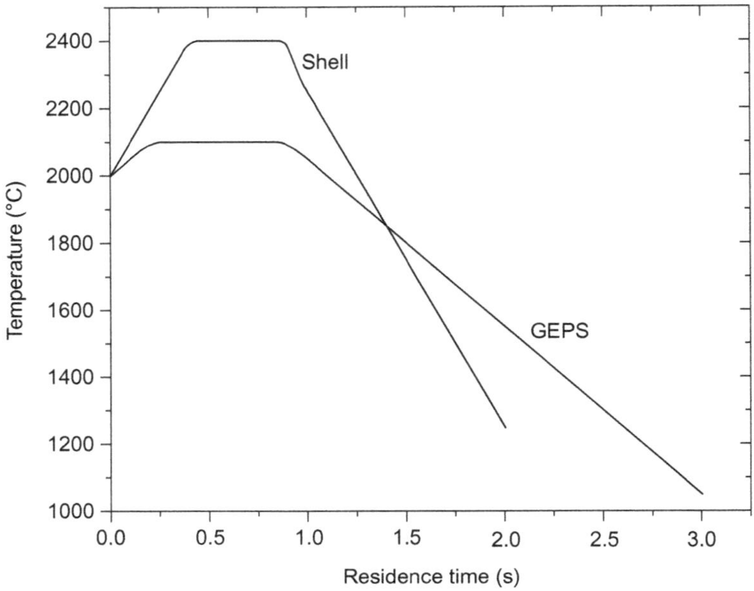

Fig. 1.5 Estimated thermal histories for *(solid curve)* Shell and *(dashed curve)* GEPS gasifiers.

CO, or H_2. The compositions entering the gasifiers, based on average values of O_2/coal and steam/coal in Table 1.2, are about 90% O_2 and 10% steam for Shell gasifiers and 45% O_2 and 55% steam for GEPS gasifiers. For these gasifiers, outlet syngas compositions can be accurately estimated as equilibrium compositions along the quench cycle. These compositions are, for Shell gasifiers, 60%–65% CO; 24%–27% H_2; 2%–4% CO_2; and 2%–5% H_2O; and, for GEPS gasifiers, they are 33%–40% CO; 23%–28% H_2; 10%–15% CO_2; and 21%–24% H_2O. The compositions do not necessarily sum to 100% because N_2 and other minor species are also present. After the different pressures for these two EF gasifiers are taken into account, the ranges of partial pressures for the gasification agents in EF gasification are the following: 1–2.2 MPa O_2; 0–1.6 MPa CO; 0–1 MPa H_2; 0–0.6 MPa CO_2; and 0.1–1 MPa H_2O. The impact of O_2 variations is independent of the other gasification agents, because combustion chemistry is much faster than gasification chemistry whenever O_2 levels exceed about 500 ppm. Even so, the ranges for the other gasification agents span an enormous operating domain.

Hence, for applications in EF gasifiers, coal heating rates range from 10^3°C/s to 10^4°C/s; gas temperatures range from 2000°C to 2400°C; and reaction times extend to a few seconds. Pressures range from 2 to as high as 8 MPa, but are usually half that or lower. Oxygen partial pressures vary from 1 to 2 MPa. Partial pressures for the gasification agents in EF gasification vary from 0 to 1.6 MPa CO; 0 to 1 MPa H_2; 0 to 0.6 MPa CO_2; and 0.1 to 1 MPa H_2O. Such variations arise spontaneously across all EF gasifiers while the O_2 and steam in the feed are consumed during char conversion; CO and H_2 accumulate as products in the syngas; and the syngas composition changes to equilibrate the water gas shift reaction (WGSR), $CO + H_2O \leftrightarrow CO_2 + H_2$. Coal sizes of interest range from PCC grinds for dry-feed systems to coarser grinds with mean sizes of 200 μm in slurry fed systems.

The transport gasifier is an extension of old fluidized catalytic cracking technology used to produce gasoline during WWII, although no catalyst is used in the coal gasification application. As seen in the layout in Fig. 1.6, the layout closely resembles a CFBC. Steam in O_2 or air is distributed into a dense bottom bed that receives the circulating solids return from the external circulation loop. Coal is fed into or above the dense bed, perhaps with a SO_2 sorbent, and with additional steam and O_2. Oxygen is consumed in a splash zone above the bed, and then the stream passes through a reducing zone in the riser where char is gasified. The stream then passes through two stages of particulate removal before it moves into a gas cleaning system. The same unit may be fired with O_2 or air, which provides some flexibility, although transport gasifiers work best with coals of the highest volatility such as lignites and subbituminous coals, because the conversion rates of such fuels are compatible with the residence times through typical riser heights.

The grind sizes, coal heating rates, and temperatures in transport gasifiers are very similar to those in CFBCs, except that fuel temperatures along the riser are much cooler once O_2 has been consumed. Grinds have mean sizes of a few millimeters. As for CFBC, the sizes of interest are those small enough to be ejected from the dense bed into the riser, which is the smaller half of the coal PSD. Larger sizes are hydrodynamically confined to the bottom bed and ultimately ground by mechanical attrition below the threshold for elutriation or withdrawn into the bottom ash drain. Coal heating rates range from 1°C/s to 100°C/s; temperatures range from 750°C to

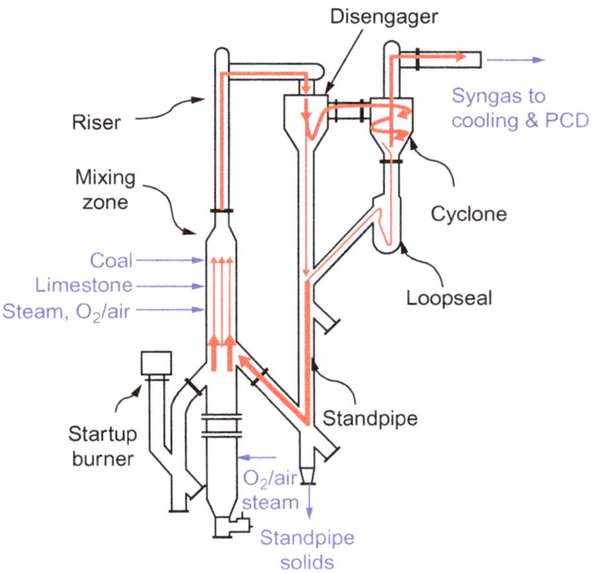

Fig. 1.6 Layout for a transport gasifier.

1000°C; and pressures are about 2 MPa. Transit times across the riser are a few seconds for gases, and as long as 10 s for entrained solids, depending on their size. Syngas compositions are widely variable, depending on the O_2/coal and steam/coal ratios, and contain some CH_4 because char conversion occurs below 1000°C. Oxygen partial pressures vary from 0 to 0.4 MPa. Partial pressures for the gasification agents vary from 0 to 0.5 MPa for CO, H_2, and CO_2; CH_4 levels remain below about 0.1 MPa; and H_2O levels vary from 0.1 to 1 MPa. Such variations arise spontaneously across transport gasifiers while the O_2 is consumed mostly in volatiles conversion. Steam in the feed is consumed and CO and H_2 accumulate in the syngas as products of char conversion; and chemistry in the gas phase reforms the syngas mixture throughout the transit time.

Fixed-bed gasifiers are the oldest gasifier designs. Coal is fed through the top of a pressure vessel and settles into a bed on a horizontal grate. The bed moves slowly under the influence of gravity as portions of the bed are consumed, sometimes with stirring to homogenize the bed voidage. Mixtures of steam in air or O_2 may be introduced from below in a countercurrent stream; or from above in a co-current stream; or into the middle where it can turn upward or downward, depending on the location of the gas outlet. For any gas flow configuration, the beds naturally partition into distinct regions for drying, devolatilization, combustion, and gasification. These gasifiers process lump coal as coarse as several centimeters in size, and impose coal residence times as long as a few hours, depending on the bed temperature. Gas residence times are also extended to tens of seconds. Many designs including the classic Lurgi gasifier have maximum bed temperatures of about 1000°C, but designs that collect ash as a molten slag run as hot as 2000°C. Coal heating rates are determined by thermal conduction through the bed supplemented in the hotter units by radiant transfer. They are very difficult to estimate because simulation results are always reported as spatial profiles through the bed, and the time coordinate for the bed is determined by the highly uncertain velocity associated with bed consumption. Pressures range from atmospheric to 6 MPa in the extreme cases, but typically range from 2.5 to 3 MPa. Syngas compositions are widely variable, depending on the O_2/coal and steam/coal ratios, and contain appreciable CH_4 if gasification zones operate below 1000°C, which is frequently the case.

For applications in fixed-bed gasifiers, coal heating rates range from 0.1°C/s to 1°C/s; bed temperatures range from 700°C to 2000°C; and reaction times extend to a few hours at the lower temperatures. Pressures range from 0.1 to 3 MPa. Oxygen partial pressures vary from 0 to 0.6 MPa. Partial pressures for the gasification agents vary from 0 to 1 MPa for CO, H_2, and CO_2; from 0 to 0.2 for CH_4; and from 0.1 to 1 MPa for H_2O. These compositions change in ways that differ among the different gas flow configurations. Coal sizes are as large as 5 cm.

The slow heating rates in fixed bed gasifiers permit chemistry among volatile products while they are escaping from the particle into the ambient gases. As explained in Chapter 4 (Section 4.1.2), this so-called secondary volatiles conversion occurs spontaneously whenever volatiles are generated so slowly that the time scale for escape to the free stream becomes comparable to the time scale for chemistry along the path for escape. Since the generation rates of volatiles diminish in proportion to reductions in the heating rate, extents of secondary conversion within particles become greater for progressively slower heating rates. Moreover, slow heating is often associated with very large characteristic dimensions of the fuel, and large dimensions are often associated with steep spatial gradients in temperature and, perhaps, pressure. These gradients will only become apparent in unsteady analyses in one or more spatial coordinates. Simply put, such multidimensional analyses are incompatible with the already large computational burdens for simulations of large-scale utilization systems. This situation will eventually be rectified by faster computers, because there is nothing missing from the suite of reaction mechanisms needed to simulate fixed bed gasifiers or, for that matter, coke ovens. But in the meantime, fixed bed gasifiers are simply too complicated to analyze with legitimate reaction mechanisms.

1.5 Operating domain for coal conversion mechanisms

This book describes the most robust reaction mechanisms that can be incorporated into simulations of the most important coal utilization technologies. We ignore coal liquefaction because comprehensive mechanisms that can depict the distinctive liquefaction behavior of individual coals remain to be formulated and validated. Stoker furnaces are omitted because this population is already small and shrinking fast because these furnaces have become uneconomical in the face of more stringent environmental regulations. Cyclone furnaces are omitted because char conversion within molten slag layers has only recently been subjected to detailed characterization work. PFBCs are omitted because they have not yet broken into the power generation market. Coke ovens and fixed bed gasifiers are beyond the validation domain because the coal heating rates are slower than about 1°C/s.

The remaining utilization technologies consume the vast majority of coal produced worldwide, and comprise PCC furnaces, CFBCs and AFBCs, and EF and transport gasifiers. Their operating domains are compiled in Table 1.3, which shows the heating rates, maximum temperatures, reaction times, pressures, grind sizes, and partial pressures of the main conversion agents. Heating rates are maximum values for the devolatilization stage. The slowest rate of interest is 1°C/s, because secondary volatiles conversion within particles occurs at slower rates. The temperatures are maximum gas temperatures; particle temperatures are often hotter. It is worth remembering that temperature profiles span several hundred degrees or more after the fuel has ignited in PCC and CFBC furnaces and EF gasifiers. The partial pressures of the gaseous reactants are maximum values. The minimum values are zero for all species except H_2O, which is always present at concentrations of at least a few percent. The gasification agents change in tandem as O_2 and steam are consumed, CO and H_2 accumulate, and reforming chemistry in the gas phase either introduces or modulates CO and CH_4 levels. Methane is appreciable only for temperatures in the vicinity of 1000°C or cooler. As broad as our operating domain of interest has become, we will see in Chapters 4–9 that direct measurements, mostly at lab scale, have already characterized most of these operating conditions. Gaps in the coverage usually arise when only a very limited range of coal quality was tested, instead of the broad range that is processed in most utilization technologies.

Table 1.3 Operating domain for coal conversion mechanisms.

Technology	Heating rate (°C/s)	T_{MAX} (°C)	Size	Partial pressures (MPa)					
				O_2	CO_2	H_2O	H_2	CO	CH_4
PCC	10^4–10^5	1700	40–300 μm	0.02	0.01	0.01	0	tr	0
Oxy-fired PCC	10^4–10^5	1700	40–300 μm	0.03	0.07	0.01	0	tr	0
AFBC	10^1–10^2	900	Several mm	0.01	0.01	0.01	0	tr	0
CFBC	10^2	950	Several mm	0.02	0.01	0.01	0	tr	0
EF gasifier	10^3–10^5	2400	40 μm–3 mm	2.0	0.6	1.0	1.0	1.6	0.06
Transport gasifier	10^2	1000	Several mm	0.4	0.5	1.0	0.5	0.5	0.1

References

Bockelie MJ, Denison MK, Chen Z, Senior CL, Sarofim AF. Proc. Pittsburgh coal conf., Pittsburgh, PA, 2003.

Tomei GL, editor. Steam its generation and use. Charlotte, NC: The Babcock and Wilcox; 2015 42nd ed.

Zheng L, Furimsky E. Comparison of Shell, Texaco, BGL, and KRW gasifiers as parts of IGCC plant computer simulations. Energy Convers Manag 2005;46(11/12):1767–79.

Fuel quality, thermophysical properties, and transport coefficients

2

Nomenclature

A	ash content in a proximate analysis, as-rec'd wt.%
Ar	Archimedes number defined in Eq. (2.12a)
C_D	drag coefficient
$C_{p,j}$	specific heat of one or more components in coal or char, cal/g K
CV	calorific value, MJ/kg
d_{bed}	size of inert fluidization solids in a fluidized bed, mm
d_{exp}	size larger than e^{-1} of a Rosin-Rammler PSD, μm or mm
$D_{i,g}$	binary mass diffusivity of species i in a gas mixture, cm^2/s
d_P	size of coal, μm or mm
E_i	Mass percentage of C, H, O, N, or S on basis i
F_{CUM}	mass percentage of a coal PSD larger than a specified size, wt.%
f_i	number fraction of element i in combustibles in coal or char
f_{PYR}	pyrite mass fraction in dry mineral matter
g	gravitational acceleration, cm^2/s
h	convective heat transfer coefficient at the external particle surface, W/m^2 K
H_{DEV}	deviation in the H-content associated with excessive inertinite levels, daf wt.%
k_m	overall mass transfer coefficient into the external surface, cm/s
m_P	particle mass, g
n_{bed}	exponent in Eq. (2.12a)
n_{RR}	spread parameter in a Rosin-Rammler PSD
Nu_P	Nusselt number for an individual particle
p	pressure, MPa
Pr	Prandtl number defined in Eq. (2.11b)
Re_{bed}	Reynolds number for an individual particle in a fluidized bed
Re_P	Reynolds number for an individual particle in dilute suspension
Sc	Schmidt number defined in Eq. (2.11c)
Sh_P	Sherwood number for an individual particle
T	temperature, K
t	time, s
U	superficial gas velocity into a fluidized bed, m/s
V_{He}	specific void volume based on He pyncnometry and Hg porosimetry, cm^3/g
v_{slip}	difference between local velocities of gas and a particle, m/s
X_{INERT}	mass fraction of inertinite in a coal

Process Chemistry of Coal Utilization. https://doi.org/10.1016/B978-0-12-818713-5.00002-2

Greek symbols

δ_p exponent in the pressure correction for the swelling factor in Eq. (2.9h)
ε actual bed voidage in a fluidized bed
ε_{mf} bed voidage under minimum fluidization conditions
λ_g thermal conductivity of the ambient gas, W/m K
μ_g viscosity of the ambient gas, cP
ρ_b bulk (or particle) density of coal, kg/m^3
Ω swelling factor
Ω_{CQ} component of swelling factor on coal quality
ν thermal diffusivity of the ambient gas, m^2/g

Subscripts

ASH ash component in coal
COMB combustibles in coal
MM minerals in coal
MOIS moisture in coal
0 initial condition
∞ ultimate, saturation condition for devolatilization

What is coal, anyway? The answer depends on the background of whom you ask. Petrographers see coal as an organic rock composed of different lithotypes resolved under reflected light. Organic chemists see it as a macromolecular network that contains just about all the common organic functional groups constructed from C, H, O, N, and S. Inorganic chemists see a few dozen familiar minerals plus scores of trace minerals and metals that span the periodic table. Engineers appreciate coals' high calorific value, but dread the damage that molten minerals can cause in furnaces. Environmentalists see a gigantic and pervasive source of primary air pollutants, hazardous trace species, and contaminants for water and soil. All these perspectives are valid but nevertheless far from definitive.

Fact is, our absolute understanding of coal constitution is woefully incomplete. And, unfortunately, the gaps in understanding are certainly not confined to the finer features of rare and inconsequential molecular structures. The stories told about the skeletal backbones of the macromolecular matrix of coal, and its diverse assortment of substituted functional groups, are compelling and corroborated by large batteries of analytical techniques. But as soon as one tries to differentiate even coarse molecular features in quantitative terms among different coal samples, he or she confronts uncertainties large enough to undermine the utility in practical applications. The only sensible response to these ambiguities is to tailor our description of coal constitution to the specific requirements of our simulation applications. From this standpoint, the appropriate question is not, "What is coal?" Rather, it is, "What features of coal structure and constitution must be explicitly resolved to simulate the distinctive behavior of individual coal samples in a particular utilization technology?"

This book addresses this question in two parts. This chapter covers the major features of coal constitution, and the surprisingly systematic ways that these features shift among coals of different geological age. Next comes the standard analyses, measurement terminology, and properties that satisfy the input requirements for simulations of utilization technology. In Chapter 5, this knowledge is used to formulate a mathematical description of any coal as a reactant in high temperature processing.

2.1 Coal rank

Residual petroleum fractions from different parts of the globe are unrelated, in that there are no unifying tendencies that connect their primary characteristics. Samples from a particular region may share many common features, but these same features for samples from other regions do not conform to any universal trends. Similarly, different forms of biomass display random variations even in their elemental compositions, again, without any underlying unification. But coal is different. Regardless of geographical origin, a particular coal sample can be easily placed on a scale called the coal rank spectrum. To a coal scientist, a sample's rank immediately calls up an array of important qualities: Calorific value, volatility, ease of ignition, agglomeration tendencies, burnout characteristics, and suitability for gasification applications are just some of the features that rank can convey. Considering the obvious utility of these features, any coal utilization specialist must gain a sound and intuitive understanding of coal rank.

First and foremost, rank denotes a sample's geological age. Age is significant because it directly connects with the process of coal maturation, which is the final stage of the coalification process. Coalification denotes the biological and physicochemical processes that convert plant matter and fossil remnants, first, into peat and, ultimately, into the different types of coal. We need not consider the early stages of coalification when most portions of the feedstocks are expelled from deposits by biological agents, and the residues are converted into resins and pitches. Rank pertains to the further maturation of these residues deep within the earth's crust under elevated pressures and temperatures. All nascent coal deposits began with fairly comparable proportions of carbon and oxygen, and sufficient hydrogen to maintain macromolecules with more aliphatic structures than condensed aromatics. Physically, nascent deposits were true colloids stabilized by numerous interactions among their most polar functional groups with the local groundwater. Maturation then preferentially eliminated oxygen and hydrogen, which initially disrupted the affinity to groundwater and eventually shifted the molecular components away from aliphatics toward planar condensed polyaromatic structures. As the planar aromatics grew, portions of adjacent macromolecules began to align into layers, which eventually expanded into nanocrystals within an amorphous glassy matrix. Further maturation expanded the crystalline portion at the expense of the glassy matrix. Once all the oxygen and most other heteroatoms had been eliminated, the planar aromatics formed lamella that stacked into crystallites in the macroscopic size range, as most of the glassy matrix was eliminated. The ultimate product of maturation is not the anisotropic, purely crystalline state of graphite, whose formation requires much more severe thermal processing than could be applied to a coal deposit. Rather, it is a disordered polycrystalline form of essentially pure carbon.

Maturation is the underlying, unifying principle for the coal rank spectrum because it generates only one ultimate product. This product is the same, regardless of the feedstocks in the nascent coal deposit, and whether the local water supply was fresh or saline, and whether the pressure was much higher than in most other coal deposits. Such factors affect the rate of maturation, and may impart distinctive features into particular coal deposits. But the distinctions are not nearly as prominent as the relentless progression toward that disordered, polycrystalline, carbonaceous solid. Rank denotes a sample's geological age because age roughly denotes the extent of maturation.

Of course, no one ever attempts to assign the geological ages of coal samples to determine their ranks. Among the many means used to denote rank, we begin with the one with the strongest connections to the course of coal maturation. Fig. 2.1 shows a coalification diagram, a.k.a. a van Krevelen diagram after the eminent coal scientist who revealed its significance. It plots the atomic ratios of H/C vs O/C for fuel samples excluding moisture and mineral matter. This rendition contains values for different forms of biomass and liquid byproducts of wood pulping called black liquor; major components of residual petroleum fractions called asphaltenes and petroleum cokes; and, of course, various coals. In these coordinates, forms of biomass and various black liquors are scattered throughout the quadrant for O/C > 0.5 and H/C > 1. Since the values do not coalesce into curves, there is no apparent correlation between the two

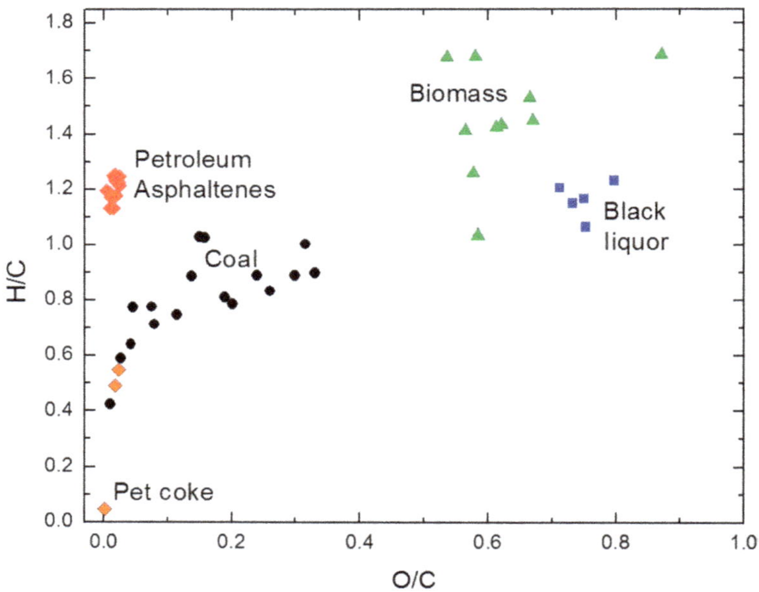

Fig. 2.1 Coalification diagram for diverse organic fuels in a condensed phase.

coordinates for these fuels or any underlying formation process. Asphaltenes occupy a very small region with hardly any oxygen and $1.1 < H/C < 1.3$. But coals conform to a band across the lower left quadrant for $0 < O/C < 0.4$ and $0.1 < H/C < 1.1$. Within this quadrant, the youngest coals appear in the near-horizontal band at the right boundary for $O/C = 0.4$. Maturation moves a sample to the left along the horizontal portion by eliminating oxygen without disrupting the hydrocarbon skeletal structure, since H/C stays the same during this stage. But further reductions in O/C below 0.2 are also accompanied by substantial reductions in the H/C ratios.

Fuel quality, thermophysical propertiesOnce O/C is reduced below about 0.04, further maturation dramatically reduces the H/C ratios of the oldest coals along the straight line from 0.5 to 0. Petroleum cokes fall along the final segment of the coalification diagram, suggesting that the later stages of coal maturation have much in common with coking and other forms of pyrolysis applied to highly aromatic organic feedstocks. Indeed, most ultimate solid residues from severe thermal processing closely resemble the disordered polycrystalline carbonaceous material at the end of the coal rank spectrum.

In Fig. 2.2, the coal rank spectrum is depicted in the plane of fixed carbon vs calorific value, which are the standard coordinates for rank assignments in the United States. Fixed carbon is one of the four elements of a proximate analysis. As described below, a proximate analysis involves heating a standard amount of coal under standardized conditions to determine the mass fractions of moisture, volatile material, carbonaceous residue, and ash. Fixed carbon is the carbonaceous residue which, in Fig. 2.2, has been adjusted to a dry, mineral-matter-free (dmmf) basis. The calorific values in the figure use a basis that also eliminates mineral matter but retains moisture. The rank spectrum progresses from samples at the far right with the lowest calorific values; upward toward the left through progressively greater levels of fixed carbon and calorific values (from samples with progressively less oxygen); and then upward toward the right through progressively greater levels of fixed carbon and lower calorific values (from samples with progressively less hydrogen).

Most important, this figure resolves the rank spectrum into the familiar labels for different coal types used throughout the coal utilization industries. The youngest coals are lignites, which are called brown coals in Australia and Eastern Europe. Low rank coals comprise lignites and both grades of subbituminous. All grades of bituminous coals are distinguished from subbituminous and the three forms of anthracites because the former melt under standardized processing conditions (like those in coke ovens), whereas the lower and higher rank coals do not. High volatile B (*hv*B) bituminous is lower in rank that *hv*A, and *hv*C is lower than *hv*B, based on calorific values. All ranks higher than medium volatile (*mv*) bituminous are collectively called low volatility coals because their fixed carbon levels are relatively very high. Medium volatile (*mv*) and low volatile (*lv*) bituminous retain the distinctive melting behavior that none of the anthracites have.

Fig. 2.2 also shows that every coal rank spans substantial ranges of fixed carbon and calorific value. This is our first clear indication that samples of the same rank usually have appreciably different properties. In turn, these differences are often responsible for appreciably different behavior during thermal processing for all but low

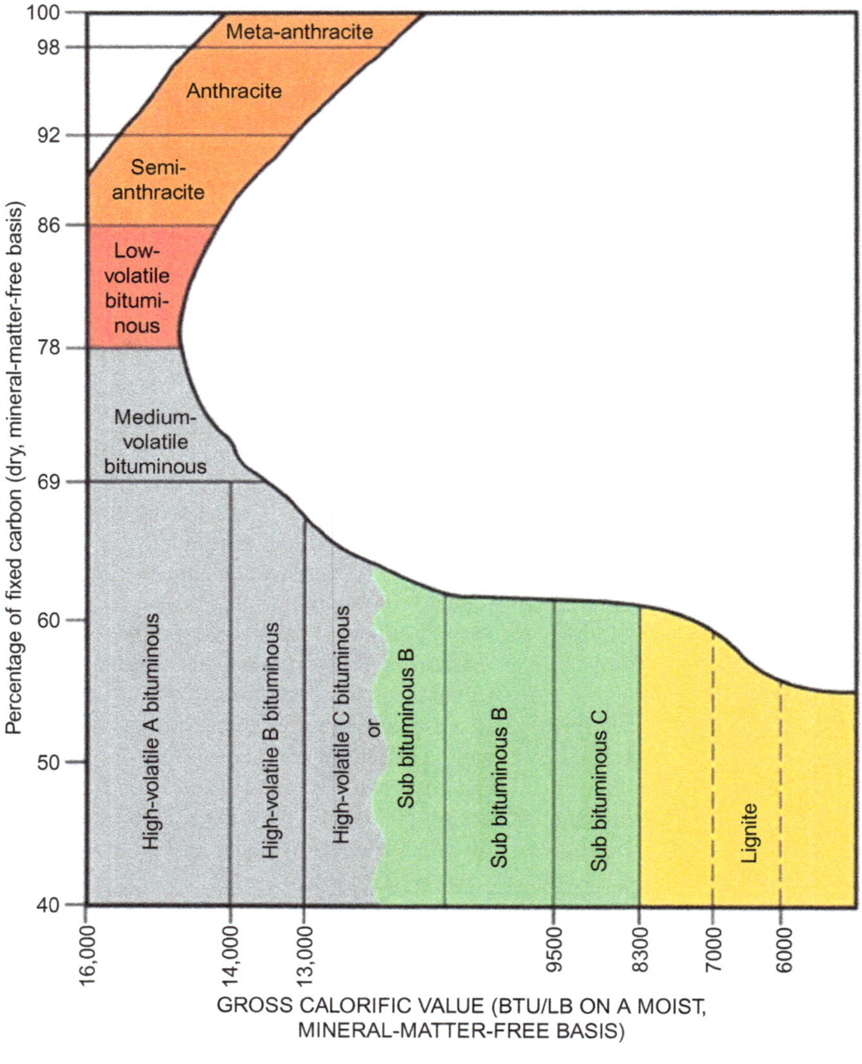

Fig. 2.2 American classification of coal rank (Schweinfurth, 2009).

volatility coals. There are many more indications to come on the variability of coal properties for the same rank, and on the implications during processing.

Figs. 2.1 and 2.2 illustrate the scientific and commercial ways to represent coal rank, but there are many other representations that perform as well. In fact, any physical property or standardized conversion index that changes monotonically for coals across the rank spectrum is a rank parameter, albeit, not necessarily among the most accurate ones. Such parameters are almost always plotted versus a sample's carbon content, preferably on a dry, ash-free basis. In-seam moisture contents, calorific values, specific internal surface areas, CaO levels in ash, and many other coal properties display the essential monotonic behavior across the rank spectrum. Fig. 2.3 illustrates this approach. The proximate volatile matter (PVM) is the mass percentage of volatile material in the same test that determines fixed carbon contents. Since the PVM diminishes for coals of progressively higher rank, it is a suitable rank parameter.

This figure also illustrates a universal quality of all rank parameters, which is that they always exhibit a loosely banded form and never conform to continuous curves, no matter how convoluted. One interpretation is that the rank spectrum can only be depicted in two dimensions, as in the planes of Figs. 2.1 and 2.2. Since Fig. 2.3 uses only C-contents to span the spectrum, we could be looking at a projection down some third axis that scatters the values in the plane of PVM vs C-content. Of course, some third axes will interpret more of the primary variations than others, and there is no reason to suppose that only two coordinates are needed. In any case, the broader questions are (i) Is it even possible to quantitatively describe the distinctive behavior of an individual coal sample in a subject technology with some fixed set of coal characteristics and, (ii) Must we settle for descriptions that depict only the average behavior for each major coal type in Fig. 2.2? From the perspective of process simulations, only the most basic coal characteristics, like C-content, are needed to correlate "average" behavior across the major coal types. But "average" performance characteristics are simply not good enough for technology management and development. Indeed, depicting the distinctive behavior of individual samples is the only means to identify

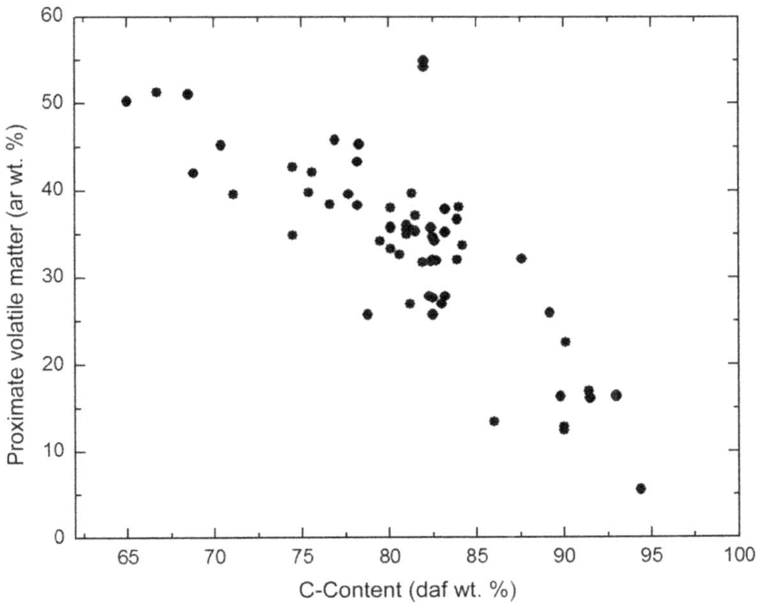

Fig. 2.3 Proximate volatile matter across the coal rank spectrum.

Table 2.1 Ranges of coal properties for the major rank labels.

Label	C-Content (daf wt.%)	Moisture (as rec'd)	PVM (daf wt.%)	O-Content (daf wt.%)
Lignite	65–70	35–55	53–63	23–30
Subbituminous	70–76	18–38	42–53	16–20
hv Bituminous	76–82	8–18	31–46	8–12
mv Bituminous	82–86	8–10	22–31	4
lv Bituminous	86–90	8–10	14–22	3
Semianthracite	90	8–10	8–14	3.5
Anthracite	92	7–9	3–8	4.5
Metaanthracite	>92	7–9	3–8	5

bad actors as well as optimal coal selections, and only sophisticated descriptions of coal constitution can depict the distinctive behavior of individual samples.

Throughout this text, a coal's C-content on a dry-ash-free basis is used as the main rank parameter. Table 2.1 compiles the typical ranges of C-content for the common coal rank labels, along with the ranges for moisture, volatile matter, and O-content. Subdesignations for subbituminous and hv bituminous coals will not be used.

2.2 Coal as a binary mixture

Even to the naked eye, coal is obviously not a homogeneous solid. Chunks of coal are striated into irregular layers around particles and bands of mineral inclusions. The most basic resolution of this structure has only two primary components, an organic matrix that holds particulate minerals. The organic matrix is actually a macromolecular network solid. We know that because coals do not dissolve completely in any solvent, but imbibe them and become swollen as only crosslinked macromolecules can. In many solvents they exhibit viscoelasticity, but may also exhibit plastic deformations at high pressures (Brenner, 1985). To interpret such behavior, the organic portion of coal is widely regarded as a covalently crosslinked macromolecular solid which also includes numerous ionic bonds and dispersive interactions, plus a substantial portion of smaller, trapped molecular fragments that can be extracted with solvents from the macromolecular matrix.

The inorganics comprise an assortment of clays, silica, alumina, calcite, and either pyrite or siderite; and minor amounts of phosphorus, alkali, and alkaline earth metals; and trace levels of dozens of other metallic species, including Hg, As, Pb, and B. Almost all the major minerals appear as inherent inclusions and extraneous particles. One notable exception is calcium in low-rank coals, which is molecularly dispersed throughout the organic coal matrix. The inherent mineral inclusions vary in size from fractions of a micron to several centimeters. When coal is pulverized for PCC furnaces, larger inclusions are freed from the coal matrix as extraneous particles. Major portions of the extraneous fraction can be eliminated by float-and-sink cleaning operations, and pyrite particles can be rejected from pulverizers before the fuel is fed into a furnace. Most of the minor inorganics like alkali and alkaline earth species are "associated" with the organic matrix; i.e., these species are coupled through weak ionic and dispersion forces to polar functional groups in the organic matrix, rather than covalently bonded. The trace metals can appear as substitution elements in major and minor minerals, or in forms associated with organics.

Certainly, for major mineral inclusions and excluded particles, the image of coal as a binary mixture of discrete inorganics within an organic matrix is an accurate description. However, this picture might also suggest that the transformations of the organic and inorganic components during thermal processing are largely independent, since particulate inclusions have relatively small contact areas with their matrix. In fact, inorganic species always catalyze both the thermal decomposition of the organic matrix and the conversion of the solid carbonaceous residue during combustion and gasification. Alkali and alkaline earth metals and Fe are the most active catalysts. The catalysis becomes stronger for coals that contain progressively more oxygen, because the polar functional groups of oxygen disperse and stabilize the inorganic catalysts throughout the organic matrix. Even so, such catalysis is probably important for all coals at all stages of their conversion into gaseous products.

This book specifically focuses on the conversion of the organic matrix throughout thermal processing. Aside from a brief overview of typical mineral compositions in the next section, we will largely ignore mineral matter transformations. In light of the strong catalysis noted above, is this perspective appropriate or fundamentally deficient? It is inappropriate from a scientific standpoint but absolutely necessary from a practical standpoint. The reason is that the overwhelming majority of tests on the conversion of the organic matrix have no information whatsoever on the levels of inorganics in the subject coal sample, except for a total ash level, so proposed reaction mechanisms cannot possibly account for any level of catalysis. Of practical necessity, these omissions are circumvented by various calibration procedures for the reactivity coefficients in the decomposition kinetics. The calibrations will be described when the reaction mechanisms are used to interpret data in later chapters. Suffice to say here that, to a large extent, these calibrations coarsely compensate for the lack of information on catalysis by inorganics during the decomposition of the organic components.

2.2.1 Minerals in coal

The physical and chemical heterogeneities of the minerals in coal normally defy detailed characterization on a routine basis. So it is not surprising that, for decades, mineral forms were ignored while researchers focused on the compositions of ashes prepared by burning coal at high temperatures. Older literature is full of debates about how the combustion transformed the mineral forms. More recently, less destructive techniques, particularly low-temperature plasma ashing, were used to prepare more representative samples of the inorganics for analysis. But the advent of computer-controlled scanning electron microscopy (CCSEM) ushered in the era when extremely detailed distributions of the compositions of pristine coal minerals can be assigned on a routine basis. The review by Skorupska and Couch (1994) of this technique is commendable for its international scope and succinct format.

For purposes of estimating mineral compositions, the major minerals in coal can be assigned to the following five groups: pyrite, illite, kaolinite, mixed aluminosilicates, and quartz. Chemical formulas and typical mass proportions are collected in Table 2.2, based on CCSEM data for over 100 bituminous coals and more than twenty American low rank coals (Huffman and Huggins, 1986). The segregation of K-bearing aluminosilicates (illite) from non-K-bearing aluminosilicates (kaolinite and mixed) is arbitrary from the standpoint of their thermal decomposition behavior. Only quartz is included on the basis of its relative abundance alone.

There are marked differences in the amounts of illite and mixed aluminosilicates in bituminous and low rank coals, but the amounts of pyrite, kaolinite, and quartz tend to be similar. The rest of the minerals in bituminous coals are calcite, siderite, and other minor species. The remainder in low rank coals includes one major form—molecularly dispersed calcium at levels ranging from 7% to 49%—plus many minor species, including iron compounds (sulfates and oxyhydroxides) and clays. Actually, the ranges of values are more meaningful than the typical levels in Table 2.2 because the sample-to-sample variability among coals of the same rank is very significant. Even coals from the same region with the same geological age can have dissimilar mineral compositions. So in lieu of consistent trends among mineral compositions

Table 2.2 Compositions of major coal minerals and ash.

		Amounts, wt.%			
		Bituminous coals		Low rank coals	
	Formula	Range	Typical	Range	Typical
Minerals					
Pyrite	FeS_2	1–27	8	1–26	7
Illite	$KAl_2(AlSi_3O_{10})(OH)_2$	2–29	14	0–12	2
Al/Si/Ca/Fe	–	5–31	17	0–22	8
Kaolinite	$Al_2Si_2O_5(OH)_8$	9–60	32	13–45	30
Quartz	SiO_2	5–44	18	7–22	15
Ash					
Silica	SiO_2		54		30
Alumina	Al_2O_3		29		15
Fe oxide	Fe_2O_3		8		10
Lime	CaO		2		20
Mg oxide	MgO		1		8
K oxide	K_2O		2		1
Na oxide	Na_2O		1		1
Titania	TiO_2		1		1
P pentox.	P_2O_5		0		0
S trioxide	SO_3		2		15

across the rank spectrum (except for Ca levels), we have only rough distinctions for low rank and bituminous coals.

The typical proportions by weight in Table 2.2 are generally consistent with CCSEM datasets in the literature for British, American, and Spanish coals (Martinez-Tarazona et al., 1992; Yu et al., 1994; Wigley and Williamson, 1997). Aluminosilicates are most abundant, followed in order by quartz, pyrite, and calcite. However, the ranges in Table 2.2 are not completely consistent with a survey of mineral compositions from 34 coals around the world, which generally reported much lower proportions of aluminosilicates and much higher proportions of quartz (Vassilev et al., 1997). Similarly, a survey of Spanish coals also reported much higher levels of quartz. The basis for these discrepancies is unclear, although Vassiliev et al. applied X-ray diffraction to plasma ashes whereas the values in Table 2.2 are based on CCSEM. It is worth remembering that mineral compositions are extremely variable for coals across the rank spectrum, and that local geological circumstances are more important than coal rank, per se.

Another caveat to Table 2.2 pertains to the association of minerals with the organic coal matrix. Any CCSEM dataset can be interpreted on several different bases, including the projected areas of mineral inclusions, total particle size, or weight fractions. Such composition distributions resolved over the PSD of pulverized grinds indicate that the largest extraneous mineral inclusions often appear as discrete mineral particles with little or no organic matter (Straszheim and Markuszeqski, 1990; Yu et al., 1994). Pyrite and calcite, in particular, tend to occur in the form of large, extraneous mineral particles along with lesser amounts of clays. So it would not be unreasonable to reduce the mass proportion of these minerals by roughly one-third to estimate the mineral composition that actually comes in intimate contact with the organic coal matrix during the initial stages of thermal treatment.

2.2.2 Normative analyses

Coals contain minerals, not ash, yet the conventional ash chemistry analysis delivers an ash composition rather than the mineral composition. Normative calculations estimate mineral compositions from the readily available compositions of high-temperature ash. The approach here assigns weight fractions to only four of the five major mineral forms in Table 2.2, because kaolinite and mixed aluminosilicates are lumped together. The Parr formula estimates the amount of mineral matter from the high temperature ash level on a dry weight basis, A_{dry}, as follows:

$$MM = 1.08 A_{dry} + 0.55 S_{tot} \tag{2.1}$$

where S_{tot} is total sulfur in dry wt.% and MM is the amount of mineral matter in coal in dry wt.%.

The following four rules were adapted from the approach of Palmer and Filby (1984): (i) Use an accurate elemental formula for illite (Srinivasachar et al., 1990) and the amount of K_2O in ash to assign the amount of illite; (ii) Assign the pyrite yield from the difference between the iron in ash as Fe_2O_3 and the iron in illite; (iii) Assign the levels of mixed aluminosilicates plus kaolinite from the difference between the Al in Al_2O_3 in ash and the Al in illite, based on an elemental formula for kaolinite; and (iv) Assign the quartz level based on the residual from the level of SiO_2 in ash and the silicon levels in illite and aluminosilicates. The equations developed from these rules are as follows:

$$\text{Wt.\%Illite} = \left[6.53 \times 10^{-3} K_2O\, A_{dry}\right] / \left(6.51 \times 10^{-2} MM\right) \tag{2.2a}$$

$$\text{Wt.\%Kaolinite} = \left[5.59 \times 10^{-3} Al_2O_3\, A_{dry} - 0.114 W_{IL} MM\right] / \left(2.09 \times 10^{-1} MM\right) \tag{2.2b}$$

$$\text{Wt.\%Pyrite} = \left[6.99 \times 10^{-3} Fe_2O_3\, A_{dry} - 0.047 W_{IL} MM\right] / \left(4.66 \times 10^{-1} MM\right) \tag{2.2c}$$

$$\text{Wt.\%Quartz} = 10^{-2} SiO_2 - 0.465 W_{KA} - 0.540 W_{PY} \tag{2.2d}$$

In these expressions, the chemical symbols denote the weight percentages of the respective components in coal ash. Also, A is the high temperature ash yield, in dry wt%; K is the level of potassium that is acid or water soluble, in wt.% dry coal; and W_{KA}, W_{IL}, and W_{PY} are the weight percentages of illite, kaolinite, and pyrite assigned from Eqs. (2.2a)–(2.2c), respectively. The percentages do not sum to 100%, because the basis is the Parr mineral matter and the normative calculations ignore calcite and various clays and iron-aluminosilicates.

The normative calculations were evaluated against CCSEM data for five American bituminous coals, including one dataset for four samples that also included the level of acid-soluble potassium, and another for a cleaned bituminous coal. Errors were largest in the aluminosilicate estimates, because the composition of this lump varies for different coals, and only the elemental formula for kaolinite was applied in the calculation. But overall, the accuracy is sufficient for many applications as, for example, when the mineral compositions are used to estimate the scavenging of minor and trace species by molten flyash particles.

2.2.3 Coals' organic matrix

One implication of observations that coals cannot completely dissolve without bond rupture is that the organic matrix has a crosslinked macromolecular network structure. Another is that nominal values for molecular weights of the matrix approach several million g/mol. But the meaning of molecular weight is ambiguous in this context because coals' macromolecular networks have no repeating monomer unit, and the degree of polymerization—the number of monomers in a polymer—is similarly ambiguous. Whenever conventional characterization tests for polymers are applied to coals, the interpretation of data inevitably bogs down under these fundamental ambiguities.

As illustrated in the hypothetical segment of a coal macromolecule in Fig. 2.4, coals' macromolecular skeleton contains condensed aromatic structures called

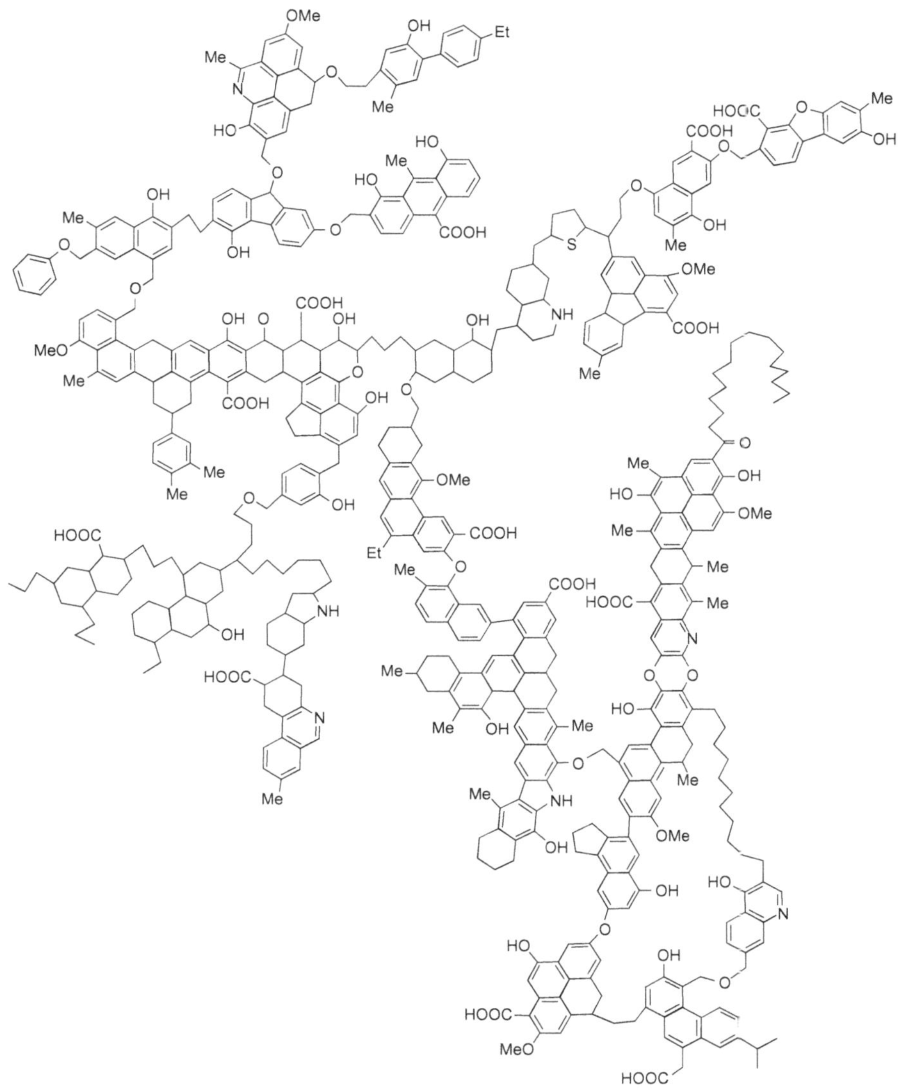

Fig. 2.4 Segment of a coal macromolecule from a *hv* bituminous coal.
Reproduced from Hatcher PG. Chemical structural models for coalified wood (vitrinite) in low rank coals. Org Geochem 1990;16:959–968 with permission from Elsevier.

nuclei in this context. In most coals, the nuclei contain 2 to 4 condensed aromatic rings each, although low volatility coals can have twice as many. There is always a distribution of size among the nuclei in any coal sample. Nuclei comprise the bulk of carbon in coal, yet they exhibit an inert, refractory character during thermal processing. They also sparsely contain nitrogen as pyridines and pyrroles, and sulfur as thiophenes. Nuclei are interconnected by many more structures than shown in Fig. 2.4 (Hatcher, 1990). These so-called "bridges" are hydroaromatics and other polymethylene chains; ether and sulfide linkages; and biphenyls and polyolefin chains. Since aromatic nuclei and hydroaromatic bridges constitute the bulk of the coal mass, this picture of coal constitution is called the "aromatic/hydroaromatic" model.

Actually, the population of bridge structures in coal is among the most uncertain aspects of coal constitution. The reason is that nuclei are substituted around their peripheries with a host of functional groups called peripheral groups, including carboxylic acids, phenols, quinones, aliphatics, olefins, methylene chains, and amines. Perhaps the most unfortunate gap in the entire characterization is that no analytical method can tag the aliphatic and chain hydrocarbon structures as either bridges or peripheral groups within useful quantitative tolerances. The allocation of oxygen into bridges and peripheral groups is similarly subject to inordinate uncertainties, although almost all oxygen is thought to appear in peripheral groups. These are crucial gaps, because bridge ruptures during processing disintegrates the organic matrix into fragments that may be expelled as products, whereas peripheral group eliminations often generate sites for additional crosslinks, which has the opposite effect on the integrity of the network structure.

Various indirect schemes have been developed to bridge these gaps. In the most conventional approach, the networks are characterized by their average number of bridges per nucleus, and by the average molecular weight between crosslinks. Numbers of bridges per nucleus convey the macromolecular architecture. For values of two or fewer, the conformation relaxes to a mixture of straight chains. For values greater than two, crosslinks become progressively more numerous and the weight between crosslinks diminishes. In principle, coal macromolecules could contain so many crosslinks that they morph into three-dimensional metallic structures in which every nucleus contains two or more crosslinks. But reported values indicate that there are rarely many more than two bridges per nucleus, so coal macromolecules are more akin to weakly crosslinked chain mixtures than to tightly bound metallic sponges. Whereas many sets of numerical values have been reported for the bridges per nucleus and weight between crosslinks, it is worth remembering that they originate in the higher moments of primary analytical signals or in interpretations based on rudimentary conceptual models, and are therefore subject to large uncertainties for any particular coal sample.

Within the organic coal matrix, minor amounts of smaller fragments called guest molecules are bound into the skeletal structure via dispersive and other long-range forces. The guest molecules are often released in solvents that swell the skeletal matrix, although it becomes difficult to distinguish decomposition fragments from true guest molecules as stronger, more reactive solvents are applied. Presumably, guest molecules constitute the earliest oils and tars from thermal processing as well.

The aromatic/hydroaromatic model describes the primary skeletal structure at the molecular scale, and also paves the way toward coal constitution at a much larger scale. The aromatic rings condensed into nuclei impart a planar character to the components in these network structures. In turn, the planar sections of various chain fragments may align into layers that are ordered across larger dimensions. These layered domains, called lamella, constitute the building blocks of carbonaceous crystallites that can be resolved into fringe images with high-resolution, transmission electron microscopy. The implications toward bona fide crystallinity are strong. But coals are not crystalline materials; rather, the organic coal matrix is a glassy solid with included crystallites. The extent and number of crystallites increase for coals that contain progressively fewer heteroatoms; e.g., anthracites, which contain all but a few percent carbon plus a little hydrogen, contain more material in the crystallites than in the glassy phase.

At an even larger scale still, coals are penetrated by pores ranging in size from nanoscale chambers within the macromolecular matrix through micropores under 2 nm, mesopores under 50 nm, and larger macropores. However, the pore system within coal is mostly irrelevant to its behavior during thermal processing. Even subbituminous and most low volatility coals soften into molten foams when rapidly heated, so the pore systems within their solid residues are completely different than the original pore systems. And among the coals which do not soften during processing, only anthracites and like-coals retain their original pore systems intact, simply because they hardly release any gases during the early stages of thermal processing so their extensive crystallites are hardly perturbed.

2.3 Thermophysical properties of coal

To this point, the presentation on the organic coal matrix emphasized its macromolecular constitution, which may have given the impression that molecular characteristics are absolutely necessary to predict the performance of particular coal samples in a subject utilization technology. This would be a serious misconception, because the macromolecular constitution of any coal sample can only be revealed with an enormous battery of analytical techniques, including several of the most sophisticated. Simulations built upon legitimate macromolecular models are too expensive, too slow, and much too elaborate for practical applications. Actually, the disconnect could not be more pronounced, because simulation specialists do not want to run any tests at all to enable their predictions. But they do have to obtain the standard coal properties, at a bare minimum. To bridge this chasm between scientific and practical imperatives, the basic question posed at the beginning of this chapter on required coal characteristics should be refined as follows: "What features of coal structure and constitution *must be derived from the standard coal properties* to simulate the distinctive behavior of individual coal samples in a particular utilization technology?" This section describes the essential coal properties for realistic process simulations, and the connections among these properties and macromolecular characteristics are presented in Chapter 5.

Complete proximate and ultimate analyses must be obtained for every coal under consideration. Calorific values can be very accurately estimated so direct measurements are often unnecessary. If any aspects of ash behavior are considered, then the standard ash chemistry analysis illustrated in Table 2.2 must also be available. The performance of all our subject utilization technologies is governed by the dynamic behavior of coal particles small enough to heat volumetrically, without steep internal gradients in temperature. But the particles usually sustain very steep gradients in the concentrations of reactant gases like O_2 and H_2. Accordingly, the most important thermophysical properties are density, specific heat, and the extent of volumetric swelling during thermal processing. The very high temperatures in most utilization technologies sustain intense thermal radiation transfer rates, so radiative properties are also required. Particle size distributions are usually first-order important. Pore size distributions may also be relevant although, surprisingly, some of the preferred mechanisms for char conversion do not require this information. This chapter specifies all necessary thermophysical and transport properties except effective diffusivities, which are discussed in standard texts on gas/solid reactions.

A proximate analysis comprises the mass percentages of moisture, ash, volatile matter, and fixed carbon, which are obtained from a series of three standardized tests. Prior to analysis, samples are often held at 10–15°C above ambient in air to release all but "residual" moisture, for which the results are reported on an "air-dried" basis. For an analysis without predrying, the results are reported "as-received." The moisture level is assigned as the weight loss in air at 100–105°C or in N_2 at 150°C. The ash level is obtained by heating a 1 g sample under air in two stages, first to 500°C and then to 750°C, during a total exposure of 4 h. The residue is assigned as the ash content. The PVM is evaluated as the weight loss corrected for moisture obtained by heating 1 g of sample in a covered platinum crucible for 7 min at 950°C. Fixed carbon levels (FC) are assigned by difference with moisture, ash, and volatile matter. Readers interested in more detail should consult ASTM D3173, D3174, and D3175.

An ultimate analysis determines the mass percentages of C, H, O, N, and S in the combustible portion excluding moisture. It uses three separate tests described in ASTM D3176, plus the ash determination from a proximate analysis. Whereas the formal definitions are straightforward, only C, H, and N levels are assigned without ambiguity. The S-determination is ambiguous because it evaluates the total S-content, and sulfur is present in both the organic and inorganic components (mostly as pyrite). Ultimate analyses are commonly regarded as the elemental compositions of the organic coal matrix, without perceptible contributions from mineral decompositions. But during the tests, mineral decompositions do generate CO_2, H_2O, SO_2, and other minor gaseous products. These products usually make negligible contributions to the C, H, N, and O inventories, but always constitute substantial or even the largest portions of the S-contents. A direct determination of the sulfide mineral levels is the only means to eliminate the ambiguities on S-contents. Pyrite levels are assigned from the Fe-ion concentrations after coal digestion in a bath of heated nitric acid.

Since there are no convenient methods to assign O-contents, they are assigned by difference, and therefore accumulate the measurement uncertainties on the other elements. These uncertainties are very small fractions of the abundant O-levels in low rank coals, but make progressively greater contributions for coals of higher rank to the extent that reported O-contents for low volatility coals often have no statistical significance at all, particularly for anthracites.

Whether the moisture level is assigned on as-received or air-dried bases, a proximate analysis should always be reported on a whole-coal basis, as it was tested. In some applications, the PVM is useful as an index of the fuel volatility, for which the analysis should be converted to the daf basis to remove the ash and moisture variations. The conversion is

$$PVM_{daf} = \frac{PVM}{1 - \frac{(A+M)}{100}} \tag{2.3}$$

where A and M denote the ash and moisture from a proximate analysis.

There is no standardized basis to report an ultimate analysis although, for our purposes, a daf basis best connects an elemental composition to the organic coal matrix. Test results are reported on a dry basis, and can be converted to a daf basis as follows:

$$E_{i,\,daf} = \frac{E_{i,dry}}{1 - \frac{A_{dry}}{100}} \quad \text{and} \quad A_{dry} = \frac{A}{1 - \frac{M}{100}} \tag{2.4}$$

where E_i denotes one of the five elements in the analysis.

Measured levels of volatile matter are plotted vs C-content on a daf basis for coals across the rank spectrum in Fig. 2.3. The C-contents span a range from 65 to 95 daf wt.%, which covers the full commercial utilization domain. The PVM-levels were mostly obtained with predried coals, and the ash levels were usually between

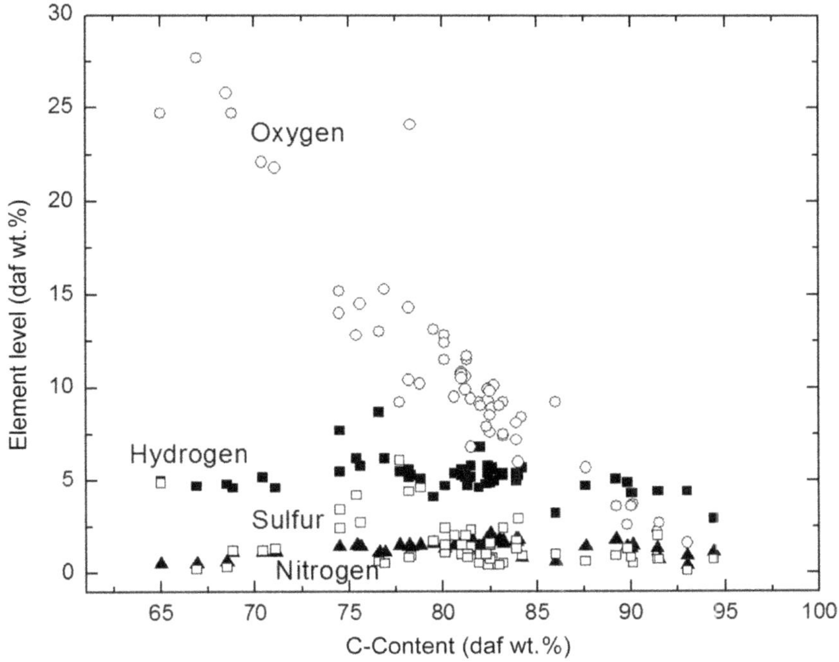

Fig. 2.5 Elemental compositions across the coal rank spectrum.

6 and 12 wt.%. The PVM diminishes continuously for coals of progressively higher rank, beginning with 50% for Australian brown coals and nearly vanishing for anthracites. The most striking feature is the sample-to-sample variability for every particular coal rank. For hv bituminous—from 80% to 84% C—the PVM levels vary from 25 to 41%, which is ±25% about a mean of 33 wt.%. Even when the extreme outliers at 82% C are ignored, and even when the thermal processing is strictly regulated, coals' sample-to-sample variability must be reckoned with in practical applications.

Elemental compositions from the ultimate analyses appear in Fig. 2.5. Both O-and S-contents display strong rank variations, whereas H- and N-contents do not. O-contents diminish in near-proportion to C-contents from almost 30 daf wt.% until they vanish in anthracites. Taken at face value, it appears that S-contents pass through a maximum through the subbituminous ranks. But actually high-S coals of any rank are used commercially. In this figure, there is one brown coal with 5% S and other low-rank coals with 2% S and a moderately high-S anthracite. The vast majority of coals processed in the developed world are cleaned to standardize S-contents, whereas coals in the developing economies may not be. So most reported S-contents reflect the pretreatment conditions rather than any inherent rank variation. Hydrogen-contents tend to scatter around 5 daf wt.% for all but the very highest ranks, whose H-contents trend lower. In this selection of coals, the levels range from 3.3 to 8.7 daf wt.%, although it is uncommon to find levels above about 6.7% in commercial applications. Even so, this variation is among the most important in the coals' thermal decomposition reaction mechanisms. N-contents range from 0.4 to 2.1 daf wt.% and exhibit no rank dependence whatsoever.

In power plants, coals are ground into the pulverized fuel (p.f.) size range in massive pulverizers. Fines are rejected via aerodynamic classification and the coarsest fractions are recycled to tightly control the portion larger than a threshold called the top size. The full PSD extends from a few microns to about 300 μm. Historically, PSDs have been measured by shaking coal samples through stacks of calibrated wire-mesh containers called sieves, and measuring the weights of the fractions between every pair of size cuts. Whereas optical methods have superseded sieve shakers in characterizing PSDs, the older terminology persists. In the United States, two cuts in the US Standard Mesh system determine the p.f. size grade. The fineness is the mass fraction of sample passing through a 200-mesh sieve, whose opening is 74 μm. The top size is the mass fraction on 50 mesh, which collects particles larger than about 300 μm. Typical p.f. feeds have 70%–80% through 200 mesh and less than 0.5% on 50 mesh. The mean size is usually about 45 μm.

These PSDs inevitably conform to a Rosin-Rammler PSD, which is

$$F_{\mathrm{cum}} = \exp\left(-\left(\frac{d}{d_{\mathrm{exp}}}\right)^{n_{\mathrm{RR}}}\right) \tag{2.5}$$

where F_{cum} is the cumulative mass fraction larger than size d; n_{RR} is the spread parameter; and d_{exp} is the size for which F_{cum} equals $1/e$. The latter two parameters are adjustable and assigned by fitting any two size fractions in the coal grind. Size distributions broaden for smaller values of the spread parameter. Typical PSDs based on

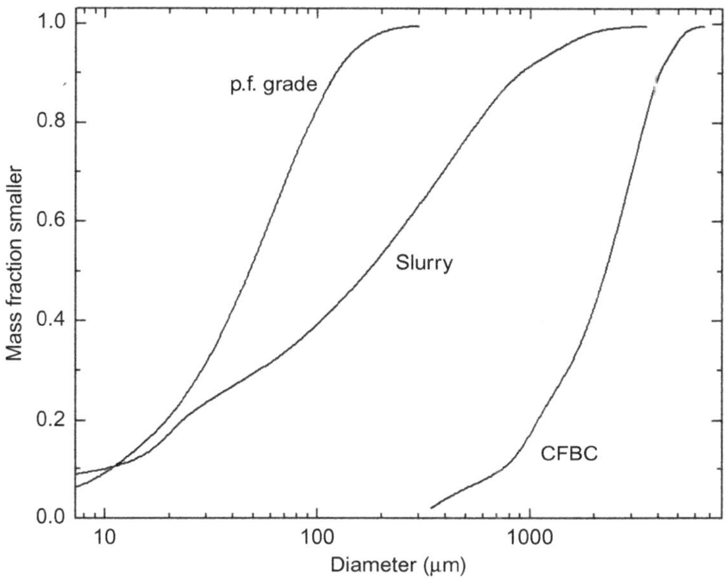

Fig. 2.6 Coal PSDs for different technologies.

Eq. (2.5) for the p.f. grade, coal-water slurries fed into entrained flow gasifiers, and CFBCs are shown in Fig. 2.6. As is typical, the mean size in the p.f. grade is 48 μm, and the spread parameter of 1.3 is within the expected range from 1 to 1.4 for this grind. The PSD for slurries is much broader to manage slurry viscosity. Whereas, the PSD for slurries and p.f. grinds are similar for small sizes, the slurry grind extends to much larger sizes, and has mean and spread values of 267 μm and 0.73, respectively. The CFBC PSD is narrowest of all, with a spread of 2.07. Grinds for fluidized systems are always much coarser than those for entrained-flow systems. This one extends from 0.5 mm to 1 cm, and has a mean size of 2.6 mm.

Calorific values are inherently important in all furnace applications, and are often among the better regression variables on many furnace performance characteristics, including emissions. As seen earlier in Fig. 2.2, calorific values are one of the two primary variables used to assign a sample's rank. Their magnitudes reflect C-contents with positive attenuation by H- and S-contents and negative attenuation by O-content. The Dulong correlation rarely gives significant deviations from measured values, unless the sources of the data are questionable. It is

$$CV = \left[14{,}544C_{\text{daf}} + 62{,}028\left(H_{\text{daf}} - \frac{O_{\text{daf}}}{8}\right) + 4050S_{\text{daf}}\right]\left[\frac{1 - \frac{(M+A)}{100}}{100}\right] \tag{2.6}$$

The constants in this rendition return calorific values in Btu/lb of as-received coal. Values in kcal/kg are obtained by multiplying the value from Eq. (2.6) by 0.5556, and in MJ/kg by using 2.326×10^{-3}. For the samples with 80–84 daf wt.% C in Fig. 2.3, the calorific values varied by ±15% about a mean value of 13,250 Btu/lb, so sample-to-sample variability within the same rank is substantial.

The only self-contained method to estimate a sample's specific heat was reported by Merrick (1983) over 30 years ago. These values remain the only alternative to direct measurements, which would never be undertaken to support simulation work. The accuracy of specific heats is not a small consideration, since specific heats typically change by a factor of two or more while devolatilization transforms a coal into residual char, and are also first-order important in the estimation of thermal histories during the early stages of any thermal processing, as demonstrated shortly.

Coal is a ternary mixture of combustibles, ash, and moisture whose components make additive contributions to the total specific heat. The following equations estimate the specific heat of any coal, $C_{P,WHOLE}$, throughout its conversion into char:

$$C_{P,WHOLE} = \left(\frac{PVM + FC}{100}\right)(1 - f_{VOL}(t))C_{P,COMB} + \frac{A}{100}C_{P,ASH} + \frac{M}{100}f_{MOIS}(t)C_{P,MOIS} \tag{2.7a}$$

$$C_{P,ASH} = 0.18 + 1.4 \times 10^{-4}(T_P, °C) \tag{2.7b}$$

$$C_{P,COMB} = R'G_1\left(\frac{380}{T_P}\right) + 2G_1\left(\frac{1800}{T_P}\right) \tag{2.7c}$$

with

$$R' = \left(\frac{1.987}{100}\right)\left[\frac{C_{daf}f_C(t)}{12} + \frac{H_{daf}f_H(t)}{1} + \frac{O_{daf}f_O(t)}{16} + \frac{N_{daf}f_N(t)}{14} + \frac{S_{daf}f_S(t)}{32.04}\right]\left(\frac{1}{1 - f_{VOL}(t)}\right) \tag{2.7d}$$

$$G_1(x) = \frac{\exp(x)}{\left[\frac{\exp(x) - 1}{x}\right]^2} \tag{2.7e}$$

The units are cal/g K or Btu/lb °F. Separate analyses specify $f_{MOIS}(t)$, the moisture fraction remaining to be evaporated at time t, and $f_{VOL}(t)$, the volatiles yield as a fraction of the original combustibles in coal, which is evaluated as (PVM + FC). The instantaneous number fractions of the original elements in coal and the nascent char, $f_i(t)$, must be specified throughout thermal processing. The instantaneous particle temperature, $T_P(t)$, is also required in kelvins, except as indicated in Eq. (2.7b). The equations to assign these quantities are presented at the end of Chapter 5. A proximate analysis specifies PVM, FC, A, and M, and an ultimate analysis specifies the elemental composition of the parent coal on a daf basis. Predicted values for $C_{P,SOLID}$ are discussed after the estimation schemes for bulk density and swelling factor.

Bulk (or particle) density describes the density within the external particle surface area, including the internal porosity, ash inclusions, moisture, and combustible matter. Strugala's estimation scheme (1994) returns realistic values and has been validated with measured densities for lignites through anthracites. The calculation sequence first estimates the mass fraction of pyrite in total mineral matter, because only pyrite's density is much greater than all other mineral forms. It then applies separate correlations for the total porosity, based on accessibility to He, and for the density of hypothetical nonporous combustible matter to evaluate the "true" density of combustibles plus mineral matter. The bulk density is then formulated as a weighted sum of the contributions for solids, porosity, and moisture. The solution for the following three simultaneous equations assigns dry mineral matter, the pyrite fraction, f_{PYR}, and the S-content in pyrite, S_{PYR}, which, in turn, specify the mineral density:

$$\mathrm{MM} = 1.113A_{dry} + 0.34S_{PYR} \tag{2.8a}$$

$$f_{PYR} = \frac{1.3(S_{PYR} - 0.30)}{\mathrm{MM}} \tag{2.8b}$$

$$S_{PYR} = \frac{\mathrm{MM}f_{PYR}}{1.871} \tag{2.8c}$$

$$\rho_{MM} = \left[\frac{1 - f_{PYR}}{2.73} + \frac{f_{PYR}}{5.10}\right]^{-1} \tag{2.8d}$$

The pyrite density is nearly double that for the rest of the mineral matter.

The true density of combustibles, ρ_{COMB}, and the total specific porosity, V_{He}, are evaluated from

$$\rho_{COMB} = \left[0.6625 - 0.5696(C_{daf})^2 + 1.0494C_{daf}\right]^{-1} \tag{2.8e}$$

$$V_{He} = 0.0368 + 16.6E - 04\,\mathrm{MM}(1 - f_{PYR}) + 0.0233S_{PYR} - \frac{\left(1 - \frac{\mathrm{MM}}{100}\right)98.118E - 04(\mathrm{PVM}_{daf})^3}{(100 - O_{daf})^2 + 0.133O_{daf} - 37.37E - 04(O_{daf})^2 + 1.0531E - 04(O_{daf})^3} \tag{2.8f}$$

The true density weights the contributions from combustibles and mineral matter, but not moisture or porosity:

$$\rho_{true} = \left[\frac{1 - \frac{\mathrm{MM}}{100}}{\rho_{COMB}} + \frac{\frac{\mathrm{MM}}{100}}{\rho_{MM}}\right]^{-1} \tag{2.8g}$$

The mineral contribution is disproportionate because most mineral densities are roughly double that for combustibles. Finally, the bulk density factors in moisture and porosity as follows:

$$\rho_b = \left[\left\{ \left(V_{He} - \frac{\frac{M}{100}}{\rho_{MOIS}} \right) + \frac{1}{\rho_{true}} \right\} \left(1 - \frac{M}{100} \right) + \frac{\frac{M}{100}}{\rho_{MOIS}} \right]^{-1} \tag{2.8h}$$

Omitting any contributions from moisture, bulk densities from this system of equations range from 1.0 to 1.5 g/cm^3, so the variations across the rank spectrum are somewhat smaller than for specific heat.

In principle, this scheme can be applied to chars throughout thermal processing by substituting char compositions for coal compositions. But in practice, the changes in bulk density throughout devolatilization are much greater than the impact of char composition on the true density of combustibles in Eqs. (2.8e), (2.8f). Any realistic analysis must also account for volumetric swelling during devolatilization, which typically reduces bulk densities by more than a factor of two during processing in the subject utilization technologies. Swelling is included in the form of a swelling factor, Ω, defined as the ratio of the ultimate size after devolatilization is complete to the original size:

$$\Omega = \frac{d_{P,\infty}}{d_{P,0}} \tag{2.9a}$$

Since swelling is driven by coal softening and volatiles release during thermal processing, the dynamics of particle expansion are expressed in terms of the instantaneous and ultimate volatiles yields, according to

$$\frac{d_P(t)}{d_{P,0}} = 1 + (\Omega - 1)\frac{V(t)}{V_\infty} \tag{2.9b}$$

where $V(t)$ is the instantaneous volatiles yield and V_∞ is the ultimate yield after extended thermal processing. As $V(t)$ approaches V_∞, the particle size ratio approaches Ω.

Swelling factors are strong functions of coal quality and pressure, and pass through maxima for *hv* bituminous coals and for pressures around 1.0 MPa. Based on a large database of values for diverse coals and pressures to 4 MPa, the following correlations were developed to evaluate Ω:

$$\Omega'_{CQ} = 8.6667 - 0.0833C_{daf} \quad \text{for } 89 < C_{daf} \le 92 \tag{2.9c}$$

$$\Omega'_{CQ} = -4.5834E - 02 + 0.0146C_{daf} \quad \text{for } 72 < C_{daf} \le 89 \tag{2.9d}$$

where Ω_{CQ}' is the reference value for the primary coal quality impacts. For C-contents greater than 92 daf wt.% or less than 72%, the swelling factor is unity. However, some coals from the Southern Hemisphere, particularly South Africa, were observed to swell by less than their C-contents would indicate (presumably due to an abundance of inertinite in the maceral mix). Accordingly, the primary estimate for Ω_{CQ}' for these coals is adjusted by

$$\Omega_{CQ} = \Omega'_{CQ} - \left(\Omega'_{CQ} - 1\right)\frac{X_{INERT} - 19.98}{100 - 19.98} \tag{2.9e}$$

where

$$X_{INERT} = 19.98 - 29.04 H_{DEV} \tag{2.9f}$$

$$H_{DEV} = H_{daf} - \left[171.9 - 7.025 C_{daf} + 0.0976 (C_{daf})^2 - 4.472E - 04 (C_{daf})^3\right] \tag{2.9g}$$

All values for X_{INERT} must be bounded between 0 and 100.

For applications at atmospheric pressure, Ω is taken as Ω_{CQ}. The final adjustment only comes into play at elevated pressures, for which the swelling factor is given by

$$\Omega = \left(\Omega_{CQ}\right)^{\delta_P} \text{ where } \delta_P = 0.7143 + 2.8570p \text{ for } 0.1 < p, \text{MPa} \leq 0.8 \tag{2.9h}$$

$$\delta_P = 3.500 - 0.625p \text{ for } 0.8 < p, \text{MPa} \leq 4.0$$

An even larger database of reported swelling factors was compiled to develop correlations that depict the maximum in swelling for heating rates around 10^4°C/s (Shurtz et al., 2011), and also covers broad ranges of coal quality and pressure (Shurtz et al., 2012).

Once a swelling factor has been specified, particle diameters are evaluated throughout thermal processing from Eq. (2.9b), and the impact of swelling on the bulk density is evaluated from

$$\rho_b = \rho_{b,0} \frac{\dfrac{m_P(t)}{m_{P,0}}}{\left[\dfrac{d_P(t)}{d_{P,0}}\right]^3} \tag{2.9i}$$

where $\rho_{b,0}$ is evaluated from Eq. (2.8h) and the diameter ratio is evaluated from Eq. (2.9b) with the swelling factor from Eq. (2.9h).

For coals across the rank spectrum, specific heats range from 0.25 to 0.45 cal/g K and tend to diminish for coals of progressively greater rank. Bulk densities range from 0.9 to 1.5 g/cm^3 and are somewhat more scattered due to the independent contributions from moisture, ash, and elemental compositions. Swelling factors for atmospheric pressure are bounded by unity and 1.25 although values greater than 1.20 are rare. For elevated pressures where swelling is maximized, the magnitude of Ω_{CQ} is cubed, so that the diameters may expand by as much as 75%.

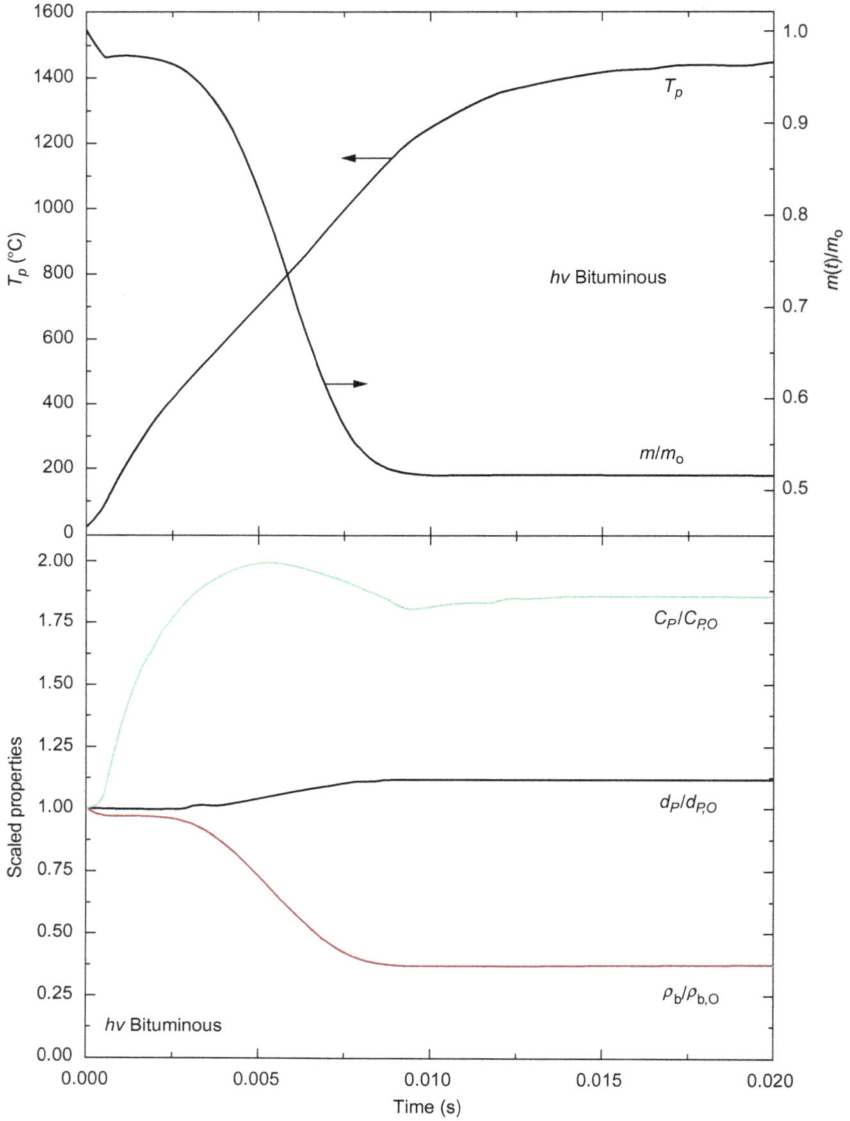

Fig. 2.7 Variation in thermal and physical properties during thermal treatment to flame temperatures.

Bear in mind that the coal quality impacts are relatively weaker than the changes in specific heat and bulk density during thermal processing. As an illustration, Fig. 2.7 shows how specific heat, size, and bulk density are estimated to change while 63 μm particles of an *hv* bituminous coal are heated to flame temperatures at atmospheric pressure. The nominal heating rate is of the order of 10^5°C/s, and the ultimate particle temperature is 1400°C. The scaled mass loss in Fig. 2.7 resolves separate stages for moisture release at very low temperatures, and devolatilization from 400°C to 1100°C. Whereas moisture release hardly perturbs the thermal and physical properties, devolatilization radically changes them. The specific heat doubles due to the joint impacts of heteroatom elimination and heating to elevated temperatures. The bulk density plummets by two-thirds due to simultaneous mass loss and swelling. Both changes are much stronger than the coal quality impacts. But the changes compensate each other in analyses for transient heating because the product of density and specific heat appears in the governing equations (cf. Section 5.7). The size change is necessarily bounded by unity and the ultimate value of the swelling factor, which gives a 12% expansion in this case.

As noted in Chapter 1, intraparticle temperature gradients tend to be unimportant for the subject utilization technologies, so thermal conductivities are unnecessary in almost all simulation applications. Readers who want to assess this issue for their particular simulation conditions can use the estimation routine for thermal conductivity developed by Atkinson and Merrick (1983).

The high processing temperatures in p.c. flames and entrained flow gasifiers promote intense radiation transfer within the coal suspension and from the suspension to the waterwalls or reactor lining. From the standpoint of heat transfer calculations for individual particles moving through these systems, the only required radiative property is the hemispherical total emittance of coal and char. For coals, this property is significantly affected by numerous absorption bands associated mostly with oxygen functional groups and the extent of aromatic crystallites in the macromolecular structure. But coals are converted to char in miniscule fractions of the total transit times through furnaces and reactors, and the radiative properties of char are much simpler than coals'. Chars' highly aromatic, carbonaceous, combustible component approaches black body behavior, but mineral inclusions are more reflective. Consequently, reported values of the hemispherical total emittance for char cluster between 0.8 and 0.9.

2.4 Transport coefficients

The most important transport coefficients are convective heat and mass transfer coefficients; drag coefficients; the binary diffusivities for all gaseous reacting species; and the effective diffusivities through a pore system in char for all gaseous reactants in char conversion. It will also be necessary to evaluate various mean gas properties to identify the flow and transport regimes, including viscosity, density, heat capacity, and thermal conductivity. Basic definitions for these coefficients may be found in the standard texts on transport phenomena.

All our subject utilization technologies process suspensions of coal in an entrainment gas stream. Entrainment streams are usually air or a gas mixture similar to air, such as the mixtures of O_2, steam, and CO_2 fed into some gasifiers. The behavior of coal-water slurries is fundamentally different within fuel injectors, but becomes much the same as air flows once the water vaporizes. Mean particle sizes range from 40 to 50 μm in furnaces to a few hundred microns in slurries to 1–20 mm in fluidized systems. Suspension loadings can be as high as 0.5 kg-coal/kg-air at the throats of burners and fuel injectors in furnaces and entrained-flow gasifiers, but the solids eventually disperse into the regime of isolated, individual particles downstream. However, the solids loadings in fluidized beds remain firmly in the dense suspension regime. The solids in CFBCs are only in dense suspension in the dense bottom bed, which is a relatively very small portion of the entire flowfield. This section first considers transport in dilute suspensions, then covers the coefficients for fluidized bed applications.

2.4.1 Transport coefficients for dilute suspensions

Since the gas properties are fairly uniform among the systems with dilute coal suspensions, the particle size determines the flow regime for convective transport. Very small particles move in Stokes flow, where the primary transport mechanisms are viscous friction, molecular conduction, and Brownian diffusion. Larger particles hold wakes that supplement the molecular transport with convective exchanges. So the first step in any transport analysis is to identify the flow regime from the Reynolds number based on the particle size, Re_P:

$$Re_P = \frac{\rho_g v_{slip} d_P}{\mu} = \frac{v_{slip} d_P}{\nu} \tag{2.10a}$$

where the gas viscosity, μ, density, ρ_g, and their ratio in the kinematic viscosity, ν, are evaluated at a mean temperature between the particle surface and a free stream condition. The slip velocity, v_{slip}, is the difference between the local velocities of the gas, v_g, and particles, v_P. It is routinely evaluated in CFD simulations with a simple balance among inertia, gravitational acceleration adjusted for buoyancy, and a drag force on an individual particle, as follows:

$$m_P \frac{dv_P}{dt} = -m_P\left(1-\frac{\rho_g}{\rho_P}\right)g - F_D = -m_P\left(1-\frac{\rho_g}{\rho_P}\right)g - \frac{1}{2}\rho_g v_{slip}^2 A_P' C_D \tag{2.10b}$$

where it is understood that the slip velocity, drag force, and gravitational acceleration, g, are vector quantities. A_P' is the projected area of the particle. The drag coefficient, C_D, is a function of Re_P. In the Stokes regime, Re_P is unity or less, for which C_D is simply $Re_P/24$. In the transitional regime, where Re_P ranges from unity to 1000, the drag coefficient is given by

$$C_D = \frac{24}{Re_P}\left[1 + 0.15 Re_P^{0.687}\right] \tag{2.10c}$$

This expression covers the full domain of Re_P for dilute coal suspensions, albeit without consideration of wall effects and deviations from sphericity. It also would need to be refined for analyses of the deposition of flyash and aerosol fumes on convective heat transfer surfaces.

In an analogous way, heat and mass transfer coefficients are developed for the Stokes regime, then extended for broader ranges of Re_P. The respective coefficients are correlated as nondimensional groups called the Nusselt number, Nu_P, and Sherwood number, Sh_P, defined as

$$Nu_P = \frac{hd_P}{\lambda_g} \quad \text{and} \quad Sh_P = \frac{k_m d_P}{D_{i,g}} \tag{2.11a}$$

where h is the mean convective heat transfer coefficient in W/m^2 K and k_m is the mean mass transfer coefficient in m/s. Average values for the gas thermal conductivity, λ_g, and binary Brownian diffusion coefficient, $D_{i,g}$, of reactant species i in the gas mixture must be specified to evaluate the transport coefficients.

The heat transfer coefficient is correlated with Re_P and the Prandlt number, Pr, as follows:

$$Nu_P = 2 + \left[0.4Re_P^{0.5} + 0.06Re_P^{0.667}\right] Pr^{0.4} \left(\frac{\mu}{\mu_S}\right)^{0.25} \quad \text{where } Pr = \frac{\mu C_{P,g}}{\lambda_g} \tag{2.11b}$$

In the Stokes regime, Nu_P relaxes to two in the limit that molecular conduction governs the heat transfer rate. The additional terms come into play for Reynolds numbers well above unity, since Pr is always very close to 0.7 for the gas mixtures of interest in coal utilization. The viscosity ratio is the ratio of gas viscosities for mean and surface conditions, which deviates from unity during transient heating and also under quasisteady combustion conditions with substantial O_2 levels.

The mass transfer coefficient is correlated with Re_P and the Schmidt number, Sc, as follows:

$$Sh_P = 2 + \left[0.552Re_P^{0.5}\right] Sc^{0.333} \quad \text{where } Sc = \frac{\nu}{D_{i,g}} \tag{2.11c}$$

In the Stokes regime, Sh_P relaxes to two in the limit that Brownian diffusion governs the mass transfer rate. The additional term comes into play for Reynolds numbers greater than unity.

2.4.2 Transport in fluidized systems

Fluidized beds process coal with very high suspension loadings and relatively coarse particle sizes, although the bulk of the solids are an inert fluidization medium such as quartz sand or agglomerated ash in a finer size range. The beds sustain at least two phases, an emulsion that contains almost all solids at volume fractions approaching one-half, and a bubble phase with hardly any solids. Under some bed operating conditions, wakes and clouds around individual bubbles entrain intermediate suspension loadings. The heat and mass transport coefficients of the coal and char particles must be specified to evaluate thermal histories and char conversion rates. In addition to the normal molecular and convective transport mechanisms, heat transfer rates are significantly accelerated by conduction during intermittent particle-to-particle contact.

One major complication from the high suspension loading is that the slip velocity introduced in Eq. (2.10a) will actually be erratic and distributed over a very broad range of values. Consequently, Re_P is not a suitable measure of the flow regime or the mean transport rates in fluidized beds. Instead of even estimating a scale for relative particle motion, emulsions are simply regarded as well-stirred reaction phases with uniform gas concentrations throughout.

The prominent role for particle-to-particle conduction carries significant uncertainties into the heat transfer coefficient for particle-to-gas heat transfer. This mechanism becomes faster for progressively larger particles and passes through a maximum for faster superficial gas velocities. For the coarse coal grinds and the smaller fluidization media fed into fluidized systems, a reasonable nominal value for h is 350 W/m^2 K. This value remains consistent with test data through sizes to several mm, and for gas velocities as fast as three times the velocity at minimum fluidization. Many correlations for Nu_P have been reported but there is little consistency among even the regression parameters. One form uses the Archimedes number, Ar, and the ratio of diameters for char and bed solids, d_{bed}, according to

$$Nu_P Ar^{-n_{bed}} = 3.539\left(\frac{d_P}{d_{bed}}\right)^{-0.257} \text{ where } Ar = \frac{\rho_g d_P^3(\rho_b - \rho_g)g}{\mu^2} \text{ and } n_{bed} = 0.105\left(\frac{d_P}{d_{bed}}\right)^{0.082} \tag{2.12a}$$

Following a series of developments in the literature over many years, the mass transfer coefficient was eventually correlated with conditions either for minimum or actual fluidization. The Sherwood number based on actual conditions is evaluated from

$$Sh_P = 2\varepsilon_{mf} + \left[0.69\left(\frac{Re_{bed}}{\varepsilon_{bed}}\right)^{0.5}\right]Sc^{0.333} \text{ where } Re_{bed} = \frac{Ud_P}{\nu} \tag{2.12b}$$

The expression contains bed voidages under minimum fluidization, ε_{mf}, and actual fluidization, ε_{bed}, conditions. The Reynolds number for the bed contains the superficial gas velocity into the bed, U, but the length scale is still the particle diameter. Its value ranges from 1 to 10 for AFBC. Alternatively, a correlation with U_{mf} and ε_{mf} within the bracketed term and the same coefficient and exponent has also been used to estimate k_m within the scatter of reported results. For small values of Re_{bed}, Sh_P relaxes to twice the minimum fluidization voidage, which is less than the Brownian diffusion limit; in other words, fluidization inhibits mass transfer.

2.5 Thermophysical properties of gases

Mitchell (1980) compiled expressions for the gas properties needed to evaluate transport coefficients for the domain of conditions in coal utilization technologies. The expressions for viscosity, thermal conductivity, and binary diffusivity in N_2 were defined as follows:

$$\mu_i = A_i T^{0.6756} \tag{2.13a}$$

$$\lambda_i = B_i T^{n_t} \tag{2.13b}$$

$$D_{i,N_2} = C_i \frac{T^{1.67}}{p} \tag{2.13c}$$

Gas viscosity and thermal conductivity are independent of pressure, and the pressure dependence in the binary diffusivity in Eq. (2.13c) is for pressure in atmospheres. The constants and exponent for the temperature dependences are collected in Table 2.3. Viscosity is in centipoise (cP), which is equivalent to mPa s and equal to 10^{-3} kg/m s. Thermal conductivity is in W/m K, which equals 0.85984 kcal/m h °C or 0.5779 Btu/ft h °F. Gas densities are evaluated with the ideal gas law.

Viscosities and thermal conductivities for gas mixtures can be evaluated with Wilke additivity, according to

$$\beta_{\text{mix}} = \sum_{i-1}^{n} \frac{x_i \beta_i}{\sum_{j=1}^{n} x_i \Phi_{ij}} \quad \text{where } \Phi_{ij} = \frac{1}{\sqrt{8}} \left(1 + \frac{M_i}{M_j}\right)^{-\frac{1}{2}} \left[1 + \left(\frac{\beta_i}{\beta_j}\right)^{\frac{1}{2}} \left(\frac{M_j}{M_i}\right)^{\frac{1}{4}}\right]^2 \tag{2.13d}$$

where β is either viscosity or thermal conductivity and x_i is the mole fraction of species i.

Table 2.3 Coefficients to evaluate gas properties.

Gas	Viscosity (cP)	Thermal conductivity (W/m k)		Binary diffusivity (cm²/s)
	$A_i \times 10^2$	$B_i \times 10^5$	n_t	$C_i \times 10^5$
CH_4	2.5100	4.652	1.1778	1.658
C_2H_2	2.5695	7.868	1.0265	1.291
CO	3.7486	30.86	0.7820	1.500
CO_2	3.6078	9.752	0.9386	1.191
H_2	1.8405	217.0	0.7681	5.525
H_2O	0.5083	3.477	1.1748	1.905
N_2	3.6974	32.19	0.7722	1.488
NH_3	3.4436	2.407	1.2642	1.849
NO	4.1530	31.41	0.7870	1.550
O_2	4.4203	29.87	0.7968	1.523

References

Atkinson B, Merrick D. Mathematical models of the thermal decomposition of coal. 4. Heat transfer and temperature profiles in a coke oven charge. Fuel 1983;62:553–61.

Brenner D. The macromolecular nature of bituminous coal. Fuel 1985;64:167.

Hatcher PG. Chemical structural models for coalified wood (vitrinite) in low rank coals. Org Geochem 1990;16:959–68.

Huffman GP, Huggins FE. Reactions and transformations of coal mineral matter at elevated temperatures. In: ACS symposium series vol. 301. Mineral matter and ash in coal. American Chemical Society; 1986. p. 100–13 [chapter 8].

Martinez-Tarazona MR, Spears DA, Palacios JM, Martinez-Alonso A, Tascon JMD. Mineral matter in coals of different rank from the Asturian Central basin. Fuel 1992;71(4):367–72.

Merrick D. Mathematical models of the thermal decomposition of coal. 2. Specific heats and heats of reaction. Fuel 1983;62:546.

Mitchell RE. A theoretical model of chemically reacting recirculating flows. Report SAND79-8236, Livermore, CA: Sandia National Laboratories; 1980.

Palmer CA, Filby RH. Distribution of trace elements in coal from the Powhatan No. 6 mine, Ohio. Fuel 1984;63(3):318–28.

Schweinfurth SP. An introduction to coal quality. In: Pierce BS, Dennen KO, editors. The national coal resource assessment overview. Reston, VA: US Dept. Interior, US Geological Survey; 2009 [chapter C].

Shurtz RC, Kolste KK, Fletcher TH. Coal swelling model for high heating rate pyrolysis applications. Energy Fuels 2011;25:2163–73.

Shurtz RC, Hogge JW, Fowers KC, Sorensen GS, Fletcher TH. Coal swelling model for pressurized high particle heating rate pyrolysis applications. Energy Fuels 2012;26:3612–27.

Skorupska NM, Couch G. Coal characterisation for predicting ash deposition: an international perspective. In: Williamson J, Wigley F, editors. The impact of ash deposition on coal fired plants, Solihull (UK). Washington, DC: Taylor and Francis; 1994. p. 137–50.

Srinivasachar S, Helble JJ, Boni AA, Shah N, Huffman GP, Huggins FE. Mineral behavior during coal combustion. 2. Illite transformations. Prog Energy Combust Sci 1990;16:293–302.

Straszheim WE, Markuszeqski R. Automated image analysis of minerals and their association with organic components in bituminous coals. Energy Fuels 1990;4(6):748–54.

Strugala A. Empirical formulae for calculating real density and total pore volume of hard coals. Fuel 1994;73(11):1781–5.

Vassilev SV, Kitano K, Vassileva CG. Relations between ash yield and chemical and mineral composition of coals. Fuel 1997;76(1):3–8.

Wigley F, Williamson J. The distribution of mineral matter in pulverized coal particles in relation to burnout behavior. Fuel 1997;76(3):1283–8.

Yu H, Marchek JE, Adair NL, Harb JN. Characterization of minerals and coal/mineral associations in pulverized coal. In: Williamson J, Wigley F, editors. The impact of ash deposition on coal fired plants, Solihull (UK). Washington, DC: Taylor and Francis; 1994. p. 361–72.

Moisture release and coal drying

3

Nomenclature

a	initial particle radius, cm
A_i	pseudo-frequency factor for release of multi- or monolayer moisture, g/cm^3 s
B_{EVAP}	transport number for evaporization
C	constant in the BET adsorption isotherm, Eq. (3.11).
$C_{p,i}$	specific heat of a region of coal plus moisture form i = b, M, m, cal/g K
D_i	mass diffusivity of multi- and monolayer moisture along pore walls, cm^2/s
E_i	activation energy for release of multi- or monolayer moisture, kJ/mol
F_M	flux of moisture through an external surface, g/cm^2 s
h	convective heat transfer coefficient at the external particle surface, cal/cm^2 s K
ΔH_i	specific enthalpy of vaporization of multi- or monolayer moisture, cal/g
ΔH_{VAP}	standard specific enthalpy of vaporization of bulk moisture, cal/g
k_i	overall mass transfer coefficient for multi- and monolayer moisture from the external surface, g/cm^2 s
m_P	particle mass, g
M_0	initial mass fraction of total moisture
n_i	power-law temperature coefficient in R_i, Eq. (3.6)
Nu_0	Nusselt number
p	pressure, MPa
q_f	external heat flux to a coal particle during drying
r	radial position, cm
r_b	position of the bulk vaporization front, cm
R_{DRY}	drying rate, g-moisture/s
R_i	rate of multi- or monolayer moisture release, g/cm^3 s
t	time, s
t_b^{MAX}	time when the bulk vaporization front arrives at the center of the particle and vanishes, s
t_M^{MAX}	time to eliminate multilayer moisture, s
t_p^{MAX}	duration of the preheating stage, s
$t_{M10\%}$	time to dry to 10 wt.% moisture, s
T	temperature, K
V_P/A_P	volume-to-surface ratio, cm
x_S	steam concentration, as a mole fraction
y_i	mass fraction of moisture of form i
$y_{i,mono}$	mass fraction of a monolayer of multi- or monolayer moisture in the BET isotherm, Eq. (3.11)
${}^{EQ}y_i$	mass fraction of multi- or monolayer moisture on the external surface in adsorption equilibrium with the drying environment

Process Chemistry of Coal Utilization. https://doi.org/10.1016/B978-0-12-818713-5.00003-4

Greek symbols

α_i thermal diffusivity of the coal particle plus moisture form i, cm^2/s
η_{DRY} drying efficiency
λ_i thermal conductivity of the coal particle plus moisture form i, cal/cm s K
λ_G thermal conductivity of the ambient gas, cal/cm s K
ρ_i density of the coal particle plus moisture form i, g/cm^3

Subscripts

b region with bulk, multilayer, and monolayer moisture
f free stream condition
M region with multilayer and monolayer moisture
m region with monolayer moisture only
0 initial condition

All coals contain moisture, and all of our subject utilization technologies remove it shortly after the coal is fed into the system. The physics that describe moisture removal from porous solids like coal are well-established and fully validated, but only for most coal types. For ranks of subbituminous and higher, moisture in coal is ordinary bulk moisture which forms a separate phase that fills the internal voids and pores in the organic coal matter, and coats exposed surfaces. Such moisture is released whenever its saturated vapor pressure exceeds the ambient partial pressure of moisture. The only complications pertain to whether or not the measured value of coal moisture accurately reflects the actual moisture content as the coal is fed into a process, and whether or not the moisture release mechanism in a process simulation will introduce numerical difficulties. Both issues are clarified in this chapter.

However, for very low rank coals such as lignites and brown coals, moisture release is akin to a chemical decomposition. These coals are true colloids whose organic matter is stabilized by ionic bonds and long-range dispersion forces across multiple layers of water (Hayashi and Li, 2004). Most of the moisture in these coals is released in a normal vaporization process. But after the bulk moisture vaporizes, substantial portions of the total moisture persist in forms bound much tighter to the coal structure. The release of this bound moisture requires significantly more energy and much longer times. Accordingly, any description of low rank coal drying must supplement the conventional treatment of bulk moisture vaporization with specialized rate expressions and supplemental vaporization endotherms. These expansions constitute the bulk of this chapter.

3.1 Moisture levels in coal

The moisture level in a coal sample is a component of a proximate analysis. It is assigned as the weight percentage of a sample released under air at 100–105°C or under N_2 at 150°C. The value is often reported on an as-received basis, which simply refers to whatever state the sample was in when it was delivered to the analytical laboratory. The sample could have been pulled from a mining operation, or a coal storage pile at a power station, or from a sealed, desiccated vessel from a research laboratory. Obviously, the moisture levels from the same sample would be different among these sources. To minimize such ambiguity, coals are routinely subjected to standardized drying conditions that eliminate portions of their total moisture contents prior to analysis, and each remaining portion is labeled with a standardized term. For example, samples subjected to air drying at 10–15°C above ambient temperature contain "residual" moisture that is often reported on an "air-dried" basis. Readers interested in the assortment of such definitions and standardized analyses should consult the testing standards, e.g., ASTM D1412, D2961, D3173, or D3302.

Standardized moisture levels are especially useful for testing at lab-scale, because the samples used in the tests can be subjected to the standardized drying conditions used in the moisture determination. In that case, a reported level of residual moisture would equal the actual moisture level of the sample fed into the laboratory apparatus. Such one-to-one correspondence is always desirable, and absolutely necessary in some testing programs. For example, when tests are run to measure the distribution of all major products formed during coal devolatilization, the amount of water formed by the chemical reaction mechanisms can only be assigned as the difference between the total amount of water in the product stream and the amount of water originally in the coal. Any uncertainty on the coal moisture level would translate directly into an uncertainty on the assigned yield of chemically formed water. To remove all ambiguities between the reported and actual moisture contents, many tests at lab-scale are simply run with dried coal samples (but this strategy is infeasible with lignites and brown coals).

At larger testing scales it becomes progressively more difficult to establish a one-to-one correspondence between reported and actual moisture contents. Pilot-scale facilities fed with coals that were prepulverized, preclassified, and stored in drums under an inert gas are straightforward, because samples taken for a moisture analysis are in the as-fed state. But whenever a coal is conveyed from a pile through a pulverizer and into a furnace or reactor, a host of uncertainties come into play. What number of samples from the pile are representative? Did it rain last night? Should samples be pulled on every test day? Obviously, the ambiguities grow with the size of the operation so that in field tests at commercial power stations, the relation between a reported moisture content and the actual moisture level in coal fed when test data were recorded becomes pure conjecture. For all these reasons, moisture levels in coals are subject to a multitude of uncertainties, and error bars on the analytical data cannot possibly convey any of the largest ambiguities in field tests.

3.2 Moisture release from most coal types

This section pertains to coal ranks of subbituminous or higher that contain only bulk moisture released by only conventional vaporization. Considering the ambiguities on reported coal moisture levels from the previous section, there is no incentive to use the most sophisticated treatments for moisture release from porous solids in process simulations of coal utilization technologies. In fact, drying treatments that resolve two stages above and below the normal boiling point often introduce discontinuities into the analysis that disrupts numerical convergence schemes. Even the most conventional vaporization model (Ko, 2005) introduces a discontinuity in the transient particle temperature that can cause computational problems. For the rapid heating rates imposed by our subject utilization technologies, it is not necessary to carefully resolve moisture release because it is always finished long before the onset of primary devolatilization.

The most expedient treatment is based on a hypothetical instantaneous dryer upstream of the coal injectors that produces dry coal and an entrainment stream that contains all the moisture that was fed with coal. Pulverizers partially dry coal via frictional heating, and realize some of the attributes of this hypothetical dryer. But this approach omits the impact of moisture release on calculated thermal histories for the particles in an injected coal suspension.

The drying stage may be explicitly resolved with the following analysis, which begins with a species balance for the moisture fraction in a particle during heatup:

$$\frac{dy_M}{dt} = -\frac{F_M}{\rho_b\left(\frac{V_P}{A_P}\right)M_0} \tag{3.1}$$

where $y_M(t)$ is the instantaneous mass fraction of moisture in the particle at time t; F_M is the moisture flux in g/cm^2 s; ρ_b is the initial bulk particle density; V_P/A_P is the volume-to-surface ratio evaluated as $a/3$ where a is the particle radius; and M_0 is the initial moisture fraction in coal. The moisture flux is evaluated as a mean value from a vaporization mechanism that is limited by heat transfer. In this way, the particle temperature continues to increase throughout moisture release, but at a slower rate due to the vaporization endotherm. Drying during heat-up to the vaporization temperature range, which begins at 90°C, is deemed to be negligible. The average moisture flux is evaluated as

$$F_M = \frac{\lambda_G Nu_0}{2a}\frac{\ln(1+B_{EVAP})}{0.51} \quad \text{where } Nu_0 = 2 + 0.6Re^{\frac{1}{2}}Pr^{\frac{1}{3}}$$
$$\text{and } B_{EVAP} = \frac{0.51\dfrac{(T_G - T_{REF})}{(T_b - T_0)}}{1 + \dfrac{\Delta H'_{VAP}}{(T_b - T_0)}} \tag{3.2}$$

where λ_G is the mean thermal conductivity of the entrainment gas; $T_G - T_{REF}$ is the mean gas temperature reduced by 125°C; $T_b - T_0$ is the difference between the boiling point and the initial fuel temperature; and $\Delta H'_{VAP}$ is the specific enthalpy of vaporization divided by the specific heat of water. In the evaluation of the Nusselt number, Nu_0, the particle Reynolds number is based on a slip velocity restricted to speeds slower than 3 m/s, and the Prandtl number is evaluated at mean gas conditions. Being based on mean conditions in the gas phase, the moisture flux is a fixed nominal value

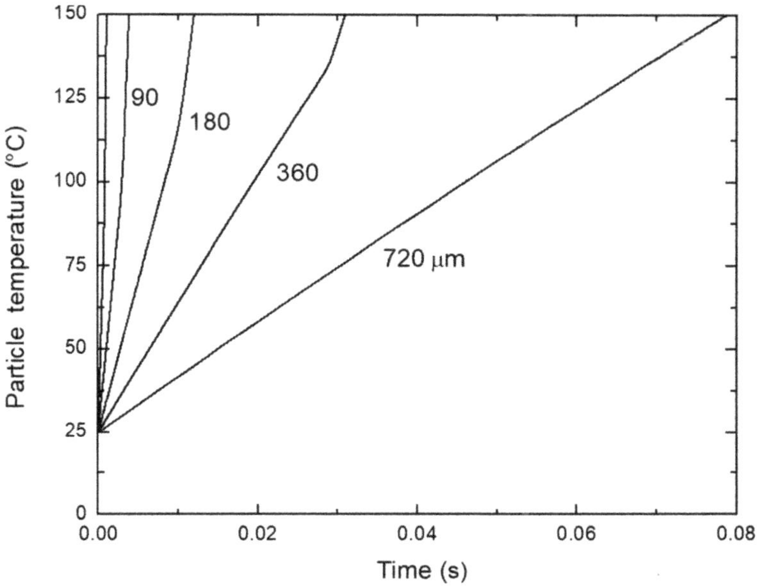

Fig. 3.1 Predicted thermal histories during release of 6.2 wt.% moisture for sizes from 44 to 720 μm.

throughout all stages of moisture release, so Eq. (3.1) can be integrated to find the instantaneous mass fraction of moisture, which is

$$y_M(t) = M_0 - \frac{F_M t}{\rho_b \left(\frac{V_P}{A_P}\right) M_0} \quad \text{where } y_M \geq 0 \tag{3.3}$$

The vaporization endotherm is expressed as $F_M \Delta H_{VAP}$, which appears as a source term in the particle energy balance developed at the end of Chapter 5 (cf. Eq. 5.17).

For the rapid heating conditions in almost all utilization technologies, this analysis spreads moisture release over a temperature range of tens of degrees, instead of imposing an isothermal stage at the normal boiling point. Consequently, the perturbations to the particle thermal histories are very modest, as seen in Fig. 3.1, and discontinuities do not arise at even the greatest moisture levels of interest. Another advantage is that Eqs. (3.2), (3.3) and the heat sink for vaporization can easily be incorporated into CFD simulations.

3.3 Moisture removal from very low rank coals

Lignites and brown coals are synonymous labels for coals of the lowest rank. They are classified as colloids because they consist of various porous particles of one phase—fulvic and humic acids, plant residues, and transformed plant fragments—dispersed in a second phase of water. This colloidal dispersion is stabilized by numerous ionic interactions between functional groups and metals in the solids with water and dissolved acids within the fine pores of the solids. Drying overcomes this colloidal stability at the expense of progressively more energy to remove progressively more water from the structure, especially during the final approach to dryness. The remainder of this chapter explains how these connections between brown coal structure and constitution affect the times for drying to a target moisture level.

Brown coals and lignites often contain more water than organic matter. The very high moisture levels are problematic because they raise transport costs, lower calorific values, and complicate handling and grinding operations. An assortment of technologies has been developed to remove most of the moisture at both mines and power plants, including mechanical rams, microwave heating, and numerous conventional heating configurations in packed and fluidized beds, as surveyed recently by Rao et al. (2015). The analysis in this chapter supports steam fluidized bed drying (SFBD) technology, although it can easily be extended to other thermal drying technologies. In SFBD, millimeter-sized fuel particles are fluidized in slightly superheated steam at near-atmospheric pressures from 120°C to 250°C to bring down the moisture levels to about 10 wt.%. Two main advantages are that the drying atmosphere is inert, which eliminates spontaneous combustion hazards, and that most of the heat of vaporization can be recovered by circulating steam from the product stream through a compressor and heat exchanger in the fluidized bed.

The technological imperatives have already spawned numerous computational schemes to predict the operating conditions needed to achieve a target dryness in the product, and to identify the best factors for process management and optimization. This work provides essential information on transport coefficients and the relevant heat transfer mechanisms, but it is limited by the universal premises that (i) the moisture in brown coals and lignites is a homogeneous bulk phase that occupies the internal porosity; and (ii) its evaporation can be described with the mathematical analyses for drying that are routinely applied to most other granular, porous materials. Unfortunately, the first premise is inconsistent with exhaustive scientific characterization of these materials and, therefore, the second premise is unsuitable. Fact is, the different forms of moisture in very low rank coals are an essential aspect of their drying behavior, without which there is little hope of describing the distinctive drying behavior of individual fuel samples.

3.3.1 Forms of moisture in very low rank coals

As explained in more detail by Allardice et al. (2004) and Yu et al. (2013), there are three distinct types of moisture in very low-rank coals: (i) Monolayer moisture fully interacts with functional groups in the coal structure via hydrogen bonds and ionic bonding complexes; (ii) Multilayer moisture forms a hydrogen bonded layer over monolayer moisture via long-range interactions; and (iii) Bulk moisture is water that volumetrically fills the bulk of the pore voidage, without chemical interactions with coal components. Monolayer and multilayer moisture are primarily stabilized by metal carboxylates, carboxylic acids, and phenolic hydroxyls in the coal structure.

Moisture on metal carboxylates is ionically bound and therefore immobile, whereas moisture on carboxylic acids and phenolic hydroxyls is hydrogen bonded and fully mobile, as are the acidic protons and the hydroxyl protons in the coal structure. A thermochemical equilibrium determines the concentrations of metal carboxylates, carboxylic acids, and the concentrations of metallic cations dissolved in the monolayer and multilayer moisture. This equilibrium may be affected by the pH of the bulk moisture, such that bulk moisture acts as a reservoir that responds to changes in the proportions of metal carboxylates and carboxylic acids.

Since the distinctive drying behavior of individual coal samples is determined by the initial mass fractions of the three types of moisture, these mass fractions are essential prerequisites for accurate predictions of drying behavior. They can be measured by a variety of means (Allardice et al., 2004), although many of these methods entail sophisticated analytical procedures and are incompatible with the pace of simulation work. Niksa and Krishnakumar (2015) developed a moisture estimation scheme based on the equilibrium moisture content (EMC) as a function of relative humidity (RH) at 30°C, which has already been reported for numerous Australian brown coal samples. Monolayer moisture is assigned as the EMC at RH of 22%, after Allardice and Evans (1971a,b). Bulk moisture is assigned from the EMC at RH of 93% and from the bed moisture, which fills interstitial voids in any aggregate coal sample. Bed moisture is based on the EMC at RH of 98%. Once the bulk and monolayer moisture fractions are calculated, the multilayer moisture is assigned by difference with the total moisture level minus bed moisture. The levels of the three estimated types of moisture were then correlated with variables from the proximate and ultimate analysis, which are the only sample-specific input requirement for the drying analysis. The regression for bulk moisture gave a (r^2) correlation coefficient of 0.91, and that for monolayer moisture was 0.78. Both regressions were most sensitive to the total moisture content and the O/H ratio.

When applied to an assortment of Australian brown coals, this estimation procedure gives broad variations in the levels of the three types of moisture among individual samples. Bulk moisture constituted 57%–80% of the total moisture; multilayer moisture constituted 19%–34%; and monolayer moisture constituted 1%–11%. These variations in the level of each form of moisture are definitely large enough to significantly affect the drying times under some, but not all, drying conditions, as demonstrated below.

During drying, the three forms of moisture are released sequentially (Allardice et al., 2004). Bulk moisture is released via evaporation and requires only the standard enthalpy of vaporization. Multilayer moisture is released by supplementing the enthalpy of vaporization with the free energy of formation of the hydrogen bonded multilayer. Monolayer moisture is released by supplementing the enthalpy of vaporization with the energy requirement to break the hydrogen bonds to phenolic hydroxyls and carboxylic acids, and to convert the metal carboxylates into an organic crosslink plus a free metallic acid. During drying at temperatures above about 150°C, metal carboxylates, carboxylic acids, and phenolic hydroxyls spontaneously decompose into CO_2, H_2O, and organic crosslinks within coal macromolecules. At temperatures above about 250°C, these decompositions become extensive enough to affect the thermochemical equilibrium that determines the concentrations of cations, carboxylic acids and carboxylates dissolved in multi- and monolayer moisture.

3.3.2 Mathematical analysis for moisture release from very low rank coals

The subject reaction system is a single, spherical particle of very low rank coal with specified proximate and ultimate analyses and particle radius, a. Initially, the particle holds specified levels of monolayer, multilayer, and bulk moisture, expressed as fractions of the total fuel mass. At some specified time, the particle is injected into a drying environment with specified uniform temperature, pressure, and steam concentration. The description of the drying environment must determine the external heat flux to the particle. For example, in a fluidized bed dryer, sufficient information would need to be provided to evaluate an overall heat transfer coefficient. The external heat flux would then be evaluated as the product of the heat transfer coefficient and the instantaneous temperature difference between the particle and fluidizing gas.

As seen in Fig. 3.2, the drying process proceeds through three regimes: (1) preheating; (2) bulk vaporization; and (3) drying under chemical reaction control. Preheating simply increases the particle temperature while the particle absorbs the external heat flux without releasing appreciable moisture. Temperature profiles are

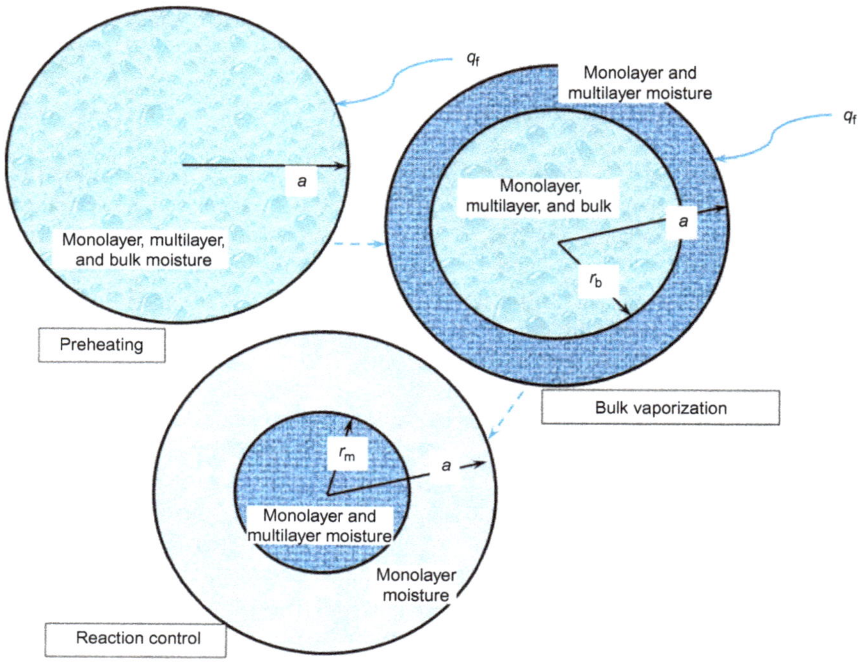

Fig. 3.2 Three regimes of very low rank coal drying.

established from the external surface to the center while the external heat flux is conducted through the coal and moisture. The preheating stage lasts as long as necessary to bring the external surface to the boiling point for the specified drying conditions, where the first bulk moisture begins to vaporize. As soon as the external surface reaches the boiling point, a bulk moisture vaporization front starts to move toward the center. The conduction flux into the front from the exterior shell provides sufficient energy to vaporize only the bulk moisture, and to continue to conduct energy toward the center. The net conduction flux across the vaporization front equals the latent heat of vaporization, so the front remains at the boiling point while it moves toward the center.

The exterior shell beyond the vaporization front continues to conduct the external heat flux, but at a slower rate due to the replacement of bulk water by steam within the internal pore system, which lowers the thermal conductivity. The steam escaping through the shell also prevents multilayer moisture from evaporating, even while the temperature through the outer layer continues to increase via heat conduction. Only after the bulk vaporization front vanishes at the particle center does multilayer moisture start to evaporate.

At this point, the mass fraction of multilayer moisture is fixed at the initial value throughout the particle. It then diminishes at the fastest rate at the external surface where the local temperature is hottest, while the drying endotherm suppresses radial temperature gradients. Eventually, the concentration of multilayer moisture vanishes from the external surface, which leaves only monolayer moisture. This region expands into an exterior shell due to the continuous conduction of the external heat flux. So there are now two domains, a core that contains a concentration profile of multilayer moisture on a uniform mass fraction of monolayer moisture and a shell that contains a profile of monolayer moisture. Since both regions have concentration profiles determined by the decomposition rates of bound moisture, this period is labeled as the Reaction Control regime in Fig. 3.2. Within the boundary between these regions, the mass fraction of multilayer moisture increases for progressively smaller radial positions, and the mass fraction of monolayer moisture remains at its initial value. On this boundary, there is no multilayer moisture but monolayer moisture remains at its initial mass fraction. Beyond this boundary, the amount of monolayer moisture diminishes for progressively greater radial positions. Note that monolayer moisture persists at the initial level until all multilayer and bulk moisture have been released from a given position. Similarly, multi- and monolayer moisture persist at their initial levels until all bulk moisture has been released from the entire particle, due to an equilibration of the mono- and multilayer moisture with the escaping steam from the bulk vaporization front. In this way, this analysis depicts the sequential elimination of the three forms of moisture.

The sequence of events in the previous section is implemented with pairs of conservation equations for energy and moisture for each of the four stages: preheating, bulk moisture vaporization, and release of multi- and monolayer moisture. All energy balances are transient equations in the radial direction of the following form:

$$\frac{\partial T_i}{\partial t} = \frac{1}{r^2}\frac{\partial}{\partial r}\alpha_i r^2 \frac{\partial T_i}{\partial r} - S_i^T \quad \text{where} \quad i = \text{b,M,m} \tag{3.4}$$

Table 3.1 Domains in time and radial extent, and moisture release terms for the four stages of moisture release

Stage	Domains		S_i^T	S_i^M
	Time	Radius		
Preheating	$0 < t < t_p^{MAX}$	$0 < r < a_P$	0	0
Bulk moisture	$t_p^{MAX} < t < t_b^{MAX}$	$0 < r < r_b$	0	0
		$r_b < r < a_P$	0	0
Multilayer	$t_b^{MAX} < t < t_M^{MAX}$	$0 < r < r_M \leq a_P$	$R_M\ (T_M, y_M)\Delta H_M/\rho_M C_{p,M}$	$R_M\ (T_M, y_M)/\rho_M$
Monolayer	$t_b{}^{MAX} < t < t_{m^{MAX}}$	$r_M < r < r_m \leq a_P$	$R_m(T_m, y_m)\ \Delta H_m/\rho_m C_{p,m}$	$R_m(T_m, y_m)/\rho_m$

where T_i is absolute temperature; t is time; r is radial position; α_i is thermal diffusivity; and S_i^T is a scaled enthalpy absorption rate for moisture release. The absolute temperature is a function of time and radial position, $T_i(r,t)$. Three subscripts label the four stages in this analysis. This is because "b" denotes regions that contain bulk, multilayer, and monolayer moisture, which arise during both preheating and bulk moisture release. Similarly, "M" denotes regions with multi- and monolayer moisture, whereas "m" denotes regions with only monolayer moisture.

The domains in time and radial position and the enthalpy absorption rates for the four stages are collected in Table 3.1. Preheating occurs throughout the entire particle, and lasts until the outer surface reaches the boiling point, which is denoted as t_p^{MAX}. Since preheating releases no moisture, there is no enthalpy requirement. Bulk moisture release partitions the particle into a shell and core, and lasts until the vaporization front reaches the center at t_b^{MAX}. Since bulk moisture is only released at the vaporization front, there is no enthalpy requirement within the shell of remaining bulk moisture. But this enthalpy requirement does factor into the following flux condition that determines the instantaneous location of the vaporization front:

$$\rho_{b,0} y_{b,0} \Delta H_{VAP} \frac{dr_b}{dt} = -\lambda_M \frac{\partial T_M(r_b, t)}{\partial r} + \lambda_b \frac{\partial T_b(r_b, t)}{\partial r} \tag{3.5}$$

where r_b is the position of the vaporization front; subscript "0" denotes the initial values for density, ρ_b, and bulk moisture weight fraction, y_b; ΔH_{VAP} denotes the standard heat of vaporization at the boiling point for the ambient conditions; and λ_i denotes thermal conductivity.

During the reaction control stage, the radial domain is divided into a core of radius r_M that contains both multi- and monolayer moisture, and a shell with only monolayer moisture. Similarly, once multilayer moisture has completely evaporated, a core of radius r_m contains only monolayer moisture, and the surrounding shell is in adsorption equilibrium with the ambient conditions.

The enthalpy absorption rates in Table 3.1 for multi- and monolayer moisture contain distinctive release rates and endotherms. The release rates are of the form

$$R_i = A_i T^{n_i} \exp\left(-\frac{E_i}{RT}\right) y_i \quad \text{where} \quad i = \text{M}, \text{m} \tag{3.6}$$

where A_i is a pseudo-frequency factor; n_i is a power-law exponent; and E_i is a desorption energy. For both forms of moisture, the assigned values are 1.5 for n_i and 4.2 kJ/mol for E_i, which are within the range of values associated with a physical adsorption/desorption processes with weak chemical interactions. The two pseudo-frequency factors are the only adjustable parameters in the analysis. Both values were set to interpret several drying datasets, and are the same for all coal samples. The vaporization endotherms for multi- and monolayer moisture were enhanced by 5% and 25%, respectively, over the normal enthalpy of vaporization, based on values reported by Allardice and Evans (1971b).

The boundary conditions on the energy balance are symmetric radial profiles at the particle center, and a convective heat flux into the external surface, according to

$$\frac{\partial T_i(0,t)}{\partial r} = 0 \quad \text{and} \quad \lambda_i \frac{\partial T_i(a,t)}{\partial r} = h(T_\text{f} - T_\text{b}(a,t)) \tag{3.7}$$

where T_f is the free stream temperature and h is the convective transfer coefficient.

Throughout the release of bulk moisture, the mass loss can be evaluated from the movement of the bulk vaporization front, according to

$$\frac{dm_\text{p}}{dt} = -\rho_{\text{b},0} y_{\text{b},0} \left(4\pi r_\text{b}^2 \frac{dr_\text{b}}{dt}\right) \quad \text{with} \quad m_\text{p}\left(t_\text{p}^\text{MAX}\right) = m_{\text{p},0} \tag{3.8}$$

where m_P is the instantaneous particle mass.

For multi- and monolayer moisture, the analysis for moisture fractions is analogous to the thermal analysis and of the form

$$\frac{\partial y_i}{\partial t} = \frac{1}{r^2} \frac{\partial}{\partial r} D_i r^2 \frac{\partial y_i}{\partial r} - S_i^\text{M} \quad \text{where} \quad i = \text{M}, \text{m} \tag{3.9}$$

where D_i are mass diffusivities; and S_i^M are the moisture release rates specified in Table 3.1 and Eq. (3.6). The diffusivities were set to very low values, because monolayer moisture is actually immobile, and surface diffusion of multilayer moisture is relatively slow. The boundary conditions on the moisture fraction equations ensure symmetry in the radial profiles through the particle center, and a convective mass flux off the external surface, according to

$$\frac{\partial y_i(0,t)}{\partial r} = 0 \quad \text{and} \quad \rho_i D_i \frac{\partial y_i(a,t)}{\partial r} = k_i\left({}^{EQ}yi(x_{\text{S},\text{f}}, T_\text{f}, p_\text{f}) - y_i(a,t)\right) \tag{3.10}$$

where k_i is a convective mass transfer coefficient and ${}^{EQ}y_i$ is the surface coverage of the moisture layer in equilibrium with the mole fraction of steam in the free stream, $x_{S,f}$, at the free-stream temperature and pressure, p_f. The surface coverages are evaluated with BET adsorption isotherms, according to

$$\frac{p_f}{{}^{EQ}yi(p_{sat}-p_f)}=\frac{1}{Cy_{i,mono}}+\frac{C-1}{Cy_{i,mono}}\frac{p_f}{p_{sat}} \tag{3.11}$$

where all parameters are based on the values reported by Allardice and Evans (1971a) with extrapolations to the hotter temperatures in commercial applications.

The analysis incorporates the thermophysical properties as functions of the three forms of moisture. Particle densities were resolved into contributions for the organic mass assigned with the correlations in Chapter 2 and the three moisture forms. Specific heats were evaluated similarly, along with contributions for moisture weighted by the mass fractions of the three forms. For thermal conductivity, the weighting factors were based on the component volume fractions, plus an estimate for thermal conductivity of the coal contribution developed by Atkinson and Merrick (1983). Convective heat and mass transfer coefficients were evaluated from the Ranz-Marshall correlation for Nusselt number, and the analog for Sherwood number.

Once the profiles of temperature and the mass fractions of multi- and monolayer moisture have been evaluated, the bulk moisture loss is evaluated from the particle mass loss as follows:

$$\begin{aligned}\frac{dm_p}{dt}&=-\int_0^a 4\pi r^2 R_M dr \quad t_M^{MAX}>t>t_b^{MAX} \quad \text{with} \quad m_p\left(t_b^{MAX}\right)\\&=m_{p,0}(1-y_{b,0})\end{aligned} \tag{3.12a}$$

$$\frac{dm_p}{dt}=-\int_0^{r_M} 4\pi r^2 R_M dr-\int_{r_M}^{a} 4\pi r^2 R_m dr \quad t>t_b^{MAX} \tag{3.12b}$$

As long as the organic coal mass does not decompose, the mass loss rate in Eq. (3.12b) equals the drying rate due to the joint elimination of multi- and monolayer moisture.

The drying efficiency, η_{DRY}, is defined from the mean mass fractions of the three forms of moisture, according to

$$\eta_{DRY}(t)=1-\frac{\langle y_b\rangle+\langle y_M\rangle+\langle y_m\rangle}{M_0} \tag{3.13a}$$

where $M_0=y_{b,0}+y_{M,0}+y_{m,0}$ and the quantities in brackets are the mean mass fractions of each form of moisture at time t. These mean values are obtained by integrating the moisture concentration profiles over the particle volume, so that

$$\langle y_i \rangle = y_{i,0} \text{ if } t < t_{\mathrm{p}}^{\mathrm{MAX}} \tag{3.13b}$$

$$\langle y_{\mathrm{b}} \rangle = \left(\frac{r_{\mathrm{b}}(t)}{a}\right)^3 y_{\mathrm{b},0};\ \langle y_{\mathrm{M}} \rangle = y_{\mathrm{M},0} \text{ and } \langle y_{\mathrm{m}} \rangle = y_{\mathrm{m},0} \text{ if } t_{\mathrm{p}}^{\mathrm{MAX}} < t < t_{\mathrm{b}}^{\mathrm{MAX}} \tag{3.13c}$$

$$y_{\mathrm{b}} = 0;\ y_{\mathrm{M}} = \frac{\int_0^{r_{\mathrm{M}}} r^2 y_{\mathrm{M}}(rt)dr}{\frac{a^3}{3}} \text{ and } y_{\mathrm{m}} = \frac{\frac{r_{\mathrm{M}}^3}{3} y_{\mathrm{m},0} + \int_{r_{\mathrm{M}}}^{a} r^2 y_{\mathrm{m}}(rt)dr}{\frac{a^3}{3}} \text{ if } t > t_{\mathrm{b}}^{\mathrm{MAX}} \tag{3.13d}$$

Once the drying efficiency has been evaluated throughout a specified drying period, the drying rate in g moisture/s can be determined by differentiation, according to

$$R_{\mathrm{DRY}}(t) = m_{\mathrm{p},0} M_0 \frac{d\eta_{\mathrm{DRY}}}{dt} \tag{3.13e}$$

3.3.3 Evaluations with reported drying histories

This section presents three case studies to demonstrate that the drying analysis can accurately depict drying histories throughout a broad domain of drying conditions. More detail on these comparisons was reported by Niksa and Krishnakumar (2015). Beeby et al. (1985) monitored steam drying at atmospheric pressure from 101°C to 178°C over extended holding times sufficient to achieve the EMCs. The Yallourn brown coal contained 64% moisture whose bulk, multilayer, and monolayer moisture fractions were 42%, 14%, and 8%, respectively. The particle size was 1 mm.

The predicted drying curve for 140°C is shown in Fig. 3.3, along with its resolution into the contributions for bulk moisture vaporization and the release of multi- and monolayer moisture. The preheating period is too short to see on this time scale. Bulk moisture is removed much faster than both other forms, as expected. At the end of the bulk vaporization period of 120 s, the residual moisture was 38 wt.%. Subsequent removals of multi- and monolayer moisture are relatively very slow, which introduces two dramatic reductions in the slope of the drying curve. About 3000 s was required to completely remove multilayer moisture, at which point the moisture content was about 18%. Further removal of monolayer moisture is even slower, such that an additional 1900 s only decreased the moisture to about 16%. About 6800 s was required to reduce the coal moisture to 10 wt.%, and over 15,000 s was needed for complete dryness.

The predicted residual particle mass fractions for all test temperatures are in excellent agreement with the measured values, as seen in Fig. 3.4. At a low superheat condition at 101°C, only bulk moisture was removed, whereas multi- and monolayer moistures were removed at the hotter temperatures. The most distinctive feature for this dataset is that, at 101°C after 1600s, there was no further advance of the bulk vaporization front toward the center, so bulk moisture vaporization ceased. The degree

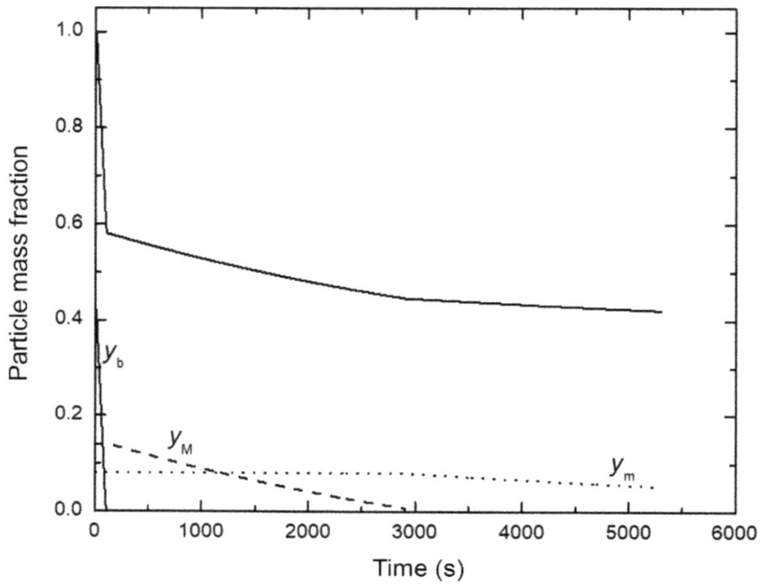

Fig. 3.3 Predicted drying curve for the test at 140°C of Beeby et al. (1985).

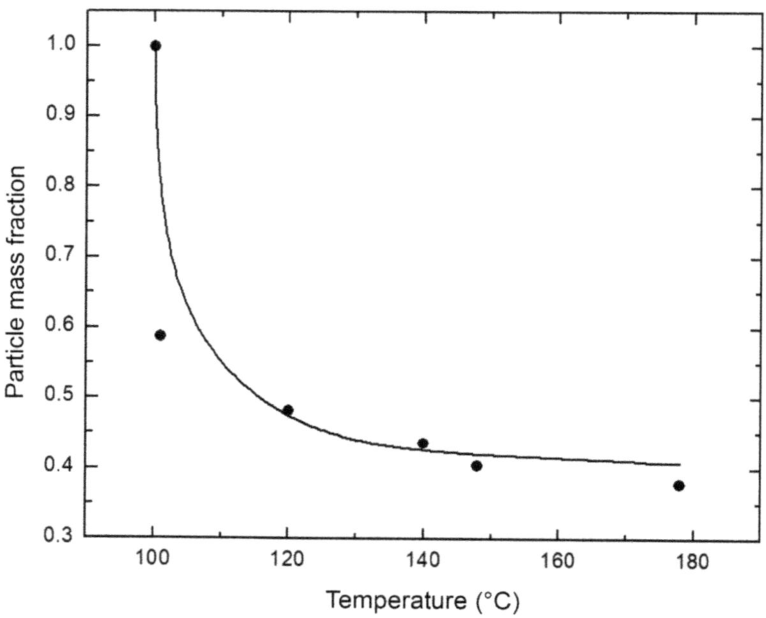

Fig. 3.4 Residual mass fractions at different temperatures for the tests of Beeby et al. (1985).

of superheat and the temperature gradients within the particle were too weak to drive the front to remove additional moisture.

In the study by Bongers et al. (1995), pressurized steam drying of low ash Loy Yang (LYLA) coal was examined in a batch autoclave from 180°C to 230°C at 1 MPa. The test coal had 62% moisture with 41% bulk; 16% multilayer; and 5% monolayer moisture. At 1 MPa, the dried coals reached their EMCs after 20 h of drying at all temperatures; increasing the residence time to 112 h did not change the EMC. The residual particle mass fractions are evaluated in Fig. 3.5. After a residence time of 20 h, the measured moisture content decreased from 62% to 5% at 222°C whereas the calculated values dropped from 62% to 1% moisture. While the predicted residual coal mass as a function of temperature is consistent with the measured values throughout most of the temperature range, the relatively steep drop in the moisture level at183°C is depicted as a more gradual removal.

The steam drying experiments by Favas and Jackson (2003) were conducted between 130°C and 350°C for 30 min after the autoclave reached the test temperature. Cases to 250°C are considered here, because the analysis does not account for the extensive loss of volatiles at hotter temperatures. The coal contained 60% moisture with 37% bulk, 18% multilayer, and 5% monolayer moisture, and was 5 mm in size. Since the test pressure was not reported, the steam pressure in simulations was adjusted continuously until the calculated moisture content of the coal after 30 min matched the measured values at 130°C and 230°C. At 130°C, the estimated degree of superheat was 2.15°C whereas at 230°C, it was 17°C. For the rest of the

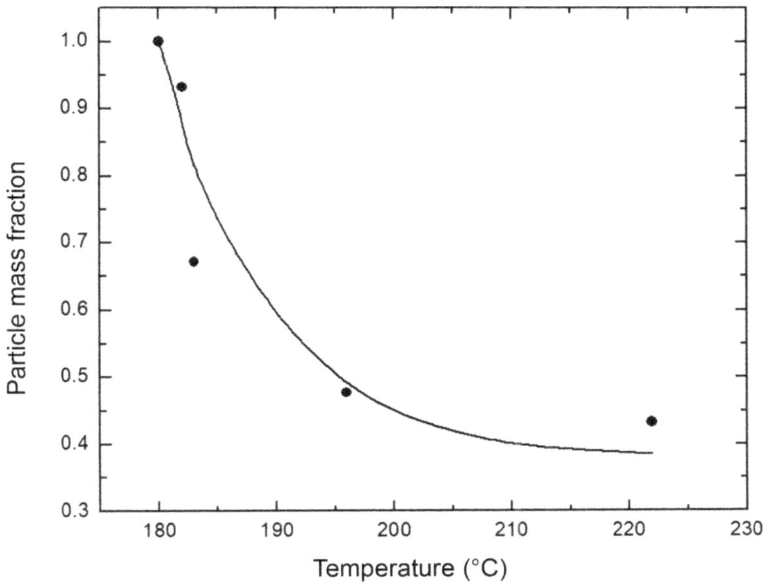

Fig. 3.5 Residual mass fractions for different degrees of superheat for the tests of Bongers et al. (1995).

temperatures, a linear interpolation was used to first calculate the degrees of superheat and then to estimate the corresponding pressure.

The predicted coal mass fractions are evaluated in Fig. 3.6. In the comparison at 250°C, the residual coal mass was corrected to represent only moisture loss because volatile coal decomposition products reduced the coal recovery to only 95.5%. Throughout the temperature range, the measured and predicted values are in excellent agreement although, for these tests, the residence time was relatively short so only bulk moisture was removed, and the rates of multi- and monolayer moisture loss did not come into play except for a marginal contribution from multilayer moisture at 250°C.

Since the analysis accurately depicts the most important aspects of drying, it can be used to identify the fuel properties and operating conditions that govern drying times and efficiencies. First consider how variations in the three forms of moisture affect drying times. Four samples that nearly cover the domain of moisture variations for Australian brown coals were compiled from the reported properties of Yallourn (YL), low ash Loy Yang (LYLA), Morewell No. 1 (MOR1), and Bowmans (BOW) coals. As seen in Table 3.2, the total bed moisture in these coals varies between 52 and 64 wt.%, based on the correlations with coal composition. Monolayer moisture varies between 2.3 and 7.2 wt.%; multilayer moisture varies between 8.9 and 20.3 wt.%; and bulk moisture varies between 36 and 43.2 wt.%. LYLA has the most

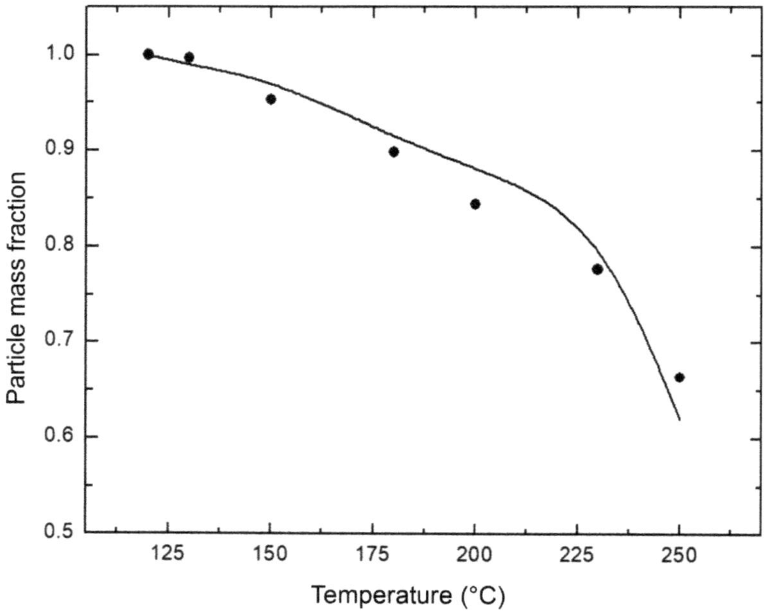

Fig. 3.6 Residual mass fractions at different temperatures for the tests of Favas and Jackson (2003).

Table 3.2 Predicted times for preheating, and to remove bulk and multilayer moisture, and to reach 10% residual moisture, and the drying efficiency for diverse brown coals.

Sample	$M_0/y_b/y_M/y_m$	t_p^{MAX} (s)	t_b^{MAX} (s)	t_M^{MAX} (s)	$t_{M10\%}$ (s)	η_{DRY} (%)
YL	64/43/14/7	42	415	6200	10,500	95
LYLA	62/39/20/2	41	383	9577	8147	93
MOR1	59/36/17/6	38	323	6856	9555	94
BOW	52/39/9/5	36	381	5715	5620	91

multilayer moisture and the least monolayer moisture; YL has the most bulk and monolayer moisture; MOR1 has the least bulk moisture; and BOW has the least multilayer moisture.

In the drying simulations, 2 mm coal particles were steam dried at a slightly sub-atmospheric pressure and 145°C. The feed steam contained 5% nitrogen. The simulations were then run to complete dryness. The times corresponding to the different stages of drying are reported in Table 3.2 for the four coals. The table reports the times to finish preheating (t_p^{MAX}); to release bulk moisture (t_b^{MAX}); and bulk plus multilayer moisture (t_M^{MAX}), and to reduce total moisture to 10 wt.% ($t_{M10\%}$). These times are cumulative and do not express drying times for individual stages. The corresponding drying efficiency, η_{DRY}, appears in the final column.

Drying times are not solely a function of the total moisture in the coal. Time $t_{M10\%}$ was longest for YL and shortest for BOW, and these coals had the highest and lowest total moisture, respectively. But the times for LYLA and MOR1 did not correlate with their total moisture contents. For all coals, preheating took negligible portions of the total drying period. Similarly, the variation in bulk moisture from 36% to 43.2% changed the bulk vaporization period by less than 100 s, which is insignificant compared to $t_{M10\%}$, and the time to eliminate bulk moisture was never more than 7% of the drying time. Consequently, variations in the bulk moisture levels for Australian brown coals do not significantly impact drying times under typical SFBD conditions. Since bulk moisture vaporization times are similar, the time for multilayer moisture removal is determined by the level of multilayer moisture, as expected. But the drying time to 10% moisture does not correlate with the monolayer moisture levels.

The comparisons between t_M^{MAX} and $t_{M10\%}$ reveal that coals with more multilayer moisture reach a target dryness faster than those with more monolayer moisture. The 10 wt.% moisture target was reached during multilayer moisture vaporization for LYLA and BOW, so that no monolayer moisture was released during these simulations; in contrast, some monolayer moisture had to be released from YL and MOR1 to reach the target. Consequently, LYLA and BOW dried faster than YL and MOR1 simply because multilayer moisture is released faster than monolayer moisture. The times to achieve a target moisture level are proportional to the total moisture in the coal, provided that the coals are first sorted into two groups, one that meets the target without releasing any of its monolayer moisture, and the other that requires the release of appreciable monolayer moisture. Within each group, drying

Table 3.3 Predicted times for preheating, and to remove bulk and multilayer moisture, and to reach 10% residual moisture for various sizes and temperatures.

Case	t_p^{MAX} (s)	t_b^{MAX} (s)	t_M^{MAX} (s)	$t_{M10\%}$ (s)
Baseline	39	294	2289	3692
0.5 mm	2	19	400	2380
5.0 mm	241	1746	3958	5566
170°C	26	180	1878	3070
120°C	64	640	3008	4670

times are proportional to total moisture. But the correlation to total moisture breaks down when samples from both groups are taken together.

Table 3.3 presents results on the impact of variations in the processing conditions on brown coal drying. In all cases, the coal contained 60% moisture allocated as 30% bulk, 10% multilayer, and 20% monolayer, and was injected into a slightly subatmospheric bed with 5% N_2 in steam. The baseline case operated with 2 mm particles at 145°C. Temperatures were varied from 120°C to 170°C, and sizes were varied from 0.5 to 5 mm. Simulations were run at all conditions until the total remaining moisture was 10%, and this target required release of about three-fourths of the monolayer moisture for the subject coal properties. The drying efficiency was 93% in all cases.

Drying times are most sensitive to variations in the particle size. Drying times more-than-doubled as size was increased from 0.5 to 5 mm. Drying times also increased as the temperature was reduced from 170°C to 120°C, but only by one-half. The results in Table 3.3 also show that the relative contributions from the stages of drying to the total drying time are strongly affected by particle size. For the smallest size, preheating and bulk moisture vaporization account for less than 1% of the drying time, whereas for the largest size, these two stages take just under one-third of the drying time. Consequently, the total drying time for the largest particle size comprises comparable contributions for the three types of moisture, but for small particles, the time to release monolayer moisture predominates.

3.3.4 General features of very low rank coal drying

To be useful for model validation work, this analysis reveals that, at a minimum, datasets must include the proximate and ultimate analyses and the mean particle size or particle size distribution of the coal; and the temperature, pressure, steam partial pressure, and residence time in every test. It also interprets the most important features of very low rank coal drying. For drying temperatures near the normal boiling point of water, bulk moisture vaporization stops even before the bulk vaporization front reaches the particle center, so that only part of the bulk moisture is removed. The drying curve and the moisture concentration profiles remain steady because the temperature gradients within the particle are too weak to remove additional bulk moisture. But for the hotter temperatures and greater degrees of superheat in many commercial drying processes, and for sizes smaller than 2 mm, both the preheating stage and the bulk vaporization stages make negligible contributions to drying times unless the particles are larger than several millimeters.

Since most of the moisture in all brown coals is bulk moisture, its rapid elimination obscures the relationship between total moisture and the time needed to achieve a target moisture level. But the drying curves of residual moisture vs. time from this analysis clearly depict the sudden deceleration in the drying rate when bulk moisture is eliminated and multilayer moisture begins to vaporize, and a second, more moderate deceleration, when multilayer moisture is eliminated and monolayer moisture begins to vaporize. Indeed, it is these two transitions, rather than the total moisture level, that determine drying times to a target moisture level.

The drying time is not solely a function of the total moisture content of the coal. For a selection of different coals, the times to achieve a target moisture level are proportional to the total moisture in the coal, but only if the coals are first sorted into two groups, one that meets the target without releasing any of its monolayer moisture, and the other that requires the release of appreciable monolayer moisture. A key aspect is whether or not the target moisture level requires removal of any monolayer moisture at all. Coals with abundant bulk and multilayer moisture can achieve targets of, say, 10% residual moisture by vaporizing only a portion of the multilayer moisture. Consequently, their drying times will be relatively modest. But coals with little bulk moisture and relatively abundant monolayer moisture will require relatively long drying times simply because monolayer moisture is always released more slowly than multilayer moisture. Hence, the distribution of the three forms of moisture in any particular coal sample is a determining factor for both drying time and enthalpy requirements.

These conclusions were demonstrated with simulations for diverse Australian brown coal samples, based on the correlation of EMCs for broad ranges of RH to estimate the initial levels of the three forms of moisture. The mathematical analysis can be applied to very low rank coals from other regions of the world without modification, although new correlations will need to be developed to estimate the moisture levels for coals from other regions, because numerous climatic and geographical factors acting on geological time scales determine the cation distributions which, in turn, affect the forms of moisture in coal beds.

References

Allardice DJ, Evans DG. The brown-coal/water system: part 1. The effect of temperature on the evolution of water from the brown coal. Fuel 1971a;50:201–10.

Allardice DJ, Evans DG. The brown-coal/water system: part 2. Water sorption isotherms on bed-moist Yallourn brown coal. Fuel 1971b;50:236–53.

Allardice DJ, Chaffee AL, Jackson WR, Marshall M. Water in brown coal and its removal. In: Li C-Z, editor. Advances in the science of Victorian brown coal. Amsterdam: Elsevier; 2004. p. 85–125.

Atkinson B, Merrick D. Mathematical models of the thermal decomposition of coal. 4. Heat transfer and temperature profiles in a coke oven charge. Fuel 1983;62:553–61.

Beeby C, Dodds C-R, Ho H, Potter OE, Yamaoka Y. Steam drying of coal. In: Proceedings of the international conference on coal science. Sydney: Pergamon Press; 1985. p. 513.

Bongers GD, Woskoboenko F, Jackson WR, Patti AF, Redlich PJ. The properties of the coal and steam condensate from pressurized steam drying. In: Pajares JA, Tascon JMD, editors. Coal science, vol. 24. Amsterdam: Elsevier Science, BV; 1995. p. 901.

Favas G, Jackson WR. Hydrothermal dewatering of lower rank coals. 2. Effect of coal characteristics for a range of Australian and international coals. Fuel 2003;82:59–69.

Hayashi J-I, Li C-Z. Structure and properties of Victorian brown coal. In: Li C-Z, editor. Advances in the science of Victorian brown coal. Amsterdam: Elsevier; 2004. p. 11–84.

Ko KK. Principles of combustion. Hoboken, NJ: John Wiley and Sons; 2005.

Niksa S, Krishnakumar B. Predicting the steam drying behavior of brown coals and lignites. Fuel 2015;159:345–53.

Rao Z, Zhao Y, Huang C. Recent developments in drying and dewatering for low rank coals. Prog Energy Combust Sci 2015;46:1–11.

Yu J, Tahmasebi A, Han Y, Yin F, Li X. A review on water in low rank coal: the existence, interaction with coal structure and effects on coal utilization. Fuel Process Technol 2013;106:9–20.

Primary devolatilization behavior

4

Nomenclature

M_n	number-average molecular weight of tar, g/mol
q	coal heating rate, °C/s
S_{ORG}	organic sulfur content of coal, wt.%
S_{PYR}	pyritic sulfur content of coal, wt.%
S_{SO4}	sulfate sulfur content of coal, wt.%
t	time, s
T	temperature, °C
W	weight loss, daf wt.%
x	fractional atomic number for S in pyrrhotite

This chapter is the first of three devoted to primary devolatilization, and introduces an organizational pattern found throughout this book. It presents the empirical basis for the important aspects of primary devolatilization and how they change with variations in coal quality and operating conditions. The next chapter develops a comprehensive reaction mechanism to quantitatively interpret these tendencies throughout the operational domain of our subject utilization technologies. A third chapter presents extensive quantitative interpretations of the empirical database with the comprehensive mechanism, and identifies areas for improvement. This same progression from an empirical basis to the requisite modeling analyses to quantitative validations is seen in the chapters on tar decomposition, volatiles conversion, and hydrogasification.

4.1 Definitions and commercial impacts

Once the fuel temperature is raised above some threshold value, depending on the coal type and heating rate, the coal matrix spontaneously disintegrates. Eventually, fragments of the original macromolecules become small enough to evaporate and escape through the fuel particle into the free stream as tar species, while lighter noncondensable gases are released both by decomposing peripheral groups and by repolymerization of fragments into larger condensed species. Collectively, tar and noncondensables are called volatiles. The disintegrating macromolecular coal matrix is ultimately reformed into an amorphous carbonaceous solid called char. After the macromolecular fragments have either been released or incorporated into char, the char expels hydrocarbon fragments and, especially, heteroatoms as additional noncondensable gases. Strictly speaking, this process is complete only when all heteroatoms have been eliminated, and only enough hydrogen remains to stabilize the carbon in char into extensive aromatic domains. But this limit is almost never achieved in commercial utilization technologies because it takes a long time at very high temperatures.

The partitioning of the coal feed into volatiles and char is crucial because volatiles are released and converted much faster than any form of char conversion. The process

Process Chemistry of Coal Utilization. https://doi.org/10.1016/B978-0-12-818713-5.00004-6

responsible for the partitioning is called "devolatilization" or "pyrolysis." In any utilization technology, devolatilization is unavoidable because its initiation temperature is always cooler than the temperature windows for all other chemical processing of coal into liquid or gaseous products. As the first stage of conversion in any utilization technology, it is important for numerous reasons. Most subbituminous and all bituminous coals form viscous melts during devolatilization at rapid heating rates, and softened, sticky fuel particles can coalesce into deposits that obstruct fuel flows through burners and fuel injectors. Volatiles ignite and stabilize coal flames on burners and fuel injectors. Volatiles combustion accounts for major portions of the total heat release, so devolatilization affects temperature fields near burner belts. The portion of coal-N released during devolatilization determines the effectiveness of aerodynamic NO_X abatement schemes. For the most effective NO_X abatement methods, the residual N in char largely determines furnace NO_X emissions. Most sulfur in coal is released during devolatilization at flame temperatures. The subsequent reactivity of char is determined during devolatilization by swelling that radically transforms the char PSD, in combination with the loss of a coal's internal pore system whenever a fuel particle melts and foams. Furnaces are sized for complete char burnout; gasifiers are sized to maximize the conversion of carbon in char.

Among all the stages of coal conversion, devolatilization is the most widely variable across the coal rank spectrum; even among different samples of the same rank, the split between volatiles and char is often markedly different. Kinetics for devolatilization rates are incorporated into most process simulations, although the total volatiles yield is the crucial characteristic. The stoichiometric requirements for volatiles combustion and reforming, and the associated enthalpy requirements are also required. But molecular species concentrations for volatiles are radically simplified or even ignored in most CFD simulations.

4.1.1 The devolatilization stage

In most utilization technologies, a coal's thermal decomposition initiates a cascade of distinct chemical processes within and around the fuel particles, for two reasons. First, the surrounding entrainment gases heat at rates comparable to those for the suspended coal particles, so volatiles continue to decompose after their release from the coal. Second, process streams almost always contain reactive gases, particularly O_2 and H_2, that accelerate the conversion of volatiles into ultimate products. The cascade begins with "primary devolatilization" which generates volatiles from chemistry within only the condensed coal phase. It is always the first process, and may initiate some or all of the conversion sequence in Fig. 4.1. This diagram shows the major products of primary devolatilization, secondary volatiles pyrolysis, volatiles combustion, and volatiles reforming. Species preceded by "+" are produced, whereas "−" denotes destruction. So-called "primary products" or "primary volatiles" comprise primary tars, noncondensable fuels (C_1-C_3 gaseous hydrocarbons (GHCs), CO, H_2, HCN, H_2S), and the gasification agents, steam and CO_2.

When neither O_2 nor H_2 is present in the surrounding atmosphere, "secondary volatiles pyrolysis" (a.k.a. "secondary pyrolysis" or "volatiles pyrolysis") converts primary volatiles into secondary pyrolysis products. The predominant transformation

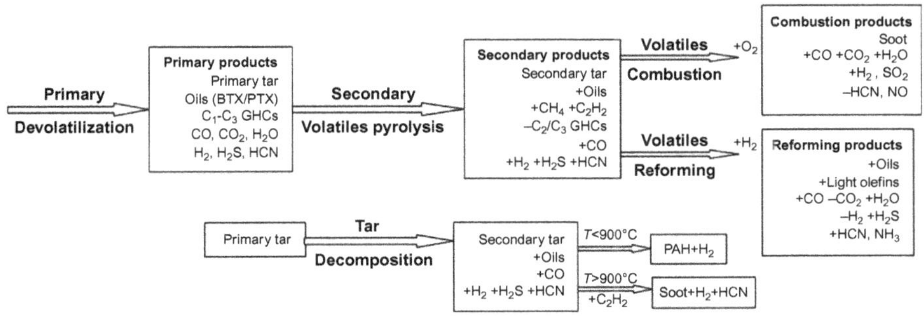

Fig. 4.1 Process chemistry within a devolatilization stage.

is "tar decomposition," which converts primary tars into secondary tars, oils, and additional noncondensables, and ultimately forms polynuclear aromatic hydrocarbons (PAH), and/or soot plus CO, H_2, H_2S, and HCN, depending on the ambient temperature. Volatiles pyrolysis also reforms the GHCs into CH_4 and C_2H_2. In the presence of O_2, "volatiles combustion" comprises the combustion of the mixtures of noncondensable fuel components in secondary volatiles, as well as the conversion of volatile-N species such as HCN and NH_3 into noxious gases and the conversion of H_2S into SO_2. In the presence of H_2 and in gasification environments, "volatiles reforming" denotes the conversion of noncondensables under reducing atmospheres, including the hydrogenation of GHCs into olefins and oils and water-gas shifting.

The major reaction processes in a devolatilization stage are primary devolatilization, tar decomposition, and other aspects of secondary volatiles pyrolysis and, if the atmosphere is reactive, volatiles combustion or volatiles reforming. As will be seen in Chapter 8, the oxidation of soot and char may compete for the available O_2 with gaseous fuel compounds, depending on the operating conditions. But soot and char gasification are relegated to different stages because their characteristic time scales are often much longer than those for any of the chemistries of devolatilization. Primary devolatilization only involves heterogeneous chemistry, by definition, but all subsequent chemical conversions of volatiles are homogeneous in the gas phase, except for a few heterogeneous reactions in which soot and char catalyze chemistry among a very limited number of volatiles. For example, NO formed during volatiles combustion can be reduced into N_2 by CO on soot and char. But this heterogeneous transformation should be considered an aspect of char oxidation, because char oxidation is the source of the requisite CO. Also, tars may deposit onto char within a coal's internal pore system, but only when heating rates are slower than those imposed in our utilization technologies of interest. So this heterogeneous transformation will be ignored.

This chapter and Chapters 5 and 6 cover primary devolatilization; Chapter 7 covers tar decomposition; Chapter 8 covers volatiles reforming and volatiles combustion; and Chapter 9 covers hydropyrolysis, tar hydrogenation, and the hydrogasification of soot and char. Collectively, these reaction mechanisms cover all the process chemistry in our subject utilization technologies (cf. Chapter 1), except char and soot conversion. This author's approach to these essential heterogeneous processes is available elsewhere (Niksa et al., 2003; Liu and Niksa, 2004).

Before we address all this chemistry, it is fair to ask, "Why resolve the devolatilization stage into a sequence of distinct chemical processes at all?" Since the time scales for primary devolatilization and volatiles conversion are comparable, is there more than academic interest to be gained by all this complexity? Shouldn't the devolatilization stage be regarded as a single chemical process? There are two very good reasons with broad practical implications to resolve the distinct chemical stages of devolatilization. First, coal constitution determines primary devolatilization behavior, but has virtually no impact on volatiles conversion. In other words, the connections among coal constitution and devolatilization behavior are readily apparent in the distributions of pristine primary products, but obscured by volatiles conversion chemistry. Indeed, the knowledge that became the basis to predict the devolatilization behavior of individual coal samples was mostly revealed by measured distributions of primary products from numerous coals.

The second reason is that nearly all the chemistry required to describe volatiles conversion was already elucidated by the combustion kinetics community before any legitimate attempts were made to unravel volatiles conversion chemistry. Once primary devolatilization and tar decomposition are factored out, volatiles conversion chemistry is completely covered by conventional homogeneous combustion chemistry among gaseous species. The oxidation of mixtures of coal volatiles abides by the mechanisms developed for the combustion of natural gas, synthesis gas, and gas mixtures from sources that have nothing to do with coal. Coal-derived gas mixtures tend to be among the most complex but can nevertheless be analyzed with conventional oxidation mechanisms. These same generalizations pertain to volatiles pyrolysis and volatiles reforming as well. But tar conversion has not yet been analyzed with the elementary mechanisms developed for the production of soot from noncondensable gas mixtures, mostly because tar structures are about as complex as coal macromolecules. So the devolatilization stage will be subdivided into distinct reaction processes as the only practical means to elucidate the connections between coal constitution and the crucial partitioning of coal into volatiles and char; and also to utilize the phenomenal knowledge base on homogeneous combustion mechanisms and kinetics in simulations of coal processing.

4.1.2 Secondary chemistry

The term "secondary chemistry" comprises tar decomposition, secondary volatiles pyrolysis, volatiles combustion, and volatiles reforming, although all four chemical processes hardly ever come into play in any particular utilization technology (CFBC being the notable exception). So secondary chemistry refers to whichever of these four chemical processes are important in a subject technology.

In abstract terms, primary devolatilization is easily distinguished from secondary chemistry: Primary devolatilization is the result of chemistry within the condensed coal phase, whereas secondary chemistry occurs in the gas phase beyond the interfacial area around the condensed phase. Even though the interface between the condensed and vapor phases is tangible, numerous ambiguities still arise in practical applications. One reason is that the size of the coal particle, per se, does not differentiate primary devolatilization from secondary chemistry. That is because the abstract boundary between the domains for primary and secondary chemistry is actually the interface between the volumetric coal phase and its internal pore system, as well as the external particle surface. Once volatiles cross the interface surrounding any portion of the coal phase and enter the internal pore system or bubbles in a coal melt, their succeeding transformations constitute secondary chemistry. So the gaseous products emitted through the external surface of a heated fuel particle can be either primary or secondary volatiles, depending on the heating rate. If they were generated so fast that their transit time to the external particle surface was too short to sustain any chemistry during transport, and if they were immediately quenched to prevent secondary chemistry in the free stream, then they would be primary products. Otherwise, they are secondary products. Of course, we already excluded the slow heating situation in our selection of commercial technologies, so only primary products are expelled through the external surface in our subject applications.

Similarly, some laboratory tests contact coal particles with a preheated gas stream. Since primary devolatilization almost always occurs while the coal is being heated to the ambient gas temperature, the volatiles are ejected into gases that are hotter than their parent particle, which sustains unregulated secondary chemistry even if the gases are inert. But other lab tests impose the opposite configuration, in which the coal is heated by thermal conduction or radiation while a cool gas stream rapidly carries the volatiles out of the hot zone. Here, the inert gas stream quenches secondary chemistry because it is cooler than the parent coal particles, so such systems generate primary products. Whenever O_2 or H_2 is fed with coal, regardless of the scale, secondary chemistry is inevitable, because these species spontaneously react with volatiles. Secondary chemistry only becomes more pronounced at progressively larger scales, because the large temperature gradients and convective mixing among reactive gas streams and the coal suspension ensure that volatiles will contact hot, reactive mixtures soon after their release from the coal suspension, and thereby sustain secondary chemistry.

Since O_2 and H_2 inevitably react with volatiles, one should also consider whether these species can alter the course of primary devolatilization by adsorbing into the condensed coal phase during decomposition, where they can directly react with coal macromolecules and their intermediates. Under some conditions, H_2 certainly can penetrate reacting coal particles and significantly affect primary devolatilization, which is why hydropyrolysis is distinguished from primary devolatilization and covered separately in Chapter 9. But O_2 cannot alter primary devolatilization, for a somewhat surprising reason. Even when the O_2 transport rate into a reacting coal particle is sufficient to overcome the outward flow of volatiles, the stoichiometric requirement for volatiles combustion is large enough to completely consume the entire O_2 flux (Lau and Niksa, 1992). In other words, O_2 reacts so fast with volatile fuel components that it will be consumed before any remains to actually adsorb into the condensed coal phase.

4.1.3 Resolution of primary devolatilization in a laboratory

From this point on through Chapter 6, our focus shifts from the devolatilization stage to primary devolatilization behavior under closely controlled conditions. As noted above, it is impossible to resolve primary devolatilization in the presence of O_2 and H_2 at any testing scale, and under inert atmospheres at pilot-scale and larger.

The fact that the chemistry of interest occurs in the condensed coal phase presents another formidable challenge. In any commercial technology, suspension loadings are too heavy to accommodate optical diagnostics, and coal is too complex to convey meaningful structural information in any spectral band. There are simply no means to monitor conversion in the condensed phase on-the-fly other than extractive sampling with rapid quenching followed by a laboratory analysis to assign conversions, such as ash tracer analyses (which are fraught with uncertainties). Moreover, devolatilization occurs in tens of milliseconds in furnaces and entrained flow gasifiers, which is too fast for extractive sampling, and many flow patterns near coal injectors are too complex for sampling probes anyway. Penetrations for sampling probes are problematic in pressurized reactors and furnaces. In fluidized systems, solids loadings are much too heavy to permit any access at all, and mixing patterns are too convoluted for extractive sampling along a time coordinate.

So all the data presented in this chapter were obtained at lab-scale under inert atmospheres. Nonetheless, these findings are relevant to coal utilization at even commercial scale, via the following three indirect connections: (1) The available laboratory database describes the yields of all major products of primary devolatilization throughout the full rank spectrum and most of the domain of operating conditions for our utilization technologies; (2) The best available reaction mechanisms interpret the entire database within useful quantitative tolerances; and (3) Predictions from even the most sophisticated devolatilization mechanism are easily incorporated into CFD simulations and process design applications for the major commercial utilization processes. The laboratory database provides the only means to stringently validate comprehensive reaction mechanisms for primary devolatilization and, once validated within useful quantitative tolerances, these mechanisms determine the rate expressions and parameters that depict primary devolatilization in process simulations of large-scale systems. Commercial applications often involve an extrapolation to faster heating rates, hotter temperatures, and shorter time scales than are routinely monitored in a laboratory, so it is important to demonstrate that the reaction mechanism is sufficiently robust for extrapolations. As convoluted as this strategy may seem, it is currently being implemented by dozens of coal technology developers worldwide, including most of the largest developers. This is because it really is the only means to accurately simulate the distinctive devolatilization behavior of individual coals that does not require laboratory support for every coal sample of interest, unless the comprehensive reaction mechanism entails sophisticated laboratory support. As emphasized in Chapter 2, the practical imperative is to accurately predict distinctive primary devolatilization behavior for individual samples based on only the proximate and ultimate analyses.

There are several prerequisites for tests that provide data that is suitable for model validation work. According to the review of coal utilization technology in Chapter 1, tests should impose heating rates substantially faster than 1°C/s. Tests with heating rates of 1000°C/s are ideal because they represent conditions closer to entrained flow conditions that can still be diagnosed within acceptable measurement uncertainties. Tests at even faster heating rates are unnecessary because today's most advanced primary devolatilization mechanisms can be extrapolated accurately from measured behavior under slower heating rates to more severe conditions. The range of test temperatures should include tests hot enough to achieve ultimate primary devolatilization yields, which are the asymptotic values achieved after extended heating periods. Either tests across a broad temperature range or multiple runs for several time increments at a uniform heating rate and temperature are needed to specify the reaction kinetics. All tests should be at a uniform pressure. When coals are entrained in suspension, the samples must be classified into narrow size fractions of no more than two standard sieve sizes, but size classification is less important when the samples are tested in batch mode, as in fluidized beds. Datasets that cover a suite of coal samples and uniform test conditions are the most valuable.

More formally, the following testing features are required of a dataset to be used to evaluate a primary devolatilization mechanism:

(1) **Coal properties**—Proximate and ultimate analyses are absolutely essential for every coal sample. Additional information from specialized analytical testing is not strictly required but may be informative. A mean particle size or PSD is essential for tests with entrained suspensions and desirable for batch tests.
(2) **Pressure**—Usually a uniform test pressure will be specified although a pressure history can also be analyzed.
(3) **Thermal history**—Sufficient information must be available to assign the temperature of the sample as a function of time throughout an entire test. This requirement may entail direct monitoring of sample temperatures, entrainment gas velocity fields, and/or particle transit times.
(4) **Impact of secondary chemistry**—Whenever volatiles are released into a flow that is hotter than the parent coal particle, volatiles will be transformed by secondary chemistry. The extent of this transformation should be monitored. The gas atmosphere must be chemically inert.
(5) **Relevant aspects of devolatilization behavior**—Total weight loss and a tar yield should always be monitored. Whereas the best datasets monitor all major products and their elemental compositions so that balances on C/H/O/N/S can be closed in individual tests, such resolution is a formidable challenge. Changes in char morphology, particularly in size and bulk density, are valuable. Tar molecular weight distributions (MWDs) were highly instrumental in advancing comprehensive reaction mechanisms but are almost never monitored any longer.

The assignment of thermal histories is, by far, the most cumbersome requirement. One would normally be inclined to monitor primary devolatilization behavior in an entrained-flow reactor (EFR), simply because this system processes coal in the p. f. grade under similar conditions to most industrial units. But as seen in the sketch in Fig. 4.2, the operating conditions in EFRs are not easy to diagnose or estimate, particularly near the coal injector where primary devolatilization occurs. In EFRs, thermal histories are determined by the initial coal temperature, a nominal particle size or PSD, an entrainment gas temperature and flow rate, a preheated gas temperature and flow rate, the intensity of mixing and particle dispersion at the injector, the reactor temperature profile, the residence time distribution, and the quench rate in the collection probe. In turn, this information must be incorporated into a heat transfer model that accounts for temperature- and composition-dependent coal thermophysical properties, convective mixing phenomena between the entrainment and preheated gas streams, particle dispersion, particle swelling and mass loss, and several heat transfer

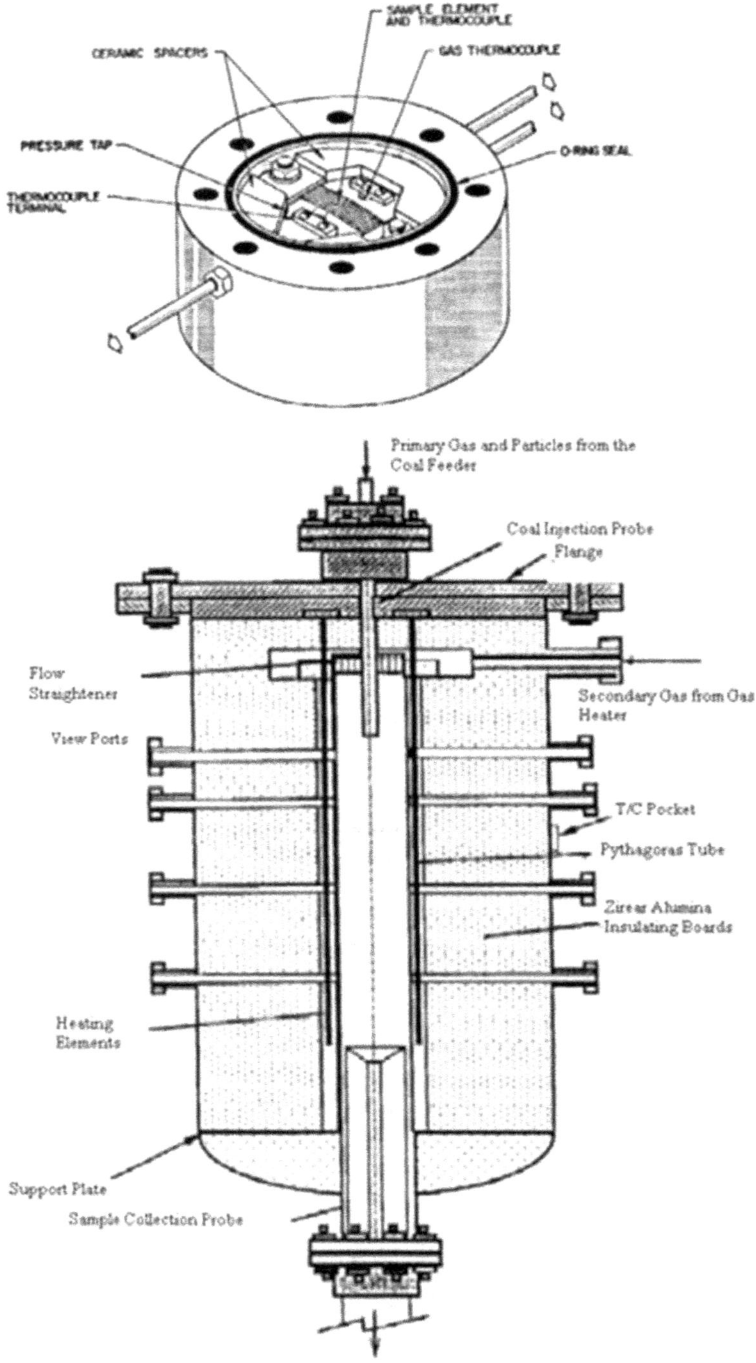

Fig. 4.2 Sketches of (bottom) an EFR and (top) a WMR.

mechanisms to assign particle thermal histories. The complexity of such calculations is well-suited to CFD simulations, although significant uncertainties in the thermal histories assigned for all EFRs cannot be eliminated.

After decades of development, there is a much simpler alternative called the "Wire Mesh Reactor (WMR)," which is also sketched in Fig. 4.2. About 10 mg of pulverized coal is pressed into or supported on a stainless steel fine wire mesh which is then mounted between the electrodes in an electrical heating circuit. In modern WMRs, the electrical power is actively controlled to heat the mesh at a prescribed uniform heating rate to a prescribed ultimate temperature for a prescribed isothermal reaction period (IRP). Dynamics are resolved by actively quenching the support at the end of the reaction period. In other words, the desired thermal history is directly imposed on the sample support, and is therefore much less ambiguous than the calculated thermal histories for EFRs. However in older systems, heating rates were not uniform, ultimate reaction temperatures were highly variable, and there was no forced quenching.

WMRs hold another distinct advantage over EFRs for primary devolatilization testing. Simply by adding a cross flow over the mesh support, primary products can be rapidly swept away from the hot sample support and recovered before they undergo secondary pyrolysis, so WMRs can easily be used to generate and characterize pristine primary devolatilization products.

Whereas WMRs are the laboratory workhorses for primary devolatilization behavior, two alternatives also satisfy the prerequisites. Curie-point pyrolyzers (CPPs) have the basic WMR configuration, but the sample supports are made of metals with graduated Curie points, which are the temperatures where a material's spontaneous magnetic and electric polarizations change to the respective induced forms (Xu and Tomita, 1987a; Wiktorsson and Wanzl, 2000). Support temperatures do not change after this transition. These systems circumvent the need to monitor reaction temperatures, and the heating rate is determined by the power delivery system and is usually not variable. Most CPPs heat samples at about 5000–10,000°C/s. The second variation, called the radiant coal flow reactor (RCFR), is a variation on an EFR. It was designed to monitor entrained coal suspensions at realistic coal loadings and heating rates without unregulated secondary volatiles pyrolysis (Chen and Niksa, 1992a). It heats entrained coal suspensions by thermal radiation from a black-body enclosure, not by a preheated gas stream. Since the entrainment stream is transparent to radiation, it can be made to remain much cooler than the suspension. Pristine products which have been quenched as soon as they were expelled can be recovered or, alternatively, the extent of secondary pyrolysis can be regulated at will. The furnace system also includes rapid quenching to resolve reaction dynamics on a 10-ms time scale, aerodynamic classification to segregate aerosol, particulate, and gaseous products, and analyses for complete product distributions. A version for pressures to 4 MPa has also been developed (Cor et al., 2000).

Lab-scale fluidized beds have also been used to obtain primary products (Tyler, 1980), albeit only for temperatures below the threshold for secondary volatiles pyrolysis. This threshold is 600°C for most coal types, but lower by about 50°C for the lowest ranks. For temperatures hotter than the threshold, secondary pyrolysis is unregulated and hard to characterize because gas residence times change whenever bed temperatures are changed.

In addition to these prerequisites on the regulation of operating conditions, the datasets must include relevant aspects of a coal's devolatilization behavior. The foremost aspect is the ultimate weight loss, on a dry-ash-free (daf) basis, which is obtained with reaction times long enough to achieve constant, asymptotic yields at elevated temperatures. Time-resolved yields are more valuable in principle, although in practice the reaction dynamics are intertwined with all the ambiguities in the assignment of thermal histories. The next most valuable characteristic is a tar yield, because tar production is associated with reaction mechanisms that are especially sensitive to variations in coal quality, heating rate, and pressure. Char characteristics are important, especially elemental compositions, particle sizes (to assign swelling factors), and both bulk and true densities. The compositions of gases and tars are also useful, provided that the extents of secondary chemistry are either made negligible or quantitatively regulated.

The datasets in this chapter will demonstrate that many important characteristics of primary devolatilization are rooted in tar production. Tar is a mixture of diverse molecular structures dispersed over a broad distribution of molecular weight that extends from about 100 to more than 1000 g/mol. Such complex mixtures present formidable challenges for quantitative recovery when coal sample sizes are limited to 10–20 mg to eliminate temperature gradients in batch reactor systems. In both batch and steady flow systems, tars will readily condense on surfaces cooler than about 100°C. And the light ends will pass through collectors operated above about 75°C. The transpiring walls, very cold collection media, centripeters, and other methods developed to manage these issues are beyond our scope. Suffice to say, it is the responsibility of any experimentalist to demonstrate that yields reported as tar yields actually comprise the complete distribution of all organic products that condense at room temperature, excluding water.

4.2 The empirical basis for primary devolatilization behavior

The remainder of this chapter presents primary devolatilization behavior in terms of measured variations with coal quality and the important operating conditions. The presentation opens with ultimate total and tar yields at atmospheric pressure across the rank spectrum, then develops the conversion dynamics from variations in heating rate and reaction time. Data for variations in pressure and particle size then carry strong implications on the roles of mass and heat transport phenomena during primary devolatilization. We then consider the compositions and primary characteristics of chars and tars. Then the discussion of the noncondensable products covers the major gases before it is focused on the precursors to noxious gases containing nitrogen or sulfur. Throughout this chapter the curves in figures simply highlight trends in the data, and are not based on any mechanistic interpretation.

Primary devolatilization moves through three stages: First, tars and noncondensables are simultaneously produced by extensive disintegrations of a coal's macromolecular matrix. Once all tar precursors have been reintegrated into a char matrix, any remaining molecular components with aliphatics and heteroatoms decompose into additional noncondensables on slower but comparable time scales. Finally, if the test imposes extended heating at or above 1000°C, thermal annealing releases additional HCN, H_2S, H_2, CO, and CH_4 on much longer time scales. Annealing is completely independent of tar production and only perturbs volatiles yields by 1–2 daf wt.%. Even so, it is a source of substantial uncertainty on reported HCN yields, particularly in interpretations of tests from different facilities that fail to account for annealing's impact at only the hottest temperatures. The bulk of the measurements in this chapter was recorded at temperatures too cool for appreciable annealing, and therefore represents only the first two stages of devolatilization. Tests that also sustained annealing will be explicitly called out.

4.2.1 Ultimate weight loss and tar yields

The coal quality impacts on ultimate weight loss and tar yields for primary devolatilization are apparent in Fig. 4.3, which shows data reported from three studies that imposed similar rapid heating rates at atmospheric pressure. The thermal histories were always severe enough to achieve the ultimate, asymptotic yields for all stages of primary devolatilization except the annealing stage. Also, all these tests were designed to eliminate secondary volatiles pyrolysis. The x-axis in Fig. 4.3 shows the C-contents of the 29 tested coals which, collectively, cover the entire rank spectrum.

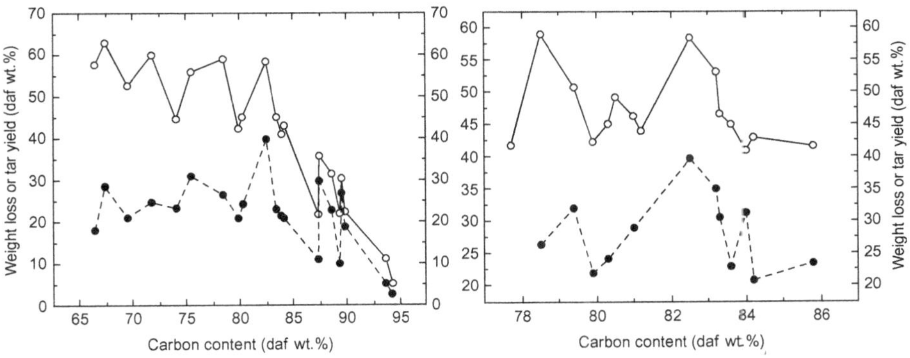

Fig. 4.3 Ultimate (○) weight loss and (●) tar yields for rapid primary devolatilization at atmospheric pressure for (left) the entire rank spectrum and (right) *hv* bituminous coals only. Reproduced with permission from Niksa S. Flashchain theory for rapid coal devolatilization kinetics. 4. Predicting ultimate yields from ultimate analyses alone. Energy Fuels 1994;8:659–70, the American Chemical Society. See Niksa S. Flashchain theory for rapid coal devolatilization kinetics. 4. Predicting ultimate yields from ultimate analyses alone. Energy Fuels 1994;8:659–70, for citations to the lab studies.

As expected from the discussion of Fig. 2.3 in Chapter 2, both weight loss and tar yields exhibit the loosely banded form of any rank parameter variation. In the left panel of Fig. 4.3, the greatest total yields from primary devolatilization are recorded with coals of the lowest rank, through C-contents of about 75 daf wt.%. The weight loss diminishes only slightly through 84% C, then falls off sharply for low volatility coals until it nearly vanishes for anthracites. Across the rank spectrum the ultimate primary devolatilization yields range from 5 to 64 daf wt.% for these particular test conditions. However, the sample-to-sample variability is substantial for every rank, and especially pronounced for lignites, subbituminous, and *hv* bituminous coals, in some cases approaching 20 daf wt.%. The same features are apparent in the tar yields, except that the greatest tar yields are usually recorded with *hv* bituminous samples. Note that the difference between total weight loss and tar yields diminishes for progressively higher ranks. In other words, tar is the most abundant volatile species, by far, from low volatility coals, and makes up about half the volatiles from *hv* bituminous and somewhat less from subbituminous coals and lignites.

The sample-to-sample variability is highlighted further in the right panel of Fig. 4.3, which shows ultimate weight loss and tar yields from *hv* bituminous coals only; again, the heating rates were about 1000°C/s and temperatures and reaction times were severe enough to achieve ultimate yields in all cases, and pressures were nearly atmospheric. The magnitudes of these variations for the same nominal coal rank are startling: Tar yields double from 20 to 40 daf wt.% and weight loss varies by 50 relative percent from 40 to 60 daf wt.%. The variations in tar yields track those in the total weight loss, suggesting that the mechanisms that produce tar are primarily responsible for the sample-to-sample variability. The noncondensable gas yields are much less variable than the tar yields, with only a few exceptions (e.g., at 84% C). Most important, these data clearly illustrate the imperative to predict the sample-to-sample variations in primary devolatilization behavior, because analyses that only give nominal average values for particular ranks are not accurate enough for quantitative simulation work. In fact, the literature contains hundreds of test data like those in Fig. 4.3, and the nominal average values for any rank can easily be evaluated as the arithmetic average on a hand-held calculator.

4.2.2 Thermal history effects

Thermal histories describe how the sample temperature changes as a function of time throughout a test. Those with the simplest forms and the smallest measurement uncertainties, by far, are for WMRs and CPPs, so all tests in this section were obtained with these reactors. As seen in Fig. 4.4, the temperature dependences for both weight loss and tar production display sigmoidal forms. The total weight loss and tar yields from primary devolatilization increase for progressively greater temperatures up to different ultimate asymptotic values, and the ultimate tar yield is recorded before the weight loss reaches its ultimate value. For this reason primary devolatilization comprises at least two stages, an initial stage where tar constitutes almost all the weight loss; and a second stage due to elimination of noncondensables from char on a similar time scale. Beyond the second stage, the thermal annealing of char releases small amounts of

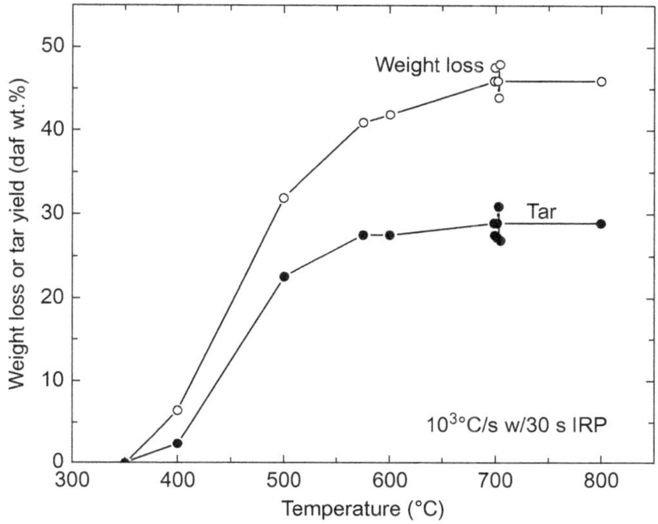

Fig. 4.4 (○) Weight loss and (●) tar yields from an *hv* bituminous coal vs ultimate sample temperature for a heating rate of 1000°C/s with 30 s IRPs.
Reproduced with permission from Gibbins JR, Kandiyoti R. The effect of variations in time temperature history on product distribution from coal pyrolysis. Fuel 1989b;68:895, Elsevier.

HCN, H_2S, and H_2, but on much longer times scales and usually at much hotter temperatures. This annealing stage cannot be recognized in datasets like those in Fig. 4.4.

The oldest notion of primary devolatilization is that every coal contains a fixed amount of "volatiles" that are released upon heating. But if this process abided by the simplest decomposition process, in which the devolatilization rate is directly proportional to the amount of material remaining to be released, the same ultimate weight loss would be recorded for every sample temperature, and that amount would equal the level of precursors in the parent coal; only the time to achieve that level would vary with temperature. One could suppose that the expected ultimate value could be achieved by imposing longer times at each temperature in the test. Specifically, how long does it take, if at all, to achieve the same ultimate yield at low temperatures as recorded for the hotter temperatures in commercial applications?

This issue was addressed long ago (Pohl and Sarofim, 1977) by heating small amounts of coal in a crucible under an inert gas for extremely long reaction times, and then monitoring the elemental compositions of the chars. Fig. 4.5 shows the ultimate weight loss from a lignite and *hv* bituminous coal for sufficient heating to achieve constant weight loss at each temperature. The heating times varied from 12 h at 375°C to 20 min at 1825°C. Compared to the temperature dependence in Fig. 4.4, the weight loss after prolonged heating is greater at the lowest temperatures, as expected. The yields in Fig. 4.5 become lower at temperatures hotter than about 550°C, due to the faster heating rate in the tests in Fig. 4.4 (as explained below) and, perhaps, to differences between the coal samples. Most important, the ultimate

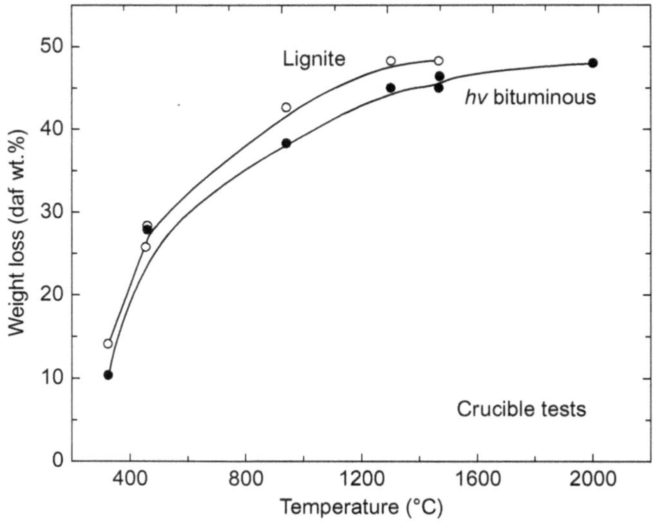

Fig. 4.5 Weight loss from (○) lignite and (●) *hv* bituminous coal after prolonged slow heating in a crucible to constant weight loss at each temperature.
Reproduced with permission from Pohl JH, Sarofim AF. Devolatilization and oxidation of fuel nitrogen. Proc Combust Inst 1977;16:491–501, Elsevier.

yields continue to display a strong temperature dependence through 1200°C. This is the first of many clear indications that a multitude of reactions are responsible for primary devolatilization, rather than a simple, first-order decomposition process. It is best to abandon the notion that coals contain fixed amounts of volatiles before proceeding further, and let the measured behavior reveal the underlying phenomenology.

Numerous complexities become apparent in the weight loss curves in Fig. 4.6 for different heating rates with and without IRPs. First, compare the curves for slow and fast heating with no IRP. For the faster heating rate, the entire decomposition process shifts toward hotter temperatures, from the onset of appreciable devolatilization through the relaxation to an ultimate yield. At the slower heating rate, primary devolatilization begins at about 350°C and relaxes to an ultimate yield at about 550°C, whereas at the faster heating rate, devolatilization begins at about 425°C (by extrapolation) and does not reach an ultimate yield without any IRP until well over 900°C. Although it is not immediately apparent in Fig. 4.6, these two curves also show that devolatilization rates increase in direct proportion to increases in the heating rate. At face value, this inference seems at odds with the observation that faster heating shifts the yields toward hotter temperatures. But in situations like this where the heating rate is strictly uniform, the following relation connects these curves to the devolatilization rate:

$$\frac{dW}{dt}=\frac{dT}{dt}\frac{dW}{dT}=q\frac{dW}{dT} \tag{4.1}$$

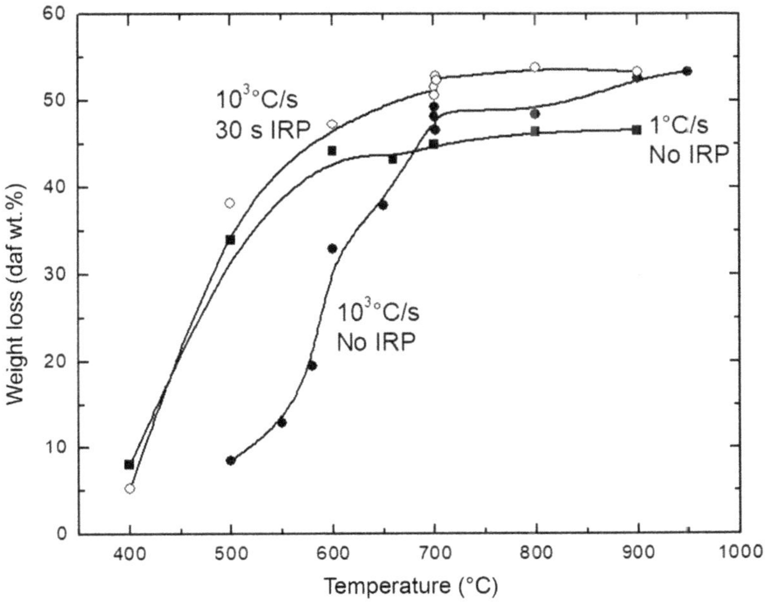

Fig. 4.6 Weight loss at 0.12 MPa from an *hv* bituminous coal vs ultimate sample temperature for a heating rate of 1°C/s with no IRP (■) and for 1000°C/s with IRPs of 0 (●) and 30 s (○). Reproduced with permission from Gibbins-Maltham J, Kandiyoti R. Coal pyrolysis yields from fast and slow heating in a wire-mesh apparatus with a gas sweep. Energy Fuels 1988;2:505, the American Chemical Society.

where W is weight loss and q is the uniform heating rate. Since the slopes in Fig. 4.6 (dW/dT) midway through devolatilization are the same for both heating rates (without IRP), Eq. (4.1) indicates that the primary devolatilization rate is much faster for the faster heating rate, in this case by three orders of magnitude. In general, primary devolatilization rates increase in proportion to increases in the heating rate for all coal types.

To return to Fig. 4.6 for a second comparison, compare the curves for the fast heating rate with and without an IRP. Adding 30 s IRP increases the yields for all temperatures up to the maximum test temperature in this dataset, where the weight loss with and without an IRP are the same. The important implication is that, for any specified heating rate, less time is required to achieve the ultimate yield for primary devolatilization at progressively hotter temperatures. For this particular heating rate of 1000°C/s, ultimate yields are obtained without any IRP for temperatures hotter than about 900°C, which is typical for any coal type. Adding a 30 s IRP lowers the temperature that achieves an ultimate yield to about 700°C, but it certainly does not eliminate the temperature dependence altogether. As noted previously, this temperature dependence can be diminished but not eliminated by extending the IRP further.

Finally, compare the curve for the fast heating rate with 30 s IRP to that for the slow heating rate with no IRP. Note that total reaction times are longer with the slow heating rate throughout all stages of primary devolatilization. Since we have already seen that a 30 s IRP achieves ultimate yields for temperatures hotter than 700°C with the faster heating rate, it is immediately apparent that faster heating rates enhance ultimate primary devolatilization yields at atmospheric pressure. In this case the enhancement is about 6 daf wt.%, which is well beyond the measurement uncertainty. Enhanced yields for progressively faster heating rates is the first definitive indication that primary devolatilization is a reaction process with competitive elements, because the simplest decomposition processes necessarily give the same ultimate yields for different heating rates.

The heating rate dependence is more prominent in Fig. 4.7, which shows ultimate weight loss and tar yields from two *hv* bituminous coals after heating at rates from 1°C/s to 1000°C/s to 700°C with a 10 s IRP at both atmospheric and elevated pressure. At atmospheric pressure, the ultimate weight loss was enhanced by about 8 daf wt.% when the heating rates were increased from 1°C/s to 1000°C/s, and all of this enhancement, if not slightly more, is apparent in the tar yields. Consequently, enhanced ultimate yields for primary devolatilization are due to an acceleration of tar production, with smaller effects on the gas production mechanisms. Note also that heating rate enhancements are only about 3 daf wt.% per order-of-magnitude increase in the heating rate. This value is typical for *hv* bituminous coals, but it can be smaller for low rank and low volatility coals. So at atmospheric pressure, accurate extrapolations from the fastest heating rates in WMR systems to the faster rates in commercial applications entail only modest adjustments to measured total and tar yields from primary devolatilization.

This tidy estimation guide carries only one major qualification: As seen in the lower panel of Fig. 4.7, total volatiles yields are not enhanced at all by faster heating rates at elevated pressure. This particular study did not report tar yields at 7 MPa, but other datasets at slightly different conditions (Table 4.1) show that the tar yields are, in fact, uniform for all heating rates at elevated pressures. So the restriction of yield enhancements for faster heating to near-atmospheric pressures does not disrupt the connection between the yield enhancement and the mechanism for tar production. We shall also see in Chapter 9 that total and tar yields are diminished by faster heating rates under elevated pressures of H_2, but that trend is a characteristic of hydropyrolysis, not primary devolatilization.

The datasets in this section clearly demonstrate that primary devolatilization is certainly not a simple, first-order decomposition process; in fact, it entails a multitude of chemical reactions, including competitive channels in the mechanism of tar production. Tar production determines weight loss during the first stage, and is the most sensitive to heating rate variations. Coal cannot possibly contain a fixed amount of "volatiles" within an inert char matrix because the proportions of volatiles and char strongly depend on temperature, heating rate, and IRP. Since the ultimate weight loss is enhanced by faster heating at atmospheric pressure, char must be a bona fide reaction product of the underlying competitive reaction scheme.

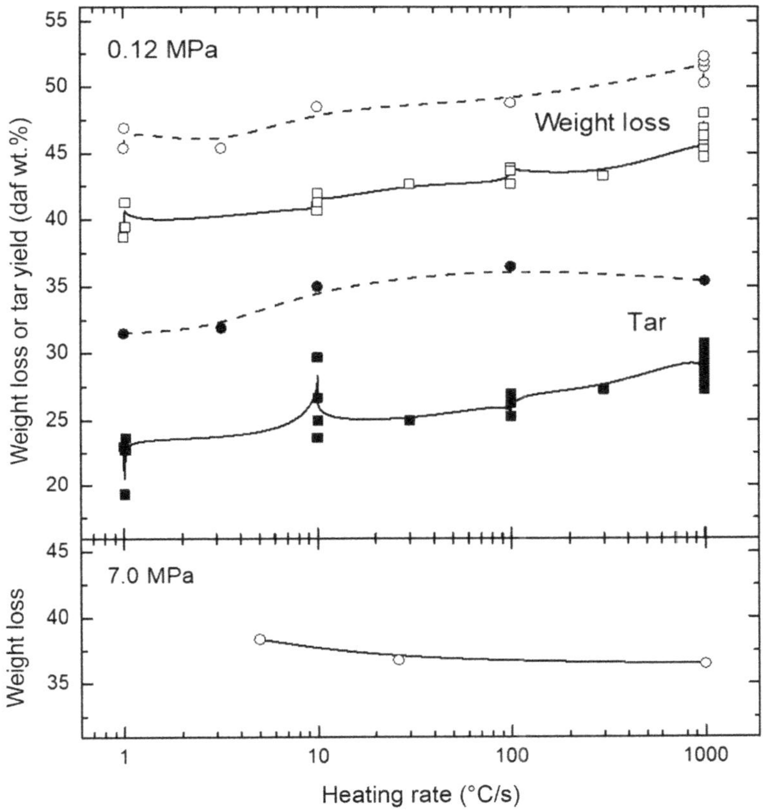

Fig. 4.7 Ultimate weight loss *(open symbols)* and tar yields *(filled symbols)* from two *hv* bituminous coals *(solid and dashed curves)* for heating at different rates to 700°C with 10 s IRP at (top) near-atmospheric pressure and (bottom) 7 MPa.
Data from Gibbins-Maltham J, Kandiyoti R. Coal pyrolysis yields from fast and slow heating in a wire-mesh apparatus with a gas sweep. Energy Fuels 1988;2:505; Gibbins J, Kandiyoti R. Experimental study of coal pyrolysis and hydropyrolysis at elevated pressures using a variable heating rate wire-mesh apparatus. Energy Fuels 1989a;3:670–77, the American Chemical Society; from Gibbins JR, Kandiyoti R. The effect of variations in time temperature history on product distribution from coal pyrolysis. Fuel 1989b;68:895, Elsevier.

4.2.3 Pressure effects

Ultimate weight loss and tar yields for pressures to 7 MPa from coals representing the three main segments of the rank spectrum appear in Fig. 4.8. Note the distinctive influence of coal quality, and the much more pronounced pressure effect in the tar yields. Ultimate weight loss from this particular Victorian brown coal is essentially independent of pressure. The bulk of the available data on low-rank coals, however, does exhibit a pressure effect, as discussed below. The weight loss from the *hv* bituminous and low volatility coals diminishes by 15% to 25%, with most of the reduction

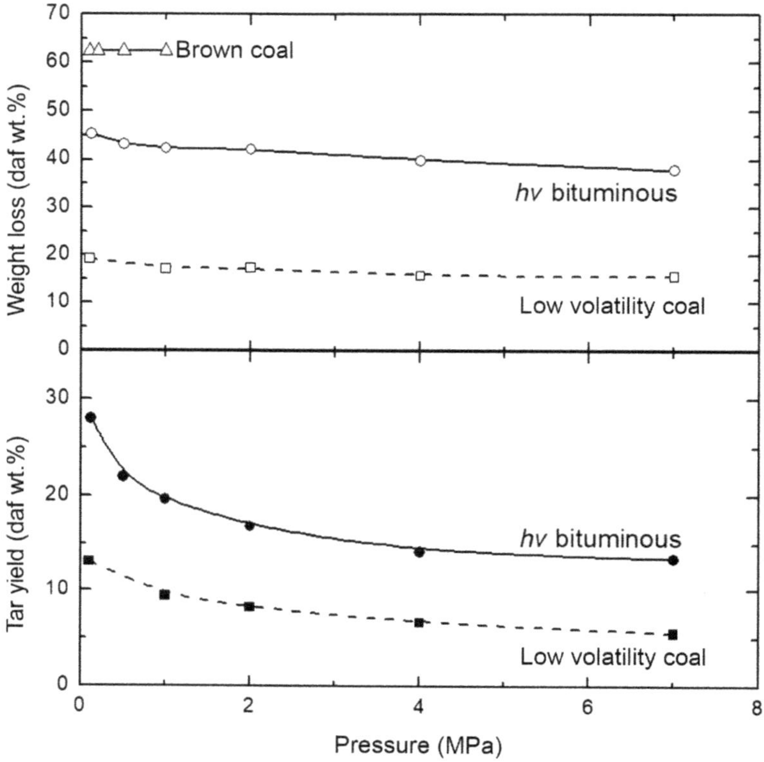

Fig. 4.8 (Top) Ultimate weight loss and (bottom) tar yields from (△) Victorian brown coal, (○, ●) *hv* bituminous, and (□, ■) *lv* bituminous for a broad pressure range.

occurring below 1 MPa. The corresponding reduction in the tar yields is much greater at roughly 50%. Since the tar yields diminish by more than the reduction in weight loss, gas yields increase for progressively higher pressures, but not by enough to compensate for the reduction in tar yields.

Among *hv* bituminous coals, there is little variation among the quantitative sensitivity of weight loss to pressure, as seen in Fig. 4.9. The ultimate weight loss among *hv* bituminous coals diminishes by approximately 2.5 daf wt.% per MPa increase in pressure, even for a suite of samples whose total and tar yields are appreciably different. This sample suite has C-contents from 78.2 to 82.6 daf wt.%, and identical thermal histories were imposed in all tests, so that pressure was the only variable operating condition. The slopes of the curves of weight loss versus pressure are nearly the same within experimental uncertainty, even while the ultimate yields at atmospheric pressure vary from 42 to 57 daf wt.%. Tar yields were not reported for all these cases, but they would probably vary by at least as much as the weight loss.

Whereas elevated operating pressures definitely affect ultimate primary devolatilization yields, they do not appreciably affect the reaction dynamics. This feature is illustrated in Fig. 4.10 with transient weight loss from an *hv* bituminous coal at three pressures. The WMR in this study featured reproducible thermal histories and a

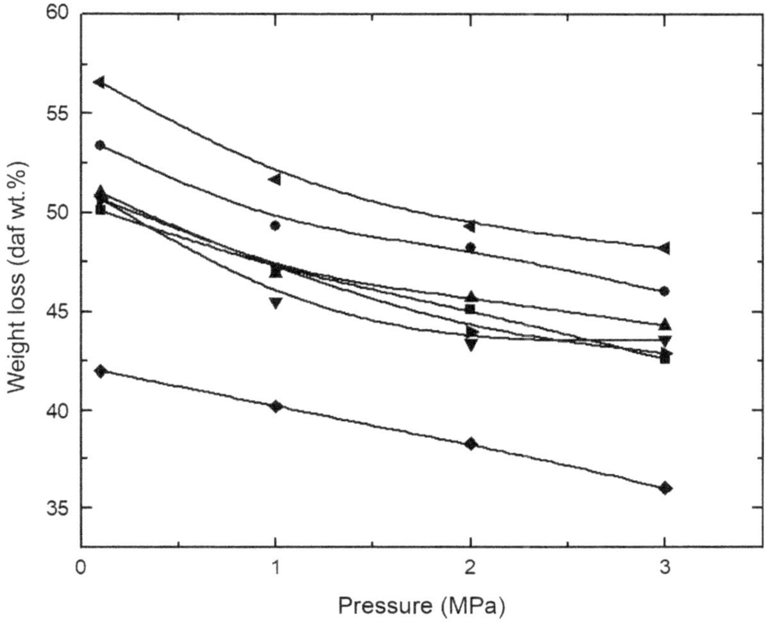

Fig. 4.9 Ultimate weight loss from *hv* bituminous coals for different pressures in a WMR for 1000°C/s to 1000°C with 10 s IRP (Messenbock et al., 1999a,b).

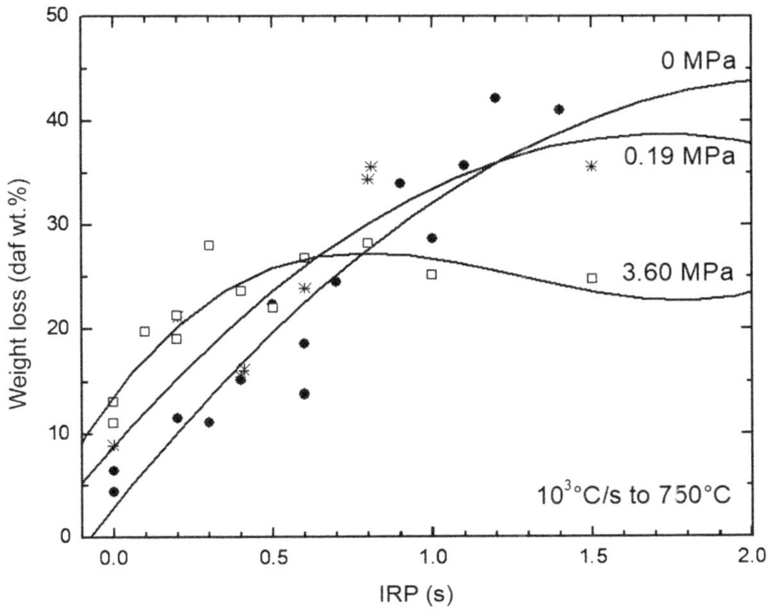

Fig. 4.10 Weight loss resolved throughout the IRPs after heatup at 1000°C/s to 750°C with an *hv* bituminous coal under (●) vacuum, (*) 0.19, and (□) 3.6 MPa (Niksa et al., 1982b).

nitrogen spray quench that could resolve reaction times into 100 ms increments. It was used to resolve the reaction dynamics at three disparate pressures, as seen in Fig. 4.10. Whereas elevated pressures definitely affect ultimate volatiles yields, they do not appear to affect the reaction dynamics. There is a discernable shift in the weight loss transients to shorter times for progressively higher pressures, which was attributed to heat transfer aspects of this particular WMR and is probably inconsequential to the apparent indication of pressure-independent devolatilization rates. Through the first 0.8 s IRP, the weight loss for all three pressures exhibits the same nominal rate, then saturates to pressure-dependent asymptotic ultimate values at longer contact times.

One demonstration that elevated pressures eliminate yield enhancements for faster heating rates appears in Fig. 4.7. The additional data (Guell and Kandiyoti, 1993) collected in Table 4.1 firmly establish this tendency, and suggest that yield enhancements due to faster heating may weaken continuously for progressively higher pressures. The data for the Linby *hv* bituminous exhibit a substantial yield enhancement due to faster heating at 0.25 MPa in both weight loss and tar yields. But at 2 MPa only the weight loss is enhanced while tar yields are nearly independent of the heating rate. At 7 MPa, weight loss from this coal is insensitive to variations in heating rate, while tar yields diminish slightly for faster heating. Among the five other coals in Table 4.1, both subbituminous coals (Pecket, Catamutum) exhibit appreciable weight loss enhancements at 7 MPa, but neither of the bituminous coals (Pit. no. 8, Longannet) shows an enhancement. The apparent enhancement with the low volatility coal (Tilmanstone) is difficult to resolve from experimental uncertainty. Notwithstanding these ambiguities, the tar yields from five of these six coals are slightly lower for 1000°C/s than for 1°C/s. Evidently, faster heating rates promote the production of intermediate compounds that are unable to vaporize at elevated pressures and therefore unable to be recovered as tar.

We are now prepared to examine the impact of coal quality at elevated pressures. Since the sample-to-sample variability is especially significant among tar yields, and since elevating the pressure suppresses tar production, one could reasonably expect

Table 4.1 Ultimate yields for various heating rates at elevated pressures from diverse coals

		Weight loss (daf wt.%)		Tar yield (daf wt.%)	
Coal	***P* (MPa)**	**1°C/s**	**10^3°C/s**	**1°C/s**	**10^3°C/s**
Linby	0.25	40.0	45.1	17.5	20.4
	2	36.8	41.9	15.0	13.7
	7	35.7	37.8	15.0	12.2
Pit. no.8	7	36.7	35.9	20.5	11.4
Pecket	7	46.8	50.9	9.5	9.3
Catamutum	7	47.3	53.6	11.6	13.8
Longannet	7	31.8	33.9	11.8	10.6
Tilmanstone	7	13.0	16.0	7.9	6.1

Reproduced with permission from Guell AG, Kandiyoti R. Development of a gas sweep facility for the direct capture of pyrolysis tars in a variable heating rate high pressure wire mesh reactor. Energy Fuels 1993;7:943–52, Elsevier.

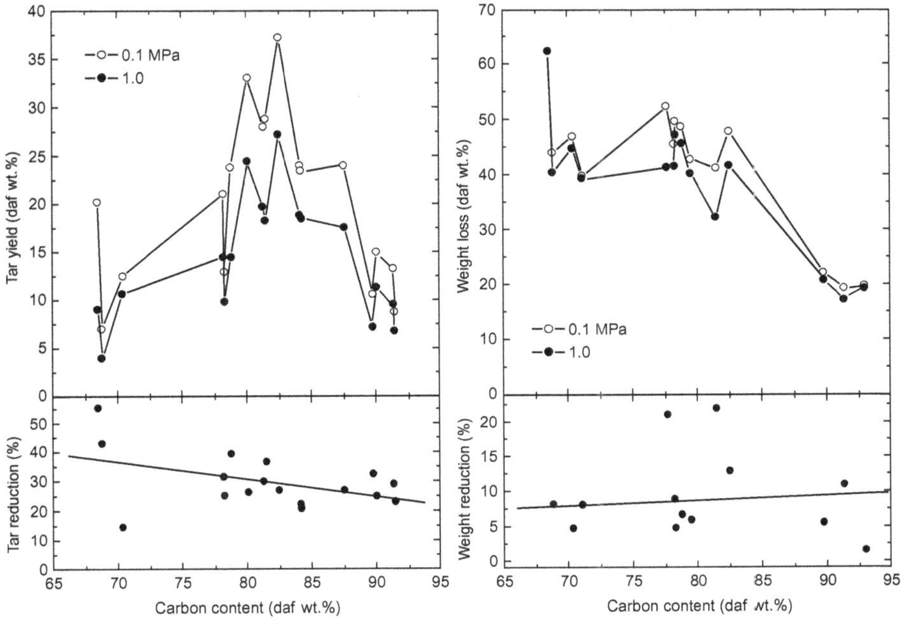

Fig. 4.11 Ultimate (left) tar yields and (right) weight loss from 7 WMR studies that imposed the same thermal histories at (○) 0.1 and (●) 1 MPa.

that tar yields and, by association, weight loss would be less sensitive to coal quality at elevated pressures than at atmospheric pressure. But the data indicate otherwise. The tests behind Fig. 4.11 imposed heating rates of 1000°C/s or faster and temperatures that achieved the ultimate yields. In each study, the same thermal histories were imposed on the same coals at 0.1 and 1 MPa. Clearly, there is a one-to-one correspondence among the tar yields at the two pressures. In other words, the sample-to-sample variability is unaffected by elevating the pressure. The percentage reduction in tar yields due to the pressure elevation (in the lower panel) diminishes from roughly 40% with lignites to 25% with low volatility coals, albeit within the considerable scatter in the data for the lowest ranks.

The weight loss associated with the tar yields in Fig. 4.11 also displays a one-to-one correspondence among the weight loss values at both pressures indicating that the sample-to-sample variability is unaffected by pressure elevations. The only exception is the coal with 77.7% C. The percentage reduction in weight loss due to the pressure elevation is essentially independent of coal quality and, with a nominal value of only 8%, the percentage reduction in weight loss is also much less than the reduction in tar yields.

With *hv* bituminous coals, the weight loss is usually about twice the tar yield, so the typical reduction in tar yield of 30% would reduce the weight loss by 15%. But the actual reduction is only half the limiting value, implying that approximately half the mass of the tar that fails to vaporize is subsequently expelled as noncondensable gases. Since the regression of the tar reduction percentages in Fig. 4.11 suggests that tar yields from the lowest ranks are reduced even more by the same pressure increase, much more than half of these tars must be cracked into gases. Tars constitute as little as a quarter of the weight loss from coals of lowest rank. If 40% of the tar fails to vaporize, then 80% of this retained tar mass must be expelled as gases to reduce the weight loss by the typical value of 8%.

4.2.4 Particle size variations

From the review of coal utilization technologies of interest in Chapter 2, entrained flow systems use coal grinds as fine as 45 μm and fluidized systems use coarser sizes to several millimeters. Such a broad size range often carries important implications for the rate limiting mechanisms in any heterogeneous reaction system, because transport phenomena become relatively more important for progressively larger sizes. During primary devolatilization, volatiles form in the condensed phase but then must cross the interface into the coal's internal pore system, move through pores or as bubbles to the external surface, and then escape from the external surface into the free stream. There are numerous transport processes that could conceivably come into play, as discussed in Chapter 5. Heat transfer rates are similarly affected by particle size variations. But, for now, the central question is, "At what particle size do transport processes mediate the underlying mechanisms of primary devolatilization, and thereby affect yields and, perhaps, product distributions?"

From a commercial standpoint, the answer is important because it would guide strategies to maximize devolatilization yields by managing coal grinds. The answer is even more important from a scientific standpoint, because it delineates test conditions that give results that solely reflect the underlying chemical reaction kinetics from those compounded by transport effects.

For sizes to just under a millimeter, all the direct measurements taken for size variations only, all other conditions the same, give the same total and tar yields. Table 4.2 compiles the data from four WMR studies on ultimate yields in daf wt.% from different size cuts at atmospheric pressure. Whereas Anthony et al. (1975) reported only ultimate weight loss, the others also reported tar yields. Bautista et al. (1986) tested three bituminous coals, including a *lv* bituminous, and Griffin et al. (1993) tested two size cuts at three heating rates. Replicate results for the same sizes address the repeatability of the reported yields.

In the results of Anthony et al. (1975), ultimate weight loss decreases by 3 wt.% for sizes from 70 to 1000 μm. This is the largest size effect ever reported, although it may not be statistically significant because it is only one-half the variation among repeated tests with 70 μm particles to determine ultimate yields at 975°C reported in other parts of this work. A least-squares regression on Suuberg's (1977) ultimate weight loss values decreases by 1.6 wt.% for sizes from 70 to 910 μm, which is also within the measurement uncertainty. The corresponding tar yields in Table 4.2 also show no systematic variation with particle size. In Bautista's study, the three coals included a *lv* bituminous sample that did not soften or swell during devolatilization. As seen in Table 4.2, neither weight loss nor tar yields from any of these coals varied for sizes

Table 4.2 Impact of variations in particle size on ultimate total and tar yields from different coal samples

Coal	Size (μm)	Ultimate yields: Tar	Ultimate yields: Total	Citations
hv Bit	70	–	47.3/47.7/47.5	Anthony et al. (1975)
	160	–	46.5	
	278	–	46.6	
	992	–	44.7/44.2	
hv Bit	67	–	47.1	Suuberg (1977)
	163	21.8	45.5/45.8/43.7	
	570	21.3/21.2	43.5/43.5/44.0	
	912	16.7/20.1	43.4/44.0/43.0/46.4/44.5	
hv Bit 1	81	27.9	40.4	Bautista et al. (1986)
	127	29.5	41.1	
	180	30.8	40.3	
hv Bit 2	81	30.4	40.7	
	127	32.4	39.3	
	180	31.1	40.4	
lv Bit	81	–	–	
	127	14.5	19.6	
	180	18.1	19.3	
hv Bit-10^1	69	25.1/26.5	44.1/45.0	Griffin et al. (1993)
	116	24.2/26.0	41.7/42.4/43.2	
hv Bit-10^3	69	30.8/31.0/33.2	47.3/47.5/48.3/50.0	
	116	26.9/29.3/30.6	48.1/46.9	
hv Bit-2 $\times 10^4$	69	32.3/34.2/34.3	51.9/53.3	
	116	26.9/28.1/29.3	50.8/52.2/53.3	

from 80 to 180 μm, except that tar yields from the *lv* bituminous were greater for the largest size. Griffin et al.'s (1993) tests at 10°C/s and 1000°C/s gave no significant size dependence for sizes from 70 to 120 μm in either total weight loss or tar yields. For 20,000°C/s, the tar yields from the smaller size cut are about 5 wt.% greater than those for the larger size, which is beyond the measurement uncertainty. But the measurements for the large size at the fastest heating rate are questionable for two reasons. First, they do not exhibit the expected tendency for enhanced tar yields for progressively faster heating rates, which is seen in the tar yields from the smaller size at all three heating rates and from the larger size at the two slower heating rates. Second, the total weight loss values from the tests with the fastest heating rate are indistinguishable for both size cuts, which is inconsistent with appreciably different tar yields. Hence, no data has been reported that conclusively demonstrates a size dependence for primary coal devolatilization for sizes to 1 mm.

These data are definitive for sizes in the p. f. grade, but they do not answer the central question. There is no doubt that, as particle size is increased without limit, transport processes will eventually govern primary devolatilization rates and, in all likelihood, affect yields and product distributions too. But the size where the transition occurs reflects the relative rates of the chemical kinetics and transport processes, which depend on both heating rate and pressure. In our subject utilization technologies, larger particles heat up slower than finer sizes, which is certainly the case in commercial fluidized technologies because they operate with much larger sizes and also much cooler temperatures than entrained flow systems. In the reaction controlled regime, devolatilization rates increase in proportion to increases in heating rate, but are independent of pressure (because both these tendencies were illustrated above with sizes too small to support transport limitations). In general, transport rates are perturbed by heating rate variations, through their relatively weak temperature dependences, and subject to pressure effects that range from negligible to strong. For example, diffusive fluxes are independent of pressure whereas convective flows are proportional to the difference between pressures within and around the particle. Consequently, the size at which transport comes into play is definitely a function of the heating rate, and may also depend on pressure.

Both dependences are clearly seen in the critical sizes for transport effects in Fig. 4.12. These critical values were based on direct measurements of volatiles yields across closely spaced size cuts, and an interpolation scheme to pinpoint the size that admitted the first mediation by transport. At atmospheric pressure, the critical size shifts toward smaller values for progressively faster heating rates, as expected. The impact of pressure is very weak at the fastest heating rate, becoming noticeable only under vacuum. But its impact expands to superatmospheric levels for the slower heating rates, and displays a tendency for smaller critical values for progressively greater pressures. Most important, the data in Fig. 4.12 show that transport effects are negligible for sizes smaller than about 200 μm for heating rates as fast as 10^4°C/s at any pressure of commercial interest. The threshold grows to 450 μm for 10^3°C/s at atmospheric pressure. For fluidized systems, heating rates are of the order of 10°C/s, and a simple proportional extrapolation of the trend for atmospheric pressure in Fig. 4.12 gives a critical value of about 3 mm. This value should be refined with data for larger sizes and slower heating rates to expand the domain of Fig. 4.12, although it appears that the transport effects on primary devolatilization in fluidized beds are unlikely to be very important. Finally, the data in Fig. 4.12 were recorded with a softening *hv* bituminous sample. Tests with a nonsoftening low rank coal gave critical sizes that were twice as large, and a weaker pressure dependence. So the threshold values for the softening coal would be very conservative estimates for nonsoftening coals.

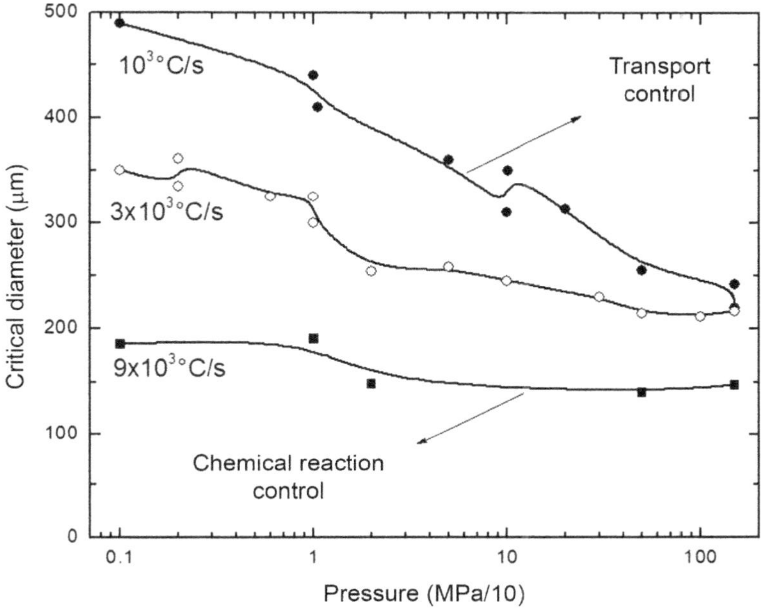

Fig. 4.12 Delineation of chemical reaction control from transport control for the rapid devolatilization of an *hv* bituminous coal for different heating rates to 800°C at different pressures.
Reproduced with permission from Wagner R, Wanzl W, van Heek, KH. Influence of transport effects on pyrolysis reaction of coal at high heating rates. Fuel 1985;64:571–73, Elsevier.

4.2.5 Elemental compositions of char

The simplest and most direct way to monitor the progress of primary devolatilization is to measure the levels of C, H, O, and N in chars throughout the process, because the chemistry is confined to the condensed phase and these elements are exclusively present in organic coal components. The clearest view is seen in Fig. 4.13, where the element retention in char is expressed as fractions of the coal's elemental contents, and plotted vs the C-content of the parent coal. All tests imposed rapid heating rates at atmospheric pressure, and the thermal histories were always severe enough to achieve the ultimate yields for all stages of primary devolatilization except the annealing stage. This omission would definitely affect the elimination of H and N, but not the other elements because O is almost completely eliminated without char annealing, and negligible levels of C-compounds are released by annealing. The entrainment stream in the flat-flame burner apparatus contained 6% O_2, but the data in Fig. 4.13 were obtained at the first sampling position after 47 ms, before the fuel ignited (Mitchell et al., 1992). This premise is corroborated by the similarities among these data and those from an RCFR (Chen and Niksa, 1992b), WMR (Suuberg, 1977), and EFR (Fletcher and Hardesty, 1992).

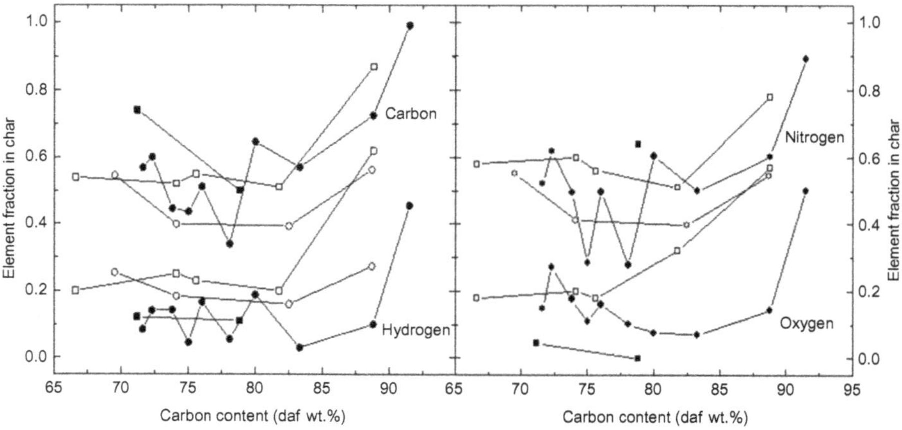

Fig. 4.13 Element fractions for C, H, O, and N in the ultimate chars from primary devolatilization in a (●) flat-flame burner, (○) RCFR, (■) WMR, and (□) EFR.

Primary devolatilization preferentially expels hydrogen and oxygen, so these element fractions in char are much lower than the fractional char mass. In contrast, the element fractions of carbon and nitrogen are comparable to the fractional char mass. The gross tendency in the coal quality impacts on element retention mimics the rank dependences for ultimate weight loss and tar yields (cf. Fig. 4.3), in that the element retentions are fairly uniform through the *hv* bituminous ranks, then increase sharply for low volatility coals. For low rank coals, the H-fractions are indistinguishable from those for bituminous coals, whereas C, O, and N are preferentially retained in the low rank coal chars. This feature suggests that tar yields are better indicators for the release of these elements than weight loss. These tendencies notwithstanding, the sample-to-sample variability is always very substantial. Note that the variations among the retentions of C, H, O, and N are the same for each individual coal sample, which suggests that some common skeletal disruption is probably responsible for the release of these elements. The variability among the O-fractions also seems to be lower than for the other elements among bituminous coals.

As seen in Fig. 4.14, the retention of sulfur in char does not abide by any common process with the other elements or, at least, such a common process is compounded by additional factors. Indeed, there is no apparent rank dependence at all, because the sample-to-sample variability is almost as great as the shift toward greater element fractions for the low volatility coals with the other four elements. As explained below in greater detail, the retention of sulfur reflects a process in common with the other elements as well as S-release during the conversion of pyrite, FeS_2, into troilite, FeS. Since pyrite levels in coals are uncorrelated with coal rank and are often larger than the levels of organic S, the variations in the retention of S in char are also uncorrelated with rank, and with the retention of the other four elements.

As seen in Fig. 4.15, faster heating promotes the release of C, H, and N, albeit not by very much for C and H. The important implication is that these three elements are

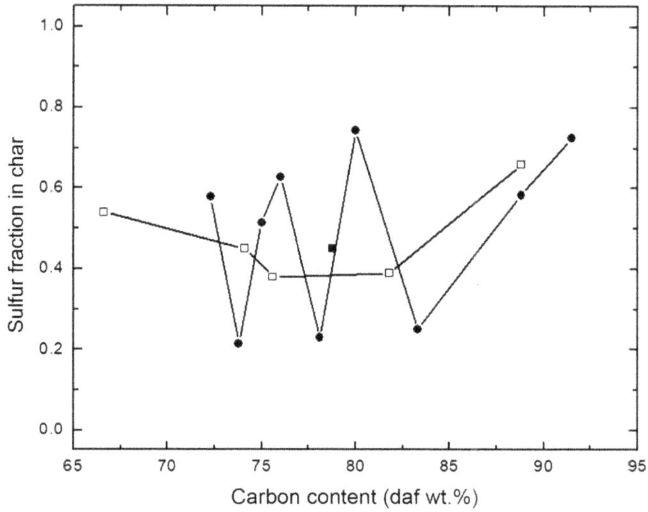

Fig. 4.14 Sulfur fractions in the ultimate chars from primary devolatilization in a (●) flat-flame burner, (■) WMR, and (□) EFR.

released throughout tar production, and are probably conveyed as components of tar into the free stream. This is not to suggest that neither O nor organic S is also shuttled away by tar; we will soon see that they are. But with O and organic S, it makes little difference whether they are released in tar or retained in the condensed phase as tar precursors that did not vaporize under particular test conditions, because all O and most organic S are expelled from the condensed phase anyway. But substantial amounts of N and some of the C and H in unvaporized tar precursors must be retained in char to account for the heating ratZe dependence.

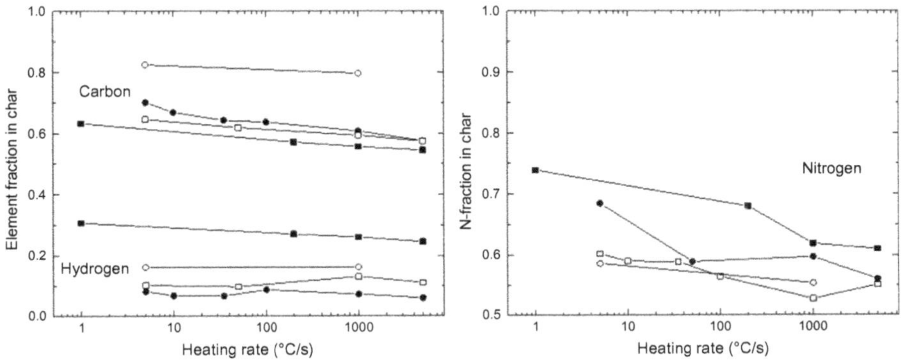

Fig. 4.15 Element fractions for (left) C, H, and (right) N in the ultimate chars from primary devolatilization in a WMR at different heating rates with (●, ■, □) three *hv* bituminous coals and (○) a *lv* bituminous coal (Cai, 1995).

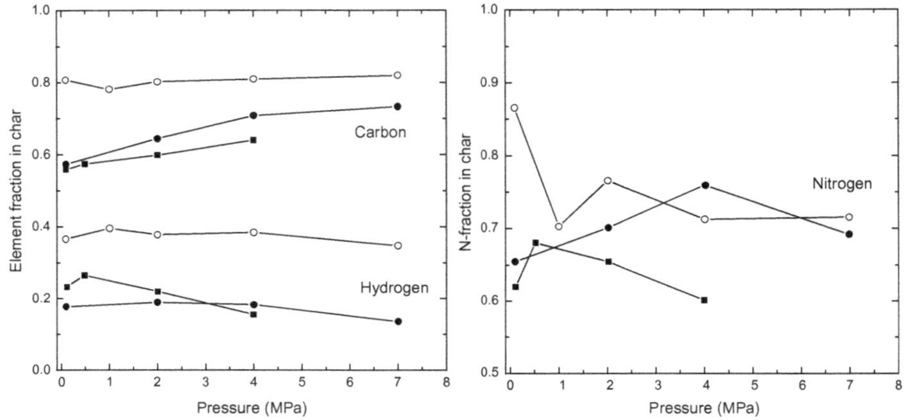

Fig. 4.16 Element fractions for (left) C, H, and (right) N in the ultimate chars from primary devolatilization in a WMR at different pressures with (●, ▪) two *hv* bituminous coals and (○) a *lv* bituminous coal (Cai, 1995).

One also expects to have more C, H, and N in chars prepared under elevated pressures for these same reasons, but this is not confirmed by the data in Fig. 4.16. Two of the three coals abide by the expectations on the C-fractions, whereas the third shows no change at all throughout the entire pressure range. The slight and consistent reductions in the H-fractions for progressively greater pressures is contrary to expectations, which suggests that chemistry among tar precursors that remain in the condensed phase is able to utilize hydrogen in the product formation channels for noncondensable gases. The N-fractions from both *hv* bituminous coals mimic the variations for greater pressures in the H-fractions, whereas values for the *lv* bituminous are distinctive.

To this point, we considered the release of each element independently, but should also consider atomic H/C ratios of char. They carry particular significance as gauges for the expansion of aromatic domains and the destruction of aliphatic components. Both at atmospheric and elevated pressures, the H/C ratios of chars fall continuously throughout devolatilization. The ultimate values are very sensitive to the severity of the imposed thermal history, especially to reaction time, because char H/C ratios diminish as H_2 is preferentially eliminated during annealing on long time scales. Typical values before the onset of annealing range from 0.30 to 0.45, which are certainly low enough to indicate predominate aromaticity in char. This tendency has been validated by direct measurements with ^{13}C NMR of significant production of aromatic rings throughout devolatilization (Miknis et al., 1988). On balance, primary devolatilization produces new aromatic rings, and the overwhelming majority end up in char.

The elemental compositions of chars presented to this point contain substantial portions of the parent coals' C, N, and S, and appreciable portions of their H. Consequently, with *hv* bituminous coals at the end of all stages of primary devolatilization except thermal annealing, a typical char composition would be 85%–90% C, 1%–2% H, and a few percent each of O, N, and S. Whereas these chars can be prepared at

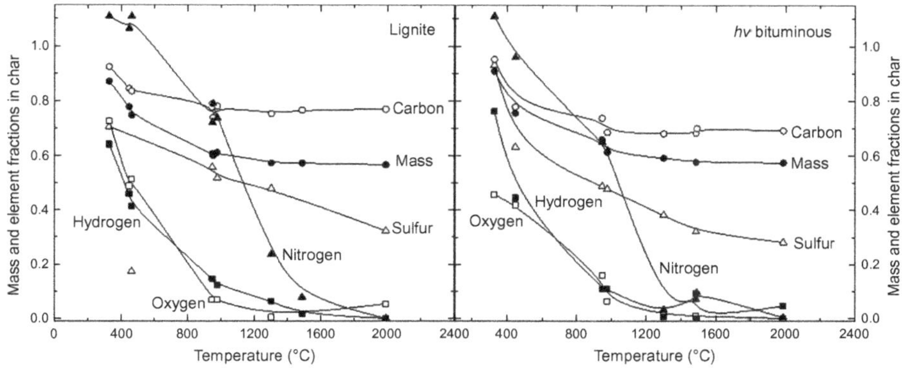

Fig. 4.17 Mass and element fractions in chars from (left) lignite and (right) *hv* bituminous coal after extended heating at various temperatures (Kobayashi, 1976).

temperatures below 1000°C in a few seconds, much more severe thermal histories are required to completely eliminate the heteroatoms via thermal annealing. The tests behind the data in Fig. 4.17 heated a crucible containing about 1 g of coal in a muffle furnace at about 1°C/s to the indicated temperature, then held temperature for at least 10 min at even the hottest temperature, then continued heating throughout an extended cool-down cycle (Kobayashi, 1976). So the duration of each test was in the tens of minutes.

Under such severe processing, all O, H, and N can be eliminated from char via thermal annealing. Note that relatively very little mass is lost while most of these heteroatoms are expelled: For both coals only about 4 wt.% is released for temperatures hotter than 1000°C. This is because CO, H_2, and HCN are the major products of annealing, along with some H_2S. Annealing also eliminates any residual organic S, but pyrite decomposition becomes frozen at the level associated with conversion into FeS in the absence of reducing agents like GHCs. So it is extremely difficult to completely eliminate all S from char under inert atmospheres. More recent tests in a molybdenum WMR with five diverse coals (Cai et al., 1998) corroborated some, but not all of these findings. The tests imposed heating rates of 5000°C/s to 1500°C with 2 s IRP. The reported H-fractions did not exceed 0.03 for any coal, and the S-fractions varied from 0.2 to 0.5, consistent with the char element fractions in Fig. 4.17. But the N-fractions also did not vanish for any sample; rather, they increased in proportion to the coals' C-content from 0.1 for a subbituminous to about 0.6 for a *lv* bituminous. So annealing will definitely eliminate nearly all O and H from char, and residual S will be present whenever appreciable portions of coal-S are in pyrite. But residual N is likely to persist in char, particularly with coals of progressively higher rank.

Thermal annealing is extremely important in most all commercial utilization technologies, but for reasons that have nothing to do with primary devolatilization. The predominant impact is on a char's intrinsic reactivity in both oxidation and gasification environments, as explained elsewhere (Niksa et al., 2003; Liu and Niksa, 2004).

With regard to primary devolatilization, annealing is inconsequential in most processing configurations. The main reason is that char conversion usually sets in via ignition before annealing expels substantial heteroatoms from char. And once char conversion begins, it becomes impossible to determine whether the heteroatomic gases were produced by the annealing stage of primary devolatilization or char conversion chemistry. In mathematical analyses, it is simpler to include the heteroatoms in the char composition and have them released at the overall rate of char conversion, so that the annealing stage of primary devolatilization is omitted from process simulations, except when the process includes extended preprocessing under reducing conditions at elevated temperatures. Notwithstanding, thermal annealing has definitely distorted the database on the elimination of heteroatoms from char under inert atmospheres, and its impact in tests should be managed by careful selection of the thermal processing conditions.

4.2.6 Physical and morphological transformations

Primary devolatilization is responsible for massive changes in the physical morphology of char that significantly affect rates of char conversion by any means, especially for coals that soften and foam during devolatilization. Char particle sizes usually expand by 25% or more, so that char volumes are often double their parent coal's. Specific surface areas become moderately greater for coals that never soften, and these chars retain distributions of micro-, meso-, and macropores that resemble their parent coals' pore size distributions. But a plastic softening stage eliminates the initial pore system, and skews a char's pore system toward macropores and, especially, macrovoids formed when a coal melt resolidifies around bubbles. Cenospheres, the near-spherical, punctured carbon balloons, are the predominant char form with highly softened and swollen chars. Whether or not a coal softens, bulk particle densities for chars plummet due to the mass loss, which is then compounded by swelling in softening coals. Like most other aspects of primary devolatilization, these transformations are different for different coal types, heating rates, and pressures.

Softening behavior is the foundation for all physical and morphological transformations during primary devolatilization. It is also a crucial aspect of coking behavior, so it has been monitored for decades with standardized performance indices, albeit at very slow heating rates. More recently, these techniques were modified to impose much faster heating rates, as reported by Yu et al. (2007). The tendencies are that faster heating rates and elevated pressures promote softening behavior. Both effects saturate for progressively more severe conditions, and have probably reached their maximum impact at about 1000°C/s and 1 MPa, respectively. So softening behavior will be maximized in any entrained coal application with the p. f. size grade, but is less pronounced in fluidized beds with coarser fuel particles.

Coal quality impacts are also first-order important but somewhat ambiguous. Certainly, coals that do not soften do not make swollen chars, so their density changes can be accurately estimated from the mass loss. Brown coals and lignites do not soften under any conditions. Conversely, all bituminous ranks soften under any operating conditions; indeed, the term "bituminous" identifies only those coal types that soften under coking conditions. But whether or not subbituminous coals swell depends on the heating rate and pressure. Subbituminous coals with progressively lower C-contents will swell for progressively faster heating rates and higher pressures. Unfortunately, there is no hard-and-fast way to delineate which particular subbituminous samples will soften under specified operating conditions from the standard coal properties. But models have been developed to address this issue, and these will be surveyed in Chapter 6.

As expected, these tendencies are mimicked in reported swelling factors. Representative data for a softening coal appear in Fig. 4.18. The impact of heating rate is explicit for rates from 0.1°C/s to almost 10^6°C/s. The coal swells by 40% at 10°C/s and by almost 60% at 10^3°C/s. Then swelling factors plummet for heating rates faster than 10^4°C/s, presumably because the release rates of gases are too fast to accumulate bubbles in the viscous melt. Changes in the porosity are similar at all but the slowest heating rates. In entrained coal applications, porosities often approach 90% with softening coals. As seen in the right panel, swelling factors pass through a maximum for progressively greater pressures. The rise to the maximum is interpreted by lower melt viscosities for progressively greater pressures as larger portions of lighter tar precursors are retained in the condensed phase. But since the asymptotic minimum tar yield as a function of pressure occurs at about 1 MPa, where swelling factors are maximized, the interpretation of the decay in swelling for higher pressures cannot invoke variations in the volatiles yields.

The heterogeneous nature of coals' physical structure is strongly reflected in the heterogeneity of char structure. Wall and coworkers developed a classification system for these variations based on three groups: (I) thin walled, hollow cenospheres; (II) multicavity crassispheres from solidified foam droplets; and (III) fusinoid particles

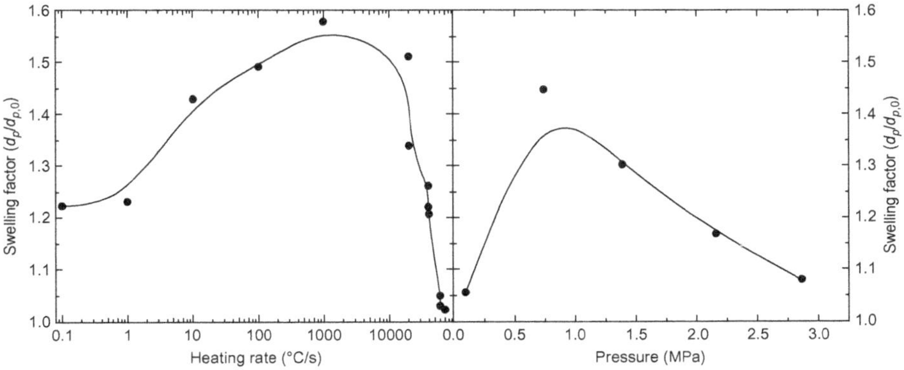

Fig. 4.18 Swelling factors as functions of (left) heating rate and (right) pressure for softening coals.
Reproduced with permission from Yu J, Lucas JA, Wall TF. Formation of the structure of chars during devolatilization of pulverized coal and its thermoproperties: a review. Prog Energy Combust Sci 2007;33:135–70, Elsevier.

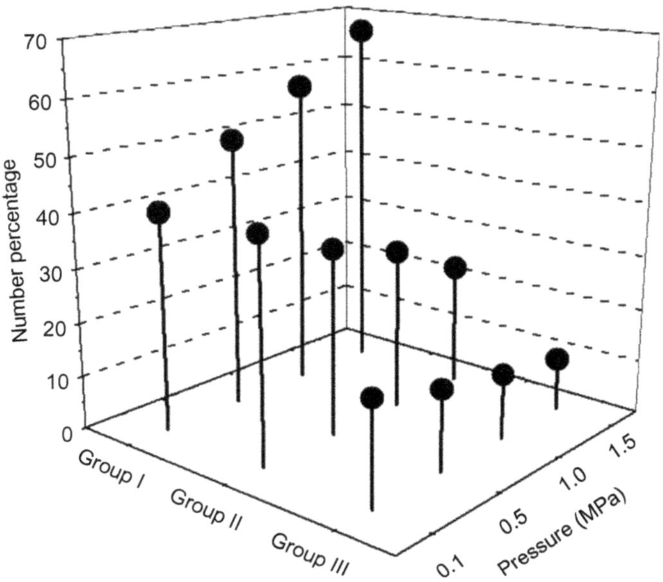

Fig. 4.19 Distributions of char types from a softening coal for various pressures. Reproduced with permission from Yu J, Lucas JA, Wall TF. Formation of the structure of chars during devolatilization of pulverized coal and its thermoproperties: a review. Prog Energy Combust Sci 2007;33:135–70, Elsevier.

that never softened (Yu et al., 2007). They monitored distributions of the three groups for various softening coals, heating rates, and pressures, one of which appears in Fig. 4.19. At any particular pressure, 80%–90% of the morphologies are either cenospheres or crassispheres. As expected for more extensive softening into a less viscous melt, the group distributions shift toward cenospheres for progressively greater pressures. Most important, Groups I and II predominate at all pressures, and both structures exhibit huge macrovoids delineated by very thin carbonaceous sheets containing micro-, meso-, and macropores. With softening coals, the vast majority of char particles in the p. f. grade have no internal pore system on the scale of their particle diameters; rather, the carbonaceous material is accessible to reactive gases through much thinner microporous domains.

Notwithstanding that char group distributions are heavily skewed toward cenospheres, the fusinoid group is the most important one whenever the goal is to accurately predict the levels of unburned carbon in flyash. The reason is that extensive microporous domains in such chars give the most hindered access to reactive gases which gives these chars relatively slow burning rates. Consequently, under conditions that completely burn out cenospheres and crassispheres, the fusinoid particles are often the only remaining form of char that can be recovered with flyash and contribute to LOI.

4.2.7 Primary tar characteristics

This section could have surveyed the monumental laboratory characterizations of primary coal tars with high resolution MS, GC-MS, HPLC, MALDI, and numerous other sophisticated analytical methods. In some applications, such as macromolecular fingerprinting or surrogate coal refining, such detailed molecular characterizations are absolutely necessary. But they are superfluous to our focus on CFD support for commercial utilization technologies, in which tar is simply an abundant intermediate that will ultimately be converted into a noncondensable fuel mixture, oils, and/or soot. As yet, there are no elementary reaction mechanisms for the secondary chemistry of mixtures as complex as tar, and we will see in Chapter 7 that they are not necessary to describe the fate of tar in commercial systems anyway. So as an intermediate and reactant, primary tar is best characterized by its average properties. From this standpoint, the most useful average properties are the elemental composition, proton and carbon aromaticities, and the molecular weight distribution, which determines a number average molecular weight. This section surveys these properties in turn.

One particular feature of a tar's elemental composition is paramount: the atomic H/C ratio. It is the easiest and therefore the best way to accurately determine whether or not secondary volatiles chemistry has affected the products of primary devolatilization in a particular test. The key tendency is illustrated in Fig. 4.20, which shows tar H/C vs fractional weight loss for tests in a RCFR (Chen and Niksa, 1992b), WMR (Solomon and Colket, 1978), and EFR (Freihaut and Proscia, 1991) with *hv* bituminous coals. The RCFR and EFR tests used the same coal sample. All values

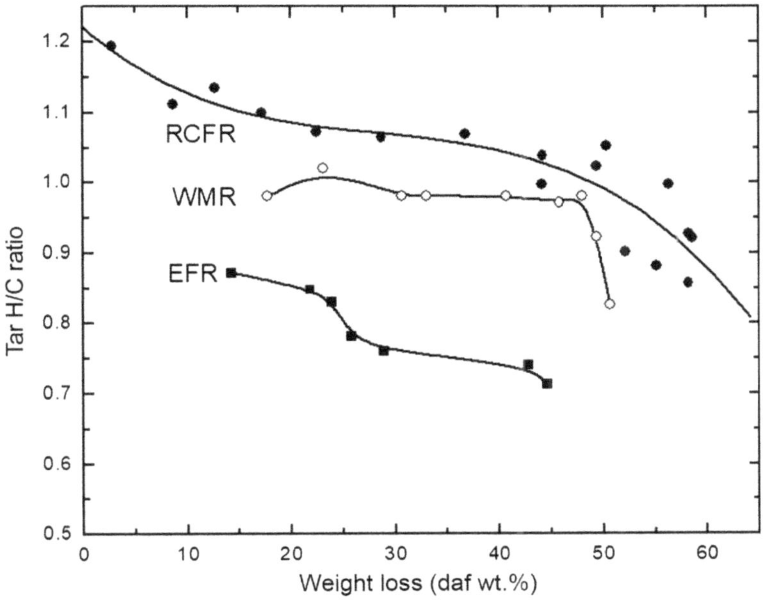

Fig. 4.20 H/C ratios of tar from *hv* bituminous coals from a (●) RCFR, (○) WMR, and (■) EFR.

exhibit the tendency for lower H/C for progressively greater extents of primary devolatilization. Since the ratios are average values for cumulative tar samples, the incremental tar additions must become much more aromatic during the course of primary devolatilization. This was verified for the RCFR tars by measuring the proton distributions and aromaticities, which showed that the tar increments became more aromatic because protons in the β and γ positions on condensed ring structures had been eliminated prior to vaporization (Chen and Niksa, 1992b). The values from the RCFR are 25% greater than from the EFR, and 10% greater than from the WMR throughout. The differences with the WMR data could also reflect a slightly lower rank fuel sample and a heating rate that was slower by an order of magnitude. But the much lower values from the EFR with the same coal reflect extensive secondary volatiles pyrolysis; in fact, when this coal was tested with regulated secondary pyrolysis in the RCFR, the tar H/C values fell to the EFR values (Chen et al., 1992).

To verify the absence of secondary volatiles pyrolysis, compare the H/C ratio of tar to the ratio for the parent coal. Table 4.3 shows H/C ratios for coals and their initial and ultimate tars from RCFR tests at atmospheric pressure. Initial tars represent only the first few daf wt.%, whereas ultimate tars make up the total tar yield. With reference to the parent coals, the tars have greater H/C ratios by 13% to 86% in the initial samples, and by 0% to 69% in the ultimate samples, although the lone coal that gave no enrichment is probably an outlier. The ratios for the initial tars give a strong negative correlation ($r^2 = 0.81$) with coal H/C. But the ultimate tar H/C ratios are essentially uncorrelated with their parent coals' ratios. Hence, with any coal type, the absence of secondary pyrolysis can be verified with ultimate tar H/C ratios that are appreciably greater than the ratios for the parent coals. In Fig. 4.20, the H/C ratio of even the initial tar sample from the EFR barely exceeds the ratio of 0.86 of its parent coal, and the ultimate ratio is much lower.

The other important inference from the data in Table 4.3 is that primary tars are considerably less aromatic than their parent coals, especially at the onset of tar production. They do become progressively more aromatic throughout primary

Table 4.3 Initial and ultimate H/C ratios of tar and their parent coal ratios for rapid devolatilization at atmospheric pressure (Chen and Niksa, 1992b; Liu et al., 2004)

		Tar H/C	
Coal-C (daf wt.%)	**Coal H/C**	**Initial**	**Ultimate**
69.5	0.86	1.21	0.99
74.1	0.86	1.11	0.92
76.3	0.99	1.12	1.03
77.8	0.96	1.27	1.10
80.4	0.71	1.32	1.20
82.3	0.69	1.11	1.06
82.5	0.82	1.19	0.90
82.6	0.71	1.09	0.71
88.7	0.68	1.17	0.84

devolatilization, but even the ultimate bulk tar sample remains much less aromatic than its parent coal. In quantitative terms, ^{13}C NMR is the best method to assign carbon aromaticities but, unfortunately, has not yet been applied to pristine tar samples unaffected by secondary tar decomposition. Proton aromaticities of primary tars recovered from a RCFR (Chen and Niksa, 1992b) increased from 0.2 to 0.4 throughout primary devolatilization, due primarily to the elimination of hydrocarbon peripheral groups at the β- and γ-positions with respect to aromatic nuclei prior to tar release from the condensed phase. In other words, aliphatic and heteroatomic peripheral groups are released from tar precursors in the condensed phase throughout primary devolatilization, which is responsible for greater aromaticities in tar released during progressively later stages of the process.

The fractional elemental compositions of ultimate tars from diverse coals for rapid primary devolatilization at atmospheric pressure appear in Fig. 4.21. There are no distinctive trends with coal quality in any of the element fractions, because the sample-to-sample variability even among coals with very similar carbon contents is enormous. Evidently, additional variations in tars' elemental compositions compound the variations among ultimate tar yields (cf. Fig. 4.3). The one-to-one correspondence among the variations in the C- and H-fractions extends across the entire rank spectrum, which is a feature seen in very few other characteristics of primary devolatilization. The N-fractions display only most of the same variations, whereas variations in the O-fractions are uncorrelated with the others. Ultimate tars are composed of 55%–85% C; 5.5%–6.5% H; 20%–45% O, with a few percent each of N and S. The abundance of oxygen in primary tar obscures detailed structural characterizations, and is also responsible for the relatively fast tar decomposition rates in secondary pyrolysis chemistry.

The conversion dynamics for tar compositions were monitored as C/H/N levels from separate RCFR tests with variable contact times with four coals (Chen and

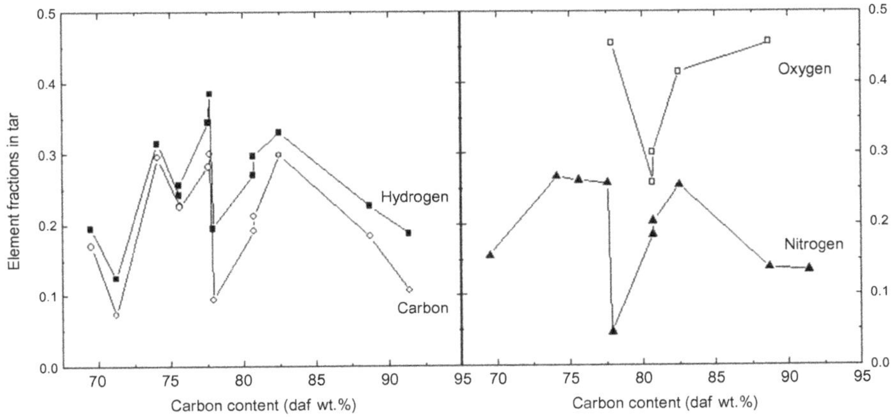

Fig. 4.21 Element fractions for (left) C and H and (right) O and N in the ultimate tars from primary devolatilization in RCFRs and WMRs at atmospheric pressure.

Niksa, 1992b) and with five coals (Liu et al., 2004). The reported element fractions increase in direct proportion to the fractional tar yield, which is the tar yield for a particular residence time normalized by the ultimate tar yield. The proportional relation indicates that there are no prominent associations between a particular element and distinct stages in the process chemistry, consistent with a form of tar shuttling in which hardly any of the molecular constituents participate in the chemistry that releases tar precursors from coal macromolecules.

As seen in Fig. 4.22, faster heating promotes the shuttling of C, H, and N with tar, albeit not by very much. This dataset is one of the very few that characterizes a heating rate dependence on tar composition, all other conditions the same. Unfortunately, it is unusual in that the tar yields for 1000°C/s and 5000°C/s were identical, which probably explains the plateau in the element partitioning for the fastest rates in Fig. 4.22. For the slower heating rates, H-shuttling is enhanced the most by faster heating, and C-shuttling the least.

One also expects to have less of the coal elements in tars prepared under elevated pressures for these same reasons, which is confirmed by the C-, H-, and O-fractions in Fig. 4.23. Estimated oxygen contents of the tars from the *hv* bituminous sample vary from 8% to 20%, which are only half to two-thirds of the values reported for

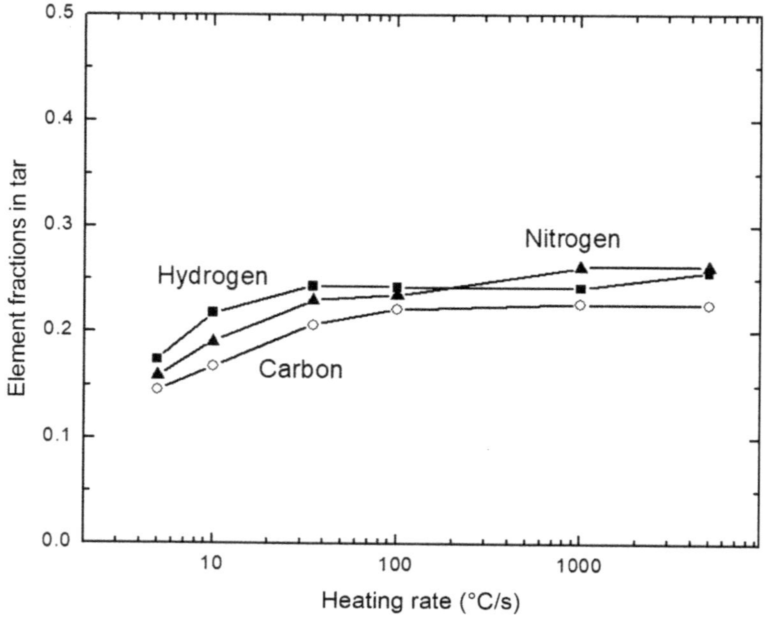

Fig. 4.22 Element fractions for C, H, and N in the ultimate tars from primary devolatilization of an *hv* bituminous coal at different heating rates at atmospheric pressure.
Reproduced with permission from Cai H-Y, Guell AJ, Dugwell DR, Kandiyoti R. Heteroatom distribution in pyrolysis products as a function of heating rate and pressure. Fuel 1993;72:321–27, Elsevier.

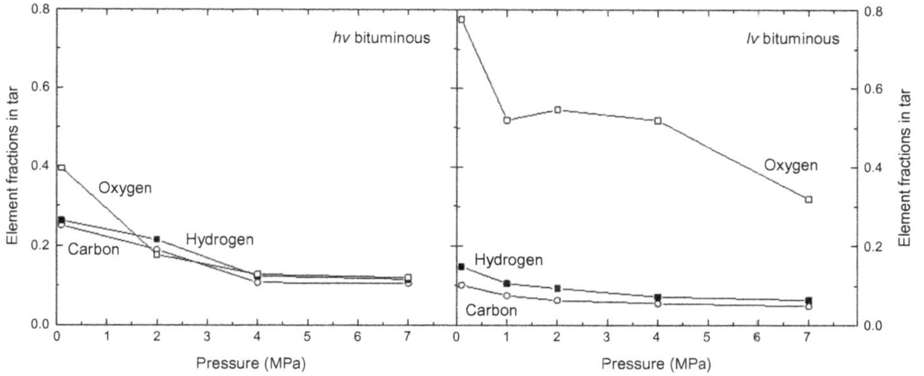

Fig. 4.23 Element fractions for C, H, and O in the ultimate tars from primary devolatilization of (left) an *hv* bituminous coal and (right) a *lv* bituminous at different pressures (Cai et al., 1993).

atmospheric pyrolysis with similar coals. This difference probably reflects the elimination of oxygen functional groups from intermediate fragments of coal molecules at elevated pressure before they were released as tar compounds. The much greater O-fractions with the *lv* bituminous in Fig. 4.23 cannot be taken at face value, because the oxygen content of this coal is so low that the measurement uncertainties on the tar-O levels are much greater.

The tendency for lower coal-H levels for progressively greater pressures is counteracted somewhat by the chemical constitution of tars at elevated pressures. In terms of the H/C ratios in Table 4.4, which are directly comparable to those in Table 4.3 for these five coals, the primary tars at elevated pressure are even more substantially enriched in hydrogen over the whole-coal values than tars prepared at atmospheric pressure. The H-enrichments are again greater for initial tars, but at elevated pressure, these H/C ratios can be double the parent coal values. Ultimately, the enrichment varies from 30% to almost 80%, which is usually greater than for tars prepared at atmospheric pressure. Consequently, primary tars generated under elevated pressures are even less aromatic than those for atmospheric pressure, and much less aromatic than their parent coals.

Table 4.4 Initial and ultimate H/C ratios of tar and their parent coal ratios for rapid devolatilization at 1.0 MPa (Manton et al., 2004)

		Tar H/C	
Coal-C (daf wt.%)	**Coal H/C**	**Initial**	**Ultimate**
76.3	0.99	1.29	1.22
77.8	0.96	1.40	1.25
80.4	0.71	1.34	1.18
82.3	0.69	1.14	1.02
82.6	0.71	1.40	1.08

Tar MWDs and M_n-values hold little practical interest but are nevertheless essential elements of modern reaction mechanisms for primary devolatilization. It would not be an overstatement to say that network depolymerization mechanisms could not have been developed before the MWDs of primary tars and their precursors in the condensed phase were measured. Unger and Suuberg (1984) first demonstrated that gel permeation chromatography (GPC) could be used to monitor the MWDs of coal tars prepared with rapid heating rates. Given the abundance of oxygen in polar functional groups in primary tars, GPC is subject to legitimate concerns about sample holdup and biasing that have been diminished but never fully resolved. But the forms of the reported MWDs and their shifts with variable operating conditions are undoubtedly correct, and will be emphasized here. Since the absolute magnitudes are not directly pertinent to our interests, there is no need to delve into the measurement uncertainties.

The relation between the MWDs of primary tars and their precursors in the condensed phase is illustrated in Fig. 4.24. In the tests, a *hv* bituminous coal was heated at 1000°C/s to a specified temperature with neither an IRP nor forced quenching, at both near-atmospheric pressure and under vacuum. The chars were extracted in tetrahydrofuran (THF) at room temperature, and the MWDs of the extracts and recovered tars were monitored with GPC. In the left panel of Fig. 4.24, the extract yields pass through a maximum at about 350°C, which is only slightly hotter than the temperature at which tar production begins. So the THF extracts are representative of the precursors to tar in the condensed phase. Tar production continues through 625°C, beyond which additional noncondensables determine the approach to ultimate weight loss at 900°C.

For the test at 550°C, the samples of both extract and tar represent substantial portions of the respective maximum yields. The three MWDs all have the form of gamma

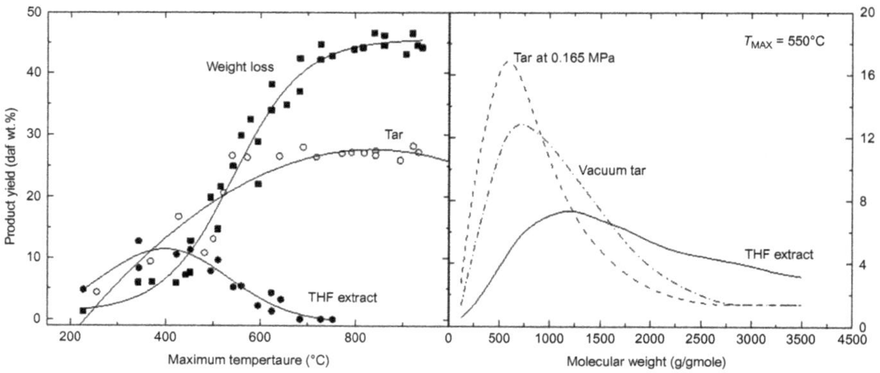

Fig. 4.24 (Left) Weight loss, tar yields, and THF extract yields from an *hv* bituminous coal heated at 1000°C/s with no IRP at 0.165 MPa; and (right) MWDs of *(solid)* THF extract and tars prepared at *(dot-dashed)* 0 and *(dashed)* 0.165 MPa for the test at 550°C.
Reproduced with permission from Unger PE, Suuberg EM. Molecular weight distributions of tars produced by flash pyrolysis of coal. Fuel 1984; 63:606–11, Elsevier.

(γ)-distributions, in which the MWD abruptly rises from a minimum value of 100 g/mol and passes through a maximum into a tail that extends to weights of several thousand g/mol. Succeeding refinements to the GPC analytical protocol showed that the maximum tar weights are probably no greater than 1500 g/mol (Oh et al., 1989; Griffin et al., 1993; Darivakis et al., 1994). Notwithstanding, γ-distributions are widely recognized throughout petrochemical refining as the characteristic form for the evaporation of heavy hydrocarbons. The extract MWD is the broadest by far, and the MWDs of near-atmospheric and vacuum tars are shifted toward lighter weights by at least a factor of two. The obvious implication of this relation is that tars are the portion of their precursors in the condensed phase that were volatile enough to evaporate under particular test conditions. The tars are skewed toward the lighter weights, compared to their precursors, simply because hydrocarbons become more volatile for progressively lower molecular weights. Yet the tar MWDs extend toward very large weights because even these hydrocarbons have finite vapor pressures. Hence, the form of tar MWDs and their relation to the MWDs of tar precursors are definitive markers for a central role for tar vaporization in the mechanism of tar production.

The other important feature in Fig. 4.24 is that the MWD of tar from vacuum is significantly heavier than that for atmospheric pressure. This feature also reflects tar vaporization, because hydrocarbon vapor pressures are functions of temperature but not pressure. So the ambient pressure determines the driving force for tar evaporation. Since a test under vacuum gives the greatest driving force, it also gives the most extensive evaporation of the heaviest tar components, which shifts the entire MWD toward heavier weights. Consequently, tar yields from tests under vacuum are appreciably greater than those at atmospheric pressure, which is consistent with the lower maximum in the MWD for vacuum tar in Fig. 4.24. The shift toward lighter tar for progressively greater pressures was corroborated by several other older WMR tests in the United States (Unger et al., 1985; Oh et al., 1989; Solomon et al., 1990), but has not received any attention since.

Unger and Suuberg (1984) also reported MWDs for portions of the total tar sample released at different temperature intervals that showed shifts toward heavier weights of a few hundred g/mol for progressively hotter temperatures, and this tendency was verified by other groups (Oh et al., 1989). This shift is easily rationalized by the positive temperature dependence in the saturated vapor pressures of heavy hydrocarbons. Tar MWDs may also shift toward heavier weights for progressively faster heating rates, although this effect is not always evident in the measured M_n-values. Griffin et al. (1993) monitored M_n of tars prepared at 800°C at atmospheric pressure with heating rates from 10 to 2×10^4°C/s. As seen in Table 4.5, the average values for both sizes show little variation through 10^3°C/s, then diminish slightly at the faster heating rates. Li et al. (1993) reported a similarly weak dependence on heating rate. Since primary devolatilization occurs at hotter temperatures for progressively faster heating rates, the heating rate effect could be an indirect reflection of the temperature dependence in the saturated vapor pressures of tar precursors, although additional factors may also contribute to its magnitude.

Table 4.5 M_n-values for rapid pyrolysis tars from a *hv* bituminous coal at atmospheric pressure from two sizes at several heating rates (Griffin et al., 1993)

	M_n (g/mol)	
Heating rate (°C/s)	**69 μm**	**115 μm**
10	261	257
	267	274
		287
200	261	
	247	
1×10^3	254	243
	235	254
5×10^3	215	
	237	
	232	
2×10^4	226	231
	242	212
		202

In what is probably the only reported characterization of tar MWDs from a diverse assortment of coals, Unger and Suuberg (1984) found substantially lighter tars from both a lignite and a *lv* bituminous coal compared to two *hv* bituminous samples.

4.2.8 Noncondensable gas yields

Quantitative resolution of the complete distribution of primary devolatilization products represents a formidable challenge and, for many different reasons, laboratory studies are usually focused on one or more portions of the distribution. These partial distributions are often valuable even though the data do not determine closures on the mass and element balances. The few datasets that close the balances on mass, C, H, and N to within ±5% in individual tests are featured in this section wherever possible. Since there are no direct determinations for the O-levels in chars and tars, O-balances necessarily reflect the accumulation of errors from independent sources and are often not closed to the same tolerance. S-balances are often impossible to evaluate, either because the specialized analyses for char-S and tar-S were omitted, or because the specialized equipment to monitor H_2S, the predominant noncondensable S-species, was omitted. Also, this section covers the major oxygenated and hydrocarbon gas species, whereas N- and S-species are considered in succeeding sections because their yields are always considerably smaller.

Xu and Tomita (1987a,b) monitored the yields of all major gaseous products from 17 coals that span the rank spectrum in a CPP at atmospheric pressure. All tests in the coal quality survey imposed a heating rate of about 3000°C/s to 765°C with an IRP of 4 s, which was shown to be sufficient to achieve ultimate primary yields. The distributions of oxygenated gases (CO_2, CO, H_2O) and GHCs (CH_4, C_2H_4 + C_2H_6, C_3H_6 + C_3H_8)

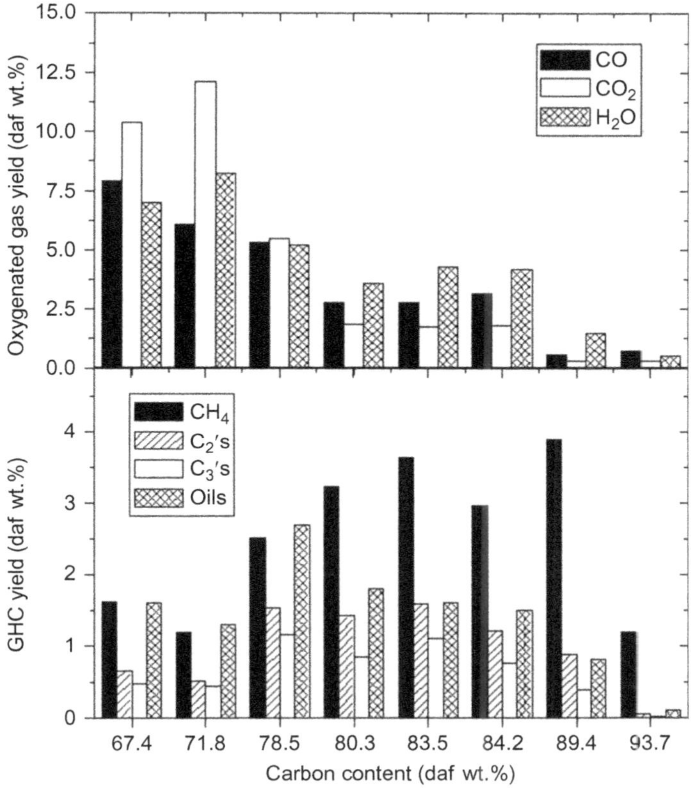

Fig. 4.25 Coal quality impacts on (top) oxygenated gas yields and (bottom) GHC yields with oils yields for rapid devolatilization at atmospheric pressure in a CPP.
Reproduced with permission from Xu W-C, Tomita A. Effect of coal type on the flash pyrolysis of various coals. Fuel 1987a;66:627–31, Elsevier.

from eight coals appear in Fig. 4.25. The lower panel also includes oils, which are mixtures of benzene, toluene, xylene, phenol, cresol, and xylenol.

Most of the noncondensable gases are oxygenated species for C-contents through 85%, which comprises all ranks except the low volatility coals. The yields of all three oxygenated gases diminish for coals of progressively higher rank, as they must, because coal-O monotonically decreases across the rank spectrum before vanishing in anthracites. This tendency is yet another aspect of primary devolatilization subject to considerable sample-to-sample variability. The yields of H_2O and CO show the fewest variations and H_2O yields are always slightly greater than CO yields, except for three lignites in the full suite of 17 coals (not shown) where these yields are the same. The CO_2 yields are the greatest of all through 72% C; then match the CO yields through 79% C; then fall off further with *hv* bituminous coals before they vanish with *lv* bituminous and anthracites.

GHC yields are lower than oxygenated gas yields through *mv* bituminous. Methane is the most abundant GHC species, by far, and its yield exceeds the sum of all C_2 and C_3 species for all coals except the three richest *hv* bituminous samples in the full suite of 17 coals whose H-contents approach or exceed 6 daf wt.%; these coals also gave the greatest tar yields. With those three coals, the CH_4 yield is only slightly lower. The trend is for greater CH_4 yields for progressively greater C-contents, until CH_4 production plummets with the anthracite. The uniformity of this trend is distinctive because the CH_4 yields do not surge for the coals whose tar yields surged. All the other GHC species except C_3H_8 change in tandem with the tar yields, and do display the surge for the three distinctive coals. In fact, all the heavier GHC species except C_3H_8 are completely correlated in their sample-to-sample variability. There are slightly more C_2's than C_3's for all coals. The C_3H_8 yields are about half those for the C_2 GHCs, and align better with CH_4 yields than with the tar/heavy GHC yields. Appreciable levels of C_4 GHCs were not detected with any coal. Oils yields mimic the variations in the tar yields. They equal the CH_4 yields through 79% C, then fall off faster than CH_4 for coals of higher rank before they vanish for anthracites. The yields of phenol and cresol are essentially the same and about double the yields of every other oil species. The H_2 yields were monitored but are not illustrated because they were 0.3 daf wt.% for C-contents through 67.4%; 0.4% through 80.3%; and either 0.4% or 0.5% for all higher C-contents.

Complete product distributions that closed mass and C/H/N balances in individual runs were reported for RCFR tests with four diverse coals at atmospheric pressure and moderately greater maximum temperatures and heating rates than the CPP tests in Fig. 4.25. As seen in Fig. 4.26, CO is the most abundant oxygenated gas except with the *lv* bituminous, and the CO_2 yields are appreciably lower than the other two. The yields of CH_4 and C_2 GHCs are comparable for all coals, whereas the C_3 yields are much lower. While the yields of all light chain GHCs in Figs. 4.25 and 4.26 are similar, the oils yields from the RCFR tests are much greater.

Noncondensable gas yields as a function of heating rate have not yet been reported, although it is already established that total gas yields are hardly perturbed for progressively faster heating rates (cf. Fig. 4.7). But detailed gas yields for different heating rates with the same coal would be needed to determine if any species are preferentially enhanced or diminished by heating rate variations. Pending such data it is reasonable to assume that all major noncondensable species yields are diminished in proportion to the percentage change in the total gas yields.

As mentioned previously, the yields of noncondensable gases are greater at higher pressures, but not by enough to compensate for the reduction in tar yields. The data in Fig. 4.27 compare gas yields from elevated and atmospheric pressures, all else the same. Moisture levels are slightly enhanced, CO yields are erratic, and CO_2 yields are usually enhanced, albeit slightly. The yields of C_3 GHCs and, especially, oils are reduced by elevated pressures. Unfortunately, the impact of elevated pressure on CH_4 and C_2 GHCs is inconsistent between the WMR and RCFR data. In the WMR data, these yields are enhanced by elevated pressures, as corroborated by other WMR datasets (Bautista et al., 1986; Griffin et al., 1993). But in the RCFR data, CH_4 and C_2 GHCs are diminished by elevated pressure. The datasets in Fig. 4.27 were

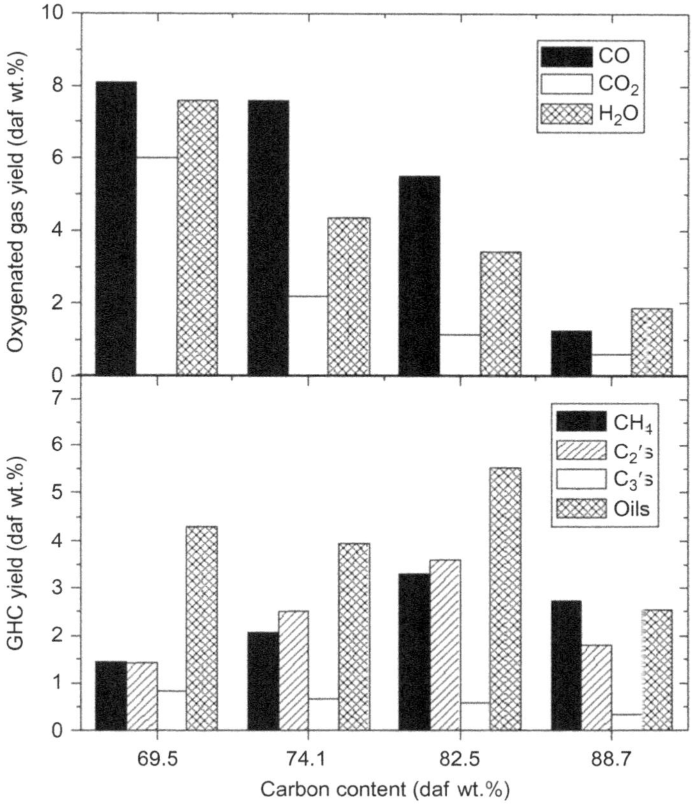

Fig. 4.26 Coal quality impacts on (top) oxygenated gas yields and (bottom) GHC yields with oils yields for rapid devolatilization at atmospheric pressure in a RCFR.
Reproduced with permission from Chen JC, Niksa S. Coal devolatilization during rapid transient heating. Part 1: primary devolatilization. Energy Fuels 1992b;6:254–64, the American Chemical Society.

obtained with two versions of an RCFR for atmospheric and elevated pressures. In principle, these facilities enable direct comparisons among any of the products at atmospheric and pressures to 4 MPa but, in practice, there are significant differences in the flowfield and, consequently, in the thermal histories for the different operating conditions. Notwithstanding, it is hard to fathom how such differences could be responsible for the large differences in the CH_4 and C_2 GHC yields in Fig. 4.27.

4.2.9 Volatile nitrogen species

Perhaps no other aspect of primary devolatilization has received as much attention as the partitioning of coal-N into volatile-N and char-N. This attention comes from tight regulations on NO_X emissions from coal fired furnaces worldwide, and the particular

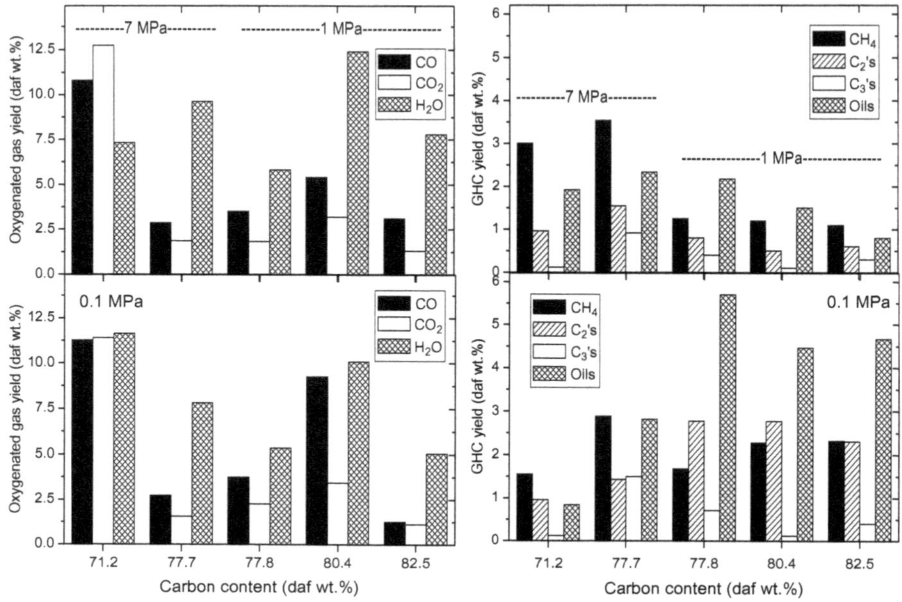

Fig. 4.27 Impact of elevated pressures on yields of oxygenated gases and GHCs (top) from a WMR at 7 MPa (Suuberg et al., 1979) and a RCFR at 1 MPa (Manton et al., 2004) and (bottom) with the same coals at atmospheric pressure (Suuberg et al., 1979; Liu et al., 2004).

interest in volatile-N comes from its prominent role in aerodynamic NO_X abatement technologies. But these imperatives carry a drawback pertaining to the release of primary N-volatiles, because the connections to flame phenomena prompt researchers to use EFRs to impose fast heating rates to very hot temperatures, where unregulated secondary volatiles pyrolysis distorts the primary volatile-N species beyond recognition. This distortion is particularly severe in the proportions of coal-N in tar and noncondensable species. The presentation in this section is based on the handful of datasets that eliminated secondary pyrolysis (unless explicitly noted) to reveal the primary release of volatile-N.

The second complication is that HCN is released as a primary product after the production of most primary tar, and as the main N-species from tar decomposition, and as a product of much slower annealing chemistry at high temperatures after the end of primary devolatilization. If secondary volatiles pyrolysis is eliminated then the second channel for HCN is also eliminated. But the first and third channels introduce a strong dependence on the total reaction time, especially in tests at temperatures hotter than about 1000°C. Consequently, there are huge variations in reported levels of residual char-N and HCN in the available database, depending on the contribution from the annealing stage. The data in this section are intended to characterize only the primary HCN release, on the time scale for primary devolatilization, with minimal contributions from annealing. Compared to many other datasets, they indicate relatively low HCN levels and high char-N levels.

Tar-N, HCN and, perhaps, NH_3 comprise volatile-N. Release of the non-condensable species is the easiest to explain because there is a direct connection between a particular bonding arrangement of a portion of coal-N and the NH_3 yield from primary devolatilization. X-ray photoelectron spectroscopy (XPS) determines three major forms of coal-N as pyrrolic-N, pyridinic-N, and quaternary-N. Kambara et al. (1995) monitored these forms throughout the rapid devolatilization of 20 coals representing ranks from brown coals through *mv* bituminous at atmospheric pressure, where the maximum temperature of 1215°C was held for about 4 s. These conditions were sufficiently severe to eliminate all quaternary-N from all the chars although, with most coals, the release was complete in 4 s at 945°C. They also monitored the yields of NH_3, HCN, and tar-N, although the proportions of tar-N and HCN were distorted by secondary pyrolysis in these tests. Evidently, the levels of NH_3 were unaffected, because they are strongly correlated with the portion of coal-N in the quaternary form, as seen in Fig. 4.28. The correlation coefficient is 0.887 with a std. dev. of 1.4%, and the proportionality constant is 0.8304. The most striking feature is that NH_3 is a relatively minor volatile-N species, never amounting to more than 12.1% of coal-N. Over half the coals released less NH_3 than 7% of coal-N, and all the coals with yields over 10% were subbituminous or lower ranks. However, it is important to realize that these NH_3 levels do not represent the NH_3 levels that come

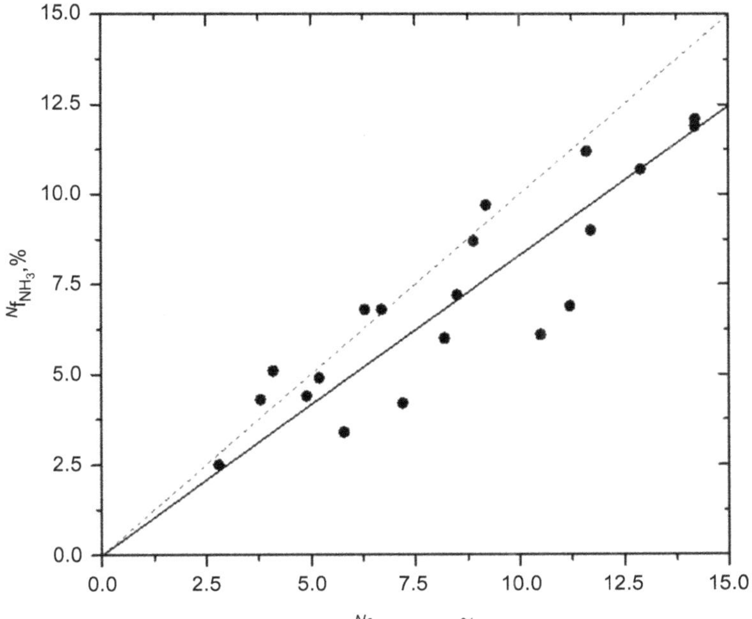

Fig. 4.28 Correlation between coal-N released as NH_3 and coal-N in the quaternary form. Reproduced with permission from Kambara S, Takarada T, Toyoshima M, Kato K. Relations between functional forms of coal nitrogen and NO_X emissions from pulverized coal combustion. Fuel 1995;74(9):1247–53, Elsevier.

into play during aerodynamic NO_X abatement, because secondary pyrolysis and volatiles combustion both generate and destroy NH_3.

Since aromatic rings are created, not destroyed, during primary devolatilization, and since the bulk of coal-N appears as pyrrolic- and pyridinic-N, it is not surprising that tar shuttling is the major release mechanism for volatile-N during primary devolatilization. This is apparent in Fig. 4.29, which shows the coal quality impacts on the ultimate fractional partitioning of coal-N into char, tar, and HCN + NH_3 in a RCFR at atmospheric pressure. The fractional tar-N levels are nearly the same as the fractional tar yields (in Chen and Niksa, 1992b,c), whereas the gaseous N-species comprise much smaller coal-N fractions for all but the subbituminous coal, which produced much more NH_3 than HCN. For the other coals, NH_3 was negligible and HCN contained 10% of coal-N or less. Both char-N and tar-N fractions are fairly uniform for all ranks through *hv* bituminous, then the tar-N falls off sharply for low volatility coals while the char-N levels surge.

The dynamics for four of the coals in Fig. 4.29 are resolved in Fig. 4.30, where the N-partitioning appears as a function of the extent of devolatilization, evaluated as fractional ultimate yield. These data clearly show that tar shuttling is essentially the only means of N-release throughout most of rapid primary devolatilization, and that gaseous N-species are released only near the end of tar production. As mentioned

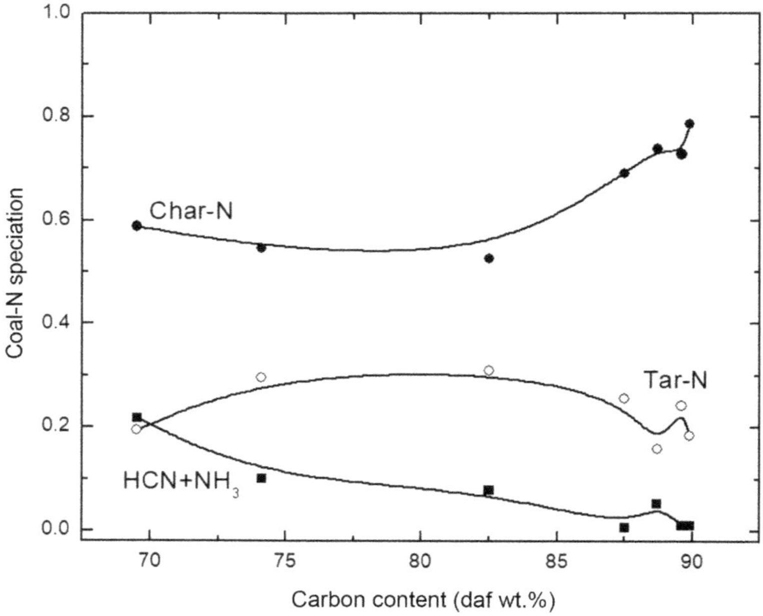

Fig. 4.29 Coal quality impacts on the partitioning of coal-N during rapid primary devolatilization in a RCFR at 0.1 MPa (Chen and Niksa, 1992c).

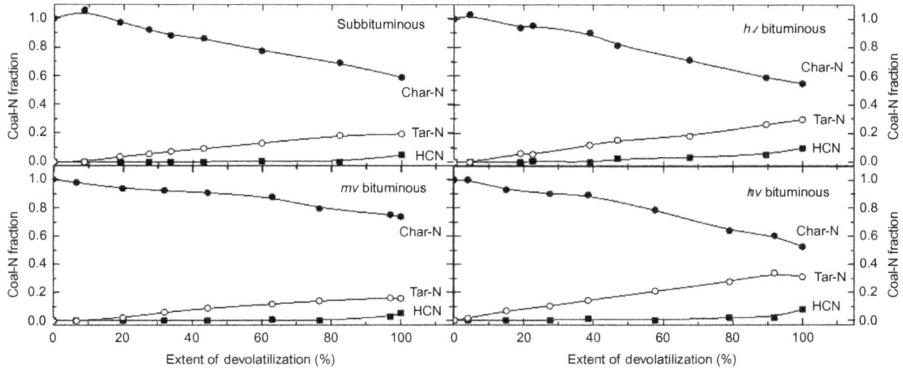

Fig. 4.30 Fractional coal-N partitioning for, in clockwise order, a subbituminous, *hv* bituminous, *hv* bituminous, and *mv* bituminous in a RCFR at atmospheric pressure. Reproduced with permission from Chen JC, Niksa S. Coal devolatilization during rapid transient heating. Part 1: primary devolatilization. Energy Fuels 1992b;6:254–264, the American Chemical Society.

above, the release of gaseous N-species persists through progressively longer reaction times and, especially, at progressively hotter temperatures, but on much slower time scales.

The central role for tar shuttling is responsible for the enhanced N-release for progressively faster heating rates in Fig. 4.31, where volatile-N denotes the sum of tar-N, HCN, and NH_3. The incremental enhancements for each order-of-magnitude increase in the heating rate are the same for both volatile-N and tar-N, within measurement uncertainties. This indicates that, unlike the precursors to the major noncondensables, the pyrrolic- and pyridinic-N retained in the condensed phase at slow heating rates that would otherwise be shuttled away in tar with faster heating is not released on the time scale of tar production. This is consistent with the direct indication in Fig. 4.30 that tar shuttling is the only means of N-release throughout all but the latest stages of primary devolatilization.

The same considerations explain why volatile-N levels diminish slightly for progressively higher pressures. The datasets in Fig. 4.32 were obtained with a WMR that imposed the same thermal history across a broad pressure range (Cai et al., 1993) to monitor ultimate primary devolatilization behavior. Levels of char- and tar-N were monitored directly, and the HCN-fraction was assigned by difference. The N-speciation for the *hv* bituminous displays the expected tradeoff between tar-N and HCN, and a slight increase in char-N for progressively greater pressures. The same tradeoff is apparent in the N-speciation for the *lv* bituminous coal, but the expected slight increase in char-N for greater pressures is probably obscured by an erroneous measurement for atmospheric pressure, and also by the relatively low tar yields from this coal sample.

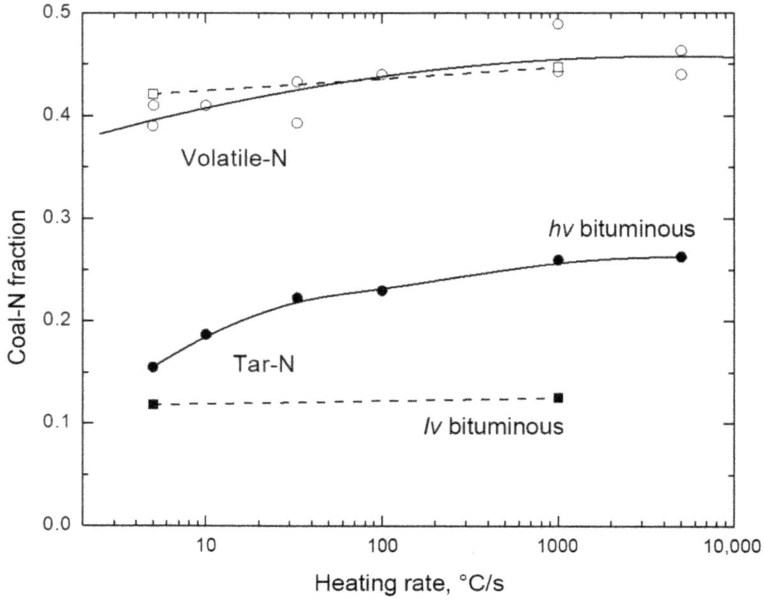

Fig. 4.31 Fractional volatile-N and tar-N from *(circles and solid curves) hv* and *(squares and dashed curves) lv* bituminous coals for various heating rates to 950°C with 5 s IRP in a WMR at atmospheric pressure.
Reproduced with permission from Cai H-Y, Guell AJ, Dugwell DR, Kandiyoti R. Heteroatom distribution in pyrolysis products as a function of heating rate and pressure. Fuel 1993;72:321–7, Elsevier.

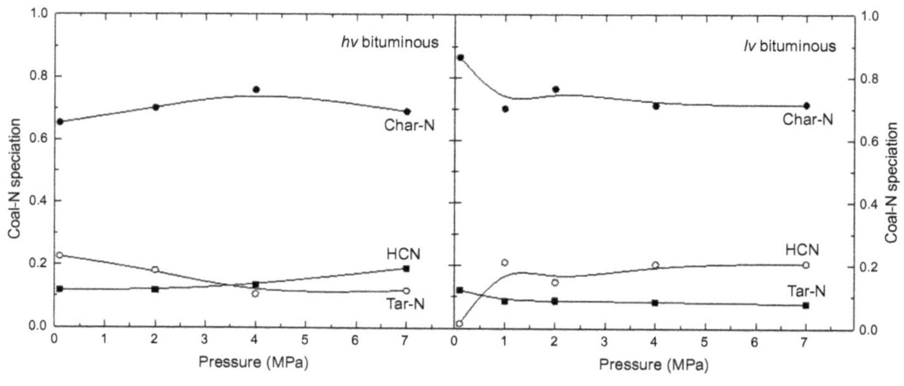

Fig. 4.32 Ultimate N-Speciation for (left) *hv* bituminous and (right) *lv* bituminous coals across a broad pressure range for 1000°C/s to 700°C with 10 s IRP in a WMR.
Reproduced with permission from Cai H-Y, Guell AJ, Dugwell DR, Kandiyoti R. Heteroatom distribution in pyrolysis products as a function of heating rate and pressure. Fuel 1993;72:321–7, Elsevier.

4.2.10 Devolatilization of coal-sulfur

Sulfur in coal is distinctive because it appears both within minerals and as functional groups in the organic coal matrix. The mineral associations are predominately pyrite, with minor contributions from the sulfates of calcium, iron, and other cations. The organic associations are broadly classified as aliphatics (mercaptans, thiols, and aliphatic sulfides) and aromatics (aromatic sulfides and thiophenes), and these two groups comprise organic-S (S_{ORG}). Total sulfur is the sum of pyritic-S (S_{PYR}), sulfate-S (S_{SO4}), and S_{ORG}. We already saw one practically important consequence of these different forms in the excessive sample-to-sample variability in the S-fractions in char (cf. Fig. 4.14). It is impossible to understand the release of volatile S-species and, especially, the retention of coal-S in char without a sure footing in the distribution of total-S among the different forms, because S_{ORG} decomposition exhibits a moderate dependence on coal quality whereas S_{PYR} decomposition is essentially the same in any coal. Consequently, the proportions of S_{ORG} and S_{PYR} are much more important than coal rank, per se.

This section briefly surveys coal-S distributions and their variation with coal rank, and the major uncertainties on reported S-distributions. In principle, we should consider only the release of organic sulfur as an aspect of primary devolatilization for parity with the release of the other four major elements from the organic coal matrix. But this is impractical because the S-content in an ultimate analysis does not distinguish mineral-S from S_{ORG}, and some components of mineral-S decompose on the time scales for devolatilization. Moreover, S_{PYR} and S_{ORG} release the same form of volatile-S, H_2S, and with most coals, most of the H_2S originates in S_{PYR} rather than S_{ORG}. The important caveat is that many coals from India, Australia, and elsewhere contain no pyrite at all, because siderite, $FeCO_3$, is their predominant Fe-mineral. Also, brown coals and lignites often have no pyrite because their iron is atomically dispersed throughout the coal matrix. In these coals most coal-S appears in the organic macromolecular structure, except for small contributions from S_{SO4}. Clearly, only datasets that accurately identify the forms of coal-S in the subject samples can be used to resolve the various contributions to volatile-S species, and to interpret residual char-S levels.

4.2.10.1 Distributions of coal-S

Distributions of the forms of coal-S are evaluated in three stages: (1) Coal-S is assigned from the SO_2 formed by complete oxidation of the coal; (2) S_{SO4} is assigned from a titration of the liquid from a heated bath of the coal in dilute HCl; and (3) S_{PYR} is assigned from the Fe-ion concentration in a bath of the HCl-washed coal in nitric acid. Then S_{ORG} is assigned by difference. The first determination has the lowest uncertainties because the reagents in the procedure are strong enough to completely oxidize all forms of coal-S. But both other analyses are unsettled by ambiguities, both in the laboratory procedures and in the interpretation of raw results. These ambiguities are beyond our scope but were carefully elaborated by Davidson (1993).

The main variation in S_{SO4} levels is that inherent mineral sulfates can be swamped by sulfates formed during weathering of the coal in stockpiles, particularly by products of pyrite oxidation at near-ambient temperatures. Notwithstanding, very few reported S-distributions have S_{SO4} at more than 10% of coal-S, and typical values are only a few percent. The main uncertainties on S_{PYR} are rooted in the assumptions that none of the Fe-minerals are removed in the HCl wash, and that all the Fe and S released in the HNO_3-wash came strictly from FeS_2, rather than any other mineral form. Also, pyrite is incorporated into coal seams as a precipitate from dissolved H_2S and ferric ion in seawater, and appears in grains as small as 10–40 μm (and as large as macroscopic chunks). The smaller grains are readily encapsulated by clays and the macromolecular matrix, and these coatings may make them impervious to acid washing. But since none of these concerns can be reconciled with any precision, there is little choice but to accept reported S_{PYR} values at face value. In coal seams, pyrite typically accounts for about half of coal-S, although most pyrite is removed prior to utilization in the developed economies, but not everywhere. So the apparent rank dependence based on reported ultimate analyses is usually an artifact of the coal cleaning process (cf. Fig. 2.5). The largest uncertainties on S-distributions pertain to S_{ORG}. Since it is specified as the difference between coal-S and S_{PYR} plus S_{SO4}, the uncertainties on all three analyses affect the assigned value.

From the standpoint of thermal stability, S-distributions must be resolved even further because the different forms of S_{ORG} exhibit grossly different kinetic behavior. This is illustrated in Fig. 4.33 by the three different temperature windows for

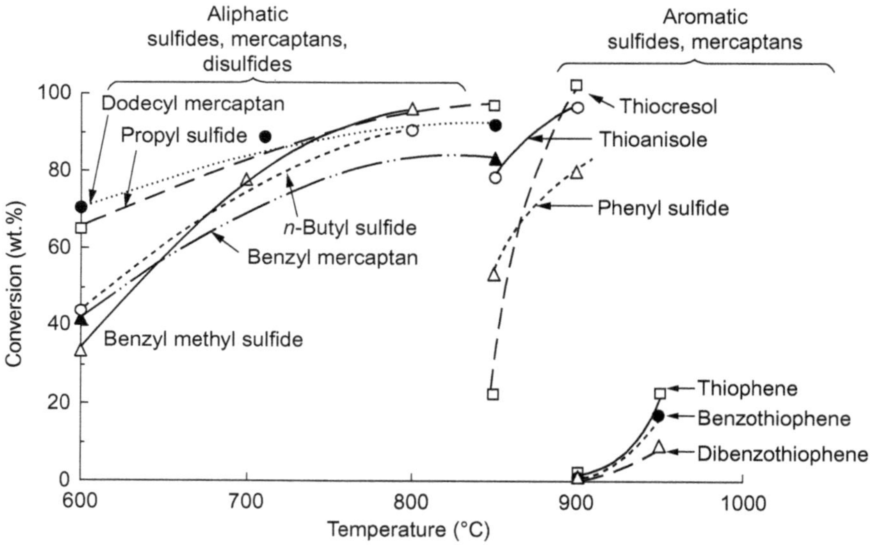

Fig. 4.33 Conversion of model S-compounds into volatile-S during rapid devolatilization with 0.5 s IRP at each temperature.

Reproduced with permission from Calkins WH. Determination of organic-sulfur containing structures in coal by flash pyrolysis experiments. Energy Fuels 1987;1:59, the American Chemical Society.

S-release from model sulfur compounds during rapid devolatilization in a WMR at atmospheric pressure. Aliphatic forms of S_{ORG} are the most reactive, by far, and display a thermal response that matches the release of the earliest primary products of coal devolatilization. The aromatic S-forms exhibit two responses, one beginning around 800°C for aromatic sulfides and mercaptans, and another beginning at 900°C for thiophenes. Clearly, volatile-S will be released from S_{ORG} throughout all stages of primary devolatilization, even during thermal annealing at the hottest temperatures (cf. Fig. 4.17). Moreover, coals with different proportions of aliphatic and aromatic S-forms will exhibit very different kinetics and yields for S-release.

The percentage of S_{ORG} as thiophene increases from 25% to 30% for lignites to 100% for anthracites in rough proportion to C-content, while aliphatic forms diminish from 50% to zero (George et al., 1991). The other aromatic forms are uniform at about 25% for all coal types. Unfortunately, there are no routine laboratory analyses that can accurately assign the forms of S_{ORG} in specific coal samples on a routine basis. This is apparent in the appreciable discrepancies among the reported distributions for the Argonne Premium Coal Samples (APCS) based on an assortment of analytical methods, including temperature programmed reduction, XPS, and XANES (Davidson, 1993). The spread in the values for aliphatic S-forms in any particular APCS sample are 15% to 20% of S_{ORG}, which is much too broad to support a thorough kinetic analysis. So it is fair to say that the tendencies in the forms of S_{ORG} along the rank spectrum have been resolved, but not the sample-to-sample variability.

Weathering is another confounding influence on the distribution of functional groups within S_{ORG}. A diverse assortment of coals exposed to air at 125°C for five days lost 40%–75% of aliphatic-S via conversion to sulfoxides, sulfones, and sulfonic acids, whereas aromatic-S was unaffected (Gorbaty et al., 1992). This treatment is severe enough to represent actual weathering in coal stockpiles only in hot climates, but it nevertheless gives an upper limit on weathering's impact on the distribution of sulfur functional groups. We shall see that this impact can definitely alter the distribution and levels of volatile-S species from primary devolatilization.

S-distributions and, especially, the distribution of the forms of S_{ORG} are crucial factors in the performance of thermal desulfurization processes that target S_{ORG}, because coals with mostly aliphatic S-forms release more of S_{ORG} at much lower temperatures. They are also responsible for the coal quality impacts on S-release during primary devolatilization presented in the next section.

4.2.10.2 Release of volatile sulfur species

During primary devolatilization, the three different forms of coal-S partially decompose into gaseous products which, unfortunately, are not distinctive. The simplest connection is that S_{SO4} decomposes into SO_2, although this process is rarely monitored during testing because most coals contain so little S_{SO4} to begin with. Sulfur dioxide is also released by the sulfones and sulfoxides created by weathering, but only as a minor decomposition product. Gorbaty et al. (1992) reported SO_2 yields of only 7%–18% of the oxidized-S from their heavily weathered coals, along with evidence for extensive conversion of oxidized-S into aromatic-S that remained within char throughout devolatilization. Whereas the SO_2 was released much faster (at much cooler temperatures) than the H_2S from unoxidized coals, weathering reduced ultimate H_2S yields by 40%–85%. These levels represent the upper limits on weathering impacts, and the tests behind them imposed slow heating to only 400°C, so the conversion of oxidized-S into aromatic-S remains to be validated for rapid heating conditions. Finally, even unweathered brown coals and lignites release SO_2 from their highly oxygenated aliphatic-S functional groups. This contribution is only a few percent of S_{ORG} with most samples, although SO_2 was the major gaseous S-species from certain lignites with high inherent pyrite levels, which are uncommon (Yani and Zhang, 2010).

The unoxidized portions of S_{ORG} release H_2S and COS during primary devolatilization, in rough proportions of at least 10 to 1. The production of COS probably involves CO_2, although it is unclear if CO_2 reacts with the S_2 from pyrite decomposition or with H_2S or with both species. The proportions of COS diminish for coals of progressively higher rank, in keeping with the tendency for less CO_2 from coals of higher rank. The decomposition of aliphatic-S mostly produces H_2S through about 1000°C, then aromatic sulfides and, ultimately, thiophenes decompose into additional H_2S at the hottest temperatures in flames. This sequence is analogous to CO production, because both CO and H_2S are released during and immediately after tar production, and also during the annealing stage on much longer time scales. All forms of S_{ORG}, pristine and weathered, are also shuttled away as tar components. So unregulated secondary tar decomposition can potentially distort distributions of volatile-S species by shifting tar-S into surplus H_2S.

As elaborated below, S_{PYR} is eventually converted into FeS plus elemental sulfur vapor, and the vapor is very rapidly converted into H_2S in the presence of GHCs and H_2. Provided that pyrite decomposition and primary devolatilization occur on comparable time scales, S_{PYR} is quantitatively converted into H_2S due to the reducing atmosphere of volatiles within the fuel particles. The ultimate H_2S yield from this source can be accurately estimated as one-half S_{PYR} because FeS is stable at even the hottest temperatures of interest unless it is exposed to H_2 or GHCs.

Coal devolatilization can also produce small amounts of carbon disulfide, CS_2, although there are two reasons to regard CS_2 as a secondary pyrolysis product. First, all tests that directly monitored this species imposed very slow heating rates and, second, CS_2 is thought to form via the reaction of COS with H_2S in the vapor phase.

We next review several datasets to establish the partitioning of coal-S among char, tar, and noncondensable gases, and how it is affected by coal quality and variations in heating rate and pressure. Then two succeeding sections present the transformations of S_{ORG} and S_{PYR} in greater detail. The literature contains an enormous database on pyrolytic desulfurization, but the vast majority of these tests were conducted with heating rates slow enough to sustain secondary tar decomposition within particles and at temperatures cooler than 600°C, usually for contact times of at least an hour. So they have little, if any, bearing on rapid primary desulfurization. The tests presented in this section feature rapid heating but may not have eliminated secondary volatiles chemistry, as noted throughout.

The ultimate partitioning of coal-S into char, tar, and gas for diverse coals appears in Fig. 4.34. The WMR tests eliminated secondary volatiles pyrolysis with a sweep

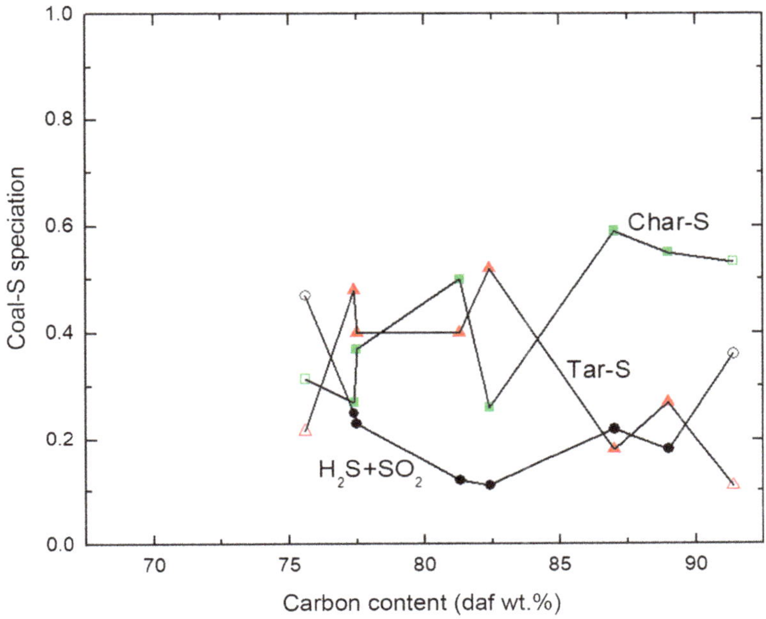

Fig. 4.34 Ultimate distribution of S-fractions in char (■), tar (▲), and gas (●) for rapid heating conditions to temperatures near 950°C from a *(open)* WMR and *(solid)* free-fall pyrolyzer.

gas, but in the free-fall pyrolyzer, falling fuel particles contacted a counter-flow of preheated N_2, and secondary tar decomposition was unregulated. These tar-S levels are minimum estimates that may not reflect the shuttling of aliphatic-S by tar if these functional groups spontaneously decomposed in the free-stream. Heating rates were estimated at roughly 5000°C/s in the free-fall pyrolyzer, and were 1000°C/s in the WMR. Unfortunately, only one subbituminous coal was tested with rapid heating rates, so data for many more low rank coal samples are needed to thoroughly characterize the coal quality impacts on S-speciation.

The most striking feature of Fig. 4.34 is that all three S-species cover exceptionally broad ranges of values, and exhibit no clear trends at all with coal quality. Both tar-S and gas-S constitute from 10% to 50% of coal-S, whereas char-S levels vary from 25% to 60%. Although many more coals should be tested at rapid heating rates to characterize the coal quality impacts, the first indication is that coal quality is only a secondary influence on the S-speciation from primary devolatilization, as expected from the uncorrelated variations in S_{ORG} and S_{PYR}.

The dynamics of coal-S partitioning among char, tar, and gas for *hv* and *lv* bituminous coals appear in Fig. 4.35, where the S-speciation is plotted vs the transit time in a free-fall pyrolyzer. For these two coals, S_{PYR} levels were 13% for the *hv* bituminous and 34% for the *lv* bituminous, and the respective S_{ORG} levels were 71% and 66%. The fact that the sum of tar-S and gas-S is less than S_{ORG}, and by a wide margin for the *lv* bituminous, obscures the relations among the forms of coal-S and their

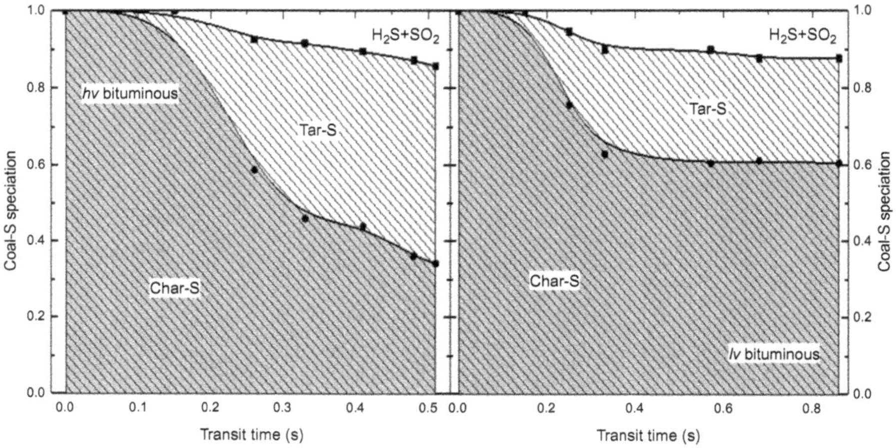

Fig. 4.35 Coal-S fractions in char, tar, and gas vs transit time through a free-fall pyrolyzer with (left) *hv* and (right) *lv* bituminous coals.
Reproduced with permission from Sugawara K, Abe K, Sugawara T, Nishiyama Y, Sholes MA. Dynamic behavior of sulfur forms in rapid pyrolysis of density-separated coals. Fuel 1995;74:1823–29, Elsevier.

transformations during primary devolatilization. Whereas the resolution of heteroatomic speciation into these three lumps is useful for coal-N, the single species lump for char-S cannot resolve the contributions from S_{PYR}, S_{SO4}, and S_{ORG}. In fact, there are no distinguishing features at any stage of primary devolatilization that resolve contributions to H_2S release from S_{ORG} and S_{PYR}, because the extremely broad thermal response of the functional groups in S_{ORG} (cf. Fig. 4.33) obscures S_{PYR}'s contribution to the H_2S yields. Pyrite decomposition will be resolved in the forms of char-S presented below.

The yields of tar-S and volatile-S for a broad range of heating rates appear in Fig. 4.36, where volatile-S was specified as one minus char-S to represent the sum of tar-S and gas-S. Tar yields (not shown) increased from 15 to 30 daf wt.% with the *hv* bituminous and from 9% to 12% with the *lv* bituminous for this range of heating rates, whereas the weight loss was enhanced by two-thirds of these amounts. Across the entire range of heating rates, volatile-S fractions are double the fractional weight loss for the *lv* bituminous, and 20% more coal-S than volatiles was released from the *hv* bituminous. The comparison reflects the much greater S_{PYR} level in the *lv* bituminous, at 45% vs 30% of coal-S, because H_2S production from pyrite is essentially independent of weight loss. Conversely, tar-S fractions are significantly lower than fractional tar yields from the *hv* bituminous, suggesting that the rupture of sulfur functional groups into H_2S may play a role in tar production as, for example, when aliphatic sulfide bridges break to reduce the size of tar precursors into their vaporization range while releasing H_2S. However, the tar-S fractions for the two coals in Fig. 4.35 are markedly greater than the estimated fractional tar yields, so this aspect remains ambiguous.

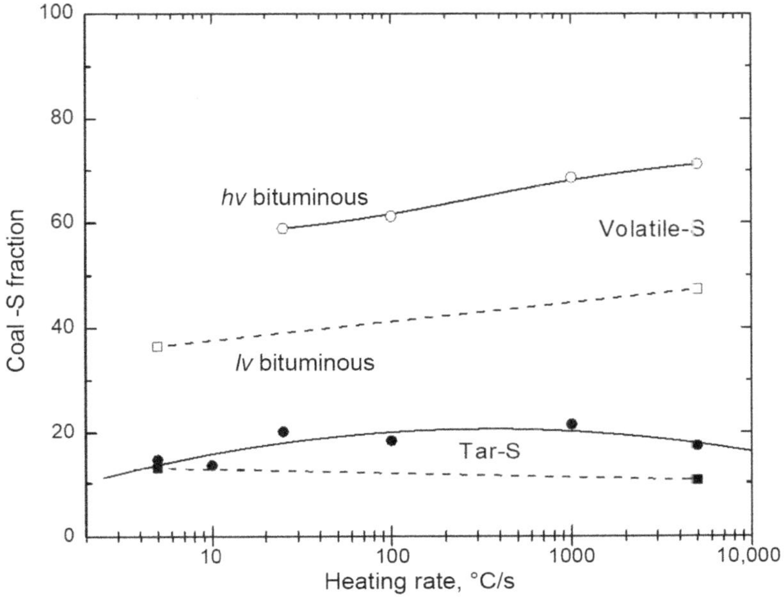

Fig. 4.36 Fractional volatile-S and tar-S from *(circles and solid curves) hv* and *(squares and dashed curves) lv* bituminous coals for various heating rates to 950°C with 5 s IRP in a WMR at atmospheric pressure.
Reproduced with permission from Cai H-Y, Guell AJ, Dugwell DR, Kandiyoti R. Heteroatom distribution in pyrolysis products as a function of heating rate and pressure. Fuel 1993;72:321–7, Elsevier.

The enhancements to volatile-S for faster heating rates match the weight loss enhancements, but neither tar-S fraction is enhanced by nearly as much as the respective tar yield; in fact, the tar-S fractions for the *lv* bituminous are diminished by faster heating. The implication is that tar precursors released late in devolatilization with the fastest heating rates are more likely to release their sulfur as H_2S before they vaporize from the condensed phase. This is consistent with the relatively faster elimination of aliphatic-S as H_2S at progressively hotter temperatures, and with the shift in tar production toward hotter temperatures for progressively faster heating rates. Aliphatic sulfides are among the most labile bridge structures in coal macromolecules. Consequently, tars produced during any stage of primary devolatilization are probably deficient in aliphatic-S, and tars produced during the later stages of devolatilization are most deficient of all. This observation remains to be verified by detailed tar characterization methods, and data like those in Fig. 4.36 for a much broader range of rank would also clarify this issue.

The rapid release of S functional groups from tar precursors also explains why volatile-S levels are insensitive to pressure variations, even while tar-S fractions plummet for progressively higher pressures. The datasets in Fig. 4.37 were obtained

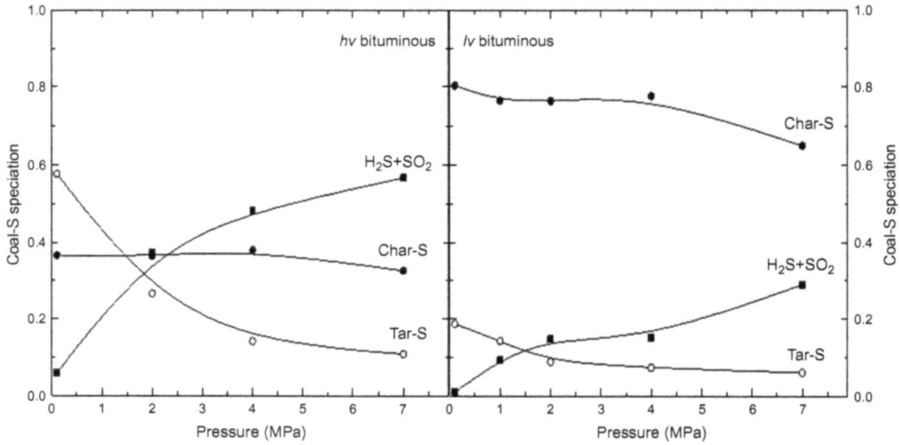

Fig. 4.37 Ultimate S-speciation for (left) *hv* and (right) *lv* bituminous coals across a broad pressure range in a WMR for 1000°C/s to 700°C with 10 s IRP.
Reproduced with permission from Cai H-Y, Guell AJ, Dugwell DR, Kandiyoti R. Heteroatom distribution in pyrolysis products as a function of heating rate and pressure. Fuel 1993;72:321–7, Elsevier.

with a WMR that imposed a uniform thermal history across a broad pressure range (Cai et al., 1993). Levels of char- and tar-S were monitored directly, and the gas-S fraction was assigned by difference. The S-speciations for both coals display the expected tradeoff between tar-S and H_2S, and a slight decrease in char-S for progressively greater pressures.

Finally, note that the tar-S levels for the *hv* bituminous samples in Fig. 4.35 from the free-fall pyrolyzer and in Fig. 4.37 at atmospheric pressure both approach 60%, whereas that level in Fig. 4.36 for 1000°C/s barely exceeds 20%. This could be a measurement artifact, because the tar collector in the tests in Fig. 4.35 was cooled with liquid N_2, and H_2S may have inadvertently condensed and dissolved into the tar sample. Or it could indicate that sulfur functional groups in tars are too reactive to remain intact during heatup to 950°C, even in a WMR with a purge gas to rapidly remove volatiles from the heated zone, but will survive heatup to 700°C. All the major tendencies for variations in both heating rate and pressure point toward extremely reactive sulfur groups on tar precursors although, once again, the measurement uncertainties must be better managed and more data on S-speciation for rapid heating rates is needed to definitively resolve these issues.

4.2.10.3 Sulfur transformations within char

Transformations of the forms of sulfur in char throughout devolatilization cannot be deduced from the distributions of gaseous products, because multiple forms of coal-S release the same gaseous S-species. But methods have been developed to directly monitor these transformations, albeit with three new potential complications. To evaluate the forms of sulfur in char, char samples are often subjected to the same three wet chemical analyses described above for coals, on the assumption that the reagents behave the same ways with the forms of sulfur in char and coal. This assumption turns out to be specious because, as already noted, pyrite is spontaneously converted into a mixture of pyrrhotite and troilite during primary devolatilization, and both forms are readily soluble in HCl (Yan et al., 2012). Since S_{SO4} levels are assigned from direct titrations, they are unaffected by the problem. But S_{PYR} is grossly underestimated, because little pyrite remains for the HNO_3 wash, and S_{ORG} is grossly overestimated if it is assigned by difference to close the S-balance. Yan et al. (2012) recommend a revised procedure that is only suitable if the chars contain no pyrite whatsoever, as happens only during the latest stages of primary devolatilization; more elaborate methods must be used for conditions that produce a mixture of Fe-sulfides including pyrite.

The second concern is a compounded version of the heterogeneous deposition within char that affects primary tars during devolatilization at slow heating rates. Since most of S_{ORG} is thiophene functional groups within aromatic nuclei, primary tars will always contain this form along with lesser amounts of aliphatic-S; consequently, tar-S and char-S levels can certainly be distorted by tar deposition within char, given sufficient transit time for volatiles before they escape the fuel particle. In addition, gaseous S_2 and H_2S released from pyrite and aliphatic S-forms can also be re-incorporated into char as additional aromatic-S, provided that the heating rate is slow. This heterogeneous deposition spontaneously coverts S_{PYR} into S_{ORG} and aliphatic-S into aromatic-S. So during slow heating, deposition of both tar and S-containing gases within char distorts the primary partitioning of the forms of coal-S. The distributions of the forms of char-S in this section were developed with heating rates fast enough to eliminate this heterogeneous deposition. However, the current database does not clearly resolve the threshold value of heating rate that delineates appreciable extents of heterogeneous deposition. The threshold of 1°C/s for tar deposition is definitely not fast enough for S-transformations in some coals, but is sufficient with others. Unfortunately, there is virtually no data that resolves this issue for moderate heating rates from 10°C/s to 100°C/s, so it is impossible to assess the impact in applications with fluidized beds and CFBCs.

Finally, H_2S is also scavenged by the Fe, Ca, and alkali metals in mineral matter to produce metallic sulfides in char, although this complication only arises with brown coals and lignites that have an abundance of finely dispersed forms of these metals, and has only been directly monitored with heating rates well below 1°C/s.

Distributions of the major forms of coal- and char-S for four diverse coals appear in Fig. 4.38. The tests imposed nominal heating rates of 5000°C/s to 960–980°C under H_2 at atmospheric pressure, and the total reaction times ranged from 400 to 500 ms, due to variations in the particle densities (Sugawara et al., 1991). The analytical data resolved S_{PYR}, S_{FeS}, and S_{ORG} in char, where S_{FeS} is the contribution from condensed pyrite decomposition products; as well as the S-partitioning into tar and gases, although individual S-gases were not monitored. The S_{SO4} levels were not monitored in the chars but they were assumed to remain unchanged throughout the tests. All contributions are expressed as mass fractions of coal-S, in mg-S/g-coal. The time scale is

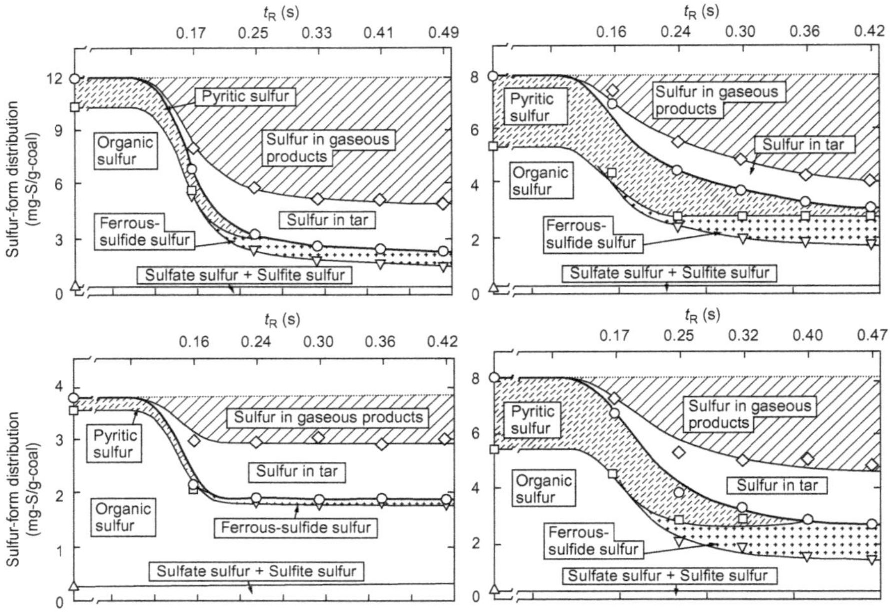

Fig. 4.38 Complete S-speciation for four coals in order of increasing rank from the upper left panel.
Reproduced with permission from Sugawara T, Sugawara K, Nishiyama Y, Sholes MA. Dynamic behavior of sulfur forms in rapid hydropyrolysis of coal. Fuel 1991;70:1091–97, Elsevier.

the transit time of particles through the free fall reactor. Note that the scales for both S-speciation and transit time vary among the panels in Fig. 4.38. In clockwise order from the upper left, the tests used two lignites, an *hv* bituminous, and a *lv* bituminous whose ultimate volatiles yields were 54.2, 55.7, 55.5, and 37.1 daf wt.%, respectively.

The impact of H_2 on the S-speciation at atmospheric pressure is probably confined to the split between tar-S and gas-S, because the time scale for hydrogenation chemistry within a condensed phase is too slow to affect primary devolatilization at the relatively rapid heating rate in these tests. In fact, the transformations of the three sulfur forms in char under N_2 exhibit the same sequence as those in Fig. 4.38, albeit with different coals (Sugawara et al., 1994, 1995).

All these distributions display substantial conversion of S_{ORG} into volatile sulfur in tar and noncondensable gases, and complete conversion of pyrite into FeS with three of the four coals; with that exceptional lignite, nearly all the pyrite was converted in the available transit time of 0.42 s. The data do not resolve whether S_{PYR} or S_{ORG} initiates S-release, although both forms decompose on very comparable time scales with all coal types. S_{ORG} continues to decompose after pyrite has been fully converted into FeS, except with the *lv* bituminous of highest rank. Extents of desulfurization are much greater than the weight loss with all four coals, and exceed 80% with one of the lignites. The contributions to tar-S are roughly comparable to the tar yields expected for these operating conditions.

Whereas half of S_{PYR} will ultimately be converted into gases with any coal type, the conversion of S_{ORG} varies with coal quality. In Fig. 4.38, 85% of S_{ORG} is converted with the coal of lowest rank vs less than half with the highest, which presumably reflects the greater proportion of aliphatic-S in the low rank coal. An even broader range of conversion appears in Table 4.6 for three density-cuts of a subbituminous coal in the same free-fall pyrolyzer (Sugawara et al., 1994). The lightest two cuts had essentially no S_{PYR}, and released two-thirds or more of S_{ORG}. The heaviest had almost half its sulfur as S_{PYR} but only released a third of S_{ORG}. The interesting feature is that the conversion of S_{ORG} does not line up with the volatiles yields. Rather, this rank dependence and, especially, the sample-to-sample variability in S_{ORG} decomposition, reflect the proportions of aliphatic- and aromatic-S, where greater proportions of aliphatic-S give progressively greater conversions of S_{ORG}. Unfortunately, the analytical methods to resolve components of S_{ORG} are too elaborate for routine testing, and are rarely reported in datasets like the ones in this section.

4.2.10.4 Devolatilization of pyritic sulfur

Pyrite appears in coal as both mineral inclusions within the combustibles and as extraneous mineral grains. During primary devolatilization, this distinction is inconsequential provided that the local atmosphere is reducing with an abundance of hydrogen. But as soon as oxygen contacts any of the intermediate products of pyrite decomposition, sulfur is released both faster and much more extensively. This section surveys the major steps for inert and reducing environments, and the more extensive mechanism for oxidizing environments is relegated to char oxidation.

Pyrite thermal decomposition moves through several intermediate mineral phases, many of which have variable compositions that are strongly affected by the concentrations of numerous species in the gas phase. So it is not surprising that strict transitions in temperature are hard to come by in the testing literature, and that the reaction products often appear to be affected by heating rate, pressure, and the overall stoichiometric ratio of the reaction system. Under inert and highly reducing conditions, the major mineral forms in the condensed mineral phase are pyrite (FeS_2), pyrrhotite (FeS_X), troilite (FeS), and metallic iron (Fe). A phase equilibria comes into play in the conversion of pyrrhotite to troilite, but the other steps are irreversible.

Table 4.6 Rapid desulfurization of three density cuts of a subbituminous coal vs volatiles yields (Sugawara et al., 1994).

Density cut	S_{ORG} (%)	X_{ORG} (%)	*W* (daf wt.%)
Lo	96	72	63
Med	97	65	56
Hi	53	32	44

Pyrite decomposes into pyrrhotite, a sulfide intermediate whose composition depends on temperature, the partial pressure of S_2 gas, and, to a lesser extent, pressure. The temperature dependence of the sulfur level for atmospheric pressure has been represented by the following polynomial (Hu et al., 2006):

$$x = -4.3738 \times 10^{-12}T^4 + 1.2034 \times 10^{-8}T^3 - 1.2365 \times 10^{-5}T^2 + 5.4779 * 10^{-3}T^3 + 1.99 \quad (4.2)$$

where the temperature is in kelvins. This expression indicates that all sulfur is eliminated at 1306°C. In actuality, the pyrite/pyrrhotite transformation is reversible as long as S_2 vapor is present. Pyrrhotite's variable composition reflects the incongruent melting of FeS_2 at 743°C into FeS_X and a sulfur-rich liquid phase. Given sufficient time under an S-deficient vapor phase, the pyrrhotite will eventually become troilite. If temperatures are progressively increased under an inert atmosphere, pyrrhotite/troilite will be reduced to metallic iron, as was observed at 1700°C (Grygleicz and Jasienko, 1992) and at 1800°C (Patrick, 1993). Under high H_2 partial pressures, metallic Fe forms at cooler temperatures.

The vast majority of tests on pyrite decomposition imposed slow heating rates. We do not consider the data for slow heating in detail, because the central issue in most slow heating tests is the re-fixation of the S_2 vapor released from pyrite into the organic coal matrix as S_{ORG}. But this work carries important implications about which gas species affect pyrite decomposition during rapid heating. The accumulation of S_2 vapor can inhibit or reverse the decomposition of pyrrhotite and troilite, whereas H_2S is not nearly as effective. So the presence of H-atoms is important, because they convert S_2 into H_2S. Hydrogen-atoms are generated during the production of gaseous GHCs during primary devolatilization, and volatile matter does promote pyrite decomposition. This effect became apparent as lower extents of desulfurization from pure FeS_2 than from FeS_2 in coal under the same test conditions (Gryglewicz et al., 1996; Yani and Zhang, 2010); as lower extents of desulfurization from FeS_2-enriched specific gravity cuts than from the pyrite in raw coal fractions (Ibarra et al., 1994a; Bonet et al., 1993); from lower extents of desulfurization from mixtures of pyrite and char than from mixtures of pyrite and volatile hydrocarbon solids (Patrick, 1993); from higher extents of desulfurization when the FeS_2-enriched fractions were blended with raw coal fractions (Ibarra et al., 1994b); and as pyrite decomposition temperatures that are lower by 100°C in whole coals than for pure pyrite (Chen et al., 1999, 2000). The promotion of H_2S production by volatiles is also responsible for the faster pyrite conversion in a bituminous coal compared to that in a subbituminous coal (Chen et al., 1999, 2000).

Perhaps the most relevant results from the slow heating literature for commercial applications are those that characterize the impact of elevated pressures. Raising the pressure of an inert atmosphere inhibits sulfur release during pyrite decomposition (Chen et al., 1999, 2000). Whereas FeS_2 was completely transformed into FeS at 950°C under N_2 at atmospheric pressure, it progressed no further than $FeS_{1.34}$ at the same temperature under 3 MPa.

Most of the available tests on pyrite transformations under rapid heating conditions imposed oxidizing conditions, although a significant fraction also covered reducing conditions to characterize slag formation under near-burner conditions. The most informative EFR tests were reported by McLennan et al. (2000), who operated their reactor at S. R.-values of 0.6 and 1.5 at 1300°C, 1450°C, and 1600°C with residence times between 1 and 2 s. The recovered flyash samples were analyzed with CCSEM, Mossbauer spectroscopy, and electron microprobe. The most important finding is that FeS/FeO eutectics formed under both oxidizing and reducing conditions, even from extraneous pyrite particles. An external reducing atmosphere sustains the eutectic in extraneous particles, whereas the locally reducing environment surrounding included pyrite particles sustains it in whole coal particles under any ambient conditions. Even for S.R.-values as low as 0.6, FeS and FeO comprised half the Fe minerals in flyash, and the rest was magnetite. This observation suggests that very high temperatures are required to completely eliminate all sulfur from pyrite under inert gases, as noted earlier. Also, thermodynamic equilibrium calculations (McLennan et al., 2000) show that, under reducing conditions (and also under locally reducing conditions within burning char particles), FeS dominates at moderate temperatures, but liquid mixtures of FeS, FeO, and Fe can form at temperatures above 950°C.

Several studies under rapid heating conditions attempted to measure extents of particle fragmentation during pyrite decomposition. At most, fragmentation is a minor effect that should probably only be applied to excluded pyrite particles, especially since the dissolution of aluminosilicates and quartz inclusions by Fe eutectics is likely to be a much more important mechanism for size changes involving pyrite inclusions.

The only tests at elevated pressure were for pure pyrolysis of the sulfur in a subbituminous coal in an EFR at 916°C for pressures from 0.7 to 6.1 MPa (Fatemi-Badi et al., 1988). These tests covered residence times from 0.1 to 1.7 s. The ultimate extents of desulfurization diminished from 45% to 10% as pressures were increased from 0.7 to 6.1 MPa and, more importantly, S_{PYR} levels increased from 0.1 to 0.3 daf wt.% over this pressure range. Hence, the suppression of S-release during pyrite decomposition at elevated pressures seen in slow heating tests has been confirmed for rapid heating conditions as well.

4.3 Summary

The devolatilization stage is the first and fastest stage of coal conversion in any commercial utilization technology, and the one most sensitive to the distinctive characteristics of individual coal samples. It is resolved into several distinct reaction processes as the only practical means to elucidate the connections between coal constitution and the crucial partitioning of coal into volatiles and char, and also to utilize the phenomenal knowledge base on combustion mechanisms and kinetics in simulations of coal processing. The devolatilization stage comprises primary devolatilization, tar decomposition and other aspects of secondary volatiles pyrolysis, volatiles reforming, and volatiles combustion. It does not include the conversion of soot and char by any means, because these solids are converted by completely different heterogeneous reaction mechanisms often on longer time scales. These time scales are so disparate for the largest sizes in coal grinds that furnaces and gasification reactors are sized to convert residual char, with little regard for the volatiles.

By definition, primary devolatilization exclusively pertains to chemistry within the condensed coal phase, whereas the other processes within the devolatilization stage are driven by homogeneous reactions in the gas phase. Collectively, the homogeneous reactions are called secondary chemistry. In general, secondary chemistry can occur within and around the coal particle during the devolatilization stage, although for the rapid heating rates in all our utilization technologies of interest, secondary chemistry occurs around particles and in the free stream that carries the coal suspensions. WMRs, CPPs, RCFRs, and fluidized beds at temperatures cooler than 550–600°C have been used to generate complete distributions of pristine primary devolatilization products. But EFRs necessarily promote uncontrolled secondary volatiles pyrolysis, and are only suitable for monitoring the total volatiles yields and char transformations associated with primary devolatilization.

The crucial characteristic of primary devolatilization is the ultimate volatiles yield, because this value delineates the portion of the coal that is converted on short time scales from the remainder that is converted much more slowly. It is apparent as the asymptotic, daf weight loss that hardly changes at hotter temperatures or with additional reaction time at temperatures hotter than about 900°C. Ultimate tar yields are recorded at less severe conditions than those that achieve ultimate volatiles yields. Accordingly, tar production determines the initial, rapid phase of primary devolatilization, then additional noncondensables are released on comparable time scales to relax the volatiles yield into the ultimate value. A third annealing stage releases small amounts of HCN, H_2S, H_2, and any remaining char-O as CO, but on much longer time scales and usually at temperatures above 1000°C. Strictly speaking, primary devolatilization is finished only when the char contains carbon with a very small amount of hydrogen to stabilize the extensive aromatic domains; plus sulfur in troilite and the most refractory thiophene structures; but no other heteroatoms. But this limit is only observed at temperatures approaching 2000°C after very long times, and is probably not achieved in any utilization technology.

Ultimate volatiles yields are different for different coal samples, heating rates, and pressures. For a specified heating rate and total reaction time, volatiles yields increase for progressively greater temperatures up to the temperature that achieves the ultimate yield. But ultimate yields, per se, are independent of temperature. They are also independent of particle size for the operating domain of our technologies of interest. Ultimate volatiles yields are comparable for brown coals, lignites, subbituminous, and *hv* bituminous samples, then diminish for low volatility coals and nearly vanish for anthracites. But, like any important coal characteristic, the sample-to-sample variability is enormous, even among samples of the same nominal rank. The variability in the ultimate tar yields is responsible for the variations in the total yields. The distinctive sample-to-sample variability in the yields from any particular sample suite is apparent at any pressure and also across any range of heating rates. Ultimate yields diminish for progressively greater pressures, especially as pressure is increased from vacuum through about 1 MPa. Thereafter, further reductions may or may not be appreciable.

The reductions in the tar yields are even greater, which indicates that noncondensables yields are enhanced at progressively greater pressures, but not by enough to compensate for the reductions in tar yields. In total, volatiles yields can be reduced by 15%–25% and tar yields are usually cut in half, and half or more of the tar precursors retained in char are released as noncondensables. Ultimate yields are also enhanced by faster heating rates, provided that pressures are below the threshold that achieves the minimum, asymptotic weight loss. At atmospheric pressure, yields are enhanced by about 3 daf wt.% per order-of-magnitude increase in heating rate with *hv* bituminous coals and somewhat less for other ranks; but at 1 MPa or greater, heating rate enhancements are negligible.

Devolatilization rates are insensitive to pressure variations and are probably the same for a specified thermal history across a broad range of pressures (although it is hard to actually impose the same thermal histories across a broad pressure range). Any mediation by transport phenomena appears to be minimal for the sizes in our technologies of interest. Devolatilization rates increase in direct proportion to increases in heating rate because faster heating shifts primary devolatilization into a broader range of hotter temperatures. With no IRP, ultimate volatiles yields are achieved at about 600°C at 1°C/s and at 900°C at 1000°C/s. Extending the IRP at every test temperature raises the yields but cannot eliminate the temperature dependence in volatiles yields altogether.

The datasets in this chapter clearly demonstrate that primary devolatilization is certainly not a simple, first-order decomposition process; in fact, it entails a multitude of chemical reactions, including competitive features in the mechanism of tar production. Tar production determines weight loss during the first stage, and is the most sensitive to variations in coal quality, heating rate, and pressure. Coal cannot possibly contain a fixed amount of "volatiles" within an inert char matrix. Rather, the proportions of volatiles and char from any coal sample strongly depend on temperature, IRP, heating rate, and pressure. Char must be recognized as a bona fide reaction product of an underlying competitive reaction scheme.

Since primary devolatilization chemistry is confined to the condensed coal phase, char compositions are especially informative. With all but the lowest rank coals, primary devolatilization eliminates nearly all coal-O and most coal-H. It also expels portions of coal-C and coal-N in rough proportion to the mass loss. However, reported releases of coal-H and coal-N are often affected by substantial and inadvertent contributions from the annealing process that should not be regarded as characteristics of rapid primary devolatilization. The release of these four elements is insensitive to rank, per se, until their retention in char surges for low volatility coals. Char desulfurization via rapid primary devolatilization reflects the proportions of S_{ORG} and S_{PYR}, although complete desulfurization has only been reported at temperatures above 1700°C, usually in the presence of H_2. Even so, ultimate extents of desulfurization are almost always much greater than fractional char yields would suggest because of the substantial contribution from S_{PYR}. Since S_{PYR} levels are almost always determined by the intensity of coal cleaning, the only legitimate rank dependence in desulfurization is in S_{ORG} conversion, due to the reduction of aliphatic-S in coals of progressively higher rank, albeit with large variations among different samples.

Primary devolatilization is also responsible for massive changes in chars' physical morphology. The crucial distinction is whether or not the condensed phase softens into a viscous melt during devolatilization. With very low-rank, nonsoftening coals and the most refractory low volatility coals, char sizes stay the same and changes in the bulk char densities and porosities simply express the mass loss. Specific surface areas become moderately greater for nonsoftening coals. With softening coals, sizes grow by as much as 50% and thereby triple the particle volumes, and porosities can approach 90%. Swelling factors increase for progressively faster heating rates up to roughly 1000°C/s, and pass through a maximum around 1 MPa for progressively greater pressures. The predominant char morphologies are cenospheres and crassispheres, both of which contain macrovoids that are segmented and bounded by thin walls penetrated by micro-, meso-, and macropores. Notwithstanding their relatively small contributions to the char morphology population, fusinoid particles that did not soften usually determine LOI levels in flyash in commercial operations with softening coals.

With all but the lowest coal ranks, primary tars are the main shuttles for heteroatoms out of the condensed coal phase. They carry as much as 40% of coal-O, and nearly all the coal-N released during primary devolatilization. Fractional tar-N levels are proportional to tar yields. There is relatively less coal-S in tar than the fractional tar yields, because much if not most of coal-S is often present as pyrite in many coals and aliphatic-S is usually expelled before tar precursors vaporize. Tar H/C ratios for the earliest tar samples can be as much as double the coal-based values, which clearly indicate that primary tars are much more aliphatic than their parent coals. Tars become more aromatic throughout primary devolatilization but H/C-values for even ultimate tar samples remain well above the coal-values. This tendency is mostly due to the preferential elimination of aliphatic carbons in the β- and γ-positions from tar precursors prior to their vaporization as devolatilization proceeds. Accordingly, tar H/C ratios provide the best means to assess the impact of secondary volatiles pyrolysis in any pyrolysis test. It can only be deemed negligible if the ultimate tar ratios are substantially greater than those for the parent coals. Finally, tar H/C ratios are also greater for tars subjected to progressively faster heating rates and higher pressures.

Tar MWDs are rarely monitored these days, but were central elements in the empirical foundation for network depolymerization mechanisms. They inevitably have the form of γ-distribution functions, which abruptly rise from about 100 g/mol through a maximum at a few hundred g/mol, and then very gradually relax to their maximum extent of 1000–1500 g/mol. This form inherently suggests that evaporation is an essential aspect of tar production which, in turn, is strongly corroborated by the relation of the MWDs of solvent extracts and their associated tars: Tar MWDs are shifted by at least a factor of two toward lighter weights than the extract MWDs, which suggests that tars are simply the portion of their precursors that were light enough to vaporize under particular test conditions. The MWDs of tar increments prepared at hotter temperatures shift toward heavier weights, consistent with the tendency for progressively faster heating rates, and also consistent with the temperature dependence in the saturated vapor pressures of heavy hydrocarbon liquids. The opposite tendency is evident for progressively greater pressures, which is consistent with a weaker driving force for tar release at progressively greater ambient pressures.

Most of the noncondensable gases from primary devolatilization are oxygenated species for all ranks except the low volatility coals. The yields of all three oxygenated gases diminish for coals of progressively higher rank, as they must, because coal-O monotonically decreases across the rank spectrum before vanishing in anthracites. The H_2O yields are comparable to or moderately greater than CO yields with high-rank coals, but usually lower than CO yields with low-rank coals. The CO_2 yields are the greatest of all through 72% C; then match the CO yields through 79% C; then fall off further with bituminous coals before they vanish with *lv* bituminous and anthracites. GHC yields are lower than oxygenated gas yields through *mv* bituminous. Methane and C_2 GHCs are the most abundant GHC species, by far, and the CH_4 yield is either comparable to or slightly exceeds the sum of all C_2 and C_3 species. The trend is for greater CH_4 yields for progressively greater C-contents, until CH_4 production plummets with anthracites. Levels of C_4 GHCs are not appreciable with any coal. Oils yields mimic the variations in the tar yields. They equal the CH_4 yields through 79% C, then fall off faster than CH_4 for coals of higher rank before they vanish for anthracites. The yields of phenol and cresol are essentially the same and about double the yields of every other oil species. The H_2 yields are always well below 1 daf wt.% unless inadvertent annealing occurred during the tests.

The yields of noncondensable gases are greater at higher pressures, but not by enough to compensate for the reduction in tar yields. Moisture levels are slightly enhanced and CO yields are slightly diminished, whereas CO_2 yields are unchanged. The yields of C_3 GHCs and, especially, oils are reduced by elevated pressures. Most but not all datasets indicate greater CH_4 yields at elevated pressures.

Despite inordinate attention to volatile-N species, there are huge variations in reported levels of tar-N, depending on the impact of secondary volatiles pyrolysis, and in residual char-N and HCN data, depending on the often inadvertent contribution from the annealing stage. This presentation focused on primary HCN release, on the time scale for primary devolatilization, with minimal contributions from annealing and none from secondary pyrolysis. Accordingly, the HCN levels are relatively low and char-N levels are high, compared to many other datasets. Tar-N, HCN and, perhaps, NH_3 comprise volatile-N. Ammonia is a relatively minor volatile-N species, never counting for more than 12% of coal-N. Most coals release less NH_3 than 7% of coal-N, and all the coals that released over 10% were subbituminous or lower ranks. The NH_3 yields are strongly correlated with the levels of quaternary-N in the parent coals from XPS spectroscopy. However, it is important to realize that these NH_3 levels do not represent the NH_3 levels that come into play during aerodynamic NO_X abatement, because secondary pyrolysis and volatiles combustion both generate and destroy NH_3. Since aromatic rings are created overall, not destroyed, during primary devolatilization, and since the bulk of coal-N appears as pyrrolic- and pyridinic-N, tar shuttling is essentially the only means of N-release throughout most of rapid primary devolatilization. Gaseous N-species are released only near the end of tar production. The central role for tar shuttling is responsible for the enhanced N-release for progressively faster heating rates. Unlike the precursors to the major noncondensables, the pyrrolic- and pyridinic-N retained in the condensed phase at slow heating rates that would otherwise be shuttled away in tar with faster heating are not released on the time scale of tar production. The same considerations explain why volatile-N levels diminish slightly for progressively higher pressures.

During primary devolatilization, S_{SO4}, S_{ORG}, and S_{PYR} partially decompose into gaseous products which are not distinctive. The simplest form to interpret is that the small amounts of S_{SO4} decompose into SO_2. Sulfur dioxide is also released by the sulfones and sulfoxides created by weathering, but with a selectivity of only 10 to 20%; the remainder is converted into aromatic-S, and severe weathering can reduce ultimate H_2S yields by up to 85%. Even unweathered brown coals and lignites release SO_2 from their highly oxygenated aliphatic-S functional groups which, with some lignites, is the predominant gaseous S-species. The unoxidized portions of S_{ORG} release H_2S and COS during primary devolatilization, in rough proportions of at least 10 to 1. The production of COS probably involves CO_2, although it is unclear if CO_2 reacts with the S_2 from pyrite decomposition or with H_2S or with both species. The proportions of COS diminish for coals of progressively higher rank, in keeping with the tendency for less CO_2 from coals of higher rank. The decomposition of aliphatic-S mostly produces H_2S through about 1000°C, then aromatic sulfides and, ultimately, thiophenes decompose into additional H_2S at the hottest temperatures in flames. This sequence is analogous to CO production, because both CO and H_2S are released during and immediately after tar production, and also during the annealing stage on much longer time scales. All forms of S_{ORG}, pristine and weathered, are also shuttled away as tar components.

The elemental sulfur vapor released by S_{PYR} conversion is very rapidly converted into H_2S in the presence of GHCs and H_2. The ultimate H_2S yield from this source at atmospheric pressure can be accurately estimated as one-half S_{PYR} because FeS is stable at even the hottest temperatures of interest unless it is exposed to H_2 or GHCs. Similarly, raising the pressure of an inert atmosphere inhibits sulfur release from pyrite.

References

Anthony DB, Howard JB, Hottel HC, Meissner HP. Rapid devolatilization of pulverized coal. Proc Combust Inst 1975;15:1303.

Bautista JR, Russell WB, Saville DA. Time-resolved pyrolysis product distributions of softening coals. Ind Eng Chem Fundam 1986;25:536.

Bonet AJ, Ibarra JV, Lazro MJ, Moliner R. Distribution of the activation energies for the thermal decomposition of organic and pyritic sulfur in coal. Fuel Process Technol 1993; 36(1/3):319–25.

Cai H. Fast pyrolysis of coals and char characterization in relation to pulverized coal combustion. [Ph. D. dissertation]. Dept. of Chemical Engr. and Chemical Technol., Imperial College of Science, Technology, and Medicine, U. London; 1995.

Cai H-Y, Guell AJ, Dugwell DR, Kandiyoti R. Heteroatom distribution in pyrolysis products as a function of heating rate and pressure. Fuel 1993;72:321–7.

Cai H-Y, Megaritis A, Messenbock L, Vasanthakumar L, Dugwell DR, Kandiyoti R. Pyrolysis of coal maceral concentrates under pf-combustion conditions: changes in heteroatom partitioning as a function of rank. Fuel 1998;77(12):1283–9.

Chen JC, Niksa S. A radiant flow reactor for high-temperature reactivity studies of pulverized solids. Rev Sci Instrum 1992a;63:2073–83.

Chen JC, Niksa S. Coal devolatilization during rapid transient heating. Part 1: primary devolatilization. Energy Fuels 1992b;6:254–64.

Chen JC, Niksa S. Suppressed nitrogen evolution from coal-derived soot and low volatility coal chars. Proc Combust Inst 1992c;24:1269–76.

Chen JC, Castagnoli C, Niksa S. Coal devolatilization during rapid transient heating. Part 2: secondary pyrolysis. Energy Fuels 1992;6:264–71.

Chen H, Li B, Zhang B. Decomposition of pyrite and the interaction of pyrite with coal organic matrix in pyrolysis and hydropyrolysis, In: Proc. tenth int. conf. on coal sci., Taiyuan, China; 1999. p. 713–6.

Chen H, Li B, Zhang B. Decomposition of pyrite and the interaction of pyrite with coal organic matrix in pyrolysis and hydropyrolysis. Fuel 2000;79(13):1627–31.

Cor J, Manton N, Mul G, Eckstrom DJ, Olsen W, Malhotra R, Niksa S. An experimental facility for the study of coal pyrolysis at 10 atmospheres. Energy Fuels 2000;14:692–700.

Darivakis GS, Howard JB, Peters WA. A rationale for heating rate and coal type effects on liquid yields and substrate morphology changes during rapid pyrolysis. Energy Fuels 1994;8:1024–32.

Davidson RM. Organic sulphur in coal. Report IEACR/60. London: IEA Coal Research; 1993.

Fatemi-Badi M, Scaroni AW, Jenkins RG. Proc Am Chem Soc Div Fuel Chem Preprints 1988;33(1):265–73.

Fletcher TH, Hardesty DR. Compilation of Sandia coal devolatilization data. Milestone Report DE92016824, Albuquerque, NM: Sandia National Laboratories; 1992.

Freihaut JD, Proscia WM. Final report on US DoE contract no. DE-AC22-89PC89759. Pittsburgh Energy Technology Center, US DoE; 1991.

George GN, Gorbaty ML, Keleman SR, Sansone M. Direct determination and quantification of sulfur forms in coals from the Argonne Premium Sample Program. Energy Fuels 1991;5:93–7.

Gorbaty ML, Keleman SR, George GN, Kwiatek PJ. Characterization and thermal reactivity of oxidized organic sulfur forms in coals. Fuel 1992;71:1255–64.

Griffin TP, Howard JB, Peters WA. An experimental and modeling study of heating rate and particle size effects in bituminous coal pyrolysis. Energy Fuels 1993;7:297–305.

Grygleicz G, Jasienko S. The behavior of sulfur forms during pyrolysis of low-rank coal. Fuel 1992;71(11):1225–9.

Gryglewicz F, Wilk P, Yperman J, Franco DV, Maes II, Mullens J, van Poucke LC. Interaction of the organic matrix with pyrite during pyrolysis of a high-sulfur bituminous coal. Fuel 1996;75(13):1499–504.

Guell AG, Kandiyoti R. Development of a gas sweep facility for the direct capture of pyrolysis tars in a variable heating rate high pressure wire mesh reactor. Energy Fuels 1993; 7:943–52.

Hu G, Dam-Johansen K, Wedel S, Hansen JP. Decomposition and oxidation of pyrite. Prog Energy Combust Sci 2006;32(3):295–314.

Ibarra JV, Bonet AJ, Moliner R. Release of volatile sulfur compounds during low temperature pyrolysis of coal. Fuel 1994a;73(6):933–9.

Ibarra JV, Palacios JM, Moliner R, Bonet AJ. Evidence of reciprocal organic matter–pyrite interactions affecting sulfur removal during coal pyrolysis. Fuel 1994b;73(7):1046–50.

Kambara S, Takarada T, Toyoshima M, Kato K. Relations between functional forms of coal nitrogen and NO_X emissions from pulverized coal combustion. Fuel 1995;74(9):1247–53.

Kobayashi H. Devolatilization of pulverized coal at high temperatures. [Sc. D. Dissertation, Dept. of Chemical Engr.]. Cambridge, MA: MIT; 1976.

Lau C-W, Niksa S. The combustion of individual particles of various coal types. Combust Flame 1992;90:45–70.

Li C-Z, Bartle KD, Kandiyoti R. Characterization of tars from variable heating rate pyrolysis of maceral concentrates. Fuel 1993;72(1):3–11.

Liu G-S, Niksa S. Coal conversion submodels for design applications at elevated pressures. Part II. Char gasification. Prog Energy Combust Sci 2004;30(6):697–717.

Liu Y-L, Malhotra R, Niksa S. Impact of pressure variations on coal devolatilization products. 1. Detailed product distributions from 0.1 MPa. Energy Fuels 2004;18:508–19.

Manton N, Cor J, Mul G, Eckstrom DJ, Malhotra R, Niksa S. Impact of pressure variations on coal devolatilization products. 1. Detailed product distributions from 1.0 MPa. Energy Fuels 2004;18:520–30.

McLennan AR, Bryant GW, Stanmore BR, Wall TF. Ash formation mechanisms during pf combustion in reducing conditions. Energy Fuels 2000;14(1):150–9.

Messenbock RC, Dugwell DR, Kandiyoti R. CO_2 and steam gasification in a high-pressure WMR: the reactivity of Daw Mill coal and the combustion reactivity of its chars. Fuel 1999a;78:781–93.

Messenbock RC, Dugwell DR, Kandiyoti R. Coal gasification in CO_2 and steam: development of a steam injection facility for high pressure WMRs. Energy Fuels 1999b;13(1):122–9.

Miknis FP, Turner TF, Ennen LW, Netzel DA. NMR characterization of coal pyrolysis products. Fuel 1988;67:1568–77.

Mitchell RE, Hurt RH, Baxter LL, Hardesty DR. Compilation of Sandia coal char combustion data and kinetic analyses. Milestone Report DE92018668, Albuquerque, NM: Sandia National Laboratories; 1992.

Niksa S, Russel WB, Saville DA. Time-resolved weight loss kinetics for the rapid devolatilization of a bituminous coal. Proc Combust Inst 1982b;19:1151.

Niksa S, Liu G-S, Hurt RH. Coal conversion submodels for design applications at elevated pressures. Part I. Devolatilization and char oxidation. Prog Energy Combust Sci 2003; 29(5):425–77.

Oh MS, Peters WA, Howard JB. Kinetics of generation and destruction of pyridine extractables in a rapidly pyrolyzing bituminous coal. Fuel 1989;65:251.

Patrick JW. Sulphur release from pyrites in relation to coal pyrolysis. Fuel 1993;72(3):281–5.

Pohl JH, Sarofim AF. Devolatilization and oxidation of fuel nitrogen. Proc Combust Inst 1977;16:491–501.

Solomon PR, Colket MB. Evolution of coal nitrogen in coal devolatilization. Fuel 1978;57:749.

Solomon PR, Serio MA, Deshpande GV, Kroo E. Crosslinking reactions during coal conversion. Energy Fuels 1990;4:42.

Sugawara T, Sugawara K, Nishiyama Y, Sholes MA. Dynamic behavior of sulfur forms in rapid hydropyrolysis of coal. Fuel 1991;70:1091–7.

Sugawara K, Tozuka Y, Kamoshita T, Sugawara T, Sholes MA. Dynamic behavior of sulfur forms in rapid pyrolysis of coals with alkali treatment. Fuel 1994;73:1224–8.

Sugawara K, Abe K, Sugawara T, Nishiyama Y, Sholes MA. Dynamic behavior of sulfur forms in rapid pyrolysis of density-separated coals. Fuel 1995;74:1823–9.

Suuberg EM. Rapid pyrolysis and hydropyrolysis of coal [Sc. D. Dissertation, Dept. of Chemical Engr.]. Cambridge, MA: MIT; 1977.

Suuberg EM, Peters WA, Howard JB. Product compositions and formation kinetics in rapid pyrolysis of pulverized coal—implications for combustion. Proc Combust Inst 1979; 17:117–30.

Tyler RJ. Flash pyrolysis of coals. Devolatilization of bituminous coals in a small fluidized-bed reactor. Fuel 1980;59:218–26.

Unger PE, Suuberg EM. Molecular weight distributions of tars produced by flash pyrolysis of coal. Fuel 1984;63:606–11.

Unger PE, Suuberg EM, Lily WD. Experimental study on mass transfer from pyrolyzing coal particles. Fuel 1985;64:966.

Wiktorsson L-P, Wanzl W. Kinetic parameters for coal pyrolysis at low and high heating rates—a comparison of data from different laboratory equipment. Fuel 2000;79:701–16.

Xu W-C, Tomita A. Effect of coal type on the flash pyrolysis of various coals. Fuel 1987a;66:627–31.

Xu W-C, Tomita A. Effect of temperature on the flash pyrolysis of various coals. Fuel 1987b;66:632–6.

Yan J, Bai Z, Zhao H, Bai J, Li W. Inappropriateness in the standard method in sulfur form analysis of char from coal pyrolysis. Energy Fuels 2012;26:5837–42.

Yani S, Zhang D. An experimental study into pyrite transformation during pyrolysis of Australian lignite samples. Fuel 2010;89:1700–8.

Yu J, Lucas JA, Wall TF. Formation of the structure of chars during devolatilization of pulverized coal and its thermoproperties: a review. Prog Energy Combust Sci 2007;33:135–70.

Further reading

Calkins WH. Determination of organic-sulfur containing structures in coal by flash pyrolysis experiments. Energy Fuels 1987;1:59.

Gibbins J, Kandiyoti R. Experimental study of coal pyrolysis and hydropyrolysis at elevated pressures using a variable heating rate wire-mesh apparatus. Energy Fuels 1989a;3:670–7.

Gibbins JR, Kandiyoti R. The effect of variations in time temperature history on product distribution from coal pyrolysis. Fuel 1989b;68:895.

Gibbins-Maltham J, Kandiyoti R. Coal pyrolysis yields from fast and slow heating in a wire-mesh apparatus with a gas sweep. Energy Fuels 1988;2:505.

Niksa S. Flashchain theory for rapid coal devolatilization kinetics. 4. Predicting ultimate yields from ultimate analyses alone. Energy Fuels 1994;8:659–70.

Niksa S, Russel WB, Saville DA. Captive sample reactor for kinetic studies of coal pyrolysis and hydropyrolysis on short time-scales. Fuel 1982a;61(12):1202–17.

Wagner R, Wanzl W, van Heek KH. Influence of transport effects on pyrolysis reaction of coal at high heating rates. Fuel 1985;64:571–3.

Reaction mechanisms for primary devolatilization

5

Nomenclature

a_i	stoichiometric coefficients for CO, CO_2, and H_2O in FLASHCHAIN®
A	normalized molar concentration of aromatic nuclei
A_i	generic pseudo-frequency factor for channel i in a global reaction scheme
A_p	external surface area of a coal particle, cm^2
A_{VP}	constant in the saturated vapor pressure of metaplast, $P^{SAT}(MW_{tj},T)$
A	generic pseudo-frequency factor in a global reaction channel
AC/Cl	number of aromatic carbons per monomer unit in a macromolecule
B	normalized molar concentration of labile bridges
B_V	blowing factor for convective heat transfer to a devolatilizing coal particle
C	normalized molar concentration of char links
C_0	initial fraction of char links in whole coal, for CBK
C_p	specific heat of coal, cal/g K
E	normalized molar concentration of recombination sites in DISCHAIN
E_a	activation energy in an Arrhenius rate constant
E_0	mean activation energy in an energy distribution function
f_a'	carbon aromaticity of a whole coal
${}^H f_a$	proton aromaticity of a whole coal
$F^b(0)$	labile bridge fraction in any fragment population
$f(E)$	energy distribution function in the DAEM
F_D	drag coefficient on a coal particle, g/cm^2 s
G	instantaneous mole fraction of noncondensables within a coal particle
h	convective heat transfer coefficient to a coal particle, cal/cm^2 K s
H	instantaneous mole fraction of tar within a coal particle
H_C	normalized mole concentration of char chain fragments in DISCHAIN
I	Increment in the integral of a rate constant over a thermal history in discrete form
k_i	Arrhenius rate constant for channel i in a global reaction scheme
L	latent heat of moisture vaporization, cal/g
m_j	normalized moles of metaplast j-mers
m_p	mass of an individual coal particle, g
M	normalized molar concentration of metaplast in DISCHAIN
M_{CL}	molecular weight of a monomer unit in whole coal in CPD
M_δ	molecular weight of a peripheral group in CPD
M_{SITE}	molecular weight of aromatics in a monomer unit in whole coal in CPD
MW_j	molecular weight of species j, g/mol
N_E^K	number of element E per coal component K
Nu	Nusselt number for convective heat transfer
P_C	constant in the saturated vapor pressure of metaplast, $P^{SAT}(MW_{tj},T)$
p_{CE}	probability that a fragment end is attached to a char link in DISCHAIN
P^{SAT}	saturated vapor pressure of a metaplast j-mer at temperature T
p_j	partial pressure of a tar j-mer

Process Chemistry of Coal Utilization. https://doi.org/10.1016/B978-0-12-818713-5.00005-8

p_M^e probability that a metaplast fragment contains bridge remnants
p_{mj} number fractions of metaplast *j*-mers in a fragment mixture
p_{mi}^b labile bridge fraction in metaplast *j*-mers
p_0 probability for links of any kind among the nuclei in whole coal
P_0 ambient pressure
q,Q heating rate, °C/s
q_i mass flux of gases from a coal particle, g/cm^2 s
r^2 statistical correlation coefficient
R universal gas constant, J/K mol
Re_p Reynolds number for a coal particle
S coal reactant in a global reaction scheme
S_{ORG} organic sulfur species in whole coal, daf wt.%
t time, s
t_{IRP} duration of the isothermal reaction period in a thermal history, s
t_j normalized moles of tar *j*-mers
t_q duration of the heating period in a thermal history, s
T temperature, K
T_P normalized concentration of tar product in DISCHAIN
T_V normal boiling point of H_2O, K
$V(t)$ instantaneous volatiles yield, daf wt.%
V^∞ ultimate volatiles yield, daf wt.%
x_j normalized moles of any *j*-mer in the condensed phase
X_{DV} fractional volatiles yield, $V(t)/V^\infty$
y_i stoichiometric coefficients in the C2SM
z constant in the saturated vapor pressure of metaplast, $P^{SAT}(MW_{tj},T)$

Greek symbols

α stoichiometric coefficient
β fraction of nuclei connected by char links, $p_0(1 - F^b(0))$
ε_p total hemispherical emittance of a coal particle
γ_{CH4} correlating parameter for mass fraction of CH_4 among GHCs
Γ_j instantaneous release rate of tar *j*-mers
η average number of N-atoms per nucleus
η_{LoT} average number of N-atoms per nucleus during N-release at low temperatures in CPD
λ_g thermal conductivity of a gas stream, cal/cm K s
ν_B scission selectivity coefficient for the bridge conversion rate, $0 < \nu_B < 1$
ν_C moles of noncondensables from spontaneous charring of a labile bridge
ρ_p bulk particle density, g/cm^3
σ standard deviation about the mean energy in an energy distribution function
σ_C coordination number in a Bethe lattice
Θ_K number of O-atoms per coal component K
Ω swelling factor

Subscripts

B labile bridge
C char link
N aromatic nucleus
S peripheral group
j degree of polymerization of a coal-derived fragment
0 initial condition

The data in the previous chapter clearly show that the impacts of all the important processing conditions are apparent in the available laboratory database. One response to this knowledge is to sharpen the focus of the testing that supports technology development programs. Since the operating conditions in a commercial process which most strongly affect ultimate volatiles yields have been revealed, one can ensure that supporting tests impose the same or very similar heating rates, temperatures, and pressures. The test results can then be input into process simulations to accurately describe the crucial partitioning of the coal into volatiles and char and, perhaps, other important characteristics of devolatilization like swelling factors and char-N levels. Some also contend that the lab-scale database demonstrates that the coal quality impacts on devolatilization behavior are so erratic that a sample's distinctive volatiles yield could never be accurately predicted anyway. If so, then the only way that technology developers can contend with samples' distinctive devolatilization behavior is to test every sample that factors into their performance guarantees, as they did throughout almost all the previous century.

Problem is, the number of coals that technology developers must manage keeps expanding, seemingly without limit. One reason is that every major technology developer has already become a global enterprise with licenses and installations around the world. Regional expertise with selected coal deposits is no substitute for being able to manage and maintain performance with all the fuels a client wants to use, now and into the foreseeable future, regardless of where that client is located. As the center of gravity for coal utilization settles into Asia, and all the major coal suppliers target the export potential into this region, the number and diversity of coal samples that some companies must consider has become enormous. It is not hard to find power plants in Japan and South Korea that process dozens of coals from different sources in a single year, and plants in India and China are also contending with much greater diversity in their fuel supplies.

This chapter and the one to follow pose an approach to manage distinctive devolatilization behavior that is much faster and far less expensive than sample-specific testing. Its central premise is that the deepest value of the laboratory database is in stringent validations of comprehensive reaction mechanisms for primary devolatilization. Quantitative comparisons among model predictions and measured behavior are paramount, because once we are convinced that a reaction mechanism accurately interprets the major trends for all the important operating conditions and also depicts the distinctive behavior of individual coal samples, then further testing becomes superfluous. The reaction mechanism can then be run as a virtual coal laboratory to provide all the necessary input to the process simulations that physical lab testing had provided, and much more information as well. This chapter develops an intellectual history for the network depolymerization mechanisms that have already satisfied these prerequisites, and the next chapter presents the quantitative validation work for the author's rendition, called FLASHCHAIN®.

5.1 Global rate expressions for primary devolatilization

It may seem odd to begin with the simplest global rate expressions for devolatilization, when we are aiming for comprehensive mechanisms with predictive capabilities. The rejection of these simple rate laws was an important step in the intellectual development of more fundamentally sound approaches. Nevertheless, global rates remain very relevant today, because they are the mathematical functions used to incorporate accurate predictions for primary devolatilization into process simulations at commercial scale. Unfortunately, that important application does not circumvent the often-substantial misunderstandings surrounding these rate expressions.

There is enormous potential for confusion whenever someone tries to represent a reaction mechanism as complex as primary coal devolatilization with a rudimentary reaction rate law. The research literature is full of inconsistent and fundamentally incorrect material on what nominal rates of devolatilization are, how they may be assigned, and how they should be used. One potential source of confusion is that specialists use the term "reaction rate" to refer to very different quantities. In its most fundamental sense, this term refers to the time rate of change of the concentration of a species that participates in a single chemical reaction, as a function of temperature, pressure, and the concentrations of all other participating species. Coal devolatilization is certainly not yet understood at the level of elementary chemical reactions, so this definition is inappropriate. In a broader sense, "reaction rate" is also used to distinguish the major components of a complex process. For example, we often want to know if chemistry or transport is limiting the overall rate of a chemical process. Even this broader meaning is unsuitable for primary devolatilization because nominal rates are definitely not equal to any of the reaction rates of the major chemical steps underlying this process. The best meaning of the term for our purposes is simple and literal: A global devolatilization rate is the time rate of change of a product yield. This definition can be applied to any group of products or to an individual species. Since it does not refer to the underlying mechanism in any way, it minimizes the potential for confusion.

Consider the nominal rate of the entire primary devolatilization process, which is often evaluated from time-resolved data on daf weight loss. The instantaneous weight loss is governed by numerous distinct chemical reaction rates as well as the statistical probabilities underlying depolymerization and crosslinking, and the conditions for tar vaporization. The interplay among these mechanisms is so complex that it is futile to try to infer anything about the chemical reaction rates from weight loss rates. Indeed, nominal devolatilization rates assigned with a comprehensive reaction mechanism are grossly different than the rates of any of the chemical reactions within that mechanism, conclusively demonstrating just how unrelated the nominal rate parameters are to the parameters in the chemical mechanisms. It is best to regard the global rates in succeeding sections as regression functions that may or may not contain a sufficient number of adjustable parameters to accurately correlate a particular selection of datasets. And even if the regressions are accurate, the magnitudes of the parameters still have no mechanistic significance whatsoever.

5.1.1 Single first-order reaction

The single first-order reaction (SFOR) rests on the long-standing premise that the combustibles in coal appear in two forms, volatiles and fixed carbon. Primary devolatilization converts the volatiles in coal to volatile products, and the fixed carbon into char. The instantaneous production rate of volatiles is proportional to the remaining volatile matter in the coal, according to

$$\frac{dV(t)}{dt} = A\exp\left(-\frac{E_a}{RT}\right)(V^{\infty} - V(t)) \tag{5.1}$$

where $V(t)$ is the instantaneous volatiles yield; V^{∞} is the hypothetical ultimate volatiles yield; A is a pseudo-frequency factor, and E_a is a pseudo-activation energy. Historically, the PVM was regarded as the ultimate amount of volatiles. Beginning in the 1960s, it became progressively clearer that volatiles yields were significantly affected by heating rate, temperature, and pressure as well. Once these dependences were recognized, V^{∞} became nothing more than an adjustable parameter which will be called the "hypothetical ultimate yield parameter." Note that, in simulation applications, V^{∞} is the crucial devolatilization characteristic, yet in the SFOR, it is an adjustable input parameter. Similarly, A and E_a are simply numbers that may faithfully mimic devolatilization kinetics over some restricted range of conditions.

Numerous sets of rate parameters for the SFOR have been reported, although most assignments did not properly account for the fact that primary devolatilization almost always occurs while the sample is being heated to its ultimate temperatures. Van Heek and coworker (Juntgen and Van Heek, 1979) developed a nonisothermal kinetic analysis with SFORs for numerous product evolution rates monitored in TGA tests at different slow heating rates. For tests with rapid heating rates, yields are much easier to monitor than evolution rates, and the best way to evaluate the rate parameters is to fit weight loss as a function of temperature during heatup. Provided that the heating rate is constant, the instantaneous volatiles yield is

$$X_{DV}(T) = 1 - \exp\left[-\left(\frac{AT}{q}\right)\frac{\exp\left(-\frac{E_a}{RT}\right)}{\frac{E_a}{RT}}\right] \tag{5.2a}$$

where $X_{DV}(T)$ is the fractional yield, $V(T)/V^{\infty}$, during heatup to temperature T; and q is the uniform heating rate (Lau and Niksa, 1993). This solution can be rearranged into the following coordinates for a modified Arrhenius diagram:

$$\ln\left[\left(\frac{q}{RT^2}\right)\ln\left[\frac{1}{1-X_{DV}}\right]\right] = \frac{E_a}{RT} - \ln\left[\frac{A}{E_a}\right] \tag{5.2b}$$

The left-hand side is evaluated from the test conditions and the measured yields, provided that some of the test conditions were severe enough to achieve an ultimate volatiles yield for each heating rate under consideration. Recall from Chapter 4 that ultimate yields are usually achieved at about 600°C for a heating rate of 1°C/s and over 900°C for 1000°C/s. When the left side of Eq. (5.2b) is plotted versus reciprocal absolute temperature, one hopes that the values form a single straight line from which A and E_a can be determined. In practice (Niksa et al., 1984), the data for each heating rate cluster into lines with different slopes, as seen below, which is an indication of an inherent limitation in the SFOR for this application. Also, the assigned activation energies tend to be very low, which was often misinterpreted as a rate limiting role for transport phenomena during primary devolatilization.

The ratio of the fractional yields at the same temperature for two different heating rates can be obtained by rearranging Eq. (5.2a) into the following form:

$$[1 - X_{DV}(q_2)] = [1 - X_E(q_1)]^{\frac{q_1}{q_2}} \tag{5.2c}$$

When q_1 is slower than q_2, this expression shows that the yield at any specified temperature at the faster heating rate must be lower than that at the slower heating rate. Consequently, the analysis demonstrates that nothing more than a single rate constant of Arrhenius form is needed to explain why the onset of primary devolatilization shifts toward hotter temperatures for progressively faster heating rates.

Eq. (5.1) can be integrated beyond the heating period to obtain the fractional yield as a function of time during an IRP, which is

$$X_{DV}(t_{IRP}) = 1 - [1 - X_{DV}(t_q)]\exp\left[-A\exp\left(-\frac{E_a}{RT}\right)t_{IRP}\right] \tag{5.2d}$$

where $X_{DV}(t_q)$ is the yield at the end of the heating period, t_q, from Eq. (5.2a) and t_{IRP} is time after t_q throughout the IRP. According to this expression, primary devolatilization achieves its ultimate volatiles yield in the limit of long IRPs at any temperature, which is fundamentally incorrect. In fact, this is also the main limitation of the SFOR as a correlating function in process simulations, because the important implication is that the SFOR can only be used with thermal histories that are severe enough to actually achieve the ultimate yield. It is fine for applications with p.c. grinds at temperatures hotter than about 1000°C, which comprise all furnaces and entrained flow gasifiers. It is also fine for fluidized bed applications provided that solids residence times are sufficient to achieve ultimate yields at the lower bed temperatures.

This stipulation will always be satisfied in fluidized bed combustors, but may not be satisfied in the coolest fluidized bed gasifiers. Also, a set of SFOR parameters should not be extrapolated to more than a single order of magnitude change in heating rate, for reasons explained in Section 6.3.1.3.

5.1.2 Distributed activation energy model

The distributed activation energy model (DAEM) retains the premise that coal is a binary mixture of volatiles and fixed carbon, but also envisions a manifold of independent, parallel SFORs as the mechanism for volatiles release. It was developed to interpret aspects of petroleum refining before Anthony and Howard (1976) used it to interpret rapid primary devolatilization behavior. The important and novel feature is that increments in the total volatiles yield are associated with increments in a continuous distribution of activation energies, according to

$$dV = V^{\infty} f(E)dE \quad \text{where } f(E) = \frac{1}{\sqrt{2\pi}\sigma}\exp\left(-\frac{(E-E_0)^2}{2\sigma^2}\right) \tag{5.3a}$$

where $dV(t)$ is an instantaneous contribution to the volatiles yield; V^{∞} is the hypothetical ultimate yield parameter; A is a pseudo-frequency factor, and E is a particular activation energy in a continuous distribution function, $f(E)$. The activation energy distribution is usually but not always the normal distribution function, which introduces σ, the standard deviation about the mean energy, E_0. The DAEM rate expression for arbitrary thermal histories is

$$\frac{dV(t)}{dt} = V^{\infty}\int_0^{\infty} A\exp\left(-\frac{E}{RT}\right)\exp\left[-\int_0^t A\exp\left(-\frac{E}{RT}\right)dt'\right]f(E)dE \tag{5.3b}$$

and the instantaneous volatiles yield is

$$V(t) = V^{\infty}\left[1-\int_0^{\infty}\exp\left[-\int_0^t A\exp\left(-\frac{E}{RT}\right)dt'\right]f(E)dE\right] \tag{5.3c}$$

Note that the DAEM contains four adjustable parameters, V^{∞}, A, E_0, and σ. As with the SFOR, the rate constant must be integrated over the thermal history and, in addition, that integral for each increment in activation energy must also be integrated over the energy distribution function. The integral in time over the thermal history has analytical solutions for both uniform heating rates (cf. Eq. 2.5a) and for an exponential approach to some ultimate reaction temperature, as arises during convective heating. In turn, these solutions have been used to obtain robust analytical solutions for the full DAEM for these thermal histories (Lau and Niksa, 1993). These solutions show that there is an autocorrelation between E_0 and A, and that only the ratio of σ to E_0 is an independent model parameter. In other words, identical DAEM predictions can be obtained with any number of pairs of E_0 and σ, provided that their ratio stays the same. Consequently, DAEM parameters have no mechanistic significance whatsoever.

In return for one additional parameter, the DAEM is able to represent the impact of temperature and heating rate variations much better than the SFOR, as seen in Fig. 5.1. The left panel shows the modified Arrhenius diagram for uniform transient heating at 0.5, 1, and 1000°C/s with *hv* bituminous coals, and the right panel shows the DAEM-based correlations for the associated transient yields. Different ultimate yield parameters for each heating rate were used to incorporate enhanced yields for faster heating into the analysis. In contrast to the single line for all heating rates expected from the SFOR, data from different thermal histories separate into a family of smooth curves with greater activation energies for faster heating rates. Most important, a single parameter set for the DAEM generates the three curves that accurately correlate these datasets, and reproduces the transient yields within the measurement uncertainties across three orders of magnitude in heating rates (Lau and Niksa, 1993).

In general, the predicted rates based on a single DAEM parameter set are accurate for at least three orders of magnitude in heating rate, and also over a very broad temperature range with arbitrary IRPs. The DAEM is especially well-suited for applications at moderate temperatures, or in two-stage processes in which a low temperature stage precedes a much hotter second stage of devolatilization. For the much slower heating rates in coke ovens, Miura simultaneously relaxed the assumptions that $f(E)$ is a normal distribution, and that A is constant (Miura, 1995). In this method, TGA datasets for three different heating rates determine complete distributions for both E and A in no prescribed functional form. In turn, the distribution functions specified this way give essentially exact fits to the rates and yields monitored in the tests for the three thermal histories, as needed to accurately forecast pressure surges during coking.

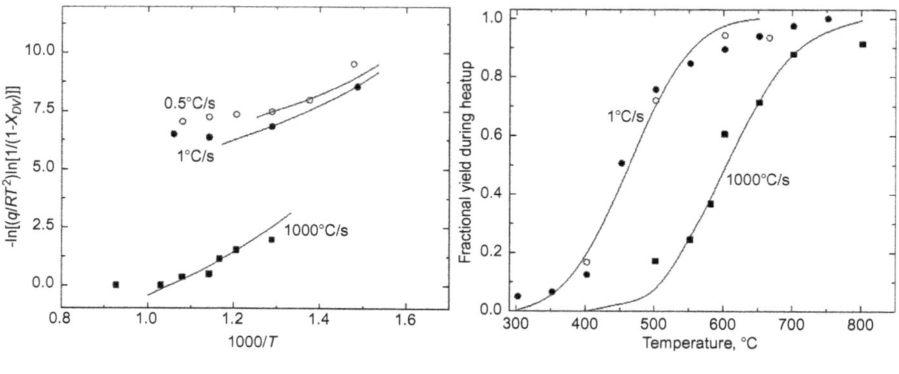

Fig. 5.1 (Left) Modified Arrhenius diagram for uniform heating rates of (●) 0.5, (○) 1, and (▪) 1000°C/s with *hv* bituminous coals and the DAEM regressions for $A = 2 \times 10^{16}$; $E_0 = 251$ kJ/mol; and $\sigma = 21.4$ kJ/mol; and (right) fractional weight loss based on this DAEM parameter set. From Lau C-W, Niksa S. Global rates of devolatilization for various coal types. Combust Flame 1993;94:294–307 with permission from Elsevier.

In another extension, Suuberg et al. (1978) specified increments in the energy distribution from specific noncondensable gas yields in measured product distributions, as seen in Fig. 5.2. The cumulative yield curve is composed of incremental additions of specific noncondensables and portions of the tar yields. Each yield increment is associated with a value of activation energy based on measured release temperatures during heatup at about 1000°C/s. Most of the individual gases make two contributions with different activation energies, suggesting independent production pathways, and CO_2 is released in three stages. In fact, the measured yields as a function of temperature during heatup for some of these species exhibited distinct jumps, consistent with disparate activation energies in multiple production channels. The most interesting finding is that the magnitudes of the yield increments are reasonably consistent with a normal distribution of activation energies, albeit within the sometimes-arbitrary subdivisions for the yields of some species.

The analysis to this point has independently specified the coal's thermal history, then evaluated DAEM parameters and instantaneous yields for that thermal history. But in process simulations, the thermal history of the coal or even its form is unknown in advance, and must be simultaneously evaluated with the DAEM prediction as the simulation advances through time increments. The following DAEM solution scheme is robust for thermal histories of any form, and has already been implemented in CFD

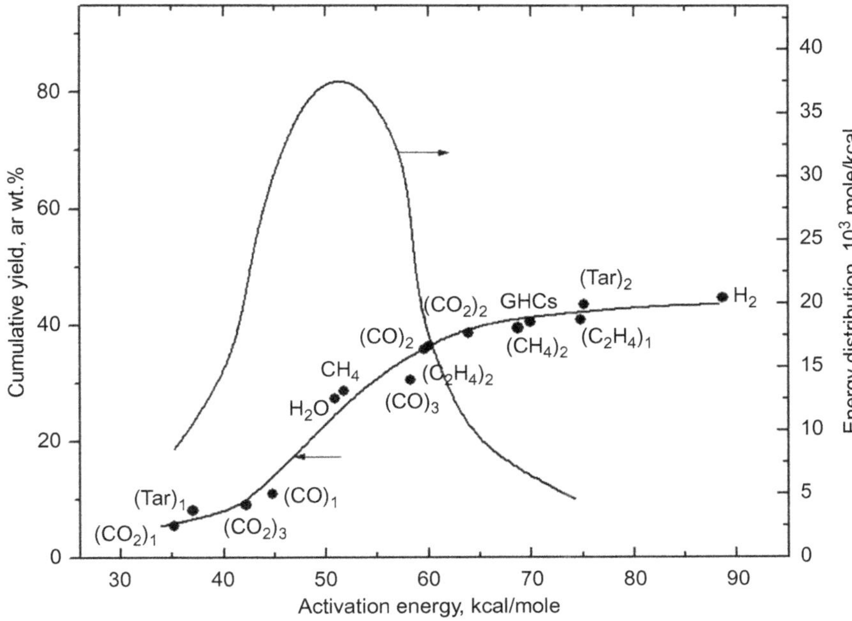

Fig. 5.2 Cumulative yield of all volatile products indicated at a particular activation energy for lignite pyrolysis at roughly 1000°C/s along with the normal distribution assigned by Suuberg et al. (1978).
Reproduced with permission from the American Chemical Society.

simulations of primary devolatilization in near-burner flame zones. In discrete form, the integral of the rate constant over the thermal history is

$$I=\sum_{i=1}^{N}\int_{t_{i-1}}^{t_i} A\exp\left(-\frac{E}{RT(t')}\right)dt' \tag{5.4a}$$

where t_0 is zero; t_N is the elapsed time t; and t_i is any time shorter than t. Each of these integrals over discrete time intervals can be evaluated with Laplace's method (Lau and Niksa, 1993) to find that

$$I=\sum_{i=1}^{N}\frac{AR}{E}\left\{\frac{T_{i+1}^2\exp\left(-\frac{E}{RT_{i+1}}\right)}{\frac{dT_{i+1}}{dt}}-\frac{T_i^2\exp\left(-\frac{E}{RT_i}\right)}{\frac{dT_i}{dt}}\right\} \tag{5.4b}$$

where T_i and dT_i/dt denote temperature and heating rate at t_i. These quantities are evaluated from a particle energy balance in discrete form, which allows the time integration to be readily combined with the integration in activation energy. The time integration is simply repeated for all discrete values of activation energy, E_j, in the second quadrature. Each contribution to the activation energy integral is

$$e_j=\frac{A}{\sqrt{2\pi\sigma}}\exp\left(-\frac{E_i}{RT_N}\right)\exp\left(-I_{j,N}\right)\exp\left(-\frac{(E_i-E_0)^2}{2\sigma}\right) \tag{5.4c}$$

where T_N is the temperature at t and $I_{j,N}$ is the time integral to time t based on E_j. The instantaneous fractional yield is obtained by summing the contributions for all activation energies with the Trapezoidal rule:

$$1-X_{\mathrm{DV}}(T(t))=e_0+2\sum_{i=1}^{M-1}e_j+e_M \tag{5.4d}$$

The quadrature in activation energy does not interfere with the time-marching as the thermal history evolves, no matter how convoluted. Contributions to the time integration in Eq. (5.4b) are stored for each E_j, so that only one term is evaluated in successive time steps. Since normal distributions can be well represented with about 30 discrete values, the computational burden for this scheme is easy to manage.

The DAEM overcomes most of the defects in a SFOR with only one additional parameter. Given yields throughout heatup for at least three heating rates and the ultimate volatiles yields for each heating rate, a parameter set can be specified to accurately depict primary devolatilization across a very broad domain of thermal histories. The DAEM accurately describes the shift in weight loss toward hotter temperatures for faster heating rates, and the apparent temperature dependence in the weight loss for thermal histories with variable IRPs. The main drawback is that the DAEM still has a hypothetical ultimate yield parameter and, therefore, cannot predict the impacts of heating rate and pressure on the ultimate volatiles yield from even a single coal.

5.1.3 Competing two-step model

The competing two-step reaction model (C2SM) for devolatilization (Kobayashi et al., 1977) is

$$S \begin{array}{l} \xrightarrow{k_1} y_1V_1+(1-y_1)C_1 \\ \xrightarrow{k_2} y_2V_2+(1-y_2)C_2 \end{array} \tag{5.5a}$$

where S is the coal reactant; $V_1(t)$ and $V_2(t)$ are instantaneous volatiles yields generated through channels 1 and 2, respectively; $C_1(t)$ and $C_2(t)$ are the respective instantaneous char yields; k_1 and k_2 are rate constants of Arrhenius form; and y_1 and y_2 are stoichiometric coefficients less than unity. This model is based on two competitive reaction channels that simultaneously convert the coal reactant into both volatiles and char. These channels are normally interpreted as low- and high-temperature pathways, and y_1 is sometimes set to match the PVM, in which case y_2 represents a yield enhancement for faster heating rates. A higher activation energy in k_2 enhances the contribution from channel 2 at faster heating rates by delaying the onset of devolatilization to higher temperatures. It is also instructive to interpret channel 1 as the tar production channel, and channel 2 as the gas formation channel, as explained below. The rate law associated with this scheme is as follows:

$$\frac{dV(t)}{dt} = \int_0^t (y_1k_1(t') + y_2k_2(t'))S(t')dt' \quad \text{where } S$$
$$= S_0 \exp\left(- \int_0^t (k_1 + k_2)dt' \right) \tag{5.5b}$$

From a conceptual standpoint, the main advantage of the C2SM is that is does not contain a hypothetical ultimate yield parameter. The instantaneous and ultimate volatiles yields are determined by the competitive kinetics which, in turn, are affected by the thermal history. In principle, it should possible to assign parameter values that predict yield enhancements for faster heating rates and temperature-dependent ultimate yields. However, this capability has never been demonstrated in practice, for example, by accurately interpreting data like those in Fig. 5.1 with a single parameter set. A more fundamental problem is apparent in Eq. (5.5b), which indicates that S, the coal reactant, vanishes at every temperature, just like the postulated coal volatiles in a SFOR. To depict yields that increase for progressively hotter temperatures, the predicted char yields must diminish for hotter temperatures, which is incorrect. In actuality, the condensed phase is a mixture of partially converted coal and char that shifts toward more char for progressively hotter temperatures.

The main practical disadvantage of the C2SM is that is contains six adjustable parameters: y_1, A_1, E_1, y_2, A_2, and E_2, which obviously have no mechanistic significance. From a practical standpoint, six parameters are too many to assign from a single devolatilization weight loss transient, so multiple test conditions are required. When total and tar yields are measured, the analysis can be reduced to four adjustable parameters by identifying channel 1 as the tar only production channel, and channel 2 as the channel for simultaneous char and gas. These connections imply that y_1 is always unity. Moreover, we can stipulate that $E_1 = E_2 + 10$ kcal/mol, to ensure that tar yields are enhanced by faster heating rather than gas yields. Then the remaining four parameters to be assigned by fitting data are A_1, y_2, A_2, and E_2.

For reasons that are hard to fathom, the C2SM is especially popular with CFD practitioners who stage furnace simulations. This group rarely discloses how the rate parameters were specified and often applies the parameters for one coal to completely different samples, which is surely erroneous. Another quantitative issue is the inadvertent heating rate dependence that arises when broad PSDs are simulated, as in all commercial applications. Whether or not the C2SM parameters were assigned with datasets for different heating rates, the ultimate yields from the C2SM for a polydisperse fuel suspension display a substantial size variation, because smaller particles are subjected to faster heating rates than larger particles. Since the vast majority of C2SM parameter sets are not based on datasets for different heating rates, the inadvertent yield enhancement in process simulations is bound to be incorrect. One illustration appears in Fig. 5.3. The parameter values were assigned from datasets from a flow reactor that imposed moderately fast heating rates to moderate temperatures (Fletcher and Hardesty, 1992). The C2SM was then extrapolated to a heating rate of 3×10^5 and 3×10^4°C/s, to show the predicted transient devolatilization histories for a broad PSD under p.f. flame conditions. Whereas the extrapolation to 3×10^4°C/s gives a reasonable yield enhancement from 53.1 to 56.6 daf wt.%, the extrapolation to the fastest rate gives a heating rate enhancement of 24.5% to 77.6% weight loss, which is several times greater than any measured enhancement. Of course, the enhancement could be diminished by specifying a lower activation energy, E_2. But the point here is that a C2SM whose parameters were specified from a dataset for a narrow range of thermal histories is bound to fail in extrapolations to the much faster heating rates in flames for a broad PSD. One obvious resolution is to use datasets for heating rates as fast as those in the application, but such data are subject to inordinate measurement uncertainties for applications in flames and entrained flow gasifiers, and this approach fails to account for the inevitable range of heating rates in any practical application with a PSD. It is much more accurate to use a SFOR that does not introduce inadvertent heating rate enhancements; or even better still, to express V^∞ as an explicit function of heating rate that matches data like those in Fig. 4.7 in the previous chapter.

The C2SM is fundamentally flawed, over-parametrized, and prone to inadvertent and implausible yield enhancements due to faster heating rates in simulations for suspension-fired coal streams. Taken together, these flaws make it one of the worst possible choices for the devolatilization submodel in process simulations.

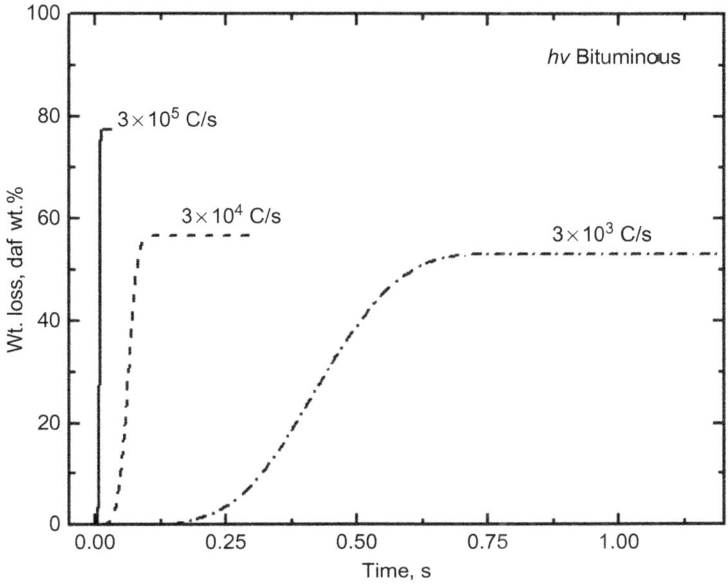

Fig. 5.3 C2SM devolatilization histories for three heating rates for parameter values assigned from flow reactor datasets by Fletcher and Hardesty (1992).

5.1.4 Multistep reaction schemes

Numerous multistep reaction schemes were proposed to interpret primary devolatilization datasets during the 1970s. The numbers of steps and adjustable parameters increased in tandem with finer resolution of the product distributions in emerging laboratory studies suggesting, perhaps, that these approaches were little more than curve fitting exercises. Here we consider only one of the simpler multistep schemes because it rectifies the main fundamental flaw in the C2SM and also invokes some of the key processes in network depolymerization mechanisms:

$$S \xrightarrow{k_1} \alpha V_1 + (1-\alpha) I \begin{cases} \xrightarrow{k_2} V_2 \\ \xrightarrow{k_3} C \end{cases} \tag{5.6}$$

The scheme contains an activation step in series with two competitive reactions. Primary devolatilization converts the coal (S) into stable volatiles (V_1) and a reactive intermediate (I) in proportion to a stoichiometric coefficient α. The intermediate subsequently partitions into a solid residue (C) and additional volatiles (V_2) based on the rate constants for these two channels. The activation energies in the three Arrhenius rate constants determine the influences of temperature and heating rate. The initial activation step eliminates the C2SM's problems with interpreting temperature-dependent volatiles yields, because the condensed phase will be a mixture of unreacted coal, intermediate, and char whose composition shifts toward char for progressively hotter temperatures. This analysis also predicts that all the coal reactant is converted at all temperatures, like the volatiles in a SFOR and the coal reactant in C2SM. But coal is converted into intermediates and, with extended IRPs at each temperature, the level of intermediates monotonically increases for progressively hotter temperatures. Consequently, the residue or char yield monotonically increases for hotter temperatures, as it should.

Transient yields throughout the heating period with different heating rates and 0 and 20 s IRPs to achieve the ultimate yields appear in Fig. 5.4. The parameters were set to reproduce the measured yields from a *hv* bituminous coal at these test conditions (Niksa et al., 1984). With only one more adjustable parameter than used in the C2SM, this scheme gives reasonably accurate sensitivity to heating rate and temperature in the transient yields during heating and in the ultimate yields. Most important, the scheme depicts these features without any hypothetical ultimate yield parameters, which demonstrates that ultimate volatiles yields across broad ranges of temperature and heating rate can be accurately reproduced with even the simplest conceivable competitive reaction process.

With benefit of hindsight, the markedly superior performance of this simple 3-step scheme is easily rationalized by its connections to more modern reaction processes for primary devolatilization. The activation step represents depolymerization of the coal macromolecule on a massive scale that converts coal into smaller and more reactive fragments (I), and also releases noncondensable gases and loosely bound tar fragments (V_1). The fate of the intermediate fragments is then determined in a competition that either makes them light enough to be released as tar plus additional noncondensables (V_2), or recombines them into a refractory char matrix (C). Faster heating hardly perturbs the depolymerization process, but it does skew the competition toward

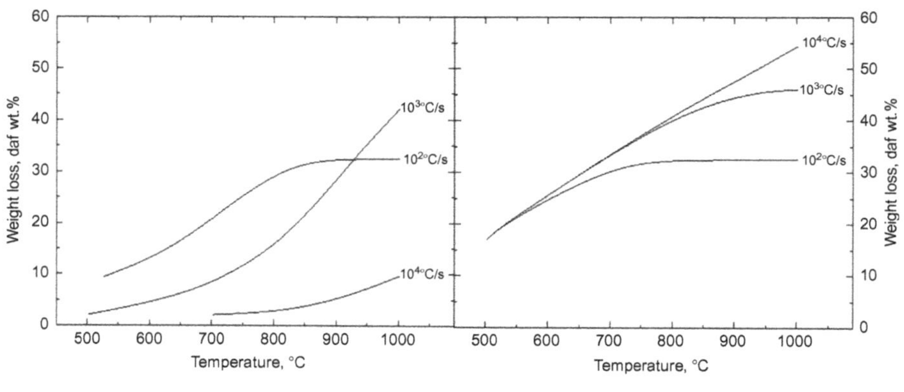

Fig. 5.4 Weight loss versus temperature from a *hv* bituminous coal for heating rates of 10^2, 10^3, and 10^4°C/s (left) with no IRP and (right) with 20 s IRP from the series/parallel reaction scheme in Eq. (5.6).

Reproduced from Niksa S, Heyd LE, Russel WB, Saville DA. On the role of heating rate in rapid coal devolatilization. Proc Combust Inst 1984;20:1445–53 with permission from Elsevier.

tar production, because k_3 has the highest activation energy of all. Bear in mind that the terminology of network depolymerization is not at all compatible with the rudimentary mathematical analysis behind the performance in Fig. 5.4. This breach will be rectified in the sections to follow.

It is also worth considering why the more complex multistep reaction schemes did not advance directly into the network depolymerization mechanisms. One reason is that none of them established the essential one-to-one correspondences among the "reactants" in coal and the chemical pathways that form the products. Any mechanism that begins with a generic coal reactant cannot possibly connect with any of the functional groups in coal that are converted during primary devolatilization, and is thereby destined to represent all aspects of devolatilization, including the coal quality impacts, through heuristic parameter adjustments. The second reason is that every step added to a scheme increases the number of adjustable parameters, and adding parameters is the opposite of insight. Since the proposed reactions were disconnected from real coal constitution, the rate parameters simply functioned as knobs to move curves around graph paper. Even if the curves closely matched measured devolatilization behavior for a particular coal, nothing could be learned in the fitting procedures to forecast how a different sample would behave. Indeed, any kinetic analysis that does not explicitly describe how coal constitution determines reactant concentrations and their decomposition rates will always degenerate into a curve-fitting exercise.

5.2 Network depolymerization reaction mechanisms

Throughout the 1970s and, especially, into the 1980s, it became progressively clearer that interpretations of data based on global reaction schemes would never circumvent the need to directly monitor the distinctive devolatilization behavior of individual coal samples in laboratory tests. This drawback prompted Solomon and coworkers at Advanced Fuel Research Inc. (AFR) and the author to connect their postulated reaction processes to particular aspects of coals' constitution and macromolecular configuration. As these connections expanded, this research adopted the mathematics to legitimately simulate coals' depolymerization into smaller fragments, and the subsequent repolymerization of intermediates into a char matrix. This advance set the stage for a new mechanism for tar production based on explicit vaporization of the lightest depolymerization fragments, without any finite-rate mass transport resistances that would introduce an incorrect dependence on the particle size. Ultimately, these novel features were coalesced into the FG-DVC and FLASHCHAIN® mechanisms, and repackaged into abridged forms in the chemical percolation devolatilization (CPD) model in a collaboration among the University of Utah, Sandia National Laboratories, and Brigham Young University.

By incorporating essential aspects of coal constitution into the kinetic analysis, these developers tried to escape the main limitations of global reaction schemes, so that (1) The rate constants would pertain to any coal, or at least vary in predictable ways for different coal samples; (2) The contributions from the various reaction mechanisms would shift for different operating conditions to automatically describe shifts in the yields and product compositions; and (3) The mechanisms would resolve the product distributions in sufficient detail for applications in synthetic fuels technologies. The ultimate goal was to circumvent hypothetical ultimate yield parameters, and formulate mechanisms that could depict the distinctive devolatilization behavior of individual coal samples; in other words, a genuine predictive capability. But the additional imperative to circumvent expensive and sophisticated laboratory support was not recognized by the developers of FG-DVC and CPD until after their mechanisms were brought to final form.

The remainder of this chapter discusses the aspects of coal constitution that found their way into the three network depolymerization mechanisms; approaches based on much more detailed chemistry are cited at the end. Model predictions and how the underlying mechanisms interpret the most important aspects of devolatilization behavior are considered separately in Chapter 6. The observations on FG-DVC and CPD pertain to the most comprehensive versions of these analyses, before numerous approximations were introduced to reduce the required laboratory support. The abridged versions with less required input are discussed separately in Section 5.4.

5.2.1 The reactants in coal and their relative reactivities

Chapter 2 introduced the most basic aspects of coals' macromolecular constitution in terms of bridge structures interconnecting aromatic nuclei into a 3D macromolecular network, with a diverse assortment of peripheral groups in isolated attachments to the nuclei. A closer examination, at the scale of larger molecular identities and functional groups, reveals an extensive hierarchy of bond strengths that imply useful information about their relative reactivities, and bond dissociation energies have been estimated for various bridge structures (Gavalas, 1982). According to the aromatic/hydroaromatic constitution model, aromatic nuclei and various alicyclics comprise the skeletal units. But alicyclics must be regarded as labile bridges because their methylene chains and ring components are the weakest of all coal structures along with sulfide linkages and, perhaps, carboxylic acids. Other functional groups with oxygen, particularly phenolic hydroxyls and ether linkages, are less reactive, as are organic sulfates and quaternary N-functionalities. The O, N, and S incorporated into aromatic nuclei are even less reactive, and the aromatic bonds within nuclei are the most refractory of all. Based on ^{13}C NMR data that showed that aromatic rings were created, not destroyed, during devolatilization (Miknis et al., 1988), aromatic nuclei can be regarded as completely refractory, and most of the heteroatoms within them will be expelled only during the annealing stage, long after primary devolatilization chemistry has run its course.

This hierarchy of reactivities is an aspect of coal constitution that must be absolutely essential to any predictive capability for primary devolatilization. While coal is heated at a uniform rate, the hierarchy determines how many of the functional groups are destroyed within a specified temperature interval which, in turn, markedly affects the fragment size distribution from the depolymerization chemistry. Surely, these connections are first-order important. The problem is that there is still no battery of analytical methods that can accurately assign complete functional group distributions for any coal. And even if such a battery could be devised, its application would be too expensive and/or uncertain and/or time consuming to use for routine coal analyses to support process simulations. Calculation schemes have been developed to estimate various mean constitution parameters from extensive lists of analytical results, and these have been connected to extensive reaction sets. But as seen below, these approaches have not yet accurately described any coal's devolatilization behavior, let alone predicted sample-specific devolatilization behavior across the rank spectrum.

The different network depolymerization models have different strategies to contend with unmeasurable functional group distributions that are worth reviewing in detail.

5.2.2 The functional group model

In their functional group (FG) model, the group at AFR adopted the interpretation by Suuberg et al. (1978) that increments in a complete product distribution are the products of specific conversion channels within a manifold of independent parallel reactions. Whereas the original DAEM associates the incremental weight loss with increments in a continuous energy distribution, the FG model applies one or multiple DAEMs to the production channel for each specific molecular product. Since there are four parameters per channel and dozens of postulated reaction channels, the FG model expanded into the antithesis of the mathematical elegance and parametric economy of the original DAEM.

Qualitatively, the FG model connected the functional group distribution in a coal to the distribution of noncondensable products, as seen in Fig. 5.5. The larger left vertical block subdivides the coal based on the yields of the indicated products: CO_2, tightly bound CO, H_2O, loosely bound CO, light GHCs, heavy GHCs, H_2, abstracted H_2, and nonvolatile carbon (or char). Abstracted H_2 is a form of coal-H that stabilizes tar fragments in the tar formation mechanism discussed below. Also, coal-N and coal-S were neglected in the model's introduction when Fig. 5.5 was presented, but were incorporated later. The area of each increment is proportional to the number fractions of the elements allocated to a particular product. The right column illustrates a hypothetical distribution of the most important functional groups in a particular sample. Even though the lengths of the segments for several functional groups appear to match the heights assigned to individual products, no measured functional group concentrations were ever used in the analysis. The one-to-one correspondence is only meant to suggest that carboxylic acids are the precursors to CO_2; ethers beget tightly bound CO; phenolic hydroxyls beget H_2O; etc., albeit in only qualitative terms. In actuality, the FG model never incorporated measurements of any of the functional group concentrations in coal; instead, the ultimate noncondensable gas yields were measured as surrogates for the functional group precursors.

In the model's introduction, the kinetic parameters were assigned to fit transient product distributions during heatup at several hundred degrees per second under vacuum. Conventional Arrhenius rate constants with two adjustable parameters each were implemented in the interpretation. The levels of the coal components, Y_i^0 in

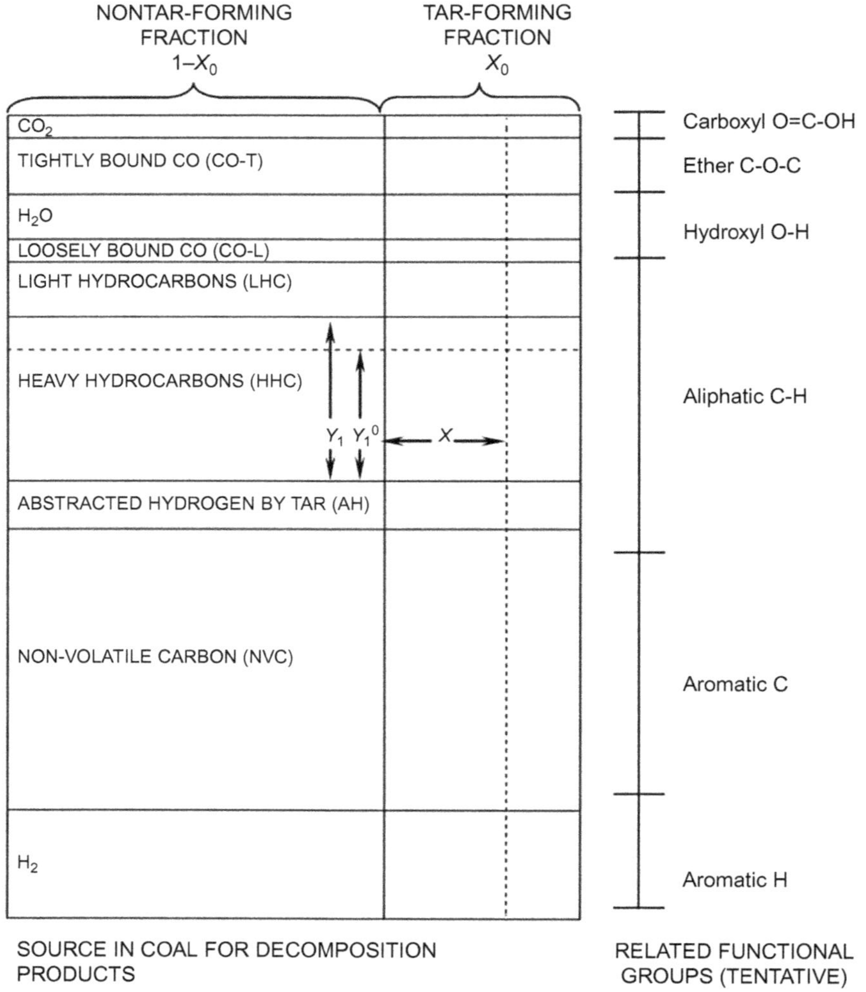

Fig. 5.5 Distribution of coal components as sources for products of devolatilization where the areas are proportional to the number percentages of each component in the coal. The implied relationship on the right between the distribution of components and the functional groups is speculative.
Reproduced from Solomon PR, Colket MB. Coal devolatilization. Proc Combust Inst 1979;17:131–43 with permission from Elsevier.

Fig. 5.5, were specified to match either the measured ultimate yields or plateaus in the evolution of the individual products. Whereas the 7 Y_i^0-values for the 7 coal components that give noncondensable products were specified from measured ultimate product yields, 14 rate parameters were adjusted to describe the approach in time to the ultimate product distribution as a function of temperature during heatup. The saving grace for all the heuristic parameter adjustments is that the product evolution from

13 different coals in rank from lignite to *lv* bituminous were reasonably well described with a single parameter set; only the variable component levels for the 7 noncondensable products were adjusted to match measured values for each coal. Eventually, this performance gave rise to the notion that the decomposition rates of each distinct functional group are the same in any coal, and it is only the concentrations of each functional group that vary from coal to coal. The irony behind this notion is that no functional groups were actually measured and incorporated into this kinetic analysis.

Moreover, coal-independent decomposition rates did not stand the test of time. In time, ever more features of the FG model became bound to data from standardized tests. Ultimately, the necessary laboratory support became truly monumental (Solomon et al., 1993), because each of the decomposition rates had been expanded into the DAEM format, and the number of subdivisions for each noncondensable product had surged. As an illustration, consider that the production of CO_2, CO, and H_2O were subdivided into three stages each, and CH_4 had two stages. So there are 27 rate parameters for the oxygenated gases plus 6 more for CH_4, plus the 11 measured yields that specify the initial coal component fractions for each stage, which gives a total of 44 input specifications to predict four noncondensable gas yields. Most important, the assigned rate parameters were no longer the same for all coal samples. When this protocol was applied to the eight APCS (Solomon et al., 1993), both the mean activation energy and std. dev. (σ) in over half the production channels were adjusted by more than 10% for different coals, although the pseudo-frequency factors were fixed. The main message in this publication is that monumental laboratory support would be required to fully specify the FG model for each coal sample.

Why was it necessary to continuously subdivide the reaction channels for each species in the FG framework over time, and expand the necessary laboratory support to monumental proportions? Because the underlying premise of multiple independent reaction channels for individual products is fundamentally wrong. This premise is almost certainly correct enough for the initial conversion of a portion of the carboxylic acid functional groups into CO_2. But that initial stage is not the only source of CO_2 in the products. And for most of the other major noncondensable products, there is not even an initial stage where some specific functional group is exclusively converted into a particular product without the influence of many other reaction intermediates. Hardly any of the GHCs, CO, and moisture are formed in independent reaction channels. In other words, just because the DAEM can describe total weight loss and the dynamics for specific products across broad domains of thermal history does not mean that multiple independent parallel reactions are legitimate reaction mechanisms for primary devolatilization. To the contrary, the inability of the FG model to evolve into any form of predictive capability is a clear indication of the limitations on independent parallel reaction models.

5.2.3 Free radical chain reaction mechanisms

Around the same time that the DAEM was being transferred from the petroleum refining community into the coal research community, another exchange was also occurring in the same direction. The free-radical chain reaction mechanisms that had dramatically advanced the chemistry of petroleum refining were being applied to coal devolatilization by Gavalas et al. (1981a, b). As in any chain reaction mechanism, a pool of radical species—CH_3-radicals, H-atoms and, in low rank coals, OH radicals—first accumulates to some quasisteady population, then de-stabilizes functional groups into intermediates, and finally stabilizes intermediates into products while continuously replenishing the radical pool. In parallel, radicals are destroyed in recombinations that may or may not yield stable products. In the coal reaction system, some of the recombination channels involve radical exchange reactions with aromatic nuclei, because unpaired electrons on polynuclear aromatics persist for much longer than typical primary devolatilization times (Lewis and Singer, 1969). Ultimately, these stable aromatic radicals are eliminated by reactions involving the radical pool or in bimolecular recombinations with other aromatic radicals. Even though the persistence of aromatic radicals is a distinctive feature of the coal reaction system, the basic progression in the chemistry of devolatilization abides by generic hydrocarbon free-radical chain reactions.

The accumulation of a radical pool is fundamentally inconsistent with multiple independent parallel reactions, because (1) the decomposition fragments of any particular functional group can contribute CH_3, H and, perhaps, OH to the pool via one or more radical exchange reactions; and (2) radicals subsequently attack and destroy many diverse functional groups much faster than unimolecular decompositions based on the bond dissociation energies. Such unimolecular decompositions do initiate radical production, but direct radical attack quickly takes over as the primary means to destroy the remaining functional groups. The fact that rapid primary devolatilization occurs while the fuel is rapidly heating accelerates the decompositions further by rapidly and progressively expanding the population of de-stabilized functional groups that decompose into additional radicals. Clearly, some of the atoms in any particular functional group can appear in the radical pool, probably in multiple molecular identities, and those that end up in a particular noncondensable product are very unlikely to have originated from a single functional group. In other words, the reactions underlying gas production are neither parallel nor independent.

The crucial distinction between free radical mechanisms and independent parallel reactions is that numerous products of free radical mechanisms display the same dynamic features during their evolution, because the dynamics of the radical pool determine the rates of most reaction channels. This congruence is clearly seen in Fig. 5.6, which shows the release rates of H_2O, CO_2, CO, and tar from a subbituminous coal during heating at 0.5°C/s (Solomon et al., 1990a). Note that the release rates of tar and the three primary oxygenated gases pass through local maxima at 18–20 min; in fact, the same coincident local maxima appear in the evolution histories for CH_4, C_2H_4, NH_3, and COS, and the only species of the nine monitored in these tests that did not exhibit this coincident maximum was SO_2, which forms from inorganic sulfates. The release rates of all these species pass through coincident maxima while the radical pool expands past a threshold that destabilizes numerous diverse functional groups at once. Some very large number of concerted chemical reactions eventually reforms molecular fragments and radicals into these eight stable products on a common time scale. Even though the detailed reaction mechanisms

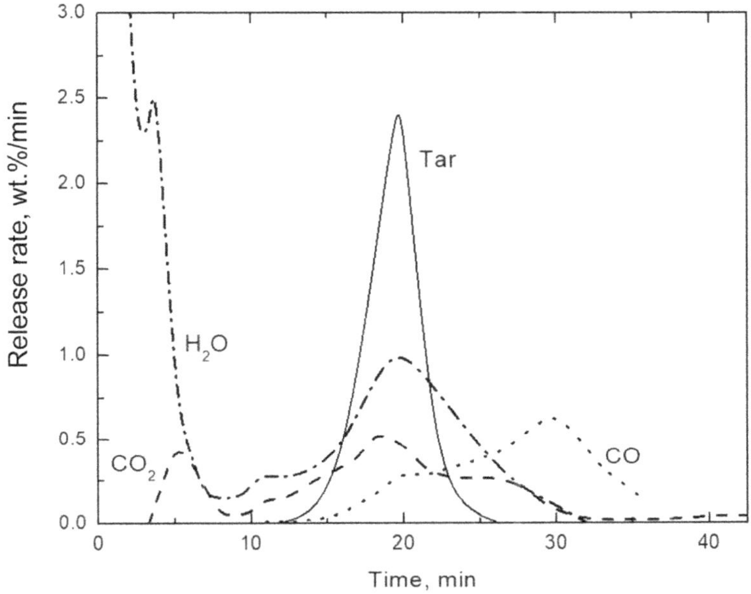

Fig. 5.6 Release rates of *(solid)* tar, *(dot-dash)* H_2O, *(dash)* CO_2, and *(dot)* CO from a subbituminous coal during heatup at 0.5°C/s in a TGA at atmospheric pressure. Reproduced from Solomon PR, Serio MA, Carangelo RM, Bassilakis R. Analysis of the Argonne premium coal samples by thermogravimetric Fourier transform infrared spectroscopy. Energy Fuel 1990a;4:319–33 with permission from the American Chemical Society.

remain obscure, it is apparent that some common free-radical chain reaction mechanism underlies tar production during primary devolatilization. This congruence is also evident in the rates reported for the seven other APCS in the tests of Solomon et al. (1990a), as well as in comparable data for several German coals from Juntgen and Van Heek (1979).

The rates in Fig. 5.6 also show two coincident maxima for H_2O and CO_2 production at 5 and 11 min, suggesting that concerted chemistry comes into play even before a massive disintegration of the macromolecular network releases tar. But except for the coincident maximum at 18 min, the production of CO is distinctive. This species is not released in common with H_2O and CO_2 prior to tar production, but instead displays two additional relative maxima once tar production starts to decline. The first of these coincides with a plateau in CO_2 production, while the second is truly distinctive. Distinctive channels like these are independent of the common reaction mechanism underlying tar production, and belong in a separate reaction mechanism for the later stages of devolatilization. As described below, this second mechanism describes the elimination of the most refractory peripheral groups and bridge fragments from coal macromolecules, as well as much slower ring-cleavages that release noncondensables during the annealing stage.

5.2.4 The coal reactant in FLASHCHAIN®

Throughout the development of the FG model, additional parallel reaction channels were continuously added to the production mechanism for each particular non-condensable product. Each parallel channel contributed an increment in the ultimate yield of that product that was specified as a hypothetical ultimate yield parameter, as a surrogate for the actual functional group concentration that participated in the channel under consideration. In the development of FLASHCHAIN®, Niksa took a diametrically opposite approach that identifies the main coal reactant as the entire population of *all* the bridges and peripheral groups in coal (Niksa, 1994). The only coal components that are excluded from the averaging are the aromatic nuclei, because nuclei are created, not destroyed, during primary devolatilization. Bridges and peripheral groups are lumped together for two reasons. First, decompositions of both types of structures sustain a common radical pool, and the evolution of the pool governs the kinetics of primary devolatilization. Since radical chain reactions coalesce the diverse peripheral groups and bridge bonding structures in any coal type into a common pool, the predominant chemical tendencies should be evident in the average chemical constitution of the entire population of radical precursors. At least that is the underlying hypothesis. The second reason is that no analytical method can quantitatively distinguish the functional groups in bridges and peripheral groups, so attempts to resolve them in a kinetic analysis are subject to inordinate uncertainties anyway.

The reactants based on this average are called "labile bridges" in FLASHCHAIN®, even though the averaging covers both bridges and peripheral groups. The atomic element ratios for bridges are evaluated from the relations in Table 5.1, which also gives the compositions of other coal structures to be discussed later. Actual bridges and peripheral groups contain only aliphatic hydrocarbon components plus heteroatomic functional groups involving oxygen, aliphatic sulfur, and quaternary nitrogen. Consequently, the atomic ratios for H/C, O/C, and S/C for bridges are evaluated for individual coal samples from only the proton-($^{H}f'a$) and carbon-(f_a') aromaticities, and the respective atomic ratios based on a conventional ultimate analysis of the whole coal, where the ratio for sulfur includes only aliphatic-S and aromatic sulfides. Moreover, both aromaticities are strongly correlated with a coal's C-content. So regressions of data in the literature are used for both quantities to circumvent a sophisticated and expensive laboratory analysis for every sample of interest (Niksa, 1994). Since quaternary nitrogen usually constitutes a small fraction of coal-N, the bridges in FLASHCHAIN® contain no nitrogen at all.

The $(H/C)_B$ and $(O/C)_B$ labile bridge ratios from these expressions for a large database of diverse coals appear in Fig. 5.7, where they have been plotted versus C-content of the whole coals along with the ratios for the whole coals. The $(H/C)_B$ ratio becomes greater for coals of progressively higher rank, even though the ratios for the whole coals diminish. Across the rank spectrum, bridge-based values increase from 1.5 to 3.0, whereas the whole coal values are quartered from unity to 0.25. The reason is that the pool of aliphatic carbon diminishes faster with rank than the pool of aliphatic hydrogen, which implies a progression in bridge structures and peripheral groups with rank from heteroatomic structures to methylene chains to greater proportions of

Table 5.1 Definitions of the reactant compositions in FLASHCHAIN®

Component	C-Number	H/C	O/C	N/C	S/C
Bridge	$\frac{(1-f_a')\mathrm{AC/Cl}}{f_a'(1-\beta^*)}$	$\left(\frac{\mathrm{H}}{\mathrm{C}}\right)\frac{(1-{}^{\mathrm{H}}f_a')}{(1-f_a')}$	$\left(\frac{\mathrm{O}}{\mathrm{C}}\right)\frac{1}{(1-f_a')}$	0	$\left(\frac{S_{\mathrm{ALIPH}}}{C}\right)\frac{1}{(1-f_a')}$
Char link	$\frac{0.45MW_B}{12+\beta^*\left(\frac{H}{C}\right)\frac{{}^{H}f_a'}{f_a'}}$	$0.45\left(\frac{\mathrm{H}}{\mathrm{C}}\right)\left(\frac{{}^{\mathrm{H}}f_a'}{f_a'}\right)\frac{1-f_a'}{1-\beta^*}$	0	0	0
Aromatic nucleus	$\frac{\mathrm{AC/Cl}}{f_a'}-(1-\beta^*)C_{\mathrm{B}}-\beta C_{\mathrm{C}}$	$\left(\frac{\mathrm{H}}{\mathrm{C}}\right)\frac{{}^{\mathrm{H}}f_a'}{f_a'}$	0	$\left(\frac{\mathrm{N}}{\mathrm{C}}\right)\frac{\mathrm{AC/Cl}}{C_A f_a'}$	$\left(\frac{S_{\mathrm{ORG}}-S_{\mathrm{ALIPH}}}{C}\right)\frac{\mathrm{AC/Cl}}{C_A f_a'}$

β^* is the number fraction of char links to nuclei in the coal.

Reproduced from Niksa S. Flashchain theory for rapid coal devolatilization kinetics. 4. Predicting ultimate yields from ultimate analyses alone. Energy Fuel 1994;8:659–70 with permission from the American Chemical Society.

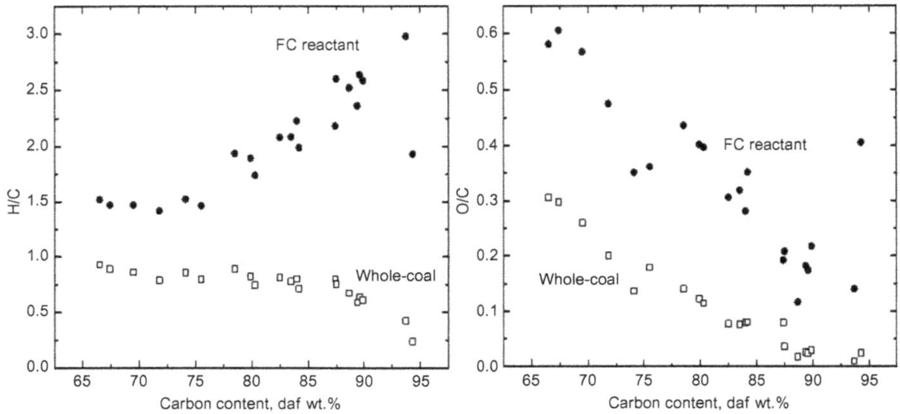

Fig. 5.7 Atomic (left) H/C and (right) O/C ratios for the bridge reactant in FLASHCHAIN® (●) compared to the whole-coal values (□).
Reproduced from Niksa S. Flashchain theory for rapid coal devolatilization kinetics. 4. Predicting ultimate yields from ultimate analyses alone. Energy Fuel 1994;8:659–70 with permission from the American Chemical Society.

methyl groups. Note also that the sample-to-sample variability in $(H/C)_B$ is much greater than in the whole coal values, which raises their potential to interpret the substantial sample-to-sample variability in most aspects of devolatilization behavior. The $(O/C)_B$ ratios in Fig. 5.7 diminish for coals of progressively higher rank, in tandem with the whole coal values, although the bridge-based values are at least twice as large. Here too the sample-to-sample variability is much greater in the bridge-based values.

The compositions of these bridge reactants are markedly different than whole coal compositions, so it is worth investigating how sample-to-sample variations in these values affect the free radical chain chemistry. Since no free radical mechanism has yet been validated for coal, the calculations in this section are based a 551-step elementary reaction mechanism for gas mixtures validated for extremely fuel-rich conditions. The bridge compositions from Table 5.1 in seven of the APCS were used to specify the levels of C_3H_8, C_6H_6, and H_2O_2 in surrogate gas mixtures with the same H/C and O/C ratios (as explained below). For this sample suite, the $(H/C)_B$ ratios increased from 1.5 to 2.8 and the $(O/C)_B$ ratios diminished from 0.52 to 0.18 to represent bridges in coals of progressively higher rank. The lignite APCS was omitted because its surrogate mixture contained benzene and peroxide without any propene, so the simulation results were not directly comparable to the other gas compositions.

In these mixtures, propene is the surrogate for methylene bridges in coal; peroxide incorporates the decomposition of hydroxyl groups into H-atoms and OH radicals at relatively low temperatures; and benzene represents more refractory hydrocarbon components. Since the $(O/C)_B$ ratios diminish and the $(H/C)_B$ ratios increase for the bridges in coals of progressively higher rank, the mixtures contain much greater amounts of peroxide and relatively more benzene than propene in surrogate mixtures for coals of progressively lower rank; in fact, the lignite mixture contains no propene and the anthracite mixture contains no benzene. The thermal decomposition was evaluated in plug flow during heatup at about 1200 °C/s to 1200°C.

As seen in Fig. 5.8, propene is converted in two stages, one at the onset of decomposition that ultimately converts 30%–90%, depending on the mixture composition, and another that begins above 850°C after a lag from the conversion level of the first stage. Conversion during the first stage becomes markedly slower and thereby shifts toward hotter temperatures for bridge reactants of progressively higher rank, while the ultimate conversion for this stage plummets. Conversion during the second stage occurs on a common time scale and becomes complete at about 1050°C for all mixtures; the apparent conversion rate for this stage slows, but only because of the large variations in the ultimate conversion levels for the first stage. In all these variations, the subbituminous and anthracite cases represent the extremes, while four of the five bituminous mixtures exhibit very similar behavior. The Ill. #6 *hv*C bituminous mixture (IL6) is an exception, but only because its H/C ratio is 20% lower and its O/C ratio is greatest of all compared to the other bituminous mixtures, and these deviations make this surrogate composition very similar to the subbituminous mixture.

The reason for these variations is apparent in the radical concentrations in Fig. 5.9, which shows the OH and CH_3 concentrations on disparate scales. The levels of H-atoms mimicked the OH concentrations, except that the maximum level for the subbituminous mixture at 600°C was only 75 ppb; O-atom concentrations were negligible

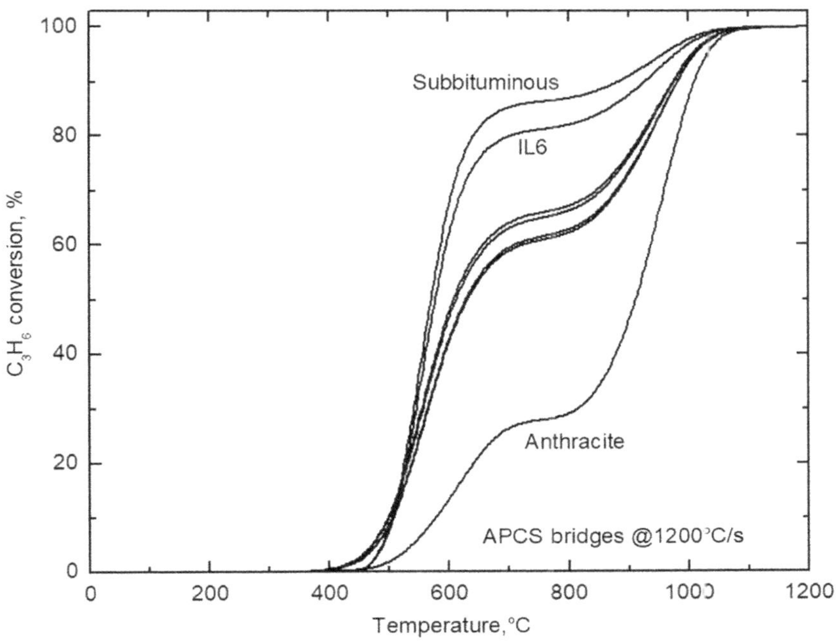

Fig. 5.8 Extents of propene conversion in surrogate gas mixtures for the bridge reactants in seven APCS during heatup at 1200°C/s.

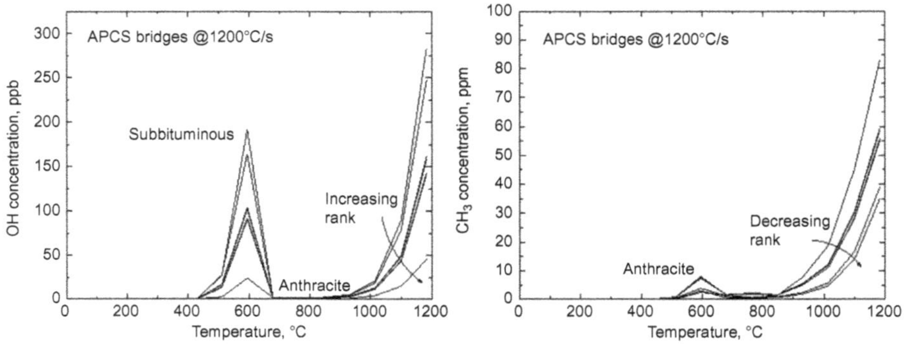

Fig. 5.9 Levels of (left) OH and (right) CH_3 in surrogate mixtures for the bridge reactants in seven APCS during heatup at 1200°C/s.

throughout. Methyl radicals are the predominant species in this radical pool throughout both conversion stages and for all surrogate mixtures. However, the more reactive hydroxyl radical makes appreciable contributions to the pool throughout the first conversion stage, but only for low-rank and *hv* bituminous mixtures. These contributions are especially strong for the mixtures based on low-rank coals, because the maximum CH_3 concentrations at 600°C are only a few ppm. Throughout both conversion stages, OH and CH_3 levels exhibit opposite variations for progressively greater rank, although OH concentrations are always inconsequential in comparison to CH_3 levels during the second conversion stage.

There are many reasons to quibble over the relevance of simulations based on surrogate gas mixtures to primary devolatilization reaction mechanisms: The omission of radical stabilization on aromatic nuclei, the much smaller assortment of surrogate fuel functional groups, and the omission of product withdrawal throughout the simulation—to name just a few concerns. But the central result that decomposition rates are accelerated by oxygen functionalities in the fuel mixture is absolutely fundamental. As such, this tendency will be apparent in the decomposition of any fossil fuel mixture and, especially, as variations in the bridge dissociation rates during the primary devolatilization of diverse coals. It clearly points toward decelerating bridge dissociation rates for coals of progressively higher rank, and undermines the notion that bridge dissociation rates are the same for all coal types.

5.2.5 Bridge conversion kinetics

Proper coal quality impacts for devolatilization kinetics can be re-created by subjecting mixtures based on the labile bridge compositions in FLASHCHAIN® to free radical chain chemistry. Before we contemplate the implications for practical applications, it is worth asking, "Why then didn't the reaction mechanism developed by the Caltech group resolve the chemistry of coal devolatilization once and for all?"

The main reason is that the analysis contained 128 kinetic parameters which were freely adjustable because (1) There are no laboratory tests that can isolate the kinetics of individual reactions in the condensed phase analogous to kinetic studies in shock tubes for chemistry in the gas phase and (2) Thermochemical estimation techniques based on well-known parameters for chemistry in the gas phase could not be transferred to chemistry in a condensed coal phase because of the strong influence of cage, gel, and other effects from species in close proximity to the reactants in the condensed phase. Consequently, the Caltech reaction mechanism still provides a solid theoretical foundation for phenomenological modeling approaches, but it never advanced to the stage of actually depicting the measured devolatilization behavior of any coal, for reasons which remain instructive.

Unfortunately, the recognition that free-radical chain mechanisms provide the means to describe the coal quality impacts on devolatilization rates does not deliver any practical means toward a tractable and quantitative analysis. The way around this important impediment is among the most elegant features of any network depolymerization mechanism. In lieu of accurate functional group distributions and the massive reaction sets they entail, bridge decompositions can be represented in a statistical way with a DAEM reaction process. In their response to a thermal history, activation energy distributions mimic the response of functional group distributions by allocating portions of coal reactants to decompose in specific temperature intervals. Of course, there is no sound basis to assert that activation energies associated with real functional group distributions are distributed as normal distribution functions. But normal distributions depict the thermal response so accurately and introduce so few adjustable parameters that it is foolhardy not to use them.

We must also acknowledge that the price for all this expedience is steep: Instead of assigning actual functional group distributions, every DAEM parameter must eventually be evaluated from some aspect(s) of coal constitution, to depict the thermal response of individual coal samples. Another noteworthy implication is that the postulated reaction "mechanisms" are, in actuality, phenomenological reaction processes because authentic chemical reactions must use actual functional groups as their reactants. Since there is no analytical method that can distinguish the functional groups that interconnect nuclei from the analogous chemical structures attached to nuclei as peripheral groups, there is no means to accurately measure bridge constitutions and concentrations in coal. So it is sensible to use an efficient probabilistic tool like the DAEM to represent the thermal response of bridge decomposition.

This approach is the one incorporated into FLASHCHAIN® for the labile bridge reactant. Niksa and Kerstein (1986, 1991) represented bridge conversion as a single DAEM process because, as explained previously, the DAEM exhibits all the correct kinetic tendencies for variations in heating rate, temperature, and reaction time. Even though a DAEM cannot depict yield enhancements for faster heating rates or pressure effects, it certainly can depict bridge decomposition kinetics. Moreover, reformulating the DAEM for bridge conversion eliminates the hypothetical ultimate yield parameter. Instead of asserting that increments in the ultimate volatiles yield are associated with increments in a continuous distribution of activation energies, we directly connect increments in the population of labile bridges to increments in the energy distribution:

$$dB = f(E)dE \quad \text{where } f(E) = \frac{1}{\sqrt{2\pi}\sigma_B} \exp\left(-\frac{(E - E_{B,0})^2}{2\sigma_B^2}\right) \tag{5.7a}$$

where B is the initial moles of bridges per unit volume of coal normalized by the initial concentration of aromatic nuclei. The normalization circumvents explicit evaluations of any of the absolute molar species concentrations, as explained elsewhere (Niksa and Kerstein, 1986, 1991), so nothing analogous to a hypothetical ultimate yield parameter appears in the analysis. The energy distribution is assumed to be Gaussian to minimize the number of parameters. Accordingly, the bridge conversion rate is given by

$$\frac{dB(t)}{dt} = -\int_0^\infty A_B \exp\left(-\frac{E_B}{RT}\right) \exp\left[-\int_0^t A_B \exp\left(-\frac{E}{RT}\right) dt'\right] f(E)dE \tag{5.7b}$$

The bridge conversion rate can also be cast in more familiar terms as an equivalent first-order reaction given by

$$\begin{aligned} \frac{dB}{dt} &= -k_B B \quad \text{where } k_B \\ &= -\frac{d}{dt} \ln\left[\int_0^\infty \exp\left[-\int_0^t A_B \exp\left(-\frac{E}{RT}\right) dt'\right] f(E)dE\right] \end{aligned} \tag{5.7c}$$

In FLASHCHAIN®, the DAEM for bridge conversion contains rate parameters A_B, $E_{B,0}$, and σ_B. In principle, these rate parameters are adjustable, because there is no way to estimate their magnitudes from scientific fundamentals, although the concept of a continuous energy distribution for the dissociation energies of bridges is corroborated by the extremely broad range of dissociation energies estimated for the postulated bridge structures in coal (Niksa and Kerstein, 1986). In practice, the mean activation energy, $E_{B,0}$, is fixed for all coals, because only the ratio of $\sigma_B/E_{B,0}$ is an independent parameter (cf. discussion of Eq. 5.3c). Values of σ_B and A_B were linearly correlated with $(O/C)_B$, based on extensive quantitative interpretations of laboratory datasets (Niksa, 1994), according to

$$\log_{10} A_B = 6.764\left(\frac{O}{C}\right)_B + 8.438 \quad \text{and} \quad \sigma_B = 61.63\left(\frac{O}{C}\right)_B + 9.38 \tag{5.8}$$

where σ_B is in kJ/mol. The bridge decomposition rate for any coal sample is specified from an ultimate analysis, and literature databases of proton- and carbon-aromaticities. Only an ultimate analysis needs to be measured for every sample.

Predicted extents of bridge conversion for the eight APCS for heatup at 1000°C/s to 1100°C appear in Fig. 5.10. These coals represent ranks from lignite through *lv* bituminous. Based on the correlations for A_B and σ_B in Eq. (5.8), bridge conversion rates in FLASHCHAIN® diminish for coals of progressively higher rank, and bridge conversion occurs within a narrower temperature range. With low rank coals, bridges are converted within a temperature range of 700°C at this heating rate; bituminous bridges in 600°C; and anthracite bridges in 400°C. These differences are most pronounced in the faster rates for the lignite and subbituminous samples, and in the slowest rate for the anthracite, whereas the five *hv* bituminous coals are assigned fairly similar bridge conversion kinetics.

Whereas FLASHCHAIN®'s bridge decomposition kinetics explicitly account for the large variations in the bridge reactant compositions across the rank spectrum, AFR's DVC component of the FG-DVC model essentially uses a single DAEM rate constant for all coal types. Their archetypal bridge structure in all coals is the ethylene linkage, so that the bridge weight is uniform at 28 g/mol, and bridge conversion kinetics were specified from model compound studies of this linkage in pure polymers of benzene and naphthalene rings, with guidance from thermochemical estimation methods (Squire et al., 1986). For the eight APCS, the mean activation energies ranged from 216 to 254 kJ/mol and the std. dev. ranged from 6.3 to 10.5 kJ/mol

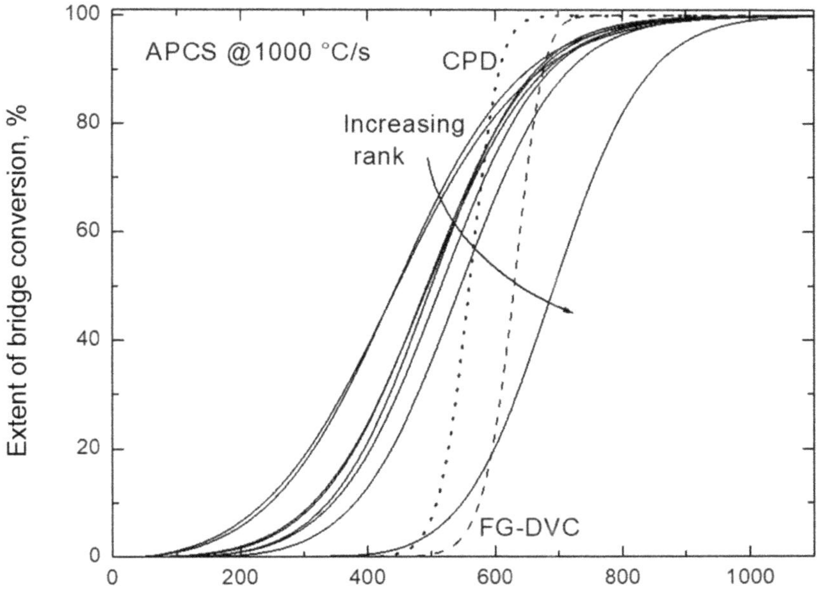

Fig. 5.10 Predicted extents of bridge conversion for 1000°C/s to 1100°C from *(solid)* FLASHCHAIN® for the eight APCS in order of increasing rank from left to right; and from *(dashed)* FG-DVC and *(dotted)* CPD for all coal types.

(Solomon et al., 1993). The frequency factor was fixed at 10^{14} s^{-1} for all coal types. Variations in the extents of bridge conversion for average parameter values are compared to the bridge conversion kinetics in FLASHCHAIN® in Fig. 5.10. Since a single surrogate bridge structure—the ethylene bridge—has been implemented for coals across the rank spectrum, the extent of bridge conversion from FG-DVC is uniform for all APCS. Due to the large activation energy and small std. dev., bridge conversion is confined to only 125°C, from 550°C to 675°C.

The developers of CPD adopted without modification AFR's mean activation energy for bridge conversion from the model compound studies, but adjusted A_B and σ_B to 2.6×10^{15} s^{-1} and 7.5 kJ/mol, respectively (Grant et al., 1989; Fletcher et al., 1990); nonetheless, CPD implements the same bridge decomposition rate for all coal samples. In Fig. 5.10, the extents of bridge conversion based on these kinetics are shifted toward cooler temperatures by 75°C compared to AFR's, but stay confined to the same narrow temperature window. It is interesting that CPD's bridge conversion kinetics are nearly identical to those for the subbituminous surrogate mixture in Fig. 5.8, whereas the kinetics in FG-DVC initially match the anthracite mixture's, but then rapidly accelerate for conversions greater than 20%.

Bridge decomposition kinetics and, especially, their dependence on coal quality, are a prominent distinguishing feature among the three network depolymerization models. FLASHCHAIN® implements distinctive rate parameters for individual coal samples and stages bridge conversion across very broad temperature ranges. Even though this approach is much simpler than legitimate free-radical chain kinetics, it incorporates the tendency for decelerating bridge conversion rates for progressively higher ranks due to variations in the bridge-based O/C ratio. AFR uses a very narrow range of rate parameters across the rank spectrum and CPD implements the same rate parameters for all coals. Both mechanisms confine bridge conversion to very narrow temperature windows, and ignore the acceleration by oxygen functional groups in coals of progressively lower rank.

5.2.6 Depolymerization of coals' macromolecular matrix and reintegration of fragments into char

Coal is a crosslinked macromolecular solid. With regard to depolymerization kinetics, the important implication is that when any particular bridge breaks, no product necessarily forms. That's because many, many linkages must break before fragments of the original coal matrix become small enough to evaporate and escape as tar. Consequently, descriptions of a coal's depolymerization into fragments, including tar precursors and tar, must account for two independent factors: (1) bridge conversion kinetics, specified in the previous section and (2) fragment statistics, which evaluate the probability that any particular bridge decomposition generates fragments of a particular size. This approach was introduced into devolatilization modeling by Gavalas and coworkers at Caltech, who developed fragment statistics for bridge dissociation based on the disintegration of a two-dimensional honeycomb network (Gavalas et al., 1981a, b). It was then implemented at AFR, in their interpretation of the decomposition of naphthalene polymers with fragment statistics for a chain of six linked naphthalene units (Solomon and King, 1984); and by Niksa and Kerstein, whose DISCHAIN model for coal devolatilization developed fragment statistics for a mixture of nominally infinite chains (Niksa and Kerstein, 1986; Niksa, 1986). In addition, DISCHAIN modeled another essential aspect of primary devolatilization for the first time: While the original macromolecular matrix is disintegrating through depolymerization, another infinite lattice called char is developing via fragment recombination reactions. Indeed, fragment statistics for devolatilization must simultaneously describe how the fragment size distribution shifts toward smaller sizes for progressively more bridge scissions, and how aromatic nuclei bound together by refractory char links accumulate into longer fragments of converted material throughout devolatilization.

Analyses to specify the fragment statistics begin by subdividing the fragment size distribution into groups based on their mobilities, and whether or not they can repolymerize into char, and whether or not they can partition into the size range for tar vaporization. This presentation uses the terminology introduced in DISCHAIN, and cites references to more complex treatments later. Reactant fragments are those that remain larger than the size range that can partition into tar, even while bridge scissions shift their size distributions toward smaller sizes. Metaplast fragments are of intermediate size and presumed to possess the mobility to form char links with reactant fragments, and to spontaneously disintegrate into the size range for tar vaporization. Tar fragments are small enough to spontaneously vaporize into the gas phase upon formation, where they are unable to recombine with fragments in the condensed phase to contribute to the char inventory.

The configurations that determine fragment statistics for this process are sketched in Fig. 5.11, where "Monomer" is used for metaplast. Prior to devolatilization, the initial coal reactant is represented as a mixture of chains of aromatic nuclei exclusively connected by labile bridges, with a peripheral group on each nucleus. Throughout devolatilization, the reactant chains disintegrate via bridge scission, which eventually produces fragments in the metaplast size class. Metaplast is subject to two competing processes. Either it attaches to a reactant fragment by forming a new refractory char link, or it spontaneously decomposes into two tar fragments. Tar immediately vaporizes into the gas phase and escapes intact into the surrounding atmosphere as a primary devolatilization product.

One essential aspect of this reaction system is that tar production is diminished by the continuous accumulation of char fragments in the reactant size class. To understand why, consider the two possible configurations for the labile bridge closest to the end of a fragment. In the most favorable configuration, that labile bridge connects only a single aromatic nucleus to the rest of the fragment. So when it breaks it forms a new metaplast fragment. This metaplast fragment does not necessarily form tar, but at least it has the potential to spontaneously decompose into tar. In the second configuration, that labile bridge connects some number of nuclei bound together by char links to the rest of the fragment. So when it breaks, an independent fragment of pure char is released from the parent fragment. Since all the linkages in the new fragment are refractory, it cannot disintegrate further into metaplast or tar.

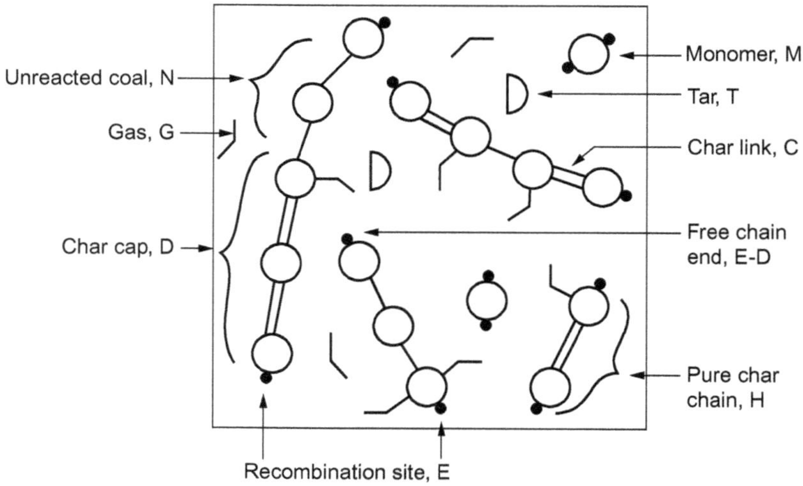

Fig. 5.11 Illustration of all reactant, intermediate, and product species in DISCHAIN. Reproduced from Niksa S, Kerstein AR. The distributed energy chain model for rapid coal devolatilization kinetics. Part I. Formulation. Combust Flame 1986;66:95–109 with permission from Elsevier.

The production rate of metaplast in this scenario is given by the following rate expression (Niksa and Kerstein, 1986):

$$\frac{dM(t)}{dt} = -2(1-B)(1-p_{\mathrm{CE}})\frac{dB}{dt} - \left(\frac{2M+E+2H_{\mathrm{C}}}{M+E+2H_{\mathrm{C}}}\right)\frac{dC}{dt} - \frac{1}{2}\frac{dT_{\mathrm{P}}}{dt} \tag{5.9}$$

where $M, B, E, H_{\mathrm{C}}, C$, and T_{P} are normalized concentrations of metaplast; labile bridges; fragment ends on all fragments with at least one labile bridge; the number of fragments containing only char links; char links; and tar fragments, respectively. The probability that the end of a fragment contains at least one char link is denoted by p_{CE}. The normalized concentrations are also probabilities so that, for example, B is the probability that two nuclei are connected by a labile bridge. The factor of $2(1-B)(1-p_{\mathrm{CE}})$ in the first term is the fragment statistic for metaplast production via bridge conversion. It expresses the probability that the last nucleus in a fragment is bound by a labile bridge, rather than a char link, which is the prerequisite for metaplast production. The factor of two appears because each fragment has two ends to consider. The product of the fragment statistic for metaplast production and dB/dt, the bridge conversion rate from the chemical kinetics, gives the production rate of metaplast via labile bridge conversion.

The fragment statistics for recombination of metaplast with any other fragment are also developed from two possible configurations. The pair of nuclei adjacent to the newly formed char link had previously been one metaplast and a reactant fragment or two metaplast fragments, which consumed one and two metaplast fragments, respectively. There are two permutations of the metaplast/reactant configuration, and two ends per metaplast fragment. All recombination sites on the larger fragments participate, and the total number of sites is the sum of all labile fragment ends plus twice the number of pure char fragments. So the numerator in this fragment statistic is the total number of recombination sites. The denominator is a conditional probability that ensures that all recombinations must involve a metaplast fragment, because the recombination of two reactant fragments is excluded based on insufficient mobility. The third term in Eq. (5.9) has no fragment statistic because, in DISCHAIN, every metaplast fragment can disintegrate into two, and only two, tar fragments.

Fragment statistics are multiplicative factors in the rates of fragment formation from simultaneous depolymerization and recombination. They are independent of the chemical reaction rates, and functions of various component concentrations and of probabilities evaluated from various component concentrations. Fragment statistics do not introduce any adjustable parameters into the kinetic analysis. By resolving transformations in the fragment size distribution throughout devolatilization, they interpret detailed features of product formation and, most importantly, pave the way for vaporization-based mechanisms of tar release.

Two of the findings from the fragment statistics in DISCHAIN with general significance appear in Fig. 5.12. The panel on the left evaluates the proportions of metaplast fragments for two limiting scenarios. One considers only bridge dissociation without formation of char and tar, and shows that the metaplast yield is not proportional to the extent of bridge conversion. After half the bridges have broken, only one-fourth of the fragments are in the metaplast size class. Metaplast production accelerates throughout bridge conversion, regardless of the chemical reaction rate for

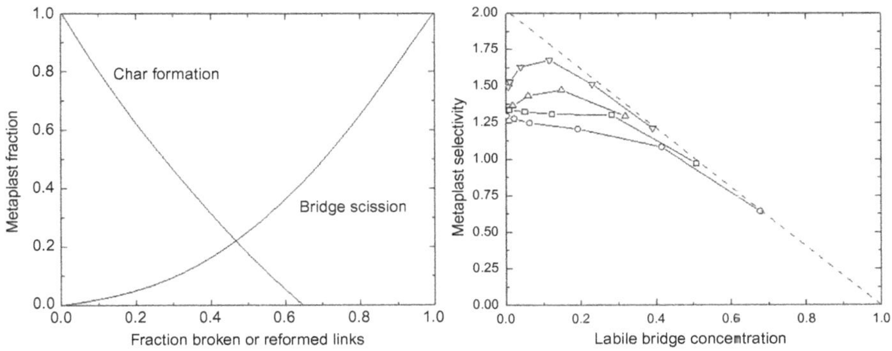

Fig. 5.12 (Left) Limiting fragment characteristics for metaplast generation by bridge dissociation and for metaplast consumption by recombination; and (right) metaplast selectivity versus bridge concentration for heatup at (○) 100, (□) 1000, (△) 10^4, and (▽) 10^5°C/s. Reproduced from Niksa S, Kerstein AR. The distributed energy chain model for rapid coal devolatilization kinetics. Part I. Formulation. Combust Flame 1986;66:95–109 with permission from Elsevier.

bridge conversion. In other words, the mathematics expresses the intuitive notion that a very large number of bridge conversions are required to produce small fragments during the depolymerization of large macromolecules.

The second limiting scenario in the left panel of Fig. 5.12 describes the repolymerization of an initial mixture of metaplast fragments into char fragments, without bridge conversion or tar production. The important result is that the number of char links needed to bind all metaplast into larger char fragments, which is evident as the value for the extent of char formation when the metaplast concentration vanishes, is much smaller than the initial number of metaplast fragments.

Both of these qualitative features may be masked in the assigned rate parameters with models that omit fragment statistics, which could cause errors to accumulate in extrapolations. Moreover, analyses that omit fragment statistics also omit one of the mechanisms responsible for enhanced tar yields with progressively faster heating rates. The fragment statistic for metaplast production via bridge conversion can be integrated to evaluate the fraction of fragments that pass through the metaplast size class at any stage of bridge conversion. This integral is depicted graphically in the right panel of Fig. 5.12 as the areas under the curves for heating rates from 100 to 10^5°C/s. Without recombinations to form char, all fragments become metaplast at all heating rates. But with charring, the cumulative population of metaplast grows from 83%–87% to 92%–96% over this range of heating rates. Of course, tar yields grow for progressively greater metaplast populations, as demonstrated in an accurate quantitative interpretation of measured weight loss from an *hv* bituminous coal for various heating rates (Niksa, 1986). One of the reasons that tar yields are greater for faster heating rates is that short reaction times to any reaction temperature accommodate consecutive, rather than simultaneous, bridge conversion and char link formation. This chronology promotes greater metaplast production for progressively faster heating rates, as follows from the fragment statistics considered here.

Since coals have crosslinked macromolecular structures in three dimensions, Kerstein and Niksa (1987) and Niksa and Kerstein (1987) then generalized the kinetic analysis in DISCHAIN for the Bethe lattice configuration shown in Fig. 5.13. Bethe lattices are mathematical constructs in two dimensions that display the crosslinking of nuclei normally found in nature in three dimensions. Closed loops are not permitted in Bethe lattices. The key parameter is the coordination number, which is the average number of connections on nuclei. Such connections are not - necessarily intact between two nuclei, so peripheral groups and vacant sites also contribute to coordination numbers. Fractional values are admitted as average values on the premise that not all nuclei necessarily have the same number of connections. The lattices in Fig. 5.13 have coordination numbers of four and six, including a single peripheral group on each nucleus and one open site on the six-coordinated lattice.

DISARAY developed chain statistics for the Bethe lattice with the same reaction scheme as DISCHAIN; in fact, DISCHAIN is simply the special case in DISARAY for a coordination number of three with one peripheral group on each nucleus, which defines infinite straight chains with one peripheral group on each nucleus.

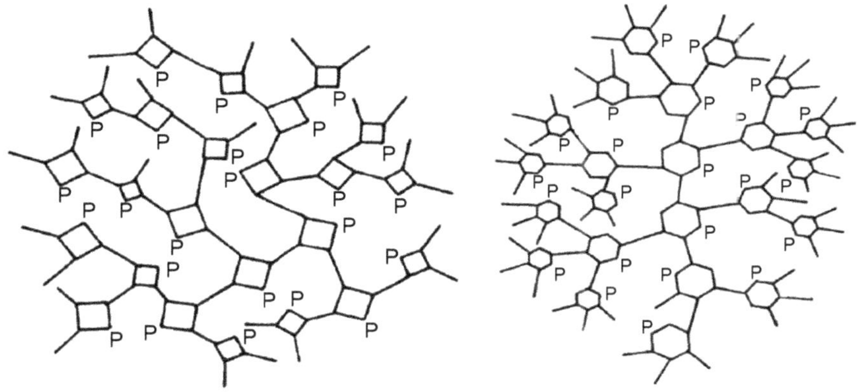

Fig. 5.13 Bethe lattices with coordination numbers of (left) four and (right) six where one connection on each nucleus contains a peripheral group, *P*.
Reproduced from Niksa S, Kerstein AR. On the role of macromolecular configuration in rapid coal devolatilization. Fuel 1987;66:1389–99 with permission from Elsevier.

Accordingly, the monomer production rate for a Bethe lattice of coordination number σ_C is given by

$$\frac{dM(t)}{dt} = -2(1-B)^{\sigma_C - 1}(1-p_{CE})\frac{dB}{dt} - \left(\frac{2M+E+2H_C}{M+E+2H_C}\right)\frac{dC}{dt} - \frac{1}{2}\frac{dT_P}{dt} \quad (5.10)$$

where the only difference with Eq. (5.9) is that the chain statistics for bridge decomposition in DISCHAIN impart a functional dependence on the bridge concentration that becomes stronger for progressively greater coordination numbers. For example, with a lattice whose coordination number is four, over half the bridges would need to break to convert 10% of the fragments into metaplast.

To better understand how fragment statistics affect predicted devolatilization behavior, consider a side-by-side comparison of the same reaction mechanism implemented with and without fragment statistics (Niksa and Kerstein, 1987). The mechanism without fragment statistics, called DISKIN, was formulated with constant stoichiometric coefficients instead of the variable coefficients in Eq. (5.10), so that every bridge conversion generated a fixed amount of metaplast throughout devolatilization. The parameters in both mechanisms were specified to fit the same datasets, and the quality of the data interpretations was very similar. The key finding appears in Fig. 5.14, which shows the weight loss during heatup at 1000°C/s to 925°C from both mechanisms. Virtually identical predictions can be obtained with DISKIN, by either increasing the frequency factor or decreasing the mean activation energy or std. dev.; and with DISARAY, by decreasing the coordination number from 4 to 2.5. With both mechanisms, the apparent temperature dependence in the weight loss would be within the measurement uncertainties, while the ultimate yields are diminished from 53 to 39 daf wt.%. Notwithstanding these similarities, these mechanisms

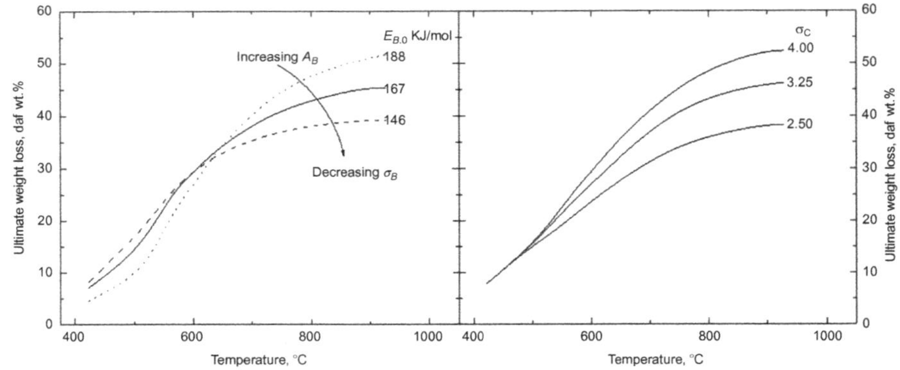

Fig. 5.14 Sensitivities of the ultimate yields (left) from DISKIN to the parameters for bridge conversion and (right) from DISARAY to the coordination number in the Bethe lattice. Reproduced from Niksa S, Kerstein AR. On the role of macromolecular configuration in rapid coal devolatilization. Fuel 1987;66:1389–99 with permission from Elsevier.

extrapolated much differently to faster and slower heating rates, whereby DISARAY gave much stronger enhancements to the tar yields for faster heating rates than DISKIN (Niksa and Kerstein, 1987).

5.2.7 Submodels for coals' macromolecular structure

Clearly, the impacts of variations in the bridge conversion kinetics and coordination number are qualitatively very similar. This exposes an awkward situation whereby a model developer can arbitrarily choose to elevate either bridge conversion kinetics or macromolecular configuration as the primary means to describe the distinctive devolatilization behavior of individual coals. In fact, the developers of FLASHCHAIN® and CPD selected the opposite extremes: In FLASHCHAIN®, the bridge decomposition kinetics cover the broad range in Fig. 5.10 to represent diverse bridge compositions in different coals while macromolecular configuration is reduced to mixtures of straight chains. This choice attributes the distinctive devolatilization behavior of individual coals to the markedly different compositions of the most reactive components, seen in the atomic ratios in Fig. 5.7, and the associated deceleration of bridge decomposition for coals of progressively higher rank in Fig. 5.8. In both CPD and FG-DVC, essentially the same bridge conversion kinetics are applied to every coal while, in CPD, four parameters pertaining to macromolecular configuration specified from ^{13}C NMR analysis represent the coal quality impacts. Assigned values for these CPD parameters appear in Fig. 5.15 for diverse coals, along with carbon aromaticities and the numbers of aromatic carbons per monomer unit.

Three of these parameters have been incorporated into all three network devolatilization models: the mass of the average monomer unit, M_{Cl}; the carbon aromaticity, f_a; and the number of aromatic carbons per monomer unit, AC/Cl. Among the three parameters used by all three models, the aromaticity increases for progressively

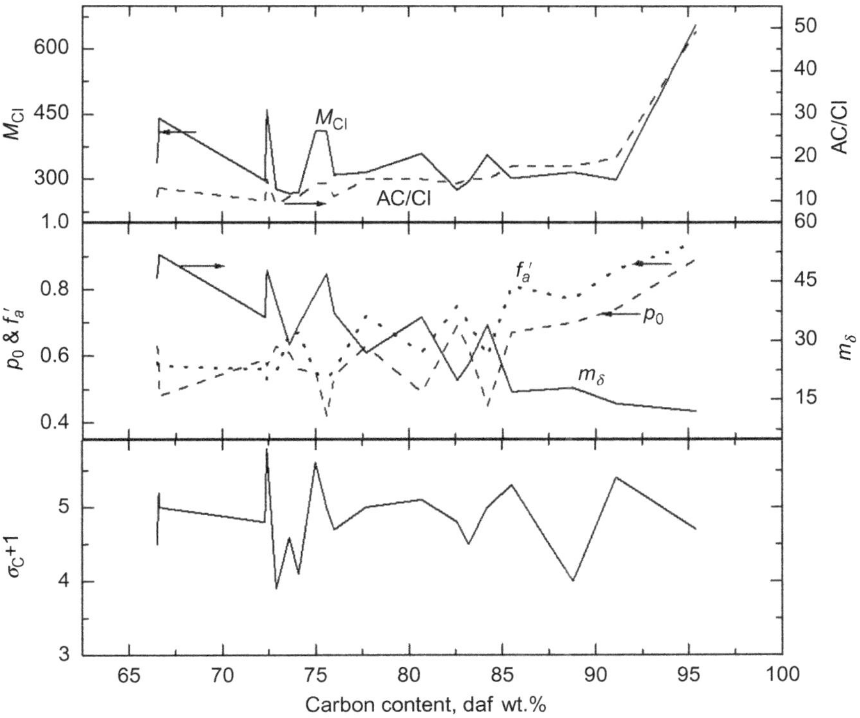

Fig. 5.15 CPD parameters evaluated from ^{13}C NMR data, including (top) M_{Cl} and AC/Cl; (middle) p_0, f_a', and m_δ; and (bottom) $\sigma_C + 1$.
Adapted from Fletcher TH, Kerstein AR, Pugmire RJ, Grant DM. Chemical percolation model for devolatilization. 3. Direct use of ^{13}NMR data to predict effects of coal type. Energy Fuel 1992;6:414–31 with permission from the American Chemical Society.

greater C-contents, albeit with appreciable sample-to-sample variability, and both the monomer mass and the aromatic carbons per monomer are fairly uniform for all ranks, except that both surge for the meta-anthracite. The additional information from ^{13}C NMR included only in CPD is the average molecular weight of peripheral groups, m_δ; the initial probability that a monomer is connected to at least one other monomer, p_0; and the coordination number, $\sigma_C + 1$. The peripheral group mass diminishes and the initial probability for connections grows for progressively greater C-contents, and both parameters track the sample-to-sample variations in f_a'. But the most striking observation of all is that the coordination number displays no rank dependence whatsoever. And since assigned coordination numbers cannot distinguish connections to peripheral groups from those to bridges, they do not directly indicate the number of bridges that must be broken, on average, to generate fragments small enough to vaporize as tar during primary devolatilization anyway.

Notwithstanding their very different foundations for the coal quality impacts, both FLASHCHAIN® and CPD describe the complete MWDs for all fragments in the condensed phase with closed-form analytical expressions based on their respective coal conformation submodels. Both approaches have been reported in elaborate mathematical detail, by Niksa and Kerstein (1991) for FLASHCHAIN®, and by Grant et al. (1989) and Fletcher et al. (1990) for CPD. FLASHCHAIN®'s submodel envisions coal as a mixture of linear chains with a broad MWD. The initial MWD is specified by adjusting p_0 to roughly match reported pyridine extract yields (Niksa, 1991b). The extracts are assumed to correspond to coal fragments with a degree of polymerization up to ten, which comprise molecular weights from 2800 to 4000 g/mol depending on coal composition. The mass fraction of this portion diminishes for progressively greater values of p_0, and vanishes at a value of unity. A database of reported extract yields exhibits fairly uniform yields for ranks through *mv* bituminous, then much lower yields for low volatility coals. So in FLASHCHAIN®, p_0 is fixed for all high-volatile coals, then surges toward unity through the low volatility coals.

CPD's conformational submodel is the Bethe lattice whose coordination number and p_0 are specified from ^{13}NMR data. As seen in Fig. 5.15, p_0 varies randomly around 0.6 through the *hv* bituminous rank, then approaches unity through the low volatility coals, whereas the coordination number is insensitive to rank. So despite the very different means to assign the initial fragment MWDs, both FLASHCHAIN® and CPD stipulate fairly uniform initial fragment MWDs through the *hv* bituminous rank, and then shift toward heavier MWDs for progressively higher ranks.

The fragment statistics in FG-DVC were first based on Monte Carlo simulations but ultimately assigned from Bethe lattices with multiple types of linkages (Solomon et al., 1990b, 1993). In the Monte Carlo simulations, coal nuclei were first assembled into a uniform mixture of chain fragments with some fixed degree of polymerization. These oligomers were then arranged to completely fill a regular rectangular lattice, and randomly interconnected by refractory linkages called crosslinks. As more crosslinks are added to the lattice, the fragment population shifts away from the uniform oligomer weight toward a broad MWD reaching much greater values. The initial proportions of crosslinks are specified to match pyridine extract yields, as in FLASHCHAIN®. Since the crosslinks are randomly distributed throughout the lattice, the number of connections per nucleus is variable, and closed loops will form whenever two or more crosslinks bind the same pair of oligomers. In principle, these features are distinctly different from the restrictions on Bethe lattices. But in practice, they are inconsequential because the average numbers of bridges per nucleus in the lattices assigned for diverse coals varied only from 2.2 to 2.5, which invokes a mixture of loosely crosslinked linear chains (Solomon et al., 1990b).

Throughout primary devolatilization, the original linkages in the oligomers break while additional crosslinks form between adjacent nuclei. Bridge scissions expand the fragment populations in the size ranges of tars, at 1–3 bound nuclei, and of solvent extracts, at 1–10 bound nuclei. Crosslinking counteracts this tendency by forming longer fragments and by reducing the probability that smaller fragments can dissociate from these larger fragments. Since the positions of the sites for scission and crosslinking are randomly selected, the original coal lattice is ultimately converted into a char lattice that may have a completely different configuration as, for example,

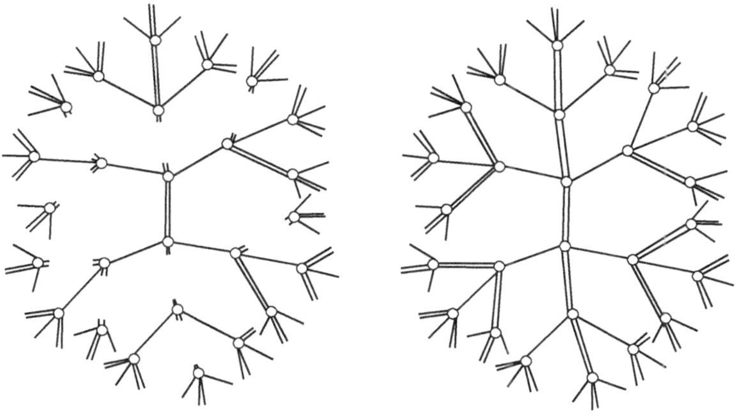

Fig. 5.16 Hybrid Bethe lattices with independent configurations for (single bond) labile bridges and (double bond) crosslinks for (left) whole coal and (right) char.
Reproduced from Solomon PR, Hamblen DG, Serio MA, Yu Z-Z, Charpenay S. A characterization method and model for predicting coal conversion behavior. Fuel 1993;72:469–88 with permission from Elsevier.

when a loosely crosslinked mixture of short chain oligomers becomes a highly branched macromolecular solid (Solomon et al., 1990b).

Once it became clear that the Monte Carlo simulations were giving fragment statistics for tars and extracts that closely resembled those from DISARAY and CPD (Solomon et al., 1990b), the AFR developers adopted a novel Bethe lattice to circumvent the inordinate computational burden of Monte Carlo simulations. As seen in Fig. 5.16, these Bethe lattices contain two distinct types of linkages, one for labile bridges and one for crosslinks. Accordingly, each nucleus in the initial coal structure contains two sites for each type. The key new feature in this approach becomes apparent by realizing that the initial coal lattice does not require that all potential linkages are actually intact. In the left panel of Fig. 5.16 for the initial coal, most of the labile bridges are intact, whereas most of the crosslinks are absent. In the right-hand rendition for char, nearly all labile bridges have been destroyed while nearly all potential crosslinks have formed. With this additional degree of freedom in the lattice, devolatilization converts the loosely crosslinked mixture of chain fragments for the coal into the highly branched lattice for char, and thereby increases the coordination number of the lattice from slightly more-than-two to four. The developers noted that these hybrid Bethe lattices better represented the fragment statistics from the Monte Carlo simulations (Solomon et al., 1993), but they did not demonstrate better predictions for the MWDs of tars and extracts.

Submodels for macromolecular structure and the associated fragment statistics are among the most similar features in FLASHCHAIN®, FG-DVC, and CPD. All three approaches use aromatic nuclei with the same composition and size. The nuclei in coal lattices are interconnected by both labile bridges and refractory char links (aka crosslinks), and not all potential connections are intact. So initially, the coal lattices are specified by the proportion of intact linkages, and also by the proportion of intact linkages that are labile bridges. Both FLASHCHAIN® and FG-DVC assign the intact linkage probability to match the fragment MWD to pyridine extract yields, and CPD assigns it from a combination of ^{13}C NMR parameters (Fletcher et al., 1992). All three approaches give moderately large portions of intact linkages for ranks from lignites through *hv* bituminous which then rise for the low volatility coals and approach unity for anthracites. However, the solvent extract data (Niksa, 1991b) and ^{13}C NMR assignments (Fletcher et al., 1992) that determine the labile bridge probabilities are widely scattered for all high volatile coals, so these assignments are subject to large uncertainties.

The labile bridge probabilities are subject to even greater uncertainties because, as previously mentioned, no analytical method is available to distinguish bridges from peripheral groups; moreover, char links in coal cannot be resolved from aromatic nuclei either. So there is no analytical method to distinguish labile bridges from refractory char links in a particular coal sample. In FG-DVC, the labile bridge fraction is specified from an interpretation of solvent swelling data based on a rudimentary theory for the elasticity of chain polymers in solution. In FLASHCHAIN®, this parameter was correlated with a coal's C-content during an early interpretation of the devolatilization behavior of about 30 coals (Niksa, 1994). As seen in Fig. 5.17, nearly

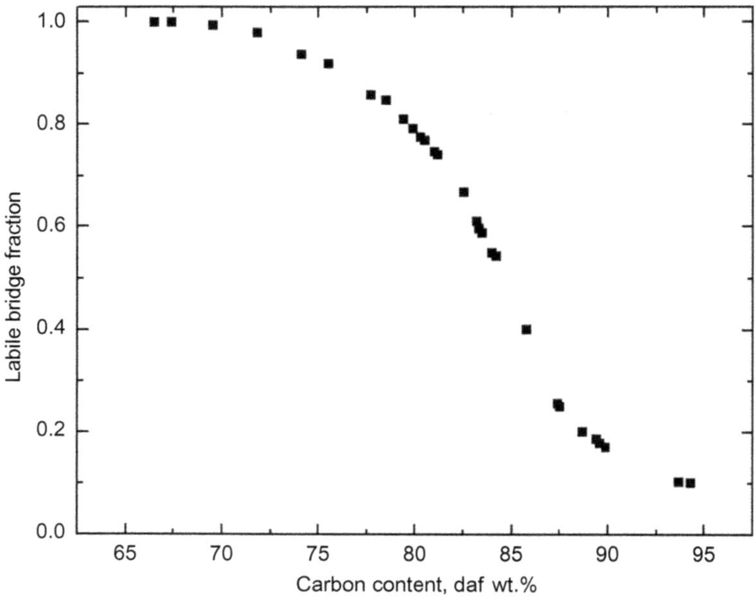

Fig. 5.17 The assigned labile bridge fraction in FLASHCHAIN® versus C-content of the whole coal.
Reproduced from Niksa S. Flashchain theory for rapid coal devolatilization kinetics. 4. Predicting ultimate yields from ultimate analyses alone. Energy Fuel 1994;8:659–70 with permission from the American Chemical Society.

all linkages are labile bridges in low-rank coals, then the fraction plummets for medium- and low-volatile bituminous coals, before it relaxes to less than 0.10 for anthracites. The general tendency for relatively fewer labile bridges in coals of progressively higher rank is fundamentally sound and a direct consequence of greater aromaticities in coals of progressively higher rank. But the postulated form in Fig. 5.17 is speculative, and it is inconceivable that this parameter does not exhibit the sample-to-sample variability with C-content seen in every other measured coal constitution parameter. In CPD the labile bridge fraction is an adjustable parameter assigned to match the predicted values to measured tar yields for different coals, and also for cases with different heating rates with the same coal. Unfortunately, neither the values specified from rubber elasticity theory for FG-DVC nor those used to tune-in tar yields in CPD have been reported for a broad selection of coals.

5.2.8 Kinetics for char link formation

One normally thinks of char as the macroscopic solid residue from primary devolatilization. But in the context of network depolymerization mechanisms, the crucial step in char formation is the formation of refractory chemical bonds between nuclei. During the initial stages of devolatilization, char links form while labile bridges are breaking, and the fragment MWD shifts toward lighter values. But as the bridges are depleted, char link formation shifts the fragment MWD toward much heavier values. At this point, smaller portions of the fragment distribution sustain tar production, while the bulk of heavier fragments accumulate into a nascent char phase. Once refractory bonds connect nuclei together in the char phase, those nuclei remain in the condensed phase, because the nuclei are also refractory. There is no other process that can break the char link or disintegrate the nucleus to release their components, except for the elimination of heteroatoms, hydrocarbons, and H_2 from nuclei via annealing at elevated temperatures. Even these annealing reactions only convert very small portions of the connected nuclei into noncondensable products, so the bulk of the char links and bound nuclei remain in the char matrix.

Char links are obviously important in partitioning nuclei among oils, tars, and char; which is to say, in partitioning the bulk of the coal mass among these major products. But what is a refractory char link? Since FG-DVC and CPD take ethylene as the archetypal labile bridge structure, the char links are simply ethylene bridges dehydrogenated into an aromatic, refractory form. According to this picture, char link formation hardly perturbs the mass of a labile bridge, via a simple dehydration reaction. In CPD, the same process coarsely describes spontaneous conversion of a labile bridge, and a separate reaction coarsely represents recombination of a fragment with the infinite lattice via a new char link.

FLASHCHAIN® envisions a much more complex chemical process, albeit without resolving the molecular details. Recall that the bridges in FLASHCHAIN® contain the aliphatics and heteroatoms from the structures that interconnect nuclei as well as peripheral groups, and that they decompose via multistep free-radical chain chemistry. In actuality, dozens of elementary chemical reactions are responsible for the decomposition of such large assortments of functional groups. From a more macroscopic perspective, one may envision two distinct outcomes for bridge conversion. Either the conversion process produces stable capping structures on both ends of the nuclei associated with the subject bridge, or it produces some refractory aromatic linkage by eliminating heteroatoms, hydrocarbons, and H_2 from the decomposing bridge functional groups. The first option is bridge scission, which reduces the fragment size without necessarily releasing appreciable amounts of noncondensables. The second is spontaneous char link production, which does not change the fragment size but does generate most of the gaseous products of primary devolatilization. In FLASHCHAIN®, the second route to char links is called spontaneous char link formation, and is diagramed as

$$x_{j-k} - x_k \longrightarrow x_{j-k} = x_k + \nu_C G \quad \text{where } \nu_C = (MW_B - MW_C)/MW_G$$

where the reactant is one of the labile bridges in a fragment with j nuclei, and the products are the newly formed char link at the same location plus ν_C moles of noncondensable products. The stoichiometric coefficient for gas production is evaluated from the molecular weights for bridges and char links which, in turn, are based on the compositions in Table 5.1 for these components. The average molecular weight for noncondensables, MW_G, is evaluated from a correlation of measured distributions of noncondensables from coals across the rank spectrum (Niksa and Kerstein, 1991). This value diminishes from 27 g/mol for lignites to 16 g/mol for anthracites.

The second mechanism in FLASHCHAIN® that forms char links is bimolecular recombination of two mobile fragments. This step increases the fragment size, and also produces noncondensable products, provided that at least one of the participating fragment ends contains a remnant of a previously broken labile bridge. Since gas release is based on the labile bridge composition in this step, the stoichiometric coefficient would be the same as for spontaneous char link production if both ends contained bridge remnants.

The association between char link formation and gas production is corroborated by experiments that showed that refractory crosslinks were incorporated into a coal's macromolecular structure whenever CO_2 and H_2O were released during the initial stages of devolatilization (Suuberg et al., 1985; Ibarra et al., 1990); and indirectly corroborated by enhanced tar yields from lignites whose carboxylic acid salts were converted with ion exchange (Tyler and Shafer, 1980). Later testing extended the connection between char link formation at intermediate stages of devolatilization to CH_4 release (Ibarra et al., 1991).

The developers of FG-DVC incorporated these findings into their analysis in a literal way: One crosslink forms for each molecule of CO_2 and CH_4 produced (Solomon et al., 1993). There are no formation kinetics for char links, per se, because the kinetics for CO_2 and CH_4 production in the FG model also determine the rate of char link formation. Like FLASHCHAIN®, CPD produces gas during the spontaneous conversion of a bridge into a char link, but with a uniform stoichiometric coefficient of two for all coal types (Grant et al., 1989). But char link production via fragment recombination in CPD produces no gas. Uniform rate parameters are used for all coal types for both spontaneous charring (Fletcher et al., 1990) and for the recombination reaction (Fletcher et al., 1992). So none of the parameters associated with char link production in CPD were connected to the production of any gas species, or to any coal-specific variation in the bridge compositions.

In FLASHCHAIN®, bridge decomposition is the common process underlying scission and char link formation, so the respective rates are evaluated with the bridge conversion rate in Eq. (5.7c) and a multiplicative scission selectivity coefficient, ν_B. This coefficient determines the contributions to scission and char link formation per bridge decomposition. Its dependence on coal quality is expressed through the correlation with the O/H ratio of bridges in Fig. 5.18. Values of $(O/H)_B$ increase for coals of progressively lower rank, so the selectivity coefficient monotonically decreases with $(O/H)_B$ because oxygen promotes crosslinking and hydrogen inhibits it by stabilizing free-radical sites on the aromatic nuclei. By virtue of its correlation with $(O/H)_B$, the selectivity coefficient exhibits the substantial sample-to-sample variability seen in the plot versus C-content in Fig. 5.18. The rate parameters for bridge conversion in Eq. (5.8) also display this variability due to their dependence on $(O/C)_B$.

The bimolecular recombination step in FLASHCHAIN® also produces gas, but only if at least one of the participating ends contains the remnants of broken bridges. The rate parameters for this step are the same for all coals, for two reasons: First, recombination is only important in *hv* bituminous coals that pass through a softened, molten stage during devolatilization, so the impact of rank on these rate constants is difficult to discern in measured yields and tar characteristics (Niksa, 1991a) and, second, very little information is available to suggest how the composition of fragment ends affects recombination rates.

FLASHCHAIN® and FG-DVC incorporate strong coal rank dependences into their rates of char link production, consistent with the reported associations among the

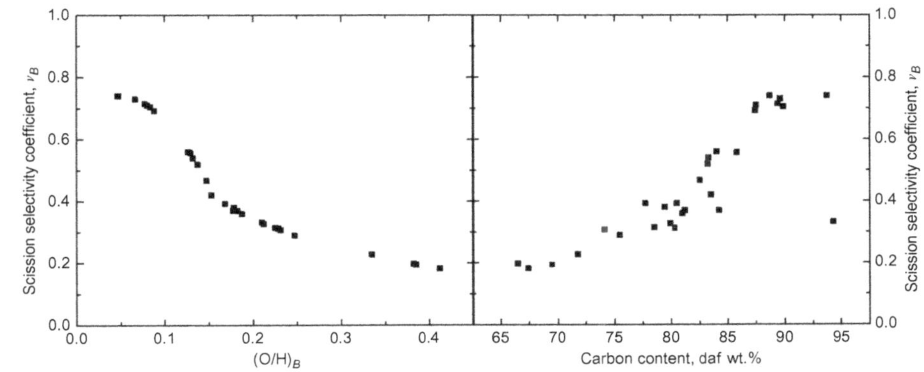

Fig. 5.18 Scission selectivity coefficient for diverse coals versus (left) $(O/H)_B$ and (right) C-content.

Reproduced from Niksa S. Flashchain theory for rapid coal devolatilization kinetics. 4. Predicting ultimate yields from ultimate analyses alone. Energy Fuel 1994;8:659–70 with permission from the American Chemical Society.

release of CO_2 and CH_4 and crosslinking in the macromolecular structures. FLASHCHAIN® uses correlations in $(O/C)_B$ and $(O/H)_B$ which, as seen in Table 5.1, are evaluated from the ultimate analyses and databases of coal aromaticities. As such, these parameters exhibit broad sample-to-sample variability, and are one of the primary means to depict the distinctive devolatilization behavior of individual fuel samples. FG-DVC uses rate constants and yields for the evolution of CO_2 and CH_4 assigned from laboratory test data. FLASHCHAIN® regards char link formation as the primary source of noncondensable products, because labile bridges comprise actual bridges plus nearly all peripheral groups, whereas FG-DVC and CPD generate relatively small amounts of gas this way, because their postulated bridge structure contain few gas precursors.

5.2.9 Mechanisms for volatiles escape

Volatiles are released volumetrically from the condensed phase during devolatilization, and must escape from a fuel particle through the external surface. This transport process is potentially important for two reasons: First, if the resistance to transport is large, the internal pressure will rise above the ambient pressure until the driving force for volatiles escape is sufficient to accommodate the volatiles flowrate. Such elevated internal pressures could conceivably become high enough to suppress tar production, as occurs with elevated ambient pressures. The second reason is that hindered volatiles transport may provide sufficient residence time within particles to sustain secondary volatiles pyrolysis chemistry, both in the vapor phase and on pore walls. Rates of secondary pyrolysis in the gas phase would also be accelerated by elevated internal pressures, through their dependences on volatiles concentrations.

Indeed, the first quantitative interpretation of lower tar yields for higher ambient pressures was based on secondary tar deposition within particles while volatiles escaped their parent particles via bulk diffusion (Anthony et al., 1975). According to this view, a portion of primary volatiles are converted into residues that remain within particles on a time scale set by the transport mechanisms for volatiles escape. Consequently, factors which promote secondary redeposition chemistry, such as the higher vapor concentrations at elevated pressures or the longer transport times in larger particles, are purported to lower yields. Over the years, transport by convective flow, bulk and Knudsen diffusion, diffusion of liquids through a melt, film-diffusion-limited evaporation from a melt, and bubble rupture and growth in a viscous melt have been analyzed in this scheme. Several of these models have correlated weight loss and tar yields versus temperature and pressure, although tar MWDs were never addressed by this approach.

Notwithstanding their impressive analytical sophistication, transport-based analyses are definitively contradicted by the absence of a dependence on particle size in the yields and product distributions for primary devolatilization. This is because it is impossible to formulate an analysis that includes any spatial gradient as the driving force for transport without simultaneously introducing a size dependence into the predicted behavior. This generalization comprises the pressure gradients that drive convective flows, and the concentration gradients that drive bulk and Knudsen diffusion. It also applies to the resistances at the external particle surfaces that govern transport rates in conventional evaporation mechanisms and bubble rupture mechanisms. Considering the profound implications of the size dependence in devolatilization behavior, readers may wish to review the data compilation in Table 4.2 that clearly shows no significant size dependences for our relevant particle sizes.

Of course, as the particle size is progressively increased, the transport resistances grow until they become rate limiting, at which point a size dependence must become apparent in the devolatilization behavior. The critical sizes for this transition have been measured, and these assignments were reproduced in Fig. 4.12. They show that the critical size shifts toward smaller values for progressively faster heating rates, as expected. Most important, they indicate that transport resistances are not rate limiting in any of our subject technology applications, because the largest particles are only used in configurations that impose the slowest heating rates.

Any truly comprehensive theory for primary devolatilization must explain the behavior under transport-limited regimes, but this book does not because it focuses on selected technological applications in which the behavior is kinetically limited. One notable exception is covered in Chapter 9, where the back diffusion of H_2 against an outward flow of volatiles is recognized as an essential element of hydropyrolysis under elevated H_2 partial pressures. In addition, FG-DVC has been incorporated into an analysis of the heat transfer to millimeter-sized particles, without finite-rate mass transport (Zhao et al., 1996).

This brings us to something of a conundrum: On the one hand, tar MWDs exhibit the functional form of γ-distributions (cf. Fig. 4.24) that confirm that tars form via evaporation of the lighter portions of multicomponent hydrocarbon mixtures. On the other, conventional analyses for multicomponent evaporation from liquid droplets impart a strong size dependence (as inverse d_p^2) on evaporation rates. What form of evaporation reconciles the form of tar MWDs with devolatilization's insensitivity to size? Niksa's Flash Distillation Analogy (FDA) is, as yet, the only mechanism for volatiles escape that accurately interprets tar MWDs without introducing a size dependence into the predicted behavior (Niksa, 1988). The phenomenology is sketched in Fig. 5.19. When steam is bubbled through a barrel of crude oil, the lightest fractions pass into the vapor and are transported away with bubbles breaking through the surface of the petroleum. But the material with high molecular weight remains in the liquid phase and condenses into coke if the temperature exceeds a certain threshold value. According to FLASHCHAIN®, coal devolatilization follows this same sequence of steps once depolymerization chemistry has disintegrated a coal's original macromolecular structure into a mixture of fragments with a broad MWD. The role of the steam is played by the noncondensable gases produced whenever labile bridges are converted into refractory char links. Tar is generated when the depolymerization fragments become small enough to vaporize into the escaping noncondensable gases. (Fragments that vaporize at processing temperatures are still heavy enough to condense into viscous liquids at room temperature.) Char forms spontaneously, and also by recombinations of heavier fragments in the condensed phase, whose further depolymerization is suppressed whenever labile connections are converted into refractory char links.

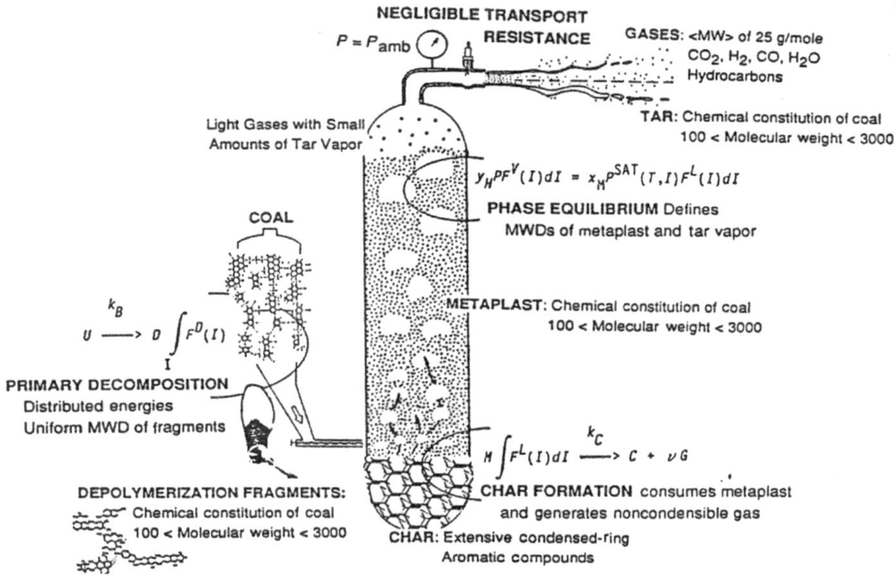

Fig. 5.19 Coal devolatilization as an equilibrium flash distillation.
Reproduced from Niksa S. Rapid coal devolatilization as an equilibrium flash distillation. AIChE J 1988;34:790–802 with permission from Wiley Interscience Publishers.

The key feature in this interpretation is that like-sized fragments in the vapor phase are in equilibrium with those in the condensed phase, so no finite evaporation rate factors into the analysis. According to the FDA, the phase equilibrium shifts to retain a larger portion of the lighter fragments in the condensed phase as the pressure is increased (Niksa, 1988). These fragments would constitute the heavy end of the tar MWD at low pressures, but remain in the condensed phase at elevated pressures. Consequently, tar prepared at higher pressures becomes lighter and the tar yield diminishes. The fragments retained in the char also contain precursors to noncondensable gases which are eventually released, so gas yields increase as the pressure is elevated, but not by enough to compensate for the retention of tar precursors. Neither finite-rate transport mechanisms nor secondary redeposition chemistry are needed to explain the pressure effect. Hence, the FDA invokes a phase equilibrium among the lighter fragments which accurately interprets the pressure effect and the functional form of tar MWDs (as shown below), and negligible transport resistances, as needed for consistency with the lack of a particle size effect.

The crux of this analysis is that internal and ambient pressures are equal because all volatiles escape at their production rates from the process chemistry. The rate of noncondensable gas production, dY_{GAS}/dt, is a summation of the rates of elimination of peripheral groups and broken bridge fragments, and of spontaneous char link formation, and of bimolecular recombination into char links, all weighted by the pertinent stoichiometric coefficients for each source term. See Niksa and Kerstein (1991) for the terms in FLASHCHAIN® that sum to an overall production rate of noncondensable gases to verify that only kinetic parameters factor into this flowrate. The rate of tar escape is also determined by the gas production rate because, in any convective flow process, species transport rates are proportional to the total escape rate and their respective mole fractions. Consequently, the molar release rate of a tar j-mer, $\Gamma_j = dY_{j,\mathrm{TAR}}/dt$, is given by

$$\Gamma_j = \left(\frac{t_j}{\sum t_j}\right)\left[H\left(\frac{1}{G}\frac{dY_{\mathrm{GAS}}}{dt}\right)\right] \tag{5.11a}$$

where H and G are the mole fractions of tar and gas within a particle; t_j is the moles of tar j-mers within the particle void space (which differs from the mole fraction in the cumulative tar sample recovered as product); and the premultiplicative factor in rounded brackets is the instaneous tar MWD, which selects only tar j-mers. The term within square brackets is the total tar flowrate.

The mole fractions of t_j are determined from the facts that (1) $H + G$ sums to unity throughout and (2) the composition of the particle void space is in phase equilibrium with the mixture of polydisperse fragments in the condensed phase. The void space consists of a mixture of noncondensables and tar fragments. The fragment concentrations are related by a generalization of Raoult's law for mixtures with continuous MWDs developed for flash distillation calculations (Cotterman and Prausnitz, 1985; Cotterman et al., 1985), according to

$$\sum_{j=1}^{J^*} p_j = x_{m_j} P^{\mathrm{SAT}}\left(T, MW_{t_j}\right) \quad \text{where} P^{\mathrm{SAT}}\left(T, MW_{t_j}\right) = P_{\mathrm{C}} \exp\left(-\frac{A_{\mathrm{VP}} MW_{t_j}^{z}}{T}\right) \tag{5.11b}$$

where p_j is the partial pressure of a t_j; x_{mj} is the mole fraction of a j-mer in the condensed phase; $P^{\mathrm{SAT}}(T,MW_{tj})$ is the saturated vapor pressure of tar precursors, called metaplast; MW_{tj} is the molecular weight of a tar j-mer; and P_{C}, A_{VP}, and z are constants fit to the yields and MWDs of tar. In the upper limit to the summation of partial pressures, J^* is a maximum degree of polymerization of the heaviest tar fragment that can vaporize, which is set to correspond to a fragment weight of at least 1500. These relations can be combined and rearranged into the following definitions for H, the instantaneous tar MWD, and the tar j-mer production rates:

$$H = \frac{P_{\mathrm{C}}}{P_0}\sum_{j=1}^{J^*}\left(\frac{m_j}{\sum_{j=1}^{\infty} m_j}\right) \exp\left(-\frac{A_{\mathrm{VP}} MW_{t_j}^{z}}{T}\right) \tag{5.11c}$$

$$\frac{t_j}{\sum_{j=1}^{J^*} t_j} = \frac{P_C}{HP_0} \frac{m_j}{\sum_{j=1}^{\infty} m_j} \exp\left(-\frac{A_{VP}MW_{t_j}^z}{T}\right) \tag{5.11d}$$

$$\Gamma_j = \left\{ \frac{\frac{P_C}{P_0} \frac{m_j}{\sum_{j=1}^{\infty} m_j} \exp\left(-\frac{A_{VP}MW_{t_j}^z}{T}\right)}{1 - \frac{P_C}{P_0} \frac{m_j}{\sum_{j=1}^{\infty} m_j} \exp\left(-\frac{A_{VP}MW_{t_j}^z}{T}\right)} \right\} \frac{dY_{GAS}}{dt} \tag{5.11e}$$

where P_0 is the internal pressure, evaluated as the ambient value. Note that the only parameters that factor into this analysis are the ambient pressure, the three parameters in P^{SAT}, and the kinetic parameters in dY_{GAS}/dt. There are no transport coefficients. The appearance of the ambient pressure in the denominators of these expressions poses problems in the interpretation of devolatilization behavior under vacuum. Actually this factor is only the first term of a series expansion in the driving force for volatiles escape (Niksa, 1988). So for vacuum applications, P_0 should be evaluated as the pressure difference across a particle, which is thought to be of the order of 0.01 MPa for sizes smaller than a few hundred microns.

The total tar production rate is given by the summation of all Γ_j; the cumulative tar yield in the free stream is obtained by integrating over all Γ_j; and the number- and mass-based MWDs of the cumulative sample are obtained by further integrations of the tar yield and instantaneous tar MWD (Niksa and Kerstein, 1991).

The predicted shift toward lighter molecular weights in the tar MWDs for progressively higher pressures appears in Fig. 5.20. As the peak sharpens and the tail contracts in the MWD, the predicted tar yield diminishes with an especially acute sensitivity to pressures from vacuum through about 0.5 MPa (Niksa, 1988). The FDA also introduces an independent basis for enhanced tar yields for progressively faster heating rates (Niksa, 1991a). Faster heating delays char link formation, which gives a greater portion of tar precursors, and also enhances tar yields by delaying metaplast production until hotter temperatures are achieved, where more of the intermediate-weight fragments have sufficient volatilities to evaporate as tar. This shift both enhances tar yields and shifts tar MWDs toward heavier values for progressively faster heating rates.

The FDA was incorporated into CPD without modification, although Fletcher et al. (1992) pointed out that the first-cut P^{SAT}-parameters in FLASHCHAIN® did not reproduce the measured vapor pressures of fractionated coal liquids. This is true, but it does not mean that the FLASHCHAIN® predictions are flawed because, as seen in the original derivation (Niksa, 1988), the predicted behavior is governed by nondimensional parameters that combine kinetic parameters with those in the saturated vapor pressure of metaplast. Consequently, it is possible to make the same predictions by adjusting various parameters in ways that keep the nondimensional groups

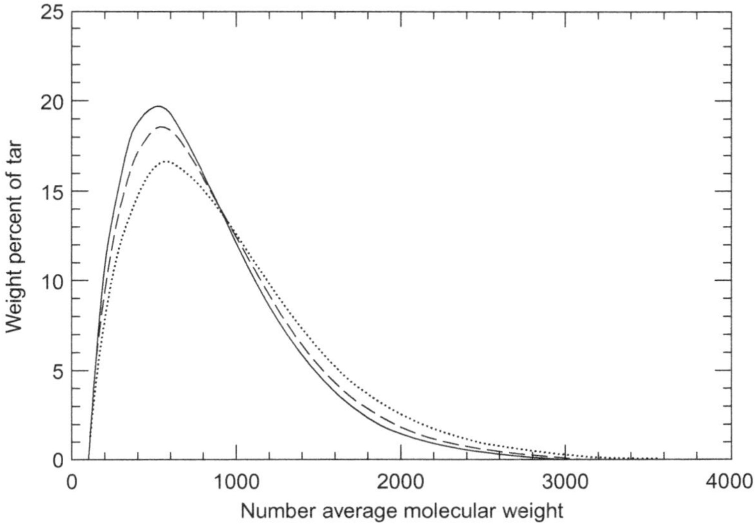

Fig. 5.20 Predicted tar MWDs for 775°C at *(dotted curve)* 0.1 MPa; *(dashed curve)* 1.0; and *(solid curve)* 3.5 MPa.
Reproduced from Niksa S. Rapid coal devolatilization as an equilibrium flash distillation. AIChE J 1988;34:790–802 with permission from Wiley Interscience Publishers.

constant. The developers of FG-DVC use a coarse treatment for the tar escape rate that also invokes equilibrium of like-sized fragments in tar and in the condensed phase (Solomon et al., 1992). One key difference is that their analysis also contains a pressure difference across the particle but, instead of evaluating this difference from a legitimate mass transfer analysis, the developers treat it as a freely adjustable parameter.

5.2.10 Tar shuttling

The FG model ushered in an important idea called tar shuttling, whose basis can be seen in Fig. 5.5 as the allocation of portions of every coal component to tar and nontar forming fractions. The nontar forming fractions are converted into noncondensable gas products, and the tar forming fractions constitute tar precursors. The tar shuttling mechanism evaluates tar compositions from the instantaneous composition of the tar forming fragments, by analogy to the distribution of a large vat of stew in a cafeteria. As soon as the stew is brought to the serving line, a serving will have the composition that left the kitchen. But in time, the lightest components will escape as vapors, and this release slowly changes the composition of succeeding servings. Similarly, the instantaneous tar composition reflects changes in the coal phase that occur prior to each incremental evaporation of tar fragments. In the FG model these changes were associated with the entire coal phase, whereas in the network depolymerization mechanisms, the tar precursors are relatively lighter molecular fragments that remain in the condensed phase, called metaplast. But the basic concept that instantaneous tar compositions are determined by representative sampling of fragments from the condensed phase is the same.

The quantitative implementation of tar shuttling is seen in the definition for the instantaneous molecular weight of a tar j-mer in FLASHCHAIN®:

$$MW_{t_j} = jMW_{\mathrm{A}} + (j-1)\frac{p^{\mathrm{b}}_{m_j}}{p_{m_j}}MW_{\mathrm{B}} + (j-1)\left(1 - \frac{p^{\mathrm{b}}_{m_j}}{p_{m_j}}\right)MW_{\mathrm{C}} + p^{\mathrm{e}}_{\mathrm{M}}(MW_{\mathrm{B}} - MW_{\mathrm{C}}) \tag{5.12}$$

where $\frac{p^{\mathrm{b}}_{m_j}}{p_{m_j}}$ gives the fraction of labile bridges; $p^{\mathrm{e}}_{\mathrm{M}}$ gives the probability that a fragment end contains the remnants of a broken bridge; and subscripts A, B, and C denote aromatic nuclei, labile bridges, and char links, respectively. The probabilities pertain to tar precursors in the condensed phase, and their values shift continuously throughout primary devolatilization as labile bridge conversion either dissociates fragments into smaller pieces or forms refractory char links. Both outcomes diminish $\frac{p^{\mathrm{b}}_{m_j}}{p_{m_j}}$. At the same time, bridge remnants are eliminated as noncondensibles, which diminishes $p^{\mathrm{e}}_{\mathrm{M}}$.

This process is illustrated in Fig. 5.21 with the bridge distributions from CPD for the rapid devolatilization of a *hv* bituminous coal. The component concentrations in the condensed phase are normalized by the initial concentration of aromatic nuclei, which relates them to the probabilities in Eq. (5.12). The labile bridge probability plummets from 13 to 18 ms, while the probabilities for char links, peripheral groups (aka side chains), and the first portion of noncondensables grow on the same time scale. This stage depicts the concerted disintegration of the coal's macromolecular structure that is responsible for tar production and the simultaneous reintegration of fragments into char. In a later stage (described below), a second portion of noncondensable gases is released on a much longer time scale.

According to a literal interpretation of Eq. (5.12), as the labile bridge fraction diminishes and the char link fraction grows, the instantaneous molecular weight of tar shifts toward smaller values, because char links are lighter than bridges. Ultimately, after all labile bridges and peripheral groups have been eliminated, the values relax to the weights of fragments of nuclei connected by only char links.[a] The levels of heteroatoms (O, N, S) and proton and carbon aromaticities for tar can also be evaluated with the same weighting by probabilities for bridges and peripheral groups for fragments in the condensed coal phase. More generally, any characteristic that can be expressed in terms of average values among the components of coal macromolecules—nuclei, bridges, char links, and peripheral groups—can be

[a] The shifts in the bridge fractions almost always occur while the coal is being heated to hotter temperatures, where flash distillation tends to release heavier fragments of tar. So for the specific quantity of molecular weight, independent factors counteract the tendency from the weighting of component molecular weights in Eq. (5.12).

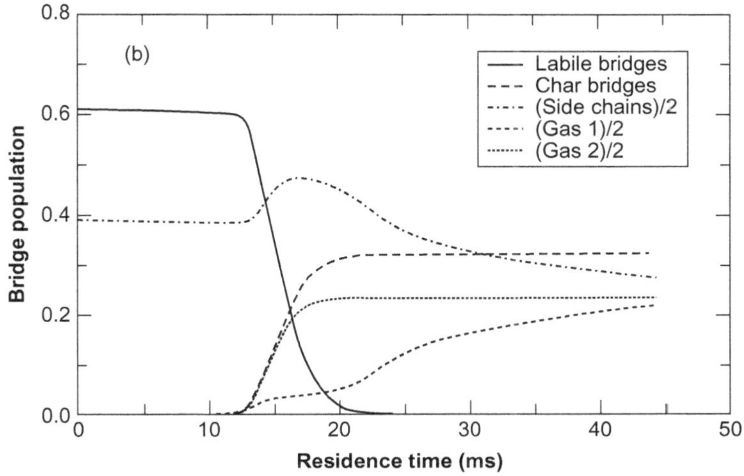

Fig. 5.21 Predicted fractions of labile bridges, char links, peripheral groups, and noncondensable gases from CPD for the rapid devolatilization of a *hv* bituminous coal. Reproduced from Fletcher TH, Kerstein AR, Pugmire RJ, and Grant DM. Chemical percolation model for devolatilization. 3. Direct use of ^{13}NMR data to predict effects of coal type. Energy Fuel 1992;6:414–31 with permission from the American Chemical Society.

evaluated for an instantaneous tar increment with weighting by the associated probabilities for fragments in the condensed phase. Then the instantaneous characteristics are integrated to determine the characteristics of cumulative tar samples.

5.3 Noncondensable gas compositions

The distributions of major noncondensable products determine the compositions of fuel vapors that react at the fastest rates in any coal utilization technology. In any practical application, the primary noncondensables are radically transformed by secondary volatiles pyrolysis before they react with O_2 or gasification agents. Even so, the starting point for any thorough analysis of the process chemistry is the distribution of primary noncondensable products. FG-DVC incorporates the manifold of multiple independent parallel reactions for major noncondensable gases in the FG submodel in Section 5.2.2. Since multiple ultimate yields for individual gases must be assigned from calibration tests with each coal sample, this analysis describes how individual gas yields are shifted for operating conditions away from the diverse calibration conditions, but does not describe distributions of noncondensables for coals without the calibration tests. CPD also incorporates the FG submodel for the main noncondensable products (Jupudi et al., 2009), except for its treatment of N-species, which is described in Section 5.3.2.

In FLASHCHAIN®, most of the aggregate gas product forms during spontaneous charring of bridges, via the elimination of broken bridges from fragment ends, and during bimolecular recombination of two fragments if at least one of the participating ends contains a bridge remnant (Niksa and Kerstein, 1991). The only rate constant in these three steps specifically associated with gas production is the one for the destruction of bridge remnants (k_G). As explained in succeeding sections, additional reactions describe the production of CO, HCN, and H_2 at elevated temperatures on much longer time scales after most bridges have been converted.

Su et al. (2015) expanded this gas production mechanism by subdividing each labile bridge into a labile connection between two nuclei plus one peripheral group. Their mechanism incorporates FLASHCHAIN®'s for the labile links and bridge remnants, plus an additional step for elimination of the new peripheral group. This step is evaluated with a new rate constant that shifts a portion of gas production toward hotter temperatures, after the original labile bridges have been converted. The chemical implication is that a portion of the labile bridge reactant is only partially converted by the radical pool into a functional group that will only decompose at elevated temperatures. Such a less reactive intermediate is plausible, considering the inevitable and concerted conversion of bridge reactants based on aliphatics with substantial levels of heteroatoms into more refractory PAH-type structures. So it is easy to imagine that the functional groups formed whenever partial bridge conversions eliminate heteroatoms are equivalent to the postulated peripheral groups in this analysis.

This expansion improved the quantitative agreement in the evaluations with a massive database, but only in the release of gases once tar formation had ended; more accurate predictions for this stage at elevated pressures were the most notable improvement (Su et al., 2015).

Gas compositions are at least as important as an aggregate gas yield because noncondensables comprise important fuels (H_2, GHCs, CO), gasification agents (H_2O, CO_2), and pollutant precursors (HCN, NH_3, H_2S). To avoid the multitude of adjustable parameters and the calibration requirement from independent parallel reactions, FLASHCHAIN® contains an accounting system that tracks the release of heteroatoms to predict the yields of individual noncondensable gases. The analysis proceeds through the following sequence of steps:

(1) Allocate the organic portion of a subject heteroatom among nuclei, labile bridges and peripheral groups, and/or char links in the parent coal.
(2) Propose reactions to describe the release of individual noncondensable gases via bridge conversion and peripheral group elimination.
(3) Propose separate reactions for gas release after bridge conversion is complete.
(4) Formulate species conservation laws that combine the contributions from all proposed reactions with tar shuttling to predict the yields of individual gases, as well as the heteroatom concentrations in tar and char.
(5) If necessary, separately describe the release of any portions of the heteroatom from inorganic compounds.

This approach directly couples the release of noncondensables to the gross disintegration of the macromolecular structure that produces tar, for consistency with the coincident maxima in Fig. 5.6 for tar and most major noncondensable products. This consistency also means that the same rate constants that describe the macromolecular disintegration also describe major portions of gas production.

5.3.1 Oxygenated gases

Coal-O is associated only with bridges and peripheral groups, because the proportions of oxygen functional groups within the aromatic components are negligible (Niksa, 1996). Similarly, the precursors to CO_2 and H_2O in clays and mineral carbonates were assumed to be negligible compared to the yields from devolatilization, which is true for the vast majority of coals. In FLASHCHAIN® bridges and peripheral groups have the same composition, so coal-O can be allocated to these components as a uniform, average value of moles-O/bridge, Θ_B. The initial value is evaluated from the coal-O level and the probabilities that specify the labile bridge fraction in the parent coal. The following sequence of reactions then converts coal-O into CO, CO_2, and H_2O, plus residual oxygen in char links:

$$\Theta_B O_B \xrightarrow{(1-\nu_B)k_B} a_1 CO + a_2 CO_2 + a_3 H_2O + \Theta_C O_C \tag{5.13a}$$

$$\Theta_B O_B \xrightarrow{\nu_B k_B} 2\Theta_S O_S \tag{5.13b}$$

$$\Theta_S O_S \xrightarrow{k_G} \frac{a_1}{2} CO + \frac{a_2}{2} CO_2 + \frac{a_3}{2} H_2O + \frac{\Theta_C}{2} O_C \tag{5.13c}$$

$$O_C \xrightarrow{k_{CO}} CO \tag{5.13d}$$

where Θ_B and Θ_C are the moles of oxygen in bridges and char links, respectively; a_1, a_2, and a_3 are molar stoichiometric coefficients; ν_B is the scission selectivity coefficient shown in Fig. 5.18; and subscript S denotes remnants of broken bridges. The first step describes the release of CO, CO_2, and H_2O during spontaneous charring of bridges at the overall rate of $(1 - \nu_B)k_B$. The second describes the exchange of bridge oxygen into peripheral groups at the bridge scission rate, $\nu_B k_B$. The third step describes the release of oxygen in bridge remnants at the generic rate for peripheral group elimination, k_G. And the fourth describes the release of the residual oxygen in char as CO, which occurs at high temperatures after most labile bridges have been converted. This step is the only one to introduce a new rate constant, k_{CO}, which is evaluated as a DAEM rate to cover the release of CO during the annealing stage. Tar-O is evaluated from a tar shuttling relation based on Θ_B and Θ_C, and the probability that tar bridges are labile; char-O is evaluated from analogous relations for all fragments in the condensed phase. Rate equations that implement this mechanism have been reported in detail (Niksa, 1996).

This approach introduces no hypothetical ultimate yield parameters for any of the three oxygenated noncondensables. It adds only the three stoichiometric coefficients plus the three rate parameters in k_{CO}. Based on validations with over two dozen diverse coal samples, the model parameters were correlated with the bridge-based O/C ratios of the parent coals, according to

$$a_2 = 0.18 + 5.37\left[\left(\frac{O}{C}\right)_B - 0.30\right]^{1.58} \tag{5.13e}$$

$$\frac{a_3}{a_2} = 1.73 + 19.11\left[0.59 - \left(\frac{O}{C}\right)_B\right]^{1.132} \tag{5.13f}$$

$$\sigma_{CO} = 143.8\left(\frac{O}{C}\right)_B - 11.76 \tag{5.13g}$$

The values of A_{CO} and E_{CO} are fixed for all coals, and a_1 is a fixed fraction of a_2.

Distributions of oxygenated products from this mechanism for a lignite and *hv* bituminous coal appear in Fig. 5.22, where each contribution is expressed as a fraction of coal-O. Primary tars contain almost one-quarter of coal-O for both coal types, but the proportions of the noncondensable gases are much more variable. Carbon dioxide predominates with the lignite, while the proportions of CO and H_2O are comparable. With the *hv* bituminous sample, H_2O predominates, while the CO yield is almost double the CO_2 yield.

5.3.2 Nitrogen products

The partitioning of coal-N among volatiles and char illustrates the case where the heteroatom is predominantly present within aromatic nuclei, since the treatments in FLASHCHAIN® and CPD omit separate steps for quaternary-N, and recognizes that

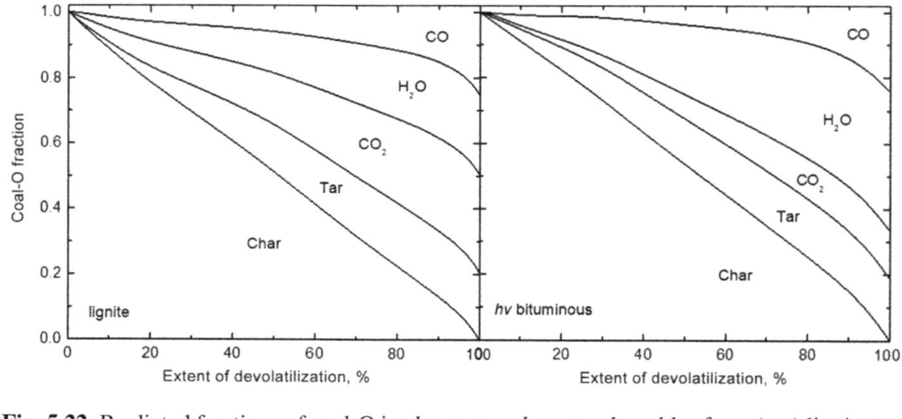

Fig. 5.22 Predicted fractions of coal-O in char, tar, and noncondensables from (top) lignite and (bottom) *hv* bituminous coals.
Reproduced from Niksa S. Flashchain theory for rapid coal devolatilization kinetics. 7. Predicting the release of oxygen species from various coals. Energy Fuel 1996;10:173–87 with permission from the American Chemical Society.

there are no inorganic N-precursors in coal. Since nuclei are refractory throughout, N-partitioning into noncondensables is unaffected by bridge conversion and peripheral group elimination. Tar shuttling determines the contribution for tar-N. The analysis for HCN evaluates an average value of moles of nitrogen per nucleus, η, from an initial value based on uniform dispersion of coal-N among all nuclei in a parent coal (Niksa, 1995). This value is diminished throughout primary devolatilization by a SFOR that produces only HCN. There are no stoichiometric coefficients, although the DAEM-rate for HCN production, k_{HCN}, introduces three rate parameters.

Whereas the mean energy and std. dev. in k_{HCN} are the same for all coal types, at 209 and 67 kJ/mol, the frequency factor was correlated with the O/N ratios of whole coals, as seen in Fig. 5.23. Lignites have O/N ratios from 20 to 25, and anthracites have values less than four. These assignments are remarkable in two respects. First, the activation energy distribution for HCN production is much broader than the ones assigned for bridge conversion in even the lowest rank coals, where bridge compositions are heavily skewed by the abundance of oxygen. Consequently, HCN is expelled over a temperature range that is broader than the ranges for any other product formation channel, in all likelihood, because physical and morphological factors determine its evolution rate. The second remarkable aspect in the HCN rate parameters is the rank dependence of A_{HCN}, which decreases by four orders of magnitude for ranks from lignite to anthracite. This tendency implies that, as the aromatic nuclei in coals of progressively higher rank become more extensive, their nitrogen

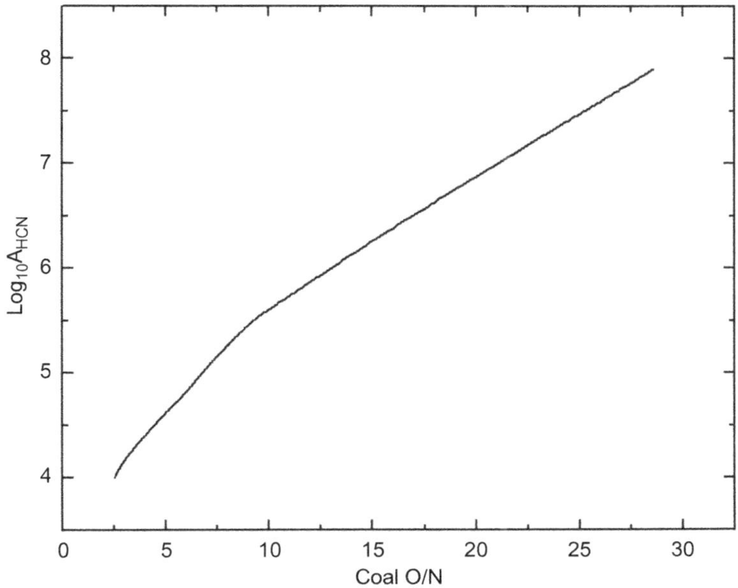

Fig. 5.23 Assigned frequency factor for HCN release versus atomic O/N ratios for whole coals. Reproduced from Niksa S. Flashchain theory for rapid coal devolatilization kinetics. 6. Predicting the evolution of fuel nitrogen from various coals. Energy Fuel 1995;9:467–78 with permission from the American Chemical Society.

becomes much harder to release. The curvature in the correlation near an O/N of 10 is probably an indication of a shift in mechanism at the transition from subbituminous to *hv* bituminous.

CPD's first submodel for coal-N release (Genetti and Fletcher, 1999) is the same as FLASHCHAIN®'s, except that coal-N was subdivided into stable and labile groups, where the stable group is much greater in low volatility coals. An additional fitting parameter to specify the initial proportions of the groups was said to improve the accuracy of the data interpretations.

A second rendition in CPD (Perry et al., 2000) aimed to explain the discrepancies between N-release from model compounds that contain postulated coal-N functional groups and N-release from actual coals. In particular, the analysis focused on the fastest stage of N-release, which occurs at temperatures cooler by a few hundred degrees than those that destabilize the nitrogen in model compounds within condensed aromatic rings. Perry et al. proposed the following three steps for N-release at low temperatures:

$$\mathrm{A-R-R'} \xrightarrow{k_1} \mathrm{A-R\bullet + R'\bullet} \tag{5.14a}$$

$$\mathrm{A-R\bullet + A-N} \xrightarrow{k_2} \mathrm{2A + HCN + Gas} \tag{5.14b}$$

$$\mathrm{A-R\bullet + R''} \xrightarrow{k_3} \mathrm{A-R-R''} \tag{5.14c}$$

where A is an aromatic nucleus without nitrogen; A-N is a nitrogen-bearing nucleus; R, R′, and R″ denote unspecified aliphatics; and • denotes a radical reaction site. The first reaction generates radical sites on nuclei; the second exchanges radicals with a nitrogen-bearing nucleus, which ultimately converts the nitrogen functional group into HCN vapor; and the third reaction eliminates the aromatic radical reaction site via recombination with another aliphatic structure.

Through a series of assumptions and substitutions, the rate of HCN production for low temperatures was evaluated as

$$\begin{aligned}\frac{d\eta_{\mathrm{LoT}}}{dt} &= -k_2[\mathrm{A}-R\bullet]\eta_{\mathrm{LoT}} = -\frac{k_2}{k_3}\left[\frac{r_1}{[\mathrm{R''}]}\right]\eta_{\mathrm{LoT}} \\ &= -k_{\mathrm{HCN}}\left[\frac{M_{\mathrm{SITE}}\dfrac{dM_{\mathrm{CL}}}{dt}}{M_{\mathrm{CL}}^2}\right]\eta_{\mathrm{LoT}}\end{aligned} \tag{5.14d}$$

where η_{LoT} is the N-release from the low temperature reactions; the chemical symbols within small brackets represent molar concentrations in the condensed phase; M_{SITE} is the mass of only the aromatic portion of an average coal monomer unit; and M_{CL} is the entire mass of a monomer. The mass of monomers changes continuously throughout devolatilization as labile bridges and peripheral groups are eliminated. This molecular weight is continuously updated with the probabilities for labile bridges and peripheral groups from the main depolymerization mechanism.

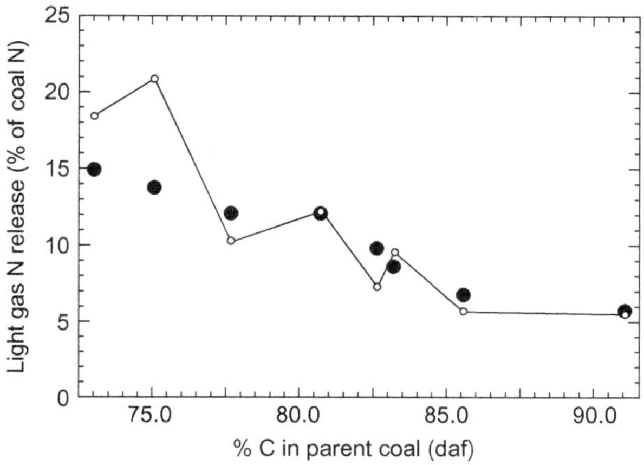

Fig. 5.24 (○) Predicted fractional HCN release from CPD for the APCS versus (●) data at 0.5°C/s to 900°C with an IRP of 3 min.
Reproduced from Perry ST, Fletcher TH, Solum MS, Pugmire RJ. Modeling nitrogen evolution during coal pyrolysis based on a global free-radical mechanism. Energy Fuel 2000;14:1094–1102 with permission from the American Chemical Society.

The main advantage in this approach is that the factor in the large square brackets on the right side of Eq. (5.14d) represents the rank dependence, rather than correlations for the parameters in k_{HCN} with coal composition. When the contribution for low temperatures was combined with a first-order reaction for high temperatures, the same rate parameters in both channels were used to interpret N-release data from a diverse assortment of eight coals as well as the APCS (Perry et al., 2000). Unfortunately, secondary pyrolysis was very extensive in the flow reactor tests with the eight diverse samples, so these measured HCN yields could not be used to evaluate the predicted HCN release. The evaluation of the predicted HCN yields in Fig. 5.24 covers devolatilization of the APCS during slow heating with extended soaking at 900°C. The predictions for coal ranks of *hv* bituminous and higher are within the measurement uncertainties, but it is hard to interpret the large discrepancies for the two coals of lowest rank, because this is the only dataset in the literature that includes the required ^{13}C NMR data for this analysis as well as measured HCN yields. Unregulated tar decomposition and the associated release of secondary HCN is also a concern in these tests, because of the slow heating rate. More validation work is needed to establish whether or not the additional channel for low temperature HCN release without rank dependent rate parameters gives significantly better accuracy than a single DAEM reaction.

5.3.3 Sulfur products

The partitioning of coal-S is the most complex among the five major coal elements because coal-S is dispersed among bridges and peripheral groups (sulfides, thiols, sulfoxides), aromatic nuclei (thiophenes), and inorganics (pyrite and sulfates).

Consequently, mechanisms for S-release have to coalesce all the mechanisms for oxygen and nitrogen, plus an independent submodel for pyrite decomposition. CPD contains no mechanism for S-release, and the one in FG-DVC is a set of independent first-order reactions in parallel that require measurements to specify their ultimate yield parameters (Solomon et al., 1993). FLASHCHAIN®'s S-release mechanism accounts for the different forms of coal-S and their dispersion among the primary coal components (Niksa, 2017). It accurately interprets S-release for any coal sample across broad ranges of temperature, heating rate, and contact time, provided that sufficient information is available to accurately specify a functional group distribution in S_{ORG}, which is rare. But the analysis did not accurately describe the partitioning of S_{ORG} among tar-S and gas-S products for most coals. Additional datasets that resolve all the major S-species along with total and tar yields throughout primary devolatilization for at least a dozen coals are needed to resolve these discrepancies. Moreover, none of the available datasets for rapid heating conditions at elevated pressures could be qualified for model validation work, so the predicted impact of pressure on S-release remains to be validated.

5.3.4 Noncondensable hydrocarbons and H_2

FG-DVC uses numerous first-order reactions in parallel to describe the approach to measured ultimate yields for C_1-C_4 GHCs and H_2 (Solomon et al., 1993). The original version of CPD does not predict GHC or H_2 yields, but an abridged version described below incorporates the FG submodel to estimate the yields of CH_4 and a lumped GHC yield for heavier hydrocarbons (Jupudi et al., 2009). The approach in FLASHCHAIN® differs from the FG submodel, and also from the generic scheme applied to coal-O and coal-N.

In FLASHCHAIN®, the total gas yield during tar production is evaluated from the reactions for bridge conversion and peripheral group elimination, whereas the instantaneous yields of CO, CO_2, H_2O, and HCN are evaluated from the release mechanisms for coal-O and coal-N. This information is first used to assign the total yields of C_1-C_3 GHCs plus H_2 by subtraction. The ultimate H_2 yield released during tar production is based on a correlation in coal-C developed from the measured yields reported by Xu and Tomita (1987) for 17 coals, which are less than 0.5 daf wt.% for most coals. The total yield of GHCs is determined by subtracting the H_2 yield from the residual gas yield.

The GHC yield is then apportioned into CH_4, C_2H_4, C_2H_6, C_3H_6, and C_3H_8 by two correlations developed from the distributions of GHCs reported by Xu and Tomita (1987). The first one correlates the mass fraction of CH_4 in the GHC mixture with a parameter γ_{CH4}, defined as

$$\gamma_{CH_4} = \frac{(1 - N_{C_{2+}})\%C}{100}\left(\frac{N_H^B}{N_C^B - a_2}\right) \tag{5.15a}$$

where $N_{C_{2+}}$ is the fractional molar yield of all GHCs except CH_4; %C is the coal's C-content; N_H^B and N_C^B are the numbers of hydrogen and carbon per bridge; and a_2

is the stoichiometric coefficient for CO_2 production in Eq. (5.13a). The second regression correlates the H/C ratio of the C_{2+} GHCs with γ_{CH4}. This value is then used to specify the average moles of carbon and hydrogen in the C_{2+} GHCs. Finally, the mass fractions for the four individual GHCs heavier than CH_4 are evaluated from balances on hydrogen and carbon with the added constraint on the H/C ratio of the mixture, and the requirement that all mass fractions sum to unity. In coals with greater-than-average H-contents, the GHC distribution shifts from alkenes toward alkanes, and in coals with exceptional H-contents, residual hydrogen may form additional H_2 during tar production.

However, most molecular H_2 forms after tar production is finished in three separate reactions that usually occur at high temperatures. The first reaction produces H_2 throughout the elimination of CO from char links, according to

$$\mathrm{A-O+A-R} \xrightarrow{k_{CO}} \mathrm{A'+CO}+\frac{1}{2}\left(\frac{\mathrm{H}}{\mathrm{C}}\right)_{\mathrm{CHAR}}\mathrm{H_2} \tag{5.15b}$$

The stoichiometry for H_2 production in this step ensures that the H/C ratio of char is unaffected by CO release from char links. The second reaction releases H_2 during the production of HCN:

$$\mathrm{A-N} \xrightarrow{k_{HCN}} \mathrm{A'+HCN}+\frac{1}{2}\mathrm{H_2} \tag{5.15c}$$

The third reaction eliminates H_2 from nuclei during thermal annealing:

$$\mathrm{A-H} \xrightarrow{k_{H_2}} \mathrm{A'}+\frac{1}{2}\mathrm{H_2} \tag{5.15d}$$

The rate constants for these reactions are in the DAEM format. FLASHCHAIN®'s submodel for the production of GHCs and H_2 introduces only a single rate constant, k_{H2}, and ensures that the predicted distribution of all noncondensable gases identically satisfies balances on carbon and hydrogen.

5.4 Abridged reaction mechanisms for practical applications

FLASHCHAIN® developed into a form that distinguished supporting information that could be evaluated from databases from the essential requirements for proximate and ultimate analyses for every particular coal sample. In contrast, CPD and, especially, FG-DVC required monumental laboratory support to specify their input data for particular fuel samples. The imperative in commercial applications to keep input data requirements to an absolute minimum eventually prompted the developers of these models to devise expedient ways to specify all input parameters.

In the abridged version of FG-DVC (Zhao et al., 1994), the input requirements stay the same, but the numerical values for individual coals are specified from interpolations that use only the proximate and ultimate analyses. The interpolation is conducted in the plane of atomic H/C versus atomic O/C of the whole coal (aka van Krevelen coordinates), which are easily evaluated from an ultimate analysis. As seen in Fig. 5.25, a triangular lattice is formed by the coordinates for nine library coals whose complete input specifications were extensively validated with data across broad domains of test conditions. Provided that the coordinates for some new coal fall within one of the triangles in the lattice, the input values are evaluated by interpolating the three sets of input specifications for the library coals that form that particular triangle. Coals whose coordinates fall beyond the lattice are analyzed with the input parameters for the nearest library coal. The same interpolation formulas are applied to every input specification.

As seen in Fig. 5.25, the ultimate tar yields based on interpolated input parameters closely match measured values for diverse coals across a broad pressure range; at only 3.4 daf wt.%, the std. error of estimation is within typical measurement uncertainties for tar yields from many different test facilities. However, the interpretations for broad ranges of heating rate and pressure displayed large discrepancies, particularly for the lowest pressures. Zhao et al. (1994) also noted that there were large discrepancies with some coals for any test conditions. They also acknowledge that this approach, being based on whole-coal compositions, only works for coals whose behavior is near-normal for each particular rank; i.e., "If the deviation is large and cannot be identified with an existing mechanism, the behavior of that coal becomes unpredictable with the current method." Tar yields and total weight loss are the only devolatilization characteristics from this interpolation scheme that have been formally evaluated in open literature.

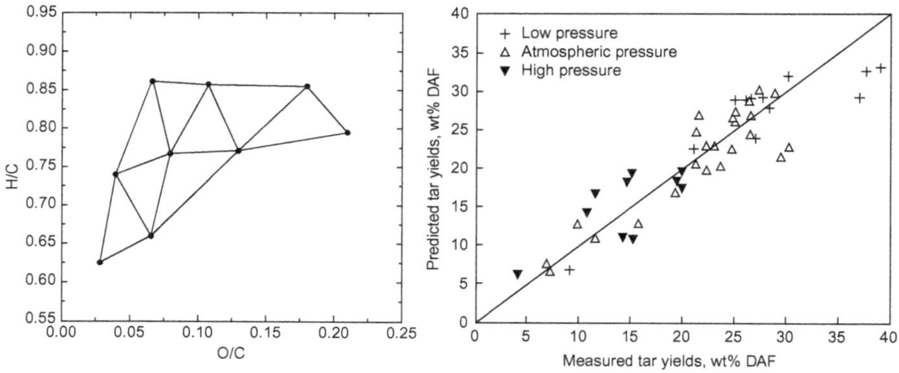

Fig. 5.25 (Left panel) Interpolation domain of (●) library coals to evaluate parameters in FG-DVC from a coal's elemental analysis; and (right) evaluation of predicted tar yields for different pressures from various coals.
Reproduced from Zhao Y, Serio MA, Bassilakis R, Solomon PR. A method of predicting coal devolatilization behavior based on the elemental composition. Proc Combust Inst 1994;25:553–60 with permission from Elsevier.

The abridged version of CPD was developed in two separate stages, one to replace the requirements for ^{13}C NMR analysis for each sample (Genetti et al., 1999) and another to adopt an abridged version of the FG submodel to estimate noncondensable gas compositions (Jupudi et al., 2009). The abridged FG submodel incorporates the manifold of independent, DAEM reactions in parallel for all major noncondensable species from the FG submodel, except that C_{2+} GHCs are treated as an aggregate lump instead of individual molecular species. The H/C and O/C ratios for ten diverse library coals were used to specify a triangular lattice on a van Krevelen diagram, so that rate parameters and ultimate gas yields for any new coal can be specified from the library of rate parameters for these reactions with the interpolation scheme developed by Zhao et al. (1994). This analysis retains the need to specify one or more ultimate yields for each gas species, although the terminology was changed from ultimate yield to "functional group/gas species source fraction" (Jupudi et al., 2009).

In the earlier abridgement, four input parameters previously specified from ^{13}C NMR analysis on individual samples were specified by nonlinear regressions of the measured values from 30 diverse coals (Genetti et al., 1999). The regression variables were the O- and C-contents of the whole coals and the PVM as terms involving multiple polynomial orders. The input parameter labels and the r^2-correlation coefficients for the regressions are collected in Table 5.2.

Except for the molecular weight of peripheral groups, the correlation coefficients are between 0.6 and 0.75, as expected for any correlation of basic structural properties based on the raw data in the proximate and ultimate analyses. It is not surprising that the lowest coefficient was obtained for the coordination number because, as seen in Fig. 5.15, this parameter displays no systematic variation whatsoever with coal rank. The uncertainties on the correlation values for p_0 and M_{CL} are not appreciably better. The much smaller uncertainties on the weight of peripheral groups are striking, considering that the range of the measured values was 4–72.

Another crucial parameter assignment is the labile bridge fraction, $F^b(0)$, which in CPD is evaluated as the difference between $p(0)$ and C_0, where C_0 is the initial fraction of intact char links in the coal macromolecule. In all previous publications on CPD before the abridgement, this parameter was freely adjusted to match the predicted values to measured tar yields for different coals, and also for cases with different heating rates with the same coal. Genetti et al., 1999 presented the following correlation for this specification:

Table 5.2 Correlation coefficients for the regressions of four CPD input parameters (Genetti et al., 1999).

Parameter	Symbol	r^2
Monomer *MW*	M_{CL}	0.72
Peripheral group *MW*	M_δ	0.94
Initial link probability	p_0	0.75
Coordination number	$\sigma_C + 1$	0.62

$$C_0 = \min[0.36, \max\{(0.118\%C - 10.1), 0.0\}] + \min[0.15, \max\{(0.014\%O - 0.175), 0.0\}] \tag{5.16}$$

This expression gives no char links whatsoever in *hv* and *mv* bituminous coals; a greater proportion of char links in coals of lowest rank, which is contrary to established tendencies in coal constitution; and a uniformly high char link fraction for ranks of *lv* bituminous and higher. The term involving the C-content vanishes for %C < 85.6 daf wt.%, then increases linearly to 0.36 through 88.6%, and maintains this value for all higher ranks. The term involving the O-content is fixed at 0.15 for %O > 23.2%, then diminishes through 12.5%, where it vanishes for all lower O-contents. Since there are no coals with more than 12.5% O that also have more than 80% C (cf. Fig. 2.5), the initial char link fraction vanishes for all *hv* and *mv* bituminous coals, which is implausible. In contrast to the monotonically decreasing labile bridge fraction for coals of progressively greater rank in FLASHCHAIN® (cf. Fig. 5.17), C_0-values from Eq. (5.16) and shown in Fig. 5.26 give bridge fractions that increase monotonically for lignites through subbituminous coals, maintain a plateau at unity for *hv* and *mv* bituminous coals, then abruptly plummet to about two-thirds for ranks of *lv* bituminous and higher.

Considering the implausible rank dependence in C_0 and the uncertain regressions for most of the input previously based on ^{13}C NMR analysis, the inability of CPD to depict the sample-specific devolatilization behavior in Fig. 5.26 is not surprising. These predictions were made with the abridged version of CPD, including the correlation in Eq. (5.16). Perhaps most striking, the predicted weight loss is nearly fixed at 50 daf wt.% for all coals with less than 85% C, in marked contrast to the variations in the measured values from 35% to 56% for this range of coal quality. The predicted tar yields are uniform from 65% to 75% C, then abruptly rise to another plateau from 75% to 85% C before they plummet through the low volatility ranks. As seen in the superposition of the char link fraction from Eq. (5.16), the plateau at zero in the char link fraction at 80%–86% C for *hv* and *mv* bituminous coals coincides with about half the range of C-contents for the maximum tar yields in Fig. 5.26. Rather than an authentic depiction of the sample-to-sample variability, the abridged CPD gives nominal average values for large segments of the rank spectrum. Since accurate ultimate yields have been reported for so many coals, such nominal values could be obtained much more easily simply by averaging the measured values for these same rank segments on an ordinary electronic calculator.

5.5 Summary comparisons among three network models

There are clear differences among the ways that coal constitution, the postulated chemical reactions and, most important, the coal quality impacts are represented in the three network depolymerization models. We are now in a position to summarize these differences, hopefully, to gain insight into how they affect the models' performance in practical applications. This is an especially important task, because none of these models account for more than a handful of coals' macromolecular

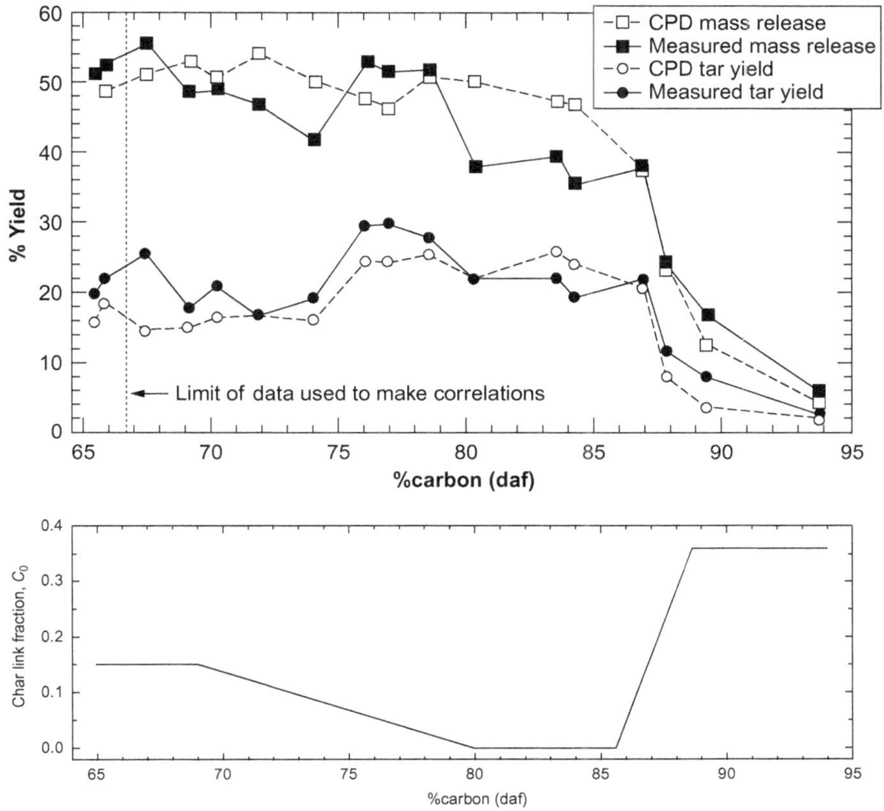

Fig. 5.26 Evaluation of the abridged version of CPD with total weight loss and tar yields from coals across the rank spectrum for rapid heating conditions.
Reproduced from Genetti D, Fletcher TH, Pugmire RJ. Development and application of a correlation of ^{13}C NMR chemical structural analyses of coal based on elemental composition and volatile matter content. Energy Fuel 1999;13:60–68 with permission from the American Chemical Society.

characteristics, and none carry little more than a passing fidelity with legitimate free radical chain reaction mechanisms. Indeed, their starting point is the average constitution of coal, with no reference at all to the significant departures from average values of just about any aspect of coal constitution in any particular coal. Rather than a firm footing in fundamentals, the validity of these models rests upon their performance in quantitative evaluations with measurements and, particularly, in the parameter adjustments required to achieve an acceptable measure of quantitative accuracy.

The different approaches to coal constitution and the chemical reaction mechanisms are outlined in Table 5.3. The label "Corr." denotes a statistical regression of a specified quantity; e. g., "Extract Corr." denotes the correlations of extract yields in FLASHCHAIN® and FG-DVC to evaluate the initial fraction of nuclei in whole coal that is connected into the macromolecular matrix. For CPD, "NMR Corr." denotes the regressions summarized in Table 5.2 in terms of the C- and O-contents and PVM of whole coals for the structural parameters evaluated from ^{13}C NMR analysis. This comparison pertains to the abridged versions of FG-DVC and CPD from Section 5.4 because these are the ones that can specify all input from a proximate and ultimate analysis, like FLASHCHAIN®.

In Table 5.3, the coal constitution parameters comprise the link probability, p_0, as well as the labile bridge fraction, the molecular weights of bridges, char links, peripheral groups, and nuclei, and the coordination number. Both FLASHCHAIN® and FG-DVC relate p_0 to reported extract yields, whereas CPD evaluates it from an NMR correlation. All the constitution parameters in CPD except the labile bridge fraction are evaluated from these correlations. This is unfortunate because, being based on whole coal properties, the correlations are only able to depict gross tendencies without much of the sample-to-sample variability (cf. the r^2 coefficients in Table 5.2). In

Table 5.3 Comparison of parameters for coal constitution and the chemical reactions in FG-DVC, FLASHCHAIN®, and CPD

Parameter	FG-DVC	FLASHCHAIN®	CPD
Coal constitution			
p_0	Extract corr.	Extract corr.	NMR corr.
Bridge fraction	Swelling corr.	%C corr.	(%C, %O) corr./ adjustable
MW_B	MW_{C2H4} for all coals	Apportion C_{AL}, H_{AL}, O, S_{ORG}	NMR corr.
MW_C	MW_B—H_2	Apportion C_{AR}, H_{AR}	NMR corr.
MW_{PG}	FG ultimate yields	n/a	NMR corr.
MW_N	NMR corr.	Apportion C_{AR}, H_{AR}, N	NMR corr.
$\sigma_C + 1$	Near-2 for all coals	2 for all coals	NMR corr.
Chemical reactions			
Bridge scission	k_T	$\nu_B k_B$	k_B,k_δ
Spont. charring	k_{CH4}, k_{CO2}	$(1 - \nu_B)k_B$	k_B,k_C
Recombination	n/a	k_R	k_R
PG elimination	FG submodel	k_G	k_G
CO_2, H_2O	FG submodel	$(1 - \nu_B)k_B$, k_R	FG submodel
CO	FG submodel	$(1 - \nu_B)k_B$, k_R, k_{CO}	FG submodel
CH_4	FG submodel	Corr. in γ_{CH4}	FG submodel
C_{2+} GHCs	FG submodel	Corr. in γ_{CH4}	FG submodel
HCN	FG submodel	k_{HCN}	$(k_{HCN})_{LoT}$, k_{HCN}
H_2S, COS	FG submodel	k_{th}, k_{py}, $\vartheta^{al}_{H_2S}$, E_{al}	n/a
H_2	FG submodel	k_{H2}	n/a
Volatiles escape	$p^{SAT}(MW_{tj}, T)$, Δp	FDA	FDA

FG-DVC, the bridge fraction is ambiguous, because the ways that labile bridges and char links are apportioned into the hybrid Bethe lattices for whole coals have never been discussed. This entry in Table 5.3 assumes that these assignments are based on rough interpretations of swelling data, as in the earlier Monte Carlo simulations. FLASHCHAIN®'s bridge fraction comes from the correlation with C-content of a whole coal in Fig. 5.17. CPD does not stipulate a labile bridge fraction, per se, but accepts the char link fraction as input, then specifies a bridge fraction as $p_0 - C_0$. These char link fractions are either adjusted to tune-in predicted tar yields to measured values; or adjusted to tune-in predicted tar yields for a broad range of heating rate; or specified from the correlation in Eq. (5.16) assigned to fit the tar yields from the dataset of Xu and Tomita (1987).

The molecular weights of nuclei, bridges, peripheral groups, and char links exhibit some of the most marked differences among these three models. In FG-DVC, the bridge weight is based on the composition of ethylene for all coals, so the weight of a char link entails only a minor adjustment for dehydrogenation. The levels of peripheral groups are estimated from measured ultimate yields for all the major non-condensable products under standardized test conditions for a set of library coals, which are then interpolated to evaluate the levels for a new subject coal. Many of the gaseous products have multiple ultimate yield parameters to represent distinct stages in the release patterns. The weight of a nucleus is calculated from databases on f_a, ${}^{H}f_a$, and AC/Cl from ^{13}C NMR.

In contrast, the component weights in FLASHCHAIN® are obtained by aggregating actual bridges and peripheral groups into a pseudo-bridge reactant species, and by allocating individual elements in whole coal to each coal component. These bridge weights are much greater than the actual weights of postulated bridge structures. Bridges contain all the aliphatic C and H, all coal-O, and and aliphatic S plus aromatic sulfides. Char links contain portions of the aliphatic hydrocarbons and residual oxygen, but no sulfur. Aromatic nuclei contain nearly all the aromatic C and H, and all coal-N, and thiophene S. The algebraic relations associated with these assignments (Niksa and Kerstein, 1991) determine the component compositions in Table 5.1. Molecular weights are obtained by summing the contributions from the five major elements. In contrast, CPD does not explicitly apportion any of the heteroatoms to particular structural components; instead, the component weights are estimated from correlations of measured values that, in turn, came from analyses of the higher-order moments of a ^{13}C NMR signal. The bridge weight is evaluated as twice the weight of peripheral groups from the NMR correlations. Consequently, variations in component molecular weights do not necessarily reflect actual variations in the levels of heteroatoms.

In FG-DVC, the coordination number is set to give approximately two ethylene bridges per nucleus for all coals. In FLASHCHAIN®, it is set to exactly two for all coals, because there are no peripheral groups in this analysis. CPD evaluates coordination numbers from an NMR correlation. These measured values must be resolved into the numbers of links per nucleus and of peripheral groups per nucleus, but that resolution has never been reported. Considering that the measured coordination numbers in Fig. 5.15, are between 4 and 6, and the number of peripheral groups are between 1 and 3 (Solum et al., 1989), the numbers of links per nucleus in CPD are probably between 2 and 3.

The treatment of heteroatoms is the distinguishing characteristic among these coal constitution submodels. FLASHCHAIN® explicitly segregates the five major elements in coal into reactive (bridges) and refractory (nuclei, char links) components, and then evaluates the necessary molecular weights from these assignments. In FG-DVC the heteroatoms are allocated to the interpolated ultimate yields for the noncondensable gas species, which essentially decouples them from the network disintegration during devolatilization. CPD ignores the heteroatoms altogether, except that coal-N is allocated to nuclei. Even the component molecular weights are assigned from the NMR correlations. Whereas the straight-chain conformation model is explicitly used in FLASHCHAIN®, FG-DVC's is only slightly more convoluted and CPD's may be more lattice-like, although this feature has not yet been described in detail.

All three models incorporate chemical reaction processes for bridge conversion—both scission and spontaneous charring—and peripheral group elimination. Scission and spontaneous charring are independent in FG-DVC. A single DAEM rate based on ethylene bridge conversion describes scission for all coals, and the rate of char link production is assigned from the release rates for CO_2 and CH_4. There is no bimolecular recombination process in this scheme. The release of peripheral groups is represented with the FG submodel, which contains one or multiple DAEM channels for each gas species.

Conversely, scission and spontaneous charring are closely coupled in both FLASHCHAIN® and CPD. FLASHCHAIN® uses one DAEM rate for bridge conversion, and a bridge selectivity coefficient, ν_B, to weight the contributions for scission and charring. Most important, the rate parameters and selectivity coefficient are explicit functions of the bridge compositions assigned from the coal constitution submodel. In this way, the kinetic parameters in FLASHCHAIN® express the sample-to-sample variability demonstrated in Fig. 5.18, and exhibit the tendency for faster bridge conversion for coals with progressively higher levels of heteroatoms seen in Figs. 5.8 and 5.10. Bimolecular recombination of fragments forms additional char links based on the same rate constant for all coal types, and peripheral group elimination converts bridge remnants on fragment ends into noncondensables with the same rate constant for all coals. The same four reactions are included in CPD, but without any connections to the variable elemental compositions of actual bridges and peripheral groups. Like FG-DVC, CPD's bridge conversion kinetics are based on a single rate of ethylene bridge conversion for all coals. Similarly, the rate constants for bridge scission (k_δ), recombination, and peripheral group elimination are the same for all coals. CPD's constitution submodel explicitly resolves peripheral groups from bridges, so even though the same label is applied to peripheral group elimination in both FLASHCHAIN® and CPD, the dispersion of peripheral groups in CPD is more realistic.

During devolatilization, bridge scissions move the fragment MWDs toward lighter sizes, while the production of new char links reduces the probability for subsequently liberating additional lighter fragments from the lattice. The analytical fragment statistics in FLASHCHAIN® and CPD retain the same lattice model throughout, whereas the hybrid Bethe lattice in FG-DVC can shift from one coordination number to another during the chemical transformations. All three approaches treat bridge scission in the same way. Char link formation in FLASHCHAIN® and CPD accounts for both spontaneous conversion of a labile bridge into a char link, which does not directly affect the fragment MWD, and bimolecular recombination of two fragments, which shifts the MWD towards heavier fragments. FG-DVC omits bimolecular recombination and only allows char links to form spontaneously between adjacent nuclei in the coal lattice. The minor differences among these three approaches are probably not responsible for any appreciable differences in the predicted yields or product compositions, including the tar and extract MWDs.

These three renditions for the fragment statistics would give similar predictions for the fragment MWDs throughout devolatilization in a hypothetical comparison in which the initial coal lattice is stipulated to have the same probabilities for intact links and labile bridges. But the three ways these important parameters are specified for individual coal samples, especially the labile bridge fractions, carry more substantial implications in practical applications.

All noncondensable products in FG-DVC and CPD are generated in the FG submodel by one or multiple DAEM reactions whose ultimate yields are specified as input data. The only exceptions are the two channels in CPD for HCN production at low and high temperatures. FLASHCHAIN® describes gas production in two distinct stages. The first stage is explicitly coupled to the kinetics for the disintegration of the macromolecular structure, based on six stoichiometric coefficients for CO, CO_2, H_2O, five GHC species, and H_2. These coefficients are evaluated as functions of the bridge compositions assigned in the coal constitution submodel. The second stage uses separate DAEM reactions to describe the release of CO and H_2 at high temperatures, after tar formation is exhausted. A separate DAEM reaction also describes HCN release, but the rate parameters are set to cover the extremely broad temperature window associated with this species. Similarly, separate DAEMs describe H_2S release from thiophene-S and pyrite over broad temperature ranges. Parameters in k_{CO} are related to bridge compositions, and those in k_{HCN} are related to the O/N ratio for the whole coal. The same k_{H2} is applied to all coals.

Both FLASHCHAIN® and CPD incorporate the FDA from Section 5.2.9 to describe volatiles release with no transport resistances whatsoever. FG-DVC incorporates a similar expression for the saturated vapor pressure of metaplast, but also contains a freely adjustable parameter that sets a higher intraparticle pressure than the ambient pressure, as actually occurs when appreciable transport resistances mediate the volatiles escape rate.

The coal constitution submodels and the chemical reaction kinetics in concert determine the coal quality impacts predicted by these models. FLASHCHAIN® emphasizes the variability of coals' chemical constitution while keeping the variations in the macromolecular conformation to a minimum. The key insight is to recognize that aromatic components are generated, not destroyed during devolatilization. Consequently, the chemical constitution that matters is the one seen after contributions from refractory aromatics have been removed from the whole coal constitution. The resulting bridge reactants are the key reaction centers and, accordingly, the kinetics for bridge conversion and the stoichiometric coefficients for gas production are based on the assigned bridge compositions, rather than any properties of the whole coal. This approach is the only one that imparts the erratic variability seen in measured product yields when plotted versus C-content to the assigned rate parameters (cf. Fig. 5.18). As such, it is the only one among the three network models that can potentially depict the distinctive devolatilization behavior of individual samples without inordinate laboratory support.

Conversely, only the chemical constitution of a whole coal factors into CPD, and even that connection is marginalized. CPD bases the coal quality impacts on the variability of macromolecular conformation while ignoring almost all variations in chemical constitution. This view regards the weights of bridges, peripheral groups, and nuclei in coal, and the connectedness among nuclei, as the main factors that determine distinctive devolatilization behavior. From a conceptual standpoint, this perspective is motivated by obvious connections among the breadth of macromolecular characteristics assigned to coals of different rank, and important processing features such as solvent extract yields, and plasticity and swelling behavior during thermal processing. But the potential in this approach was undermined by the correlations that circumvent the required laboratory support with ^{13}C NMR. The NMR correlations used to evaluate the structural input parameters in CPD factor in only the C- and O-contents and PVM values of a whole coal, so they cannot possibly express the sample-to-sample variability seen in the raw NMR assignments. This loss of fidelity is clearly evident in the contrast between the variability in measured structural parameters in Fig. 5.15 with the meager correlation coefficients in Table 5.2 for the same structural parameters. Moreover, even the raw coordination numbers do not exhibit any coherent rank dependence (cf. Fig. 5.15), so it is hard to fathom how they could possibly be used to interpret the strong rank dependences in reported devolatilization behavior.

Incorporating the FG model for the release of gas species into CPD does not strengthen the basis for coal quality impacts, because this approach relies upon interpolated values of the ultimate gas yields (aka functional group/gas species source fraction) based on measured gas yields for the library coals. Neither these hypothetical ultimate yields nor any of the DAEM rate parameters were related to any structural or constitution properties, beyond the H/C and O/C ratios of the whole coals.

The abridged version of FG-DVC contains the thinnest basis for coal quality impacts among these three models, because hardly any of the model parameters have been related in any systematic, functional way to continuous variations in coal rank. The canonical set of ultimate gas yields and the associated kinetic rate parameters were evaluated by fitting test data for the library coals, and are now specified for new coals with the triangular interpolation scheme. The library coals were selected to represent portions of the coal rank spectrum and, consequently, the predictions based on parameters from the interpolation scheme can only regenerate some average of the behavior of the coals that form a particular triangle. Do such averages reflect the variability of coal constitution along the rank spectrum? It is difficult to fathom how they could do that when the H/C and O/C ratios of the whole coals are the only means to distinguish a particular sample in this analysis. According to the developers, this approach only works for coals whose behavior is near-normal for each particular rank; i.e., some average of the behavior of three of the library coals.

The DVC submodel also omits continuous variations with coal rank in the parameter specifications because its coal analogs contain no char links and all bridges are evaluated as ethylene links with the same decomposition kinetics. Initial link probabilities, p_0, are adjusted to match a database of extract yields, as in FLASHCHAIN®, and the weights and compositions of nuclei reflect a database of f_a, $^{H}f_a$, and AC/Cl values, as in FLASHCHAIN® and CPD. But the variations in the rate parameters for bridge scission for the library coals are minimal, so shifts in the kinetics for char link production from shifts in k_{CO2} and k_{CH4} are the only appreciable rank dependences in this reaction mechanism.

All three network models can be used to interpret measured yields and product distributions. With CPD the predictions can be tuned-in to data by adjusting C_0 for tar yields, and the gas species source fractions for noncondensables. With FG-DVC, the adjustments for gas species are the same and the internal pressure parameter directly connects to tar yields. With FLASHCHAIN®, there are no adjustable parameters, which is a disadvantage in data-fitting exercises but a marked advantage in supporting process simulations. Of course, the various stoichiometric and selectivity coefficients can always be adjusted from their values assigned from correlations, as needed to tune-in predictions to particular laboratory datasets.

5.6 Legitimate chemical reaction mechanisms

By this point, readers from the scientific community must surely be desperate for connections between devolatilization modeling and realistic chemistry. Of course, chemists in the coal science community have been unraveling detailed chemical reactions for decades, and their accumulated knowledge has already been synthesized into detailed chain reaction mechanisms for primary devolatilization. The first chain reaction mechanism ever devised for devolatilization is still useful as a theoretical framework to relate coal constitution to rates, yields, and product distributions (Gavalas, 1982). More recent expansions of devolatilization chemistry subjected an elaborate model of the constitution and macromolecular structure of a *hv* bituminous coal to a ReaxFF reactive force field to monitor changes in bonding and fragment MWDs during devolatilization at 1725°C (Castro-Marcano et al., 2014). The predicted product distributions were reasonably complete but not particularly accurate. The predicted sequence of functional group decompositions was generally consistent with previously estimated bond dissociation energies, including a prominent role for sulfides during the initiation of primary devolatilization. But the computational burden of several months computing time to simulate the first 250 picoseconds of exposure time seems likely to remain prohibitive for quite some time.

One can make a compelling argument that such detailed treatments of the chemistry will never predict an accurate product distribution from any coal under any test conditions, because so many of coals' molecular features and properties cannot be monitored or estimated within useful quantitative tolerances. But detailed chemistry can certainly identify the dominant reaction channels and, hopefully, the functional groups that supply the most effective promoters into the radical pool. Provided that the most important reaction species can be accurately monitored, the prerequisites are in hand to select coal constitution parameters that better represent the coal quality impacts in devolatilization behavior. Bear in mind that the current basis for the coal quality impacts are atomic ratios for hypothetical bridge reactants in DAEM depolymerization/recombination reactions. So the potential for refinements that improve the quantitative accuracy of predicted primary devolatilization yields is substantial.

5.7 Particle thermal histories

This chapter has already developed the concepts, chemical reactions, and physical phenomena incorporated into the network devolatilization mechanisms, and identified all associated rate parameters and how they are specified. It would be superfluous to reiterate the conservation equations that must be solved to obtain the predictions for specified coal properties and operating conditions with these models. However, in addition to the rate equations associated with the reaction mechanisms, conservation equations for the mass and enthalpy changes of individual particles throughout devolatilization must also be solved simultaneously, to obtain the thermal histories that determine heating rates, ultimate temperatures, and contact times under realistic processing conditions. This section presents these equations, as the final preliminary before the predicted devolatilization behavior from FLASHCHAIN® is evaluated in Chapter 6.

The simplest situation arises in modern WMRs and CPPs. Provided that only a few milligrams of pulverized coal are loaded onto the mesh, the coal's thermal history can be assigned as the temperature program imposed on the mesh, which also means that thermal histories are independent of particle size for these systems. Actively controlled systems impose strictly uniform heating rates from a specified initial temperature to the ultimate reaction temperature. Users also specify the duration of the IRP as the elapsed time after the coal reached the ultimate temperature. These input values appear in Fig. 5.27 for three thermal histories that have progressively slower heating rates. The heating rate is uniform in all cases but at different magnitudes. The initial coal temperature is the same. History No. 1 has the fastest heating rate, Q_1, and also the hottest ultimate reaction temperature, T_1^{Ult}. The heating rate in History No. 2 is faster than that in History No. 3; i.e., $Q_2 > Q_3$. But their ultimate temperatures are the same; i.e., $T_2^{\mathrm{Ult}} = T_3^{\mathrm{Ult}}$.

Whereas heating rates and ultimate temperatures for these thermal histories are straightforward, the definitions of reaction times are potentially confusing, because only the times beyond the end of the heating period factors into IRPs. The total reaction time is the sum of the heating period, t_Q, and the IRP, t_{IRP}. The IRP extends from the end of the heating period, not the beginning of the test. Accordingly, Histories Nos. 1 and 2 have the same heating periods, but there is no IRP for History No. 2 because it ends as soon as the fuel is brought to the ultimate temperature. Fig. 5.27 shows three

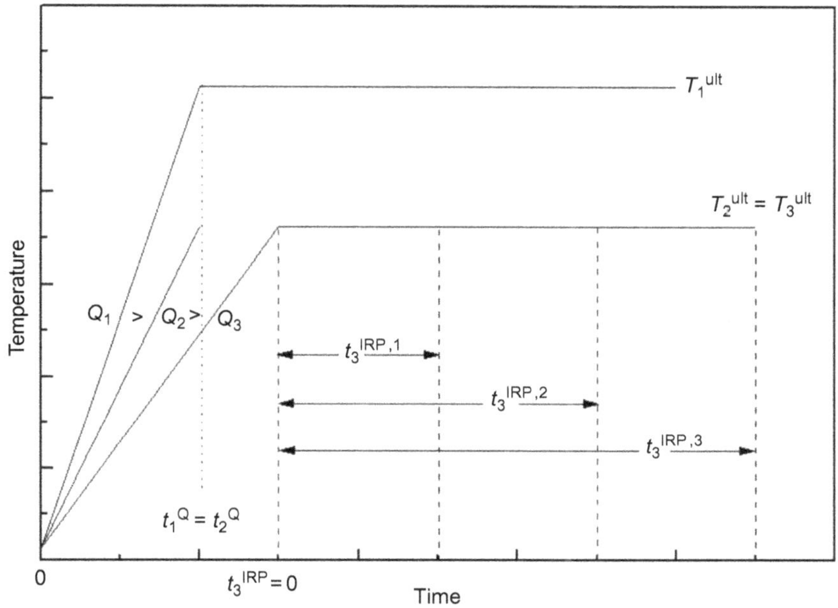

Fig. 5.27 Three typical thermal histories for WMR tests.

IRPs for History No. 3. They all begin at the end of the heating period, which is labeled as $t_3^{IRP} = 0$. Each successive period is extended by the same time increment.

Thermal histories for EFRs, fluidized beds, and other entrained flow reactors are subject to much greater uncertainties than those for WMRs. The main ambiguities pertain to convective mixing at the particle injector that mixes preheated gases with the primary coal entrainment stream. But the estimated thermal histories can be undermined by numerous imperfections in these systems which can only be avoided with an uncommon attention to detail. In most cases, the simulated behavior will only approximate the performance of actual laboratory systems. But the degree of correspondence cannot be generalized, except to observe that dynamic devolatilization behavior is much more sensitive to many minor aspects of design and implementation than are ultimate yields and product distributions. This section describes the assumptions underlying calculated thermal histories for EFRs so that they can be assessed against the features of any particular laboratory system.

In the most common configuration of laminar-flow EFRs, an axisymmetric laminar flow is preheated to the temperature of interest and conditioned into a flow with uniform velocity as it enters a heated reaction zone. At the top of the reaction zone, a pulverized fuel stream is entrained onto the flow axis through a cooled injector, and permitted to react until it is extracted into quench probes. In kinetic rate determinations, residence times are varied by adjusting the distance between the injector and the sampling point. Designs for stream temperatures to 1900°C with conversion histories resolved into tens of milliseconds have been reported.

All calculations to assign particle thermal histories are idealized in several aspects. Most importantly, the thermal history of the coal particles is assigned with an energy balance for individual particles. This approach ignores the impact of two-phase mixing phenomena at the coal injector, which is certainly important for even the most dilute suspensions. In real furnaces, temperatures through the coal jet usually vary both radially and axially. Such nonuniform temperature fields cannot be diagnosed directly. They can only be analyzed with complex heat transfer models subject to very large uncertainties because two-phase fluid mechanics determines the heat transfer rates. It is hard to find an EFR in which the time scale for heating the entrainment gas was comparable to or shorter than the time scale for individual particle heating. Typically, the gas heating time is about an order of magnitude longer than the particle heating time. Consequently, heating rates based on particle heating time scales are significantly faster than what is actually imposed in these furnaces. Compounding matters, even a solid fuel that has been finely pulverized devolatilizes while it is being heated to typical processing temperatures, and the uncertainties in calculated thermal histories have definitely obscured devolatilization rate assignments.

Notwithstanding these important concerns, the thermal histories for EFRs are invariably based on the heating of individual particles while they are entrained in a large excess of entrainment gas with variations in the gas temperature and velocity and the wall temperature along the channel length. The flow regime is not specified explicitly, but usually determines the correlation for heat transfer. The heat transfer rate from gas to particles is evaluated from Nusselt number correlations that depend on the Reynolds number that, in turn, is proportional to slip velocity.

The particle energy conservation equation balances the particle's thermal capacitance against the net heat transfer rate, which includes contributions from enthalpy changes during moisture loss and devolatilization, convection, and radiation, according to

$$m_{\mathrm{p}} C_{\mathrm{p}} \frac{dT_{\mathrm{p}}}{dt} = A_{\mathrm{p}}(q_{\mathrm{m}} \Delta H_{\mathrm{M}} + q_{\mathrm{v}} \Delta H_{\mathrm{V}}) - hA_{\mathrm{p}}(T_{\mathrm{p}} - T_{\mathrm{G}}) - \sigma \varepsilon_{\mathrm{p}} A_{\mathrm{p}}(T_{\mathrm{p}}^4 - T_{\mathrm{W}}^4) \tag{5.17}$$

where

$$q_{\mathrm{v}} = \frac{(\rho_0 d_{\mathrm{p}})}{6}\left(\frac{dY_{\mathrm{TAR}}}{dt} + \frac{dY_{\mathrm{GAS}}}{dt}\right)$$

$$q_{\mathrm{M}} = \frac{h}{C_{\mathrm{p,H_2O}}} \ln(1 + B_{\mathrm{M}}) \quad \text{with } B_{\mathrm{M}} = \frac{C_{\mathrm{p,H_2O}}(T_{\mathrm{B}} - T_{\mathrm{V}})}{L + C^{\mathrm{L}}_{\mathrm{p,H_2O}}(T_{\mathrm{V}} - T_{\mathrm{O}})}$$

$$h = \frac{Nu\lambda_{\mathrm{g}}}{dp} \frac{B_{\mathrm{V}}}{1 - e^{B_{\mathrm{V}}}} \quad \text{with } B_{\mathrm{V}} = d_{\mathrm{p}} \Sigma \frac{C_{\mathrm{p},i} q_i}{\lambda_i}$$

In these relations, the particle characteristics are denoted by m_{p} for mass; A_{p} for external surface area; d_{p} for diameter; ρ_{p} for bulk density; C_{p} for specific heat; and ε_{p} for total hemispherical emittance. The temperatures are T_{p} for the particle; T_{G} for the entrainment stream; T_{W} for the surroundings; and T_{V} for the boiling point of H_2O. Subscript 0 denotes initial values. The reaction enthalpies are denoted by

ΔH_M for moisture loss and by ΔH_V for devolatilization. The volatiles flux is determined by the sum of the release rates for gases and tars from a devolatilization reaction mechanism, dY_{TAR}/dt and dY_{GAS}/dt. The gas production rate, dY_{GAS}/dt, represents the sum of the release rates of all predicted noncondensable primary devolatilization products (H_2, CH_4, C_2H_4, C_2H_6, C_3H_6, C_3H_8, CO, CO_2, H_2O, HCN, and H_2S). The moisture evaporation flux is based on a conventional transport number that depends on the heat capacities of water, both liquid and vapor, and the latent heat of vaporization, L. The convective heat transfer coefficient, h, accounts for the flow regime and particle size of interest through the Nusselt number, Nu. For fluidized systems, Nu-correlations that also account for supplemental heat transfer due to particle-to-particle contact should be used. The convective coefficient is also proportional to the average thermal conductivity of the gas stream, λ_g. The so-called blowing factor involving B_V accounts for the reduction in heat transfer due to the outward flow of volatiles.

According to the energy balance, the heating rate is determined by the temperatures of the reactor and entrainment gas, and the initial temperature and the mean diameter of coal particles.

The particle mass changes continuously throughout devolatilization according to the following mass balance:

$$\frac{dm_p}{dt} = -A_P(q_M + q_V) \tag{5.18}$$

where

$$m_p = \frac{\pi \rho_p d_p^3}{6}$$

$$\frac{d_p}{d_{p,0}} = 1 + (\Omega - 1)\frac{V(t)}{V_\infty} \quad \text{and} \quad \Omega = \frac{d_{p,\infty}}{d_{p,0}}$$

$$\rho_p = \rho_{p,0}\frac{\dfrac{m_p}{m_{p,0}}}{\left[\dfrac{d_p}{d_{p,0}}\right]^3}$$

Once the particle mass is resolved into a product of the density and volume, diameter changes are based on a swelling factor, Ω, and the instantaneous and ultimate volatiles yields.

The particle momentum balance accounts for inertia, gravity, and drag, according to

$$m_p\frac{dv_p}{dt} = -m_p\left(1 - \frac{\rho_g}{\rho_p}\right)g - F_D = -m_p\left(1 - \frac{\rho_g}{\rho_p}\right)g - \frac{1}{2}\rho_g v_{slip}^2 A_p' C_D \tag{5.19}$$

where

$$Re_p = \frac{\rho_g v_{slip} d_p}{\mu}$$

where it is understood that the slip velocity, drag force, and gravitational acceleration, g, are vector quantities. A_p is the projected area of the particle. The drag coefficient, C_D, is a function of Re_p, the Reynolds number for particles based on slip velocity, v_{slip}. All thermophysical properties and transport coefficients were presented in Chapter 2.

The energy balance incorporates fuel- and temperature-dependent thermophysical properties, the enthalpy requirement for moisture loss, the influence of convective blowing on gas-to-particle heat transfer, and radiant transfer to the channel walls in the limit of a very large enclosure. The gray-body, total hemispherical emittance of coal is fixed for all temperatures and coal compositions.

Sample thermal histories assigned from this analysis for EFR simulations of coal devolatilization for different particle sizes appear in Fig. 5.28. These cases have the same temperatures for coal at the injection point, the entrainment gas, and the channel wall, and the same total reaction time. The channel wall is significantly hotter than the gas stream. The particle sizes were doubled in four successive cases. Each thermal history exhibits its maximum heating rate (equal to the slope of the temperature history) at time zero, then the rate falls continuously as the coal temperature eventually approaches an ultimate value between the temperatures of the gas and wall. Both of the smaller particles reach their ultimate values in the specified residence time, but neither of the two larger sizes does. For longer times than shown, the ultimate temperatures for the larger sizes become slightly hotter than for smaller sizes because the convective heat flux from the gas to the coal is inversely proportional to the particle size. Consequently, smaller particles lose heat to the gas faster than larger particles, so their ultimate temperatures remain closer to the gas temperature.

The acute size-dependence in the assigned thermal histories is the basis for yet another intrinsic source of ambiguity in interpreting EFR test data with these

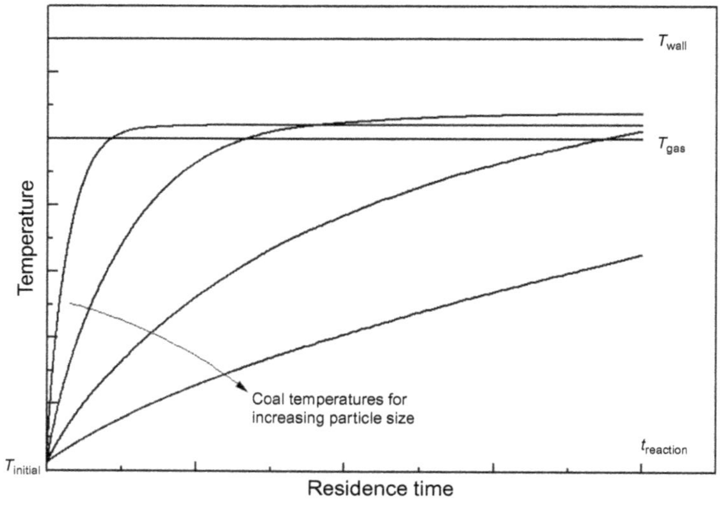

Fig. 5.28 Thermal histories for EFRs tests with four particle sizes.

simulations. No matter what classification method is applied, pulverized coal cannot be prepared into monodisperse grinds. Even the best methods generate size distributions that are broad enough to have significantly different thermal histories associated with the different sizes. Twofold variations in heating rate are common, although ultimate reaction temperatures are nearly uniform. Simulations for a suite of sizes do provide the necessary input for more detailed analyses that assess the range of thermal histories with any PSD of interest.

References

Anthony DB, Howard JB, Hottel HC, Meissner HP. Rapid devolatilization of pulverized coal. Proc Combust Inst 1975;15:1303.

Anthony DB, Howard JB. Coal devolatilization and hydrogasification. AIChE J 1976;22:625.

Castro-Marcano F, Russo MF, van Duin ACT, Matthews JP. Pyrolysis of a large-scale molecular model for Illinois no. 6 coal using ReaxFF reactive force field. J Anal Appl Pyrolysis 2014;109:79–89.

Cotterman RL, Bender R, Prausnitz JM. Phase equilibria for mixtures containing very many components. Development and application of continuous thermodynamics for chemical process design. Ind Eng Chem Process Des Dev 1985;24:194.

Cotterman RL, Prausnitz JM. Flash calculations for continuous or semicontinuous mixtures using an equation of state. Ind Eng Chem Process Des Dev 1985;24:434.

Fletcher TH, Kerstein AR, Pugmire RJ, Grant DM. Chemical percolation model for devolatilization. 2. Temperature and heating rate effects on product yields. Energy Fuel 1990;4:54–60.

Fletcher TH, Hardesty DR. Compilation of Sandia coal devolatilization data. In: Milestone report DE92016824. Albuquerque, NM: Sandia National Laboratories; 1992.

Fletcher TH, Kerstein AR, Pugmire RJ, Grant DM. Chemical percolation model for devolatilization. 3. Direct use of ^{13}NMR data to predict effects of coal type. Energy Fuel 1992;6:414–31.

Gavalas GR, Cheong PH, Jain R. Model of coal pyrolysis. 1. Qualitative development. Ind Eng Chem Fundam 1981a;20:113–22.

Gavalas GR, Jain R, Cheong PH. Model of coal pyrolysis. 2: Quantitative formulation and results. Ind Eng Chem Fundam 1981b;20:122–32.

Gavalas GR. Coal pyrolysis. Coal Sci. Technol. vol. 4. Amsterdam: Elsevier Scientific Publishing Co; 1982.

Genetti D, Fletcher TH. Modeling nitrogen release during devolatilization on the basis of chemical structure of coal. Energy Fuel 1999;13:1082–91.

Genetti D, Fletcher TH, Pugmire RJ. Development and application of a correlation of ^{13}C NMR chemical structural analyses of coal based on elemental composition and volatile matter content. Energy Fuel 1999;13:60–8.

Grant DM, Pugmire RJ, Fletcher TH, Kerstein AR. Chemical model of coal devolatilization using percolation lattice statistics. Energy Fuel 1989;3:175–86.

Ibarra JV, Cerero I, Garcia M, Moliner R. Influence of cross-linking on tar formation during pyrolysis of low-rank coals. Fuel Process Technol 1990;24:19–25.

Ibarra JV, Moliner R, Gavilan P. Functional group dependence of cross-linking reaction during pyrolysis of coal. Fuel 1991;70:409–13.

Juntgen H, Van Heek KH. An update of German non-isothermal coal pyrolysis work. Fuel Process Technol 1979;2:261–93.

Jupudi RS, Zamansky V, Fletcher TH. Prediction of light gas composition in coal devolatilizatin. Energy Fuel 2009;23:3063–7.

Kerstein AR, Niksa S. Polymer scission with irreversible reattachment: A kinetic model of pyrolysis with char formation. Macromolecules 1987;20:1811–8.

Kobayashi H, Howard JB, Sarofim AE. Coal devolatilization at high temperatures. Proc Combust Inst 1977;16:411–25.

Lau C-W, Niksa S. Global rates of devolatilization for various coal types. Combust Flame 1993;94:294–307.

Lewis IC, Singer LS. Electron spin resonance of stable aromatic radical intermediates in pyrolysis. Carbon 1969;7:93–9.

Miknis FP, Turner TF, Ennen LW, Netzel DA. NMR characterization of coal pyrolysis products. Fuel 1988;67:1568–77.

Miura K. A new and simple method to estimate f(E) and $k_0(E)$ in the distributed activation energy model from three sets of experimental data. Energy Fuel 1995;9:302–7.

Niksa S, Heyd LE, Russel WB, Saville DA. On the role of heating rate in rapid coal devolatilization. Proc Combust Inst 1984;20:1445–53.

Niksa S, Kerstein AR. The distributed energy chain model for rapid coal devolatilization kinetics. Part I. Formulation. Combust Flame 1986;66:95–109.

Niksa S. The distributed energy chain model for rapid coal devolatilization kinetics. Part II. Transient weight loss correlations. Combust Flame 1986;66:111–9.

Niksa S, Kerstein AR. On the role of macromolecular configuration in rapid coal devolatilization. Fuel 1987;66:1389–99.

Niksa S. Rapid coal devolatilization as an equilibrium flash distillation. AIChE J 1988;34:790–802.

Niksa S, Kerstein AR. Flashchain theory for rapid coal devolatilization kinetics. 1. Formulation. Energy Fuel 1991;5:647–65.

Niksa S. Flashchain theory for rapid coal devolatilization kinetics. 2. Impact of operating conditions. Energy Fuel 1991a;5:665–73.

Niksa S. Flashchain theory for rapid coal devolatilization kinetics. 3. Modeling the behavior of different coals. Energy Fuel 1991b;5:673–83.

Niksa S. Flashchain theory for rapid coal devolatilization kinetics. 4. Predicting ultimate yields from ultimate analyses alone. Energy Fuel 1994;8:659–70.

Niksa S. Flashchain theory for rapid coal devolatilization kinetics. 6. Predicting the evolution of fuel nitrogen from various coals. Energy Fuel 1995;9:467–78.

Niksa S. Flashchain theory for rapid coal devolatilization kinetics. 7. Predicting the release of oxygen species from various coals. Energy Fuel 1996;10:173–87.

Niksa S. Flashchain theory for rapid coal devolatilization kinetics. 8. Modeling the release of sulfur species from various coals. Energy Fuel 2017;31:4925–38.

Perry ST, Fletcher TH, Solum MS, Pugmire RJ. Modeling nitrogen evolution during coal pyrolysis based on a global free-radical mechanism. Energy Fuel 2000;14:1094–102.

Solomon PR, King H-H. Tar evolution from coal and model polymers: theory and experiments. Fuel 1984;63:1302–11.

Solomon PR, Serio MA, Carangelo RM, Bassilakis R. Analysis of the Argonne premium coal samples by thermogravimetric Fourier transform infrared spectroscopy. Energy Fuel 1990a;4:319–33.

Solomon PR, Hamblen DG, Yu Z-Z, Serio MA. Network models of coal thermal decomposition. Fuel 1990b;69:754–63.

Solomon PR, Serio MA, Suuberg EM. Coal pyrolysis: experiments, rates, and kinetic mechanisms. Prog Energy Combust Sci 1992;18:133–220.

Solomon PR, Hamblen DG, Serio MA, Yu Z-Z, Charpenay S. A characterization method and model for predicting coal conversion behavior. Fuel 1993;72:469–88.

Solum MS, Pugmire RJ, Grant DM. ^{13}C solid-state MNR of Argonne premium coals. Energy Fuel 1989;3:187–93.

Squire KR, Solomon PR, Carangelo RM, DiTaranto MB. Tar evolution from coal and model polymers. 2. The effects of aromatic ring sizes and donatable hydrogens. Fuel 1986;65:833–43.

Su P, Shien J, Ling L. Depolymerization model for coal devolatilization: bridges and side chains as the reaction centers. Energy Fuel 2015;29:2162–76.

Suuberg EM, Peters WA, Howard JB. Product composition and kinetics of lignite pyrolysis. Ind Eng Chem Process Des Develop 1978;17:37.

Suuberg EM, Lee D, Larsen JW. Temperature dependence of crosslinking processes in pyrolysing coals. Fuel 1985;64:1668–71.

Tyler RJ, Shafer HNS. Flash pyrolysis of coal: influence of cations in the devolatilization behavior of brown coals. Fuel 1980;59:487.

Xu W-C, Tomita A. Effect of coal type on the flash pyrolysis of various coals. Fuel 1987;66:627–31.

Zhao Y, Serio MA, Bassilakis R, Solomon PR. A method of predicting coal devolatilization behavior based on the elemental composition. Proc Combust Inst 1994;25:553–60.

Zhao Y, Serio MA, Solomon PR. A general model for devolatilization of large coal particles. Proc Combust Inst 1996;26:3145–51.

Further reading

Solomon PR, Colket MB. Coal devolatilization. Proc Combust Inst 1979;17:131–43.

Quantitative interpretations of primary devolatilization behavior

6

Nomenclature

A	generic pseudo-frequency factor in a SFOR
C	numerical constant in the expression for coal melt viscosity
E_a	activation energy in an Arrhenius rate constant
E_η	exponent in the expression for coal melt viscosity
$<k>$	SFOR-based Arrhenius rate constant for devolatilization
k_E	numerical constant in the expression for coal melt viscosity
M_n	number-average molecular weight of tar, g/mol
n_b	number of bubbles within a molten coal particle
P_b	uniform internal pressure within a molten coal particle
P_0	ambient pressure, MPa
R	universal gas constant, J/K mol
r_b	bubble radius within a molten coal particle
r_p	outer shell radius of a molten coal particle
r_{p0}	outer shell radius of a molten coal particle at ambient pressure
S_W	critical wall strength for rupture of a bubble in molten coal
t	time, s
T	temperature, K
T^*	upper temperature limit in the expression for coal melt viscosity
$V(t)$	instantaneous volatiles yield, daf wt.%
V^∞	ultimate volatiles yield, daf wt.%
X_{DV}	fractional volatiles yield, $V(t)/V^\infty$

Greek symbols

ϕ_C	critical value of ϕ_S for a softened coal phase with infinite viscosity
ϕ_M	weight fraction of metaplast in molten coal
ϕ_S	volume fraction of solids in a molten coal phase
η	viscosity of molten coal
σ	surface tension of molten coal

As explained in Section 4.1.3, the conditions are so extreme in entrained flow coal utilization technologies that measured devolatilization behavior can only be incorporated into process simulations via validated reaction mechanisms, because extrapolations are inevitable. So the best strategy to accurately depict primary devolatilization in process simulations is to (1) compile a large database that describes the impact of coal quality and all important operating conditions on the yield and products of primary devolatilization; (2) formulate a comprehensive reaction mechanism that interprets the database within useful quantitative tolerances; (3) use rudimentary rate

Process Chemistry of Coal Utilization. https://doi.org/10.1016/B978-0-12-818713-5.00006-X

expressions to incorporate the predictions from the devolatilization mechanism into computational fluid dynamics (CFD) simulations and process design applications for a subject commercial technology.

The measurements in Chapter 4 constitute a suitable validation database, and the three network depolymerization mechanisms in Chapter 5 are viable comprehensive reaction mechanisms. This chapter covers the validation of a comprehensive reaction mechanism and the transfer of predictions from a comprehensive mechanism into process simulations. It also emphasizes quantitative interpretations for the impacts of coal quality and all operating conditions. However, the presentation is limited to FLASHCHAIN®, the author's mechanism, simply because it could easily be used to prepare new results to illustrate all the most important tendencies throughout our operating domain. This flexibility was particularly useful in preparing results that show how yields and product distributions shift for variations in particular test conditions, without regard for the availability of data on each particular tendency. Comparable data interpretations based on CPD and FG-DVC may be available in the literature cited in Chapter 5.

6.1 Quantitative model validations with data

Formal publications invariably emphasize the theoretical foundations and comprehensive scope of any of the network depolymerization mechanisms. Yet validation of the model predictions in comparisons with data is the crucial prerequisite for practical applications. Whether or not a model contains every feature ever claimed to be important in devolatilization behavior, it is the quantitative accuracy across the entire operating domain in applications that makes an analysis truly comprehensive. The irony is that as more features are incorporated into an analysis, the mechanism becomes less able to accurately interpret data across a broad operating domain, because every new feature inevitably introduces additional adjustable parameters. Reality is certainly not irrelevant; but simply because some process has been observed to occur during devolatilization does not mean that the process must be included in a comprehensive model. For example, bituminous coals soften into melts and foam during devolatilization. Yet none of the depolymerization mechanisms incorporate the physics of foaming because this and every other transport mechanism do not mediate devolatilization rates under commercial operating conditions. The imperatives are to retain only essential mechanisms in an analysis and to hold the number of model parameters to an absolute minimum. Before CPD incorporated the FG submodel to estimate noncondensable gas compositions, CPD had many fewer parameters than FG-DVC, and FLASHCHAIN® still has many fewer parameters than both other models.

However, this imperative is important only while a developer is finalizing the means to assign every model parameter from the proximate and ultimate analyses. The abridged version of FG-DVC evaluates every parameter by interpolation in a plane of coordinates that is assigned from an ultimate analysis. So, in principle, this version contains no adjustable parameters that are specified by fitting predictions to test data for standardized test conditions. Similarly, the abridged version of CPD replaced the requirement for ^{13}C NMR analysis with correlations based on a proximate and ultimate analysis. FLASHCHAIN® was never abridged because it never required any standardized testing for individual fuel samples. The absence of testing requirements carries an enormous advantage over FG-DVC and CPD in quantitative validations, because the proximate and ultimate analyses have always been the only sample-specific information required to predict a coal's devolatilization behavior. Consequently, the entire backlog of published test data could be brought into FLASHCHAIN® validations while the model parameters were being related to particular coal properties, because even the oldest test reports almost always include these standard coal properties. It would be difficult, if not impossible, to discern the functional relationships to assign model parameters from bridge-based atomic ratios with data on only a handful of different coals. But, as illustrated in Fig. 5.18 for the bridge scission selectivity coefficient, these functional relationships become immediately apparent when the behavior of dozens of diverse coals can be factored into the analysis.

In contrast, the original form of FG-DVC required monumental lab support, as elaborated in Chapter 5, but none of the older test reports included these data. The model could only be tuned-in to data for specific coal samples at the pace of the supporting analytical work. Similarly, the original form of CPD required ^{13}C NMR analysis on every sample, yet such characterization data was omitted from essentially all tests on devolatilization outside of the developer's laboratory. Since both abridged versions resort to statistical regressions in terms of standard, whole-coal properties, they cannot possibly depict the distinctive devolatilization behavior of individual coal samples unless at least one, and perhaps more, parameters are tuned-in to test data. Indeed, there is no good reason to expect accurate sample-specific predictions from any sort of statistical regression with the values in a proximate or ultimate analysis.

6.1.1 Blind model evaluations with data

A blind evaluation is one in which a testing team provides to the model developer the proximate and ultimate analyses for all samples and a listing of the test conditions but not the measured yields of any of the devolatilization products. The distinguishing features are that the model developer gets no information on the measured yields before making the predictions, and the test team gets no information on the predictions before running the tests. In technical engineering research such as this, the opportunities for genuine drama are few and far between, yet blind evaluations of model predictions with test data are bound to become dramatic when the developer meets the test team in public so that all parties can compare the predicted and measured values, without even the possibility of unregulated parameter adjustments to tune away any discrepancies.

FLASHCHAIN® has been subjected to blind evaluations since the mid-1990s. The evaluations in Fig. 6.1 and Table 6.1 are especially comprehensive because the same sample suite was subjected to tests in two EFRs and a WMR. In Fig. 6 1, the measured ultimate yields from an EFR vary from 47 to 67 daf wt.%, whereas the predicted yields

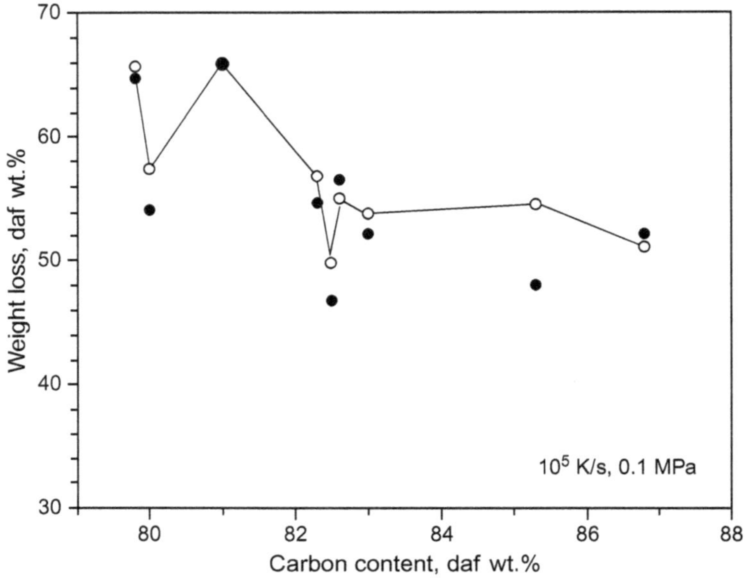

Fig. 6.1 Blind evaluation of (○) predicted ultimate yields from FLASHCHAIN® with (●) unpublished EFR data recorded by Dr. Alan Thompson of Rolls Royce Combustion Ltd.

Table 6.1 Blind evaluation of FLASHCHAIN® with unpublished ultimate yields from a WMR operated by Dr. L. Muzio of Fossil Energy Research Corp. and from an EFR operated by Dr. W. Gibb of Powergen, Ltd

	WMR		EFR	
Label	**Predicted**	**Measured**	**Predicted**	**Measured**
TyBl	44.7	41.1	57.6	61.6
Atc	40.6	37.4	50.8	51.6
Dryt	46.9	45.8	59.2	62.6
LCer	45.9	45.8	58.1	62.1
Ashl	45.8	42.6	60.5	68.9
ShCr	35.9	31.6	48.2	48.4
NRiv	47.0	44.7	58.5	64.7

extend from 50% to 67%. The predicted yields are within a measurement uncertainty of 4 wt.% for 8 of the 9 samples. Even among the four samples whose C-contents vary by only 1 wt.% from 82% to 83% C, the predictions depict the markedly different yields from individual samples within measurement uncertainties. The tests in Table 6.1 covered seven of the same samples in a WMR with a heating rate of about 1000°C/s and also in a second EFR. The predicted yields for the WMR tests are within measurement uncertainties for all samples except, perhaps, the ShCr coal. The measured range of values extends from 31.6 to 45.8 daf wt.%, versus a predicted range from 35.9% to 47%. The two coals that gave the lowest yields are correctly identified, as are the three that gave the highest yields. In the evaluation with data from the second EFR, the predicted yields are within measurement uncertainties for 5 of 7 coals. The discrepancy for Ashl coal is hard to rationalize because, in this dataset, it gave the greatest yield of all, whereas in both other datasets, its yield is in the middle of the pack. That for NRiv coal is probably another reflection of the slight underprediction for all these coals, albeit only in the evaluation with this particular EFR dataset.

A second blind evaluation used a dataset for three low-rank coals in an EFR operated at 1 MPa, and was published in the open literature (Matsuoka et al., 2003). As seen in Table 6.2, the evaluation of ultimate yields also included blind predictions from FG-DVC and CPD at three furnace temperatures. The predicted ultimate yields from FLASHCHAIN® are nearly exact at both the hottest temperatures, but high by 10 daf wt.% for 600°C. Both other models give accurate yields at the hottest reactor temperature, but underpredict the yields at both cooler temperatures. Whereas the discrepancies with FG-DVC are only 5 daf wt.%, those with CPD reach 16% for the coolest temperature.

This blind evaluation of the FLASHCHAIN® predictions is expanded to the complete product distributions from three coals at 800°C in Table 6.3. The presence of soot in the measured products is a clear indication of unregulated secondary volatiles pyrolysis in these tests. It is minimal for the first subbituminous, but becomes substantial for both coals of lower rank, especially the lignite. The predicted products represent only primary devolatilization, and the consequences of omitting secondary pyrolysis were not characterized. The predicted yields of char, tar, and gas are nearly exact for all three coals, except that the measured soot yields represent reductions in the actual yields of primary tar. The predicted GHC distribution shows the correct tendencies for lower yields of progressively heavier species, and from coals of progressively lower rank. Hydrogen yields are accurate for the first subbituminous, but more-than-double the measured yields for both lower rank coals, which could reflect the omission of secondary pyrolysis in the simulations. The CO, CO_2, and

Table 6.2 Blind evaluation of FLASHCHAIN®, FG-DVC, and CPD with ultimate yields in daf wt.% from an EFR operated at 1 MPa with a subbituminous coal

	Predicted			
T **(°C)**	**FLASHCHAIN®**	**FG-DVC**	**CPD**	**Measured**
600	52	37	26	42
700	53	46	45	51
800	54	51	52	55

Reproduced from Matsuoka A, Ma Z-X, Akiho H, Zhang Z-G, Tomita A, Fletcher TH, Wojtowicz MA, Niksa S. High-pressure coal pyrolysis in a drop tube furnace. Energy Fuels 2003;17:984–90 with permission from the American Chemical Society.

Table 6.3 Blind evaluation of FLASHCHAIN® predictions with product distributions from an EFR operated at 800°C and 1 MPa

	Subbit		Subbit		Lignite	
	Predicted	**Measured**	**Predicted**	**Measured**	**Predicted**	**Measured**
Char	47.9	45.3	51.1	48.4	48.4	47.4
Soot	0	1.4	0	4.4	0	7.3
Tar	16.4	17.2	10.2	12.1	9.7	9.6
Gas	33.9	32.8	36.6	36.2	39.5	38.9
CH_4	6.6	5.7	2.6	3.7	2.8	3.7
C_2's	4.4	5.8	1.2	2.9	1.2	2.9
C_3's	2.5	0.7	0.7	0.5	0.7	0.5
H_2	1.4	1.2	1.8	0.7	1.6	0.9
CO	6.6	7.8	12.3	9.9	14.9	10.0
CO_2	4.8	5.2	9.7	8.0	10.3	8.0
H_2O	7.6	6.4	8.3	10.5	8.0	12.0

Reproduced from Matsuoka A, Ma Z-X, Akiho H, Zhang Z-G, Tomita A, Fletcher TH, Wojtowicz MA, Niksa S. High-pressure coal pyrolysis in a drop tube furnace. Energy Fuels 2003;17:984–90 with permission from the American Chemical Society.

H_2O yields are reasonably accurate for all three coals, except for the overpredicted CO yield and underpredicted H_2O yield from the lignite.

The blind evaluations in this section fairly represent the performance of FLASHCHAIN® in many other comparable evaluations. Ultimate weight loss and the gross partitioning of coal into gas, tar, and char tend to be within the measurement uncertainties for the vast majority of samples. In other words, FLASHCHAIN® accurately predicts the distinctive ultimate yields from individual samples based only the proximate and ultimate analyses without any parameter adjustments whatsoever. The allocation of coal-O among gas, tar, and char is usually accurate, but the predicted proportions of CO, CO_2, and H_2O are subject to greater discrepancies even while they exhibit the proper tendencies for samples across the rank spectrum. Similarly, total GHC yields are accurate, simply because they are evaluated by difference from the total gas yield and the aggregate yield of oxygenated gases. But the GHC distribution is frequently subject to larger discrepancies. Hydrogen yields from primary devolatilization are very low for all coal types, so unregulated secondary volatiles pyrolysis in the tests tends to be the main source of discrepancies in evaluations of predicted values.

Blind evaluations such as these are important milestones in demonstrating the predictive capabilities needed in practical applications, because they provide the most stringent evaluation of model predictions along with the firmest assurance against adjustments to any of the model parameters. By definition, all forms of tuning are excluded. Even so, model developers may gain the greatest benefits of all, because blind evaluations provide the clear, unbiased assessments of nascent predictive capabilities they need to make the predictions even more accurate.

6.1.2 In-house evaluations of coal quality impacts

The two formal evaluations in this section were staged by the author after the correlations for all model parameters were brought to final form. The evaluation with measured ultimate total and tar yields for primary devolatilization in Fig. 6.2 covers the entire rank spectrum. The FLASHCHAIN® predictions appear as the open symbols connected by line segments, where the lines appear in the figure only to guide one's eye over the point-by-point comparisons to convey the main tendencies with coal rank. This CPP heated the samples at 3000°C/s to 767°C with a 4 s IRP at atmospheric pressure. Characterization tests established that the tests imposed sufficient thermal severity to attain ultimate yields. This is the same dataset used to tune-in the abridged form of CPD in Section 5.4 (cf. Fig. 5.26).

Among the 17 coals tested, the predicted ultimate weight loss is within the measurement uncertainties in 14 cases, the exceptions being the samples with 75, 80, and 87% C. The predictions exhibit the tendency for fairly uniform ultimate yields for lignites and subbituminous coals, with a gradual reduction through the *hv* and

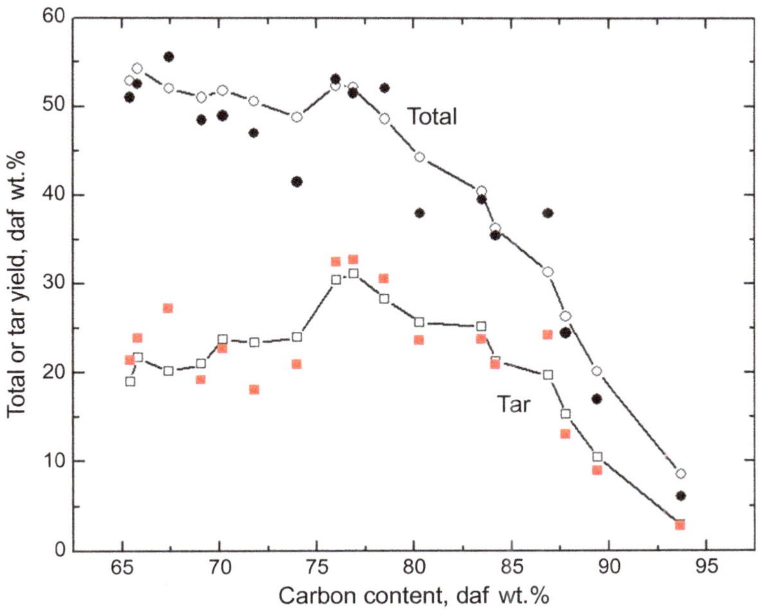

Fig. 6.2 Evaluation of predicted (○) ultimate weight loss and (□) tar yields for rapid primary devolatilization at atmospheric pressure with measurements from Xu and Tomita (1987). Reproduced from Niksa S. Predicting the evolution of fuel nitrogen from various coals. Proc Combust Inst 1994a;25:537–44. Niksa S. Flashchain theory for rapid coal devolatilization kinetics. 4. Predicting ultimate yields from ultimate analyses alone. Energy Fuels 1994b;8:659–70. Niksa S. Flashchain theory for rapid coal devolatilization kinetics. 5. Interpreting rates of devolatilization for various coal types and operating conditions. Energy Fuels 1994c;8:671–79 with permission from the American Chemical Society.

mv bituminous ranks, before the yields plummet for low volatility coals. Of course, the sample-to-sample variability is substantial throughout, and especially so for *hv* bituminous coals. The measured range of yields through 85% C is 35–56 daf wt.%, versus a predicted range from 36 to 54%. The predicted tar yields are within measurement uncertainties for all but the coals with 65, 72, and 87% C. The predictions exhibit the tendency for a shallow maximum for subbituminous coals, with comparable tar yields from low-rank, *hv* bituminous, and *mv* bituminous coals (seen in this particular dataset but not others), before the tar yields plummet for low volatility coals. The measured range of tar yields through 85% C is 19–33 daf wt.%, versus a predicted range from 19–32%. The FLASHCHAIN® predictions almost always indicate the correct change in total and tar yields for each increment along the axis for C-content, although the magnitudes of these changes are frequently underpredicted.

Even though Fig. 6.2 contains only a modest number of samples, it clearly shows that nominal average yields cannot accurately depict the measured yields for even a single rank; i.e., the sample-to-sample variability is appreciable in every instance where multiple samples of the same rank were subjected to the same test conditions. This variability is magnified in the evaluation with the suite of *hv* and *mv* bituminous coals in Fig. 6.3. This dataset combined total and tar yields from numerous tests that

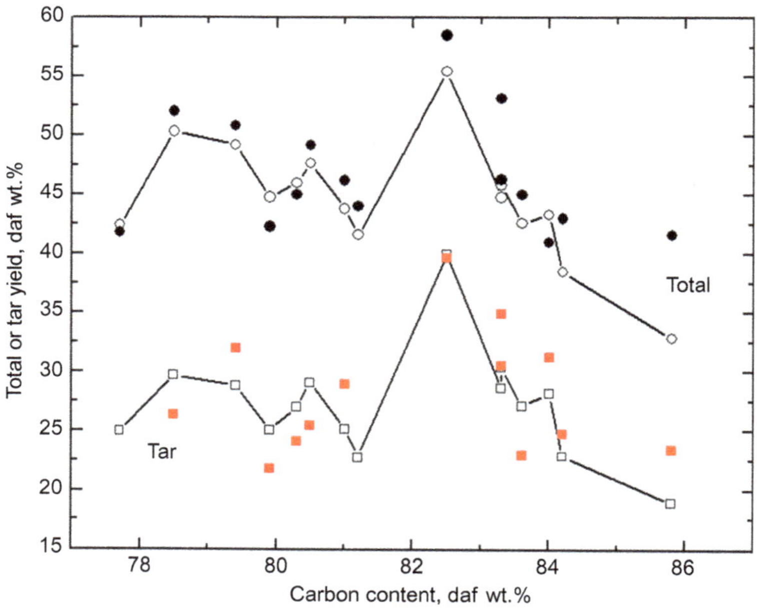

Fig. 6.3 Evaluation of predicted (○) ultimate yields and (□) tar yields for rapid primary devolatilization of *hv* and *mv* bituminous coals at near-atmospheric pressure (see Niksa, 1994a for citations for these tests).
Reproduced from Niksa S. Predicting the evolution of fuel nitrogen from various coals. Proc Combust Inst 1994a;25:537–44 with permission from the American Chemical Society.

imposed rapid heating to temperatures hot enough to achieve ultimate yields at near-atmospheric pressure. Despite the directly comparable operating conditions and the narrow range of coal rank, measured total yields range from 40 to 60 daf wt.%, and tar yields double from 20 to 40 daf wt.%. The FLASHCHAIN® predictions capture the ranges of variation in weight loss and tar yields as well as the sample-to-sample variability within experimental uncertainty of ±4 daf wt.% in all but one of the cases at 83.2%, one of the cases at 84% C and another at 86% C.

Models that can only assign nominal yields for different ranks are superfluous, because the database of available devolatilization tests is extensive enough to specify nominal yields for each rank directly by arithmetic averaging of measured values. Of course, the larger drawback is that nominal yields for each rank fail to represent the very significant impact of the sample-to-sample variability in coals' devolatilization behavior, which is clearly large enough to affect the performance and emissions from any coal utilization technology.

6.2 Interpretations for primary devolatilization behavior

There is little point to reiterating the catalogue of validation studies for FLASHCHAIN®, because formal evaluations with data for about 300 coals from all geographical regions worldwide are already in open literature. Citations to this work are collected in Table 6.4 for studies made after the assignments of model

Table 6.4 Citations for validation work on FLASHCHAIN®

Niksa S, Lau C-W. Global rates of devolatilization of various coal types. Combust Flame 1993;94:293–307
Niksa S. Predicting the evolution of fuel nitrogen from various coals. Proc Combust Inst 1994a;25:537–44
Niksa S. Flashchain theory for rapid coal devolatilization kinetics. 4. Predicting ultimate yields from ultimate analyses alone. Energy Fuels 1994b;8:659–70
Niksa S. Flashchain theory for rapid coal devolatilization kinetics. 5. Interpreting rates of devolatilization for various coal types and operating conditions. Energy Fuels 1994c;8:671–79
Niksa S. Flashchain theory for rapid coal devolatilization kinetics. 6. Predicting the evolution of fuel nitrogen from various coals. Energy Fuels 1995;9:467–78
Niksa S. Flashchain theory for rapid coal devolatilization kinetics. 7. Predicting the release of oxygen species from various coals. Energy Fuels 1996;10:173–87
Niksa S. Predicting the devolatilization behavior of any coal from its ultimate analysis. Combust. Flame 1995a;100:384–94
Niksa S, Liu G-S, Hurt RH. Coal conversion submodels for design applications at elevated pressures. Part I. Devolatilization and char oxidation. Prog Energy Combust Sci 2003;29 (5):425–77
Niksa S, Fujiwara N. Predicting the combustion kinetics of Chinese coals. In: Seventh int symp on coal combust, Harbin, PR China; July 17–20, 2011
Niksa S. Flashchain theory for rapid coal devolatilization kinetics. 8. Modeling the release of sulfur species from various coals. Energy Fuels 2017;31:4925–38

parameters were at or very close to final form. These papers can be downloaded from researchgate.net under the name, "Stephen Niksa."

Instead, this section is devoted to interpretations from FLASHCHAIN® for all the major aspects of primary devolatilization covered in Chapter 4, in the same topical order. The case studies shown without data points were prepared with a commercial computer package that contains FLASHCHAIN® called PC Coal Lab®. Most results were prepared with v.5.0 PC Coal Lab®. Cases with data were prepared with earlier versions of FLASHCHAIN®, although most of these results are indistinguishable from those from v.5.0. Cases without data are based on the properties of the APCS, usually with the West Virginia *hv*A bituminous sample representing the *hv* bituminous rank. When additional coals were needed to illustrate the variability of the coal quality impacts, the simulations were based on measured behavior for actual coal samples and the plots contain measured values, whenever possible.

6.2.1 Thermal history effects

Fig. 6.4 shows the transient weight loss for two heating rates, with a pair of isothermal reaction periods for the faster heating rates, like the actual tests in Fig. 4.6 but with coal properties that give appreciably lower ultimate yields. The three transients exhibit the correct sigmoidal functional form. The onset of devolatilization is delayed until much hotter temperatures with the faster heating rate, and the devolatilization rate

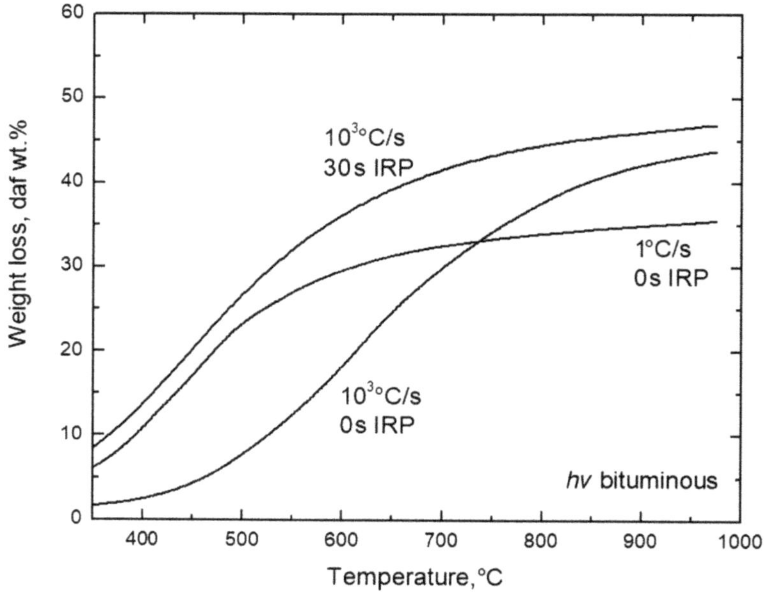

Fig. 6.4 Weight loss from an *hv* bituminous coal at *(dashed curve)* 1°C/s with 0s IRP, and at *(solid curves)* 10^3°C/s with 0 and 30 s IRP at atmospheric pressure.

increases in proportion to changes in the heating rate, as it should. Although the ultimate yield at 1°C/s is reached around 600°C, an ultimate yield greater by several percent is achieved around 900°C at the faster heating rate. Adding 30 s IRPs to the tests for 1000°C/s shifts the transient weight loss toward much cooler temperatures, but only slightly perturbs the ultimate yield, in accord with actual test results.

These thermal history effects are rooted in the chemical heterogeneity of the bonding structures in coal. Yet despite the multitude of actual chemical bonds, only a single distribution of activation energies for the bridge conversion processes is needed to depict the influences of temperature, heating rate, and reaction time. Applying the same energy distribution to bridge scission and spontaneous char link production minimizes adjustable parameters with no apparent drawbacks.

The major shifts in the reaction products and intermediates in the condensed phase are illustrated in Fig. 6.5 for two heating rates without any IRP. The figure shows the three size groups of fragments in the condensed phase. Metaplast weights extend from 125 to a value between 1400 and 2000 g/mol, depending on coal composition. Intermediate fragment weights can be double those values, and reactant weights extend to infinity in theory, but rarely exceed 10,000 g/mol once devolatilization begins. Initially, this coal contains about 10% metaplast, 20% intermediate fragments, and 70% in the reactant size class. During heatup at both rates, the reactant fragments rapidly disintegrate into intermediate fragments and metaplast, starting well before either tar or gas is released. The delayed release of products reflects the impact of fragment statistics. At the slower heating rate, scissions begin to convert reactant into intermediate fragments and metaplast around 200°C. Through heatup to 400°C, metaplast reaches its maximum level and then diminishes in tandem with tar release. For both heating rates, essentially all reactant fragments are converted out of the reactant size class, so the disintegration of the bulk of the largest coal fragments does not explain the substantially enhanced tar yield for the faster heating rate.

Gas and tar are released on very similar time scales, although tar production ends before gas production. Ultimately, slower heating enhances gas yields by about 20%,

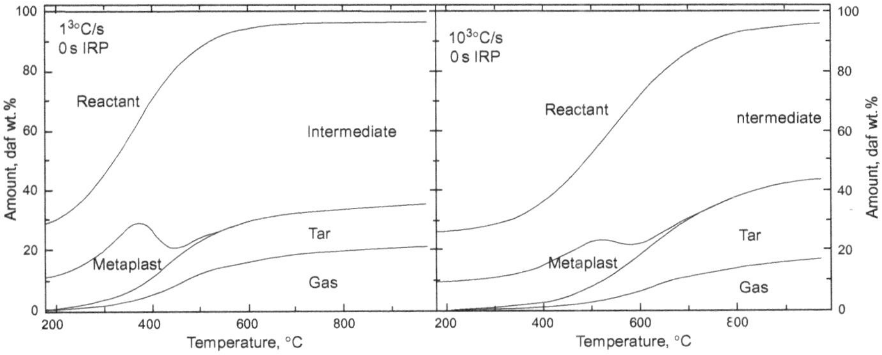

Fig. 6.5 Cumulative amounts of three fragment classes, tar, and gas during heatup of an *hv* bituminous coal at (left) 1 and (right) 10^3°C/s with 0 s IRP at atmospheric pressure.

but cuts tar yields in two. As a result, ultimate weight loss falls by almost 10 daf wt.%. Coupling gas formation to two charring reactions—recombination and spontaneous conversion—captures the selectivity to tar formation at low temperatures and short reaction times, and to gas formation at high temperatures, which is a characteristic feature of *hv* bituminous coals. Bimolecular recombination forms most char links. It completely eliminates metaplast, so that intermediate fragments are the major component in the ultimate char phase.

The key to understanding how faster heating rates enhance tar yields is in the competition for metaplast between recombination into larger intermediate fragments versus flash distillation out of the condensed phase as tar fragments. It is not determined by bridge scission, since the maximum metaplast levels are 50% greater for the slow heating rate. This greater level of tar precursors does not actually enhance tar yields because it arises at temperatures that are too cool to rapidly vaporize metaplast into tar. Instead, the bulk of metaplast recombines into larger intermediate fragments that remain in the condensed phase during slow heating. At the faster heating rate, even though the maximum metaplast level is lower, it is reached at a much hotter temperature, where the saturated vapor pressure of metaplast is much greater. Consequently, a much smaller portion of the metaplast inventory remains in the condensed phase as intermediate fragments, and tar yields are enhanced. The surprising feature is that a greater portion of the original nuclei in coal become metaplast at some point in devolatilization at slower heating rates, yet the greater maximum metaplast levels do not give greater tar yields, as demonstrated in earlier simulation studies (Niksa, 1991). It is interesting that an extended quasisteady state arises during rapid heating, whereby the metaplast inventory nearly vanishes after only about half the ultimate tar yield is reached. Once the inventory has been depleted, bimolecular recombination occurs at a negligible rate, so metaplast fragments are vaporized into tar essentially as fast as they are generated by scissions of larger nonvolatile fragments.

The ultimate total and tar yields in Fig. 6.6 cover a four order-of-magnitude range of heating rate at 0.1 and 7 MPa. The predicted ultimate weight loss exhibits the 3 daf wt.% enhancement per order-of-magnitude increase in heating rate reported for *hv* bituminous coals at atmospheric pressure (cf. Fig. 4.7). The enhancements to tar yields are slightly greater, as they should be, because gas yields diminish slightly for progressively faster heating rates. The predicted weight loss and tar yields also exhibit the much smaller enhancements at elevated pressure which, in this case, are about half as large as those at atmospheric pressure. This increment is greater than reported enhancements (cf. Table 4.1).

6.2.2 Pressure effects

FLASHCHAIN® omits both tar deposition from the gas phase and mass transport limitations which had formerly been proposed as the mechanistic basis for the impact of ambient pressure variations. Instead, it relies on the shifting phase equilibrium to provide a more compelling explanation: As the pressure is increased, the competition for metaplast between recombination into larger intermediate fragments versus vaporization into tar shifts to retard the vaporization channel, which reduces tar yields.

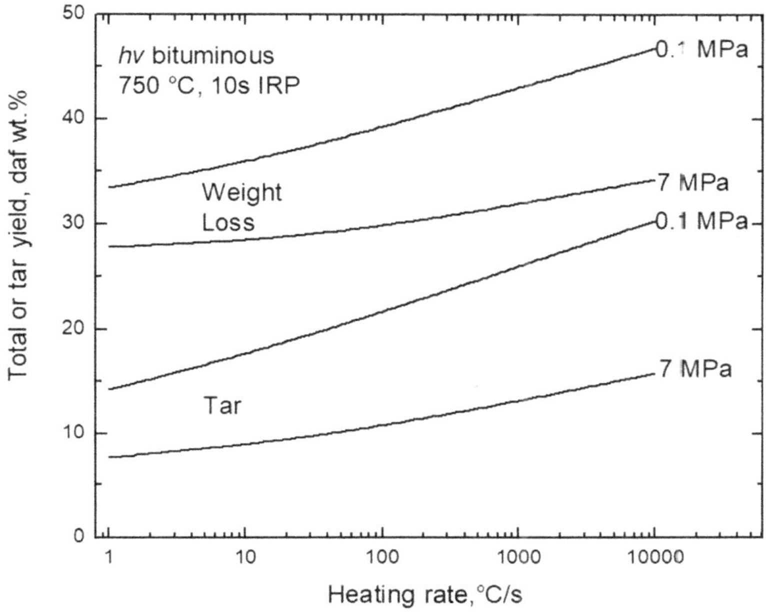

Fig. 6.6 Weight loss and tar yields from an *hv* bituminous coal for different heating rates to 750° C with a 10 s IRP at (solid curves) 0.1 MPa and (dashed curves) 7 MPa.
Reproduced from Niksa S, Liu G-S, Hurt RH. Coal conversion submodels for design applications at elevated pressures. Part I. Devolatilization and char oxidation. Prog Energy Combust Sci 2003;29:425–77 with permission from Elsevier.

This mechanism strongly influences tar yields, especially at pressures to several atmospheres.

The FDA accurately interprets ultimate total and tar yields across an enormous pressure range. The evaluation in Fig. 6.7 covers nearly the entire range of bituminous coal properties for rapid heating conditions under pressures from 0.1 to 7 MPa. Volatile yields fall very rapidly as pressures are increased to 1 MPa, then diminish with much weaker sensitivity as pressures are increased further. Tar yields are more sensitive to pressure than weight loss, because enhanced gas yields mediate the impact on total volatiles yields. The predicted weight loss is within measurement uncertainties over the full pressure range for *hv* and lv bituminous samples, and within 5 wt.% for a second *hv* bituminous for pressures to 1 MPa. The predicted tar yields are within experimental uncertainty for all three coals, provided that pressures are higher than 0.5 MPa. But the predicted tar yields are too low by at least 5 wt.% for 0.1 MPa.

The fragment distributions in Fig. 6.8 illustrate the shifts in the fragment size classes within the condensed coal phase at elevated pressures. According to the FDA, the phase equilibrium retains a larger portion of metaplast at elevated pressures, which is clearly seen in the greater maximum metaplast level at 500°C at 7 versus 0.1 MPa in Fig. 6.8. Accordingly, recombination converts larger portions of metaplast into

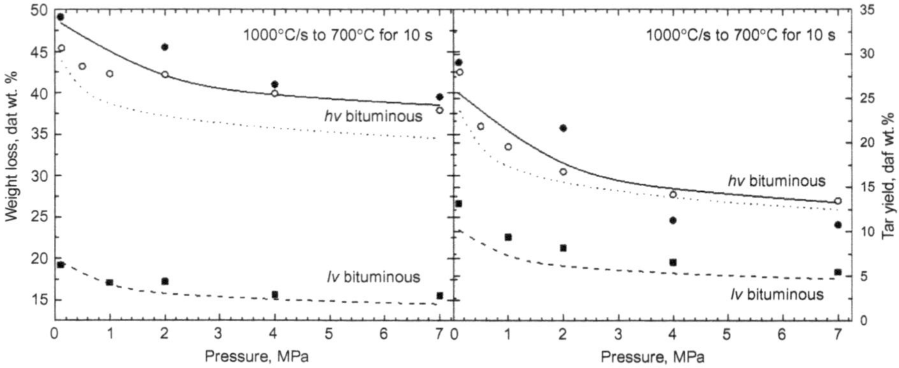

Fig. 6.7 Evaluation of predicted (left) weight loss and (right) tar yields from three diverse bituminous coals for heating at 10^3°C/s to 700°C with 10 s IRPs.
Reproduced from Niksa S, Liu G-S, Hurt RH. Coal conversion submodels for design applications at elevated pressures. Part I. Devolatilization and char oxidation. Prog Energy Combust Sci 2003;29:425–77 with permission from Elsevier.

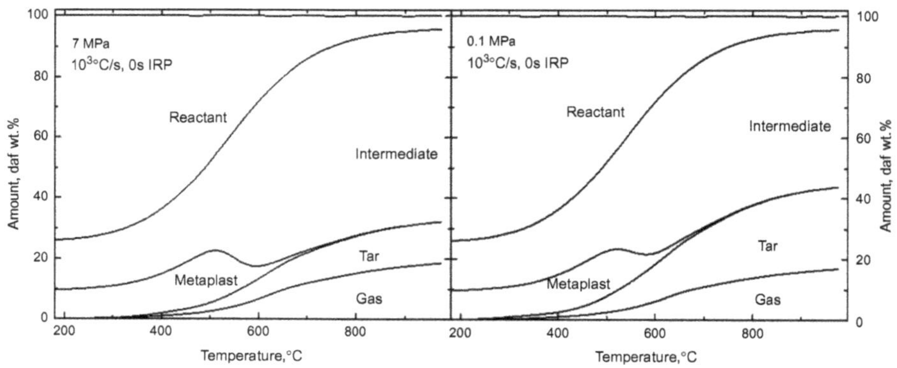

Fig. 6.8 Cumulative amounts of three fragment classes, tar, and gas from *hv* bituminous coal during heatup at 10^3°C/s at (left) 7 and (right) 0.1 MPa.

intermediate fragments, and thereby diminishes tar yields for elevated pressures. Metaplast fragments continue to disintegrate further via bridge scission so that, at even the highest pressures, some fragments become small enough to vaporize as tar. Consequently, tar MWDs shift toward lighter weights for progressively higher pressures (as seen below in Section 6.2.5).

In contrast to transport/redeposition schemes, FLASHCHAIN® also predicts devolatilization rates which are nominally independent of pressure, as seen in Fig. 6.9, which replicates the actual test conditions in Fig. 4.10. The predictions are consistent with this particular dataset and other data which show a common approach during the transient stage to different asymptotic yields for pressures from vacuum to tens of atmospheres. Basing the product release rates on the chemical reaction kinetics,

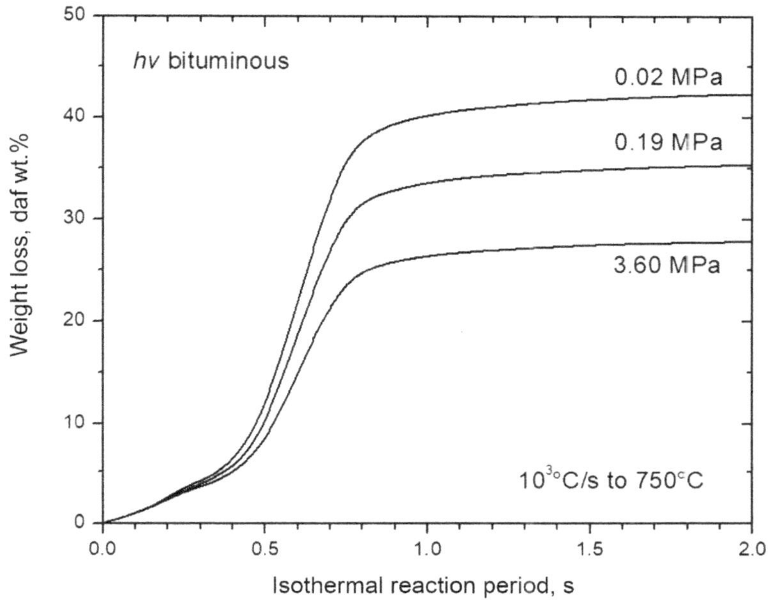

Fig. 6.9 Predicted weight loss resolved throughout the IRPs after heatup at 1000°C/s to 750°C with an *hv* bituminous coal under *(solid curve)* vacuum, *(dashed)* 0.19, and *(dotted)* 3.6 MPa. Reproduced from Niksa S, Liu G-S, Hurt RH. Coal conversion submodels for design applications at elevated pressures. Part I. Devolatilization and char oxidation. Prog Energy Combust Sci 2003;29:425–77 with permission from Elsevier (see Fig. 4.10 for the corresponding data).

not transport, ensures nominally equal devolatilization rates for all pressures, in accord with the available measurements.

The same interpretation rationalizes the smaller yield enhancements for progressively faster heating rates at elevated pressures in Fig. 6.6. Even though rapid heating increases the temperature for flash distillation of metaplast into tar vapor, at elevated pressure more tar precursors remain in metaplast, which eventually recombine into larger fragments that remain in the condensed phase.

Finally, the predictions for the entire rank spectrum at atmospheric and elevated pressures appear in Fig. 6.10. The measured yields appeared previously in Fig. 4.11, and are now shown with the FLASHCHAIN® predictions that appear as plus signs connected by line segments. The test results shown as open symbols give consistently lower yields than reported by others for directly comparable test conditions (Niksa et al., 2003). Accordingly, nearly all the most serious discrepancies are with these data. Notwithstanding, there are no systematic deviations in any portion of the rank spectrum and, most important, the sample-to-sample variability is as evident in the predictions for 1 MPa as they are at 0.1 MPa, as they should be. The one-to-one correspondence in the distinctive devolatilization behavior of individual samples at 0.1 and 1 MPa is even clearer in Fig. 6.11, which shows the comparable evaluation of ultimate weights loss for only *hv* bituminous coals.

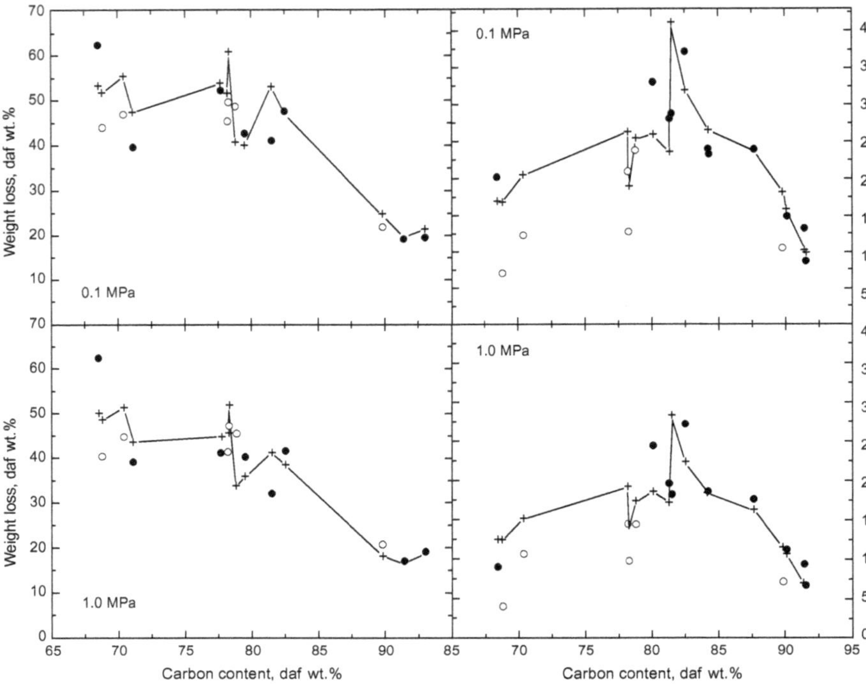

Fig. 6.10 Evaluation of predictions with (left) ultimate weight loss and (right) tar yields from diverse coals under rapid heating conditions at (top) 0.1 and (bottom) 1 MPa. Predictions appear as plus signs connected by line segments. Data as open symbols are subject to greater measurement uncertainties.
Reproduced from Niksa S, Liu G-S, Hurt RH. Coal conversion submodels for design applications at elevated pressures. Part I. Devolatilization and char oxidation. Prog Energy Combust Sci 2003;29:425–77 with permission from Elsevier.

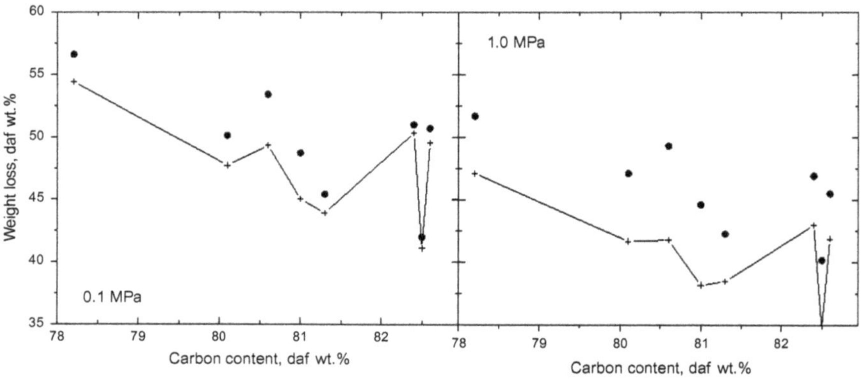

Fig. 6.11 Evaluation of predicted ultimate weight loss from *hv* bituminous coals under rapid heating conditions at (left) 0.1 and (right) 1 MPa. Predictions appear as plus signs connected by line segments.
Reproduced from Niksa S, Liu G-S, Hurt RH. Coal conversion submodels for design applications at elevated pressures. Part I. Devolatilization and char oxidation. Prog Energy Combust Sci 2003;29:425–77 with permission of Elsevier.

6.2.3 Particle size effects

Since FLASHCHAIN® omits all transport mechanisms, the predictions are completely insensitive to variations in particle size, as appropriate for the grind size distributions used in all our subject coal utilization technologies. Only the scaling for negligible transport resistances in FLASHCHAIN® is consistent with the reported lack of a particle size dependence at or above atmospheric pressure (cf. Table 4.2). Both pressure-independent devolatilization rates and the absence of particle size effects present problems for transport-based models that have not yet been reconciled.

6.2.4 Elemental compositions of char

Primary devolatilization preferentially expels hydrogen and oxygen, so these element fractions in char are much lower than the fractional char mass, whereas the element fractions of carbon and nitrogen are somewhat greater than the fractional char mass. Both tendencies are clearly apparent in Fig. 6.12, which shows the predicted element fractions in char for the 8 APCS after thermal processing to attain ultimate devolatilization yields. The gross tendency in the coal quality impacts on element retention inversely mimics the rank dependences for ultimate weight loss and tar yields, in that the element retentions are fairly uniform through the *hv* bituminous rank, then increase sharply for the lv bituminous in the APCS. The sample-to-sample variability is appreciable in the simulation results, although there are too few coal samples in this series to see if it covers the range of measured values (cf. Fig. 4.13). The predicted variations among the retentions of C, H, and N are the same for each individual coal sample, because a common macromolecular disruption is responsible for the release of these elements. The variability among the O-fractions is noticeably lower than for the other elements among all coals, simply because most chars contain little oxygen once their

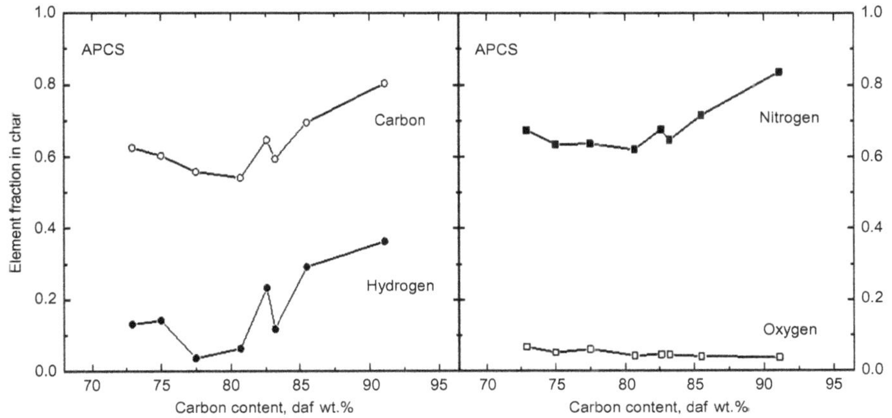

Fig. 6.12 Predicted element fractions in char from the APCS for 10^3°C/s to 975°C with no IRP at atmospheric pressure.

ultimate yields have been released. However, the reported retention of half of coal-O by an anthracite (in Fig. 4.13) could be an indication that stable oxygenated bonding structures survive primary devolatilization, at odds with the near-complete elimination for this coal type in Fig. 6.12. Unfortunately, only a handful of very high-rank coals have been monitored this way, and these samples usually contain so little oxygen that measurement uncertainties obscure the dynamics of oxygen release during devolatilization.

FLASHCHAIN® interprets these tendencies by allocating all coal-O and most coal-H to the bridge reactant, which releases all its oxygen and most hydrogen as non-condensable products during spontaneous char link production and via the elimination of bridge remnants on fragment ends. Both reaction processes extend from the onset of primary devolatilization through tar production and beyond. Nitrogen and carbon are preferentially retained in char because all coal-N and most coal-C is bound into aromatic nuclei, which remain refractory throughout tar production, and decompose relatively slowly throughout succeeding stages. Consequently, tar shuttling is the primary release mechanism for C and N.

These same allocations interpret the element retention for variations in heating rate and pressure in Figs. 6.13 and 6.14, respectively. More C, H, and N are released from char with progressively faster heating rates, in accord with reported measurements (cf. Fig. 4.15), while more O is retained in char. The preferential release inversely mimics the enhancements to tar yields for faster heating, corroborating the predominant role for tar shuttling in the devolatilization mechanism. The opposite prediction for oxygen

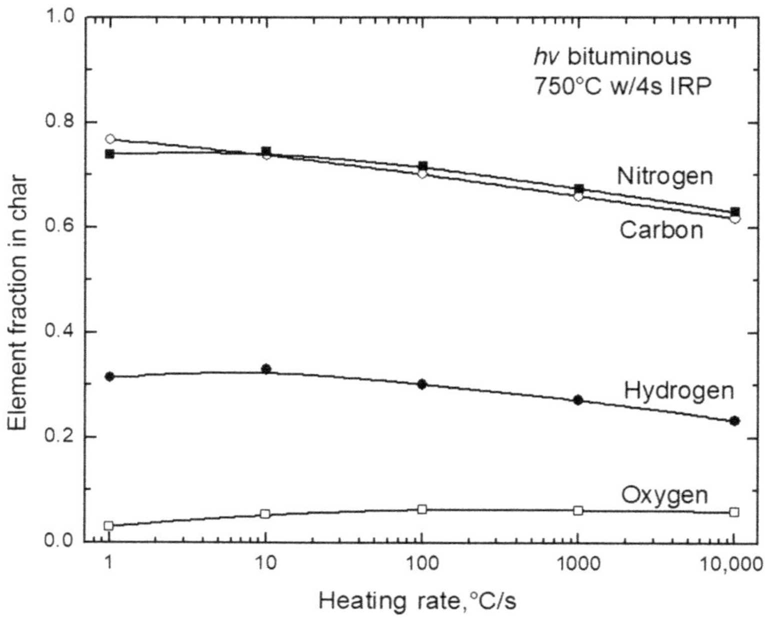

Fig. 6.13 Predicted element fractions in char from *hv* bituminous coal after heating at different rates to 750°C with 4 s IRP at atmospheric pressure.

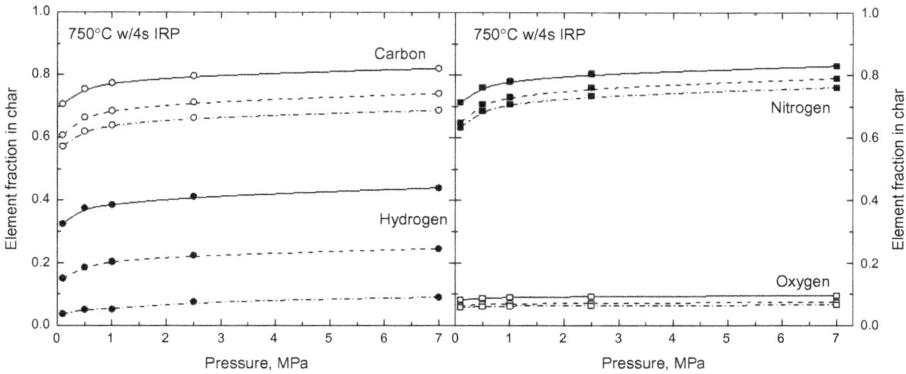

Fig. 6.14 Predicted element fractions in char from *(dot-dash)* Illinois No. 6 and *(dash)* Pittsburgh No. 8 *hv* bituminous coals and a *(solid) mv* bituminous coal for different pressures at 10^3°C/s to 750°C with 4 s IRP.

seems confusing, because whether coal-O is shuttled away in tar or expelled in noncondensables, it will be released from the char. This tendency probably reflects the much longer total reaction times for progressively slower heating rates in this simulation study. Unfortunately, no measurements are available to evaluate this aspect of the predictions.

The pressure effect is illustrated with predicted char element fractions for three bituminous coals in Fig. 6.14. In the predictions, char retains more C, H, and N for progressively higher pressures, and these char element fractions increase for coals of progressively higher rank. The tar precursors retained in char at elevated pressures contribute to the inventories of these elements. The retention of C and N are somewhat more sensitive to pressure than hydrogen retention, in accord with reported char compositions (cf. Fig. 4.16), although the measured char-H retention diminishes slightly for progressively higher pressures. This is because a greater proportion of the H in retained tar precursors is released as noncondensables, but nearly all the retained C and N are refractory. Oxygen retention is insensitive to pressure variations for all but the lowest pressures, again, because any coal-O retained in tar precursors will be released as noncondensables anyway.

The char compositions presented to this point reflect tar production and the stage of relatively fast gas production immediately afterward. The results in Fig. 6.15 feature extended heating to 2000°C to also cover the annealing stage. The 30 s IRP at all simulation temperatures obscures the faster devolatilization rate for the lignite, so that the element retention profiles for both coals are the same, except that those for C and N saturate to different ultimate levels. All elements except N are rapidly released below 900°C before reaching a saturation limit that persists through the hottest temperatures. Oxygen is completely eliminated by 900°C, whereas hydrogen reaches an asymptotic limit of only several percent of coal-H. Similarly, carbon retention reaches a saturation level between 60% and 70% at the same temperature. This level is indistinguishable from char-C fractions at the end of primary devolatilization because the carbon released as gases during annealing is imperceptible. These features are seen in

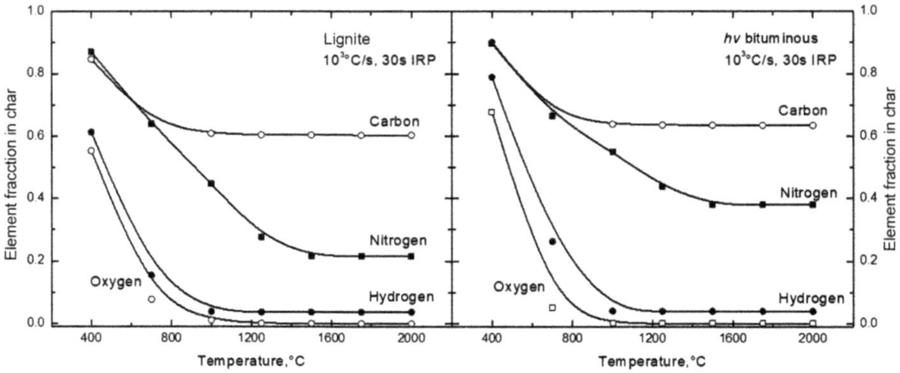

Fig. 6.15 Predicted element fractions in char from (left) lignite and (right) *hv* bituminous coal for heating at 10^3°C/s to different temperatures with 30 s IRP at atmospheric pressure.

reported char compositions across the same temperature range (cf. Fig. 4.17). In contrast, nitrogen retention does not reach its saturation level until 1400–1500°C. Ultimate char-N levels are appreciably different for the two coal types, and both are substantially lower than those at the end of the first two stages of primary devolatilization. The comparable data in Fig. 4.17 show complete elimination of N from char at 2000°C, although this discrepancy may reflect the scarcity of measured char compositions through such high temperatures.

Ultimate oxygen distributions appear in Fig. 6.16 for a lignite and an *hv* bituminous coal, where the x-axis denotes the fractional ultimate weight loss, as a scale for the

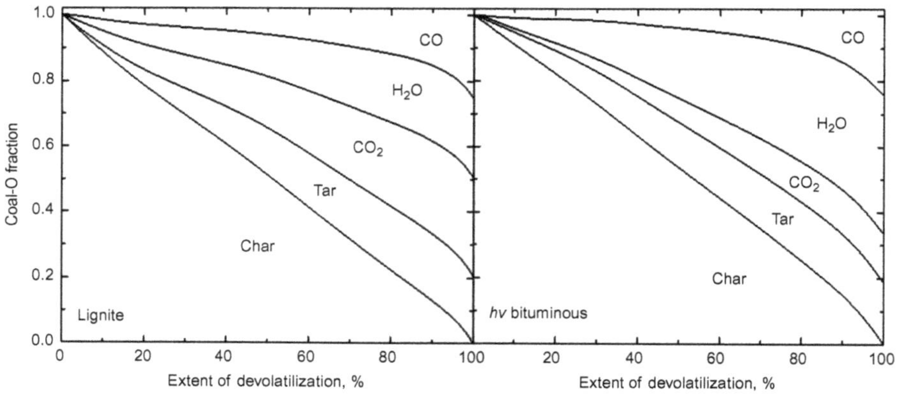

Fig. 6.16 Cumulative distributions of coal-O fractions among major oxygenated products during heatup at 1000°C/s to 1275°C for (top) lignite and (bottom) *hv* bituminous coal. The x-axis indicates the percentage of the ultimate weight loss based on the predicted transient yields.
Reproduced from Niksa S. Flashchain theory for rapid coal devolatilization kinetics. 7. Predicting the release of oxygen species from various coals. Energy Fuels 1996;10:173–87 with permission from the American Chemical Society.

extent of primary devolatilization. The most striking feature in both distributions is the near-linear inverse proportionality between residual char-O and extent of devolatilization. If they were not subject to such large measurement uncertainties, fractional char-O levels would be excellent measures of the extent of devolatilization whenever direct measurements are infeasible, and would certainly be preferable to ash tracer analyses for EFR testing at elevated temperatures. Fully devolatilized chars contain little, if any, oxygen, except chars from anthracites.

6.2.5 Primary tar characteristics

The factors that determine the elemental compositions of primary tars are even more straightforward than those responsible for char compositions, since, by definition, primary tars do not reflect the elimination of heteroatoms during secondary pyrolysis, and tar shuttling is entirely responsible for tar compositions. According to FLASHCHAIN®, the compositions of tar precursors in the coal phase change throughout devolatilization as bridges spontaneously decompose into char links and bridge remnants on fragment ends are eliminated as noncondensable gases. The precursor compositions are unaffected by flash distillation into tar vapors. Consequently, successive incremental contributions to the cumulative tar product have fewer heteroatoms and aliphatic hydrocarbon components, so tar compositions saturate toward those of PAH compounds.

As seen for the four *hv* bituminous samples in the APCS in Fig. 6.17, predicted tar H/C ratios for primary tars remain fairly uniform during the early stages, then fall off sharply during the later stages of tar production. For reference, the coal H/C ratios are 0.77 for three of these four coals, and 0.86 for the more volatile Blind Canyon APCS sample. FLASHCHAIN® correctly predicts very substantial H-enrichments in the early tars from all samples, and appreciable H- enrichments in the ultimate tars from three of these four coals.

The predicted tar compositions are considerably less aromatic than their parent coals, as they should be. It is interesting that the predicted ultimate tar H/C ratios for the three coals with the same H/C ratios are essentially uncorrelated; this disconnect to coal H/C ratios is also evident in reported tar compositions (cf. Table 4.3). This erratic character reflects the different bridge compositions assigned to coals of even the same nominal rank, and differences in the assigned char link compositions for particular samples. Bridge decompositions into char links eliminate aliphatic hydrogen and carbon from tar precursors which, in turn, diminishes H/C ratios of the cumulative tar sample.

FLASHCHAIN® also predicts higher H/C ratios for tar prepared at elevated pressures. For example, for the Pittsburgh No. 8 APCS, tar H/C ratios increased from 0.86 for 0.1 MPa to 0.92 for 7 MPa.

Predicted element fractions of tar for the APCS appear in Fig. 6.18 for ultimate primary tar samples. The predicted fractions of coal-C, -H, and -N vary from 0.10 to 0.35, consistent with the reported range (cf. Fig. 4.21). Oxygen fractions vary from 0.1 to 0.2, which is only about half the measured range from 0.25 to 0.45. The predicted element fractions for these eight coals exhibit a broad maximum in all four element fractions for *hv* bituminous coals, but this feature probably reflects the small

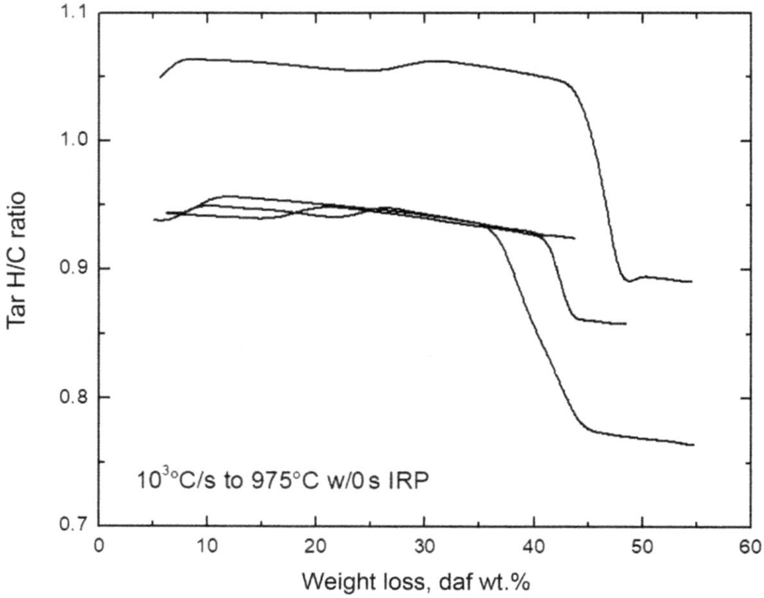

Fig. 6.17 Predicted atomic H/C ratio of tar throughout devolatilization of 4 *hv* bituminous coals from the APCS for heating at 10^3°C/s to 975°C with no IRP at atmospheric pressure.

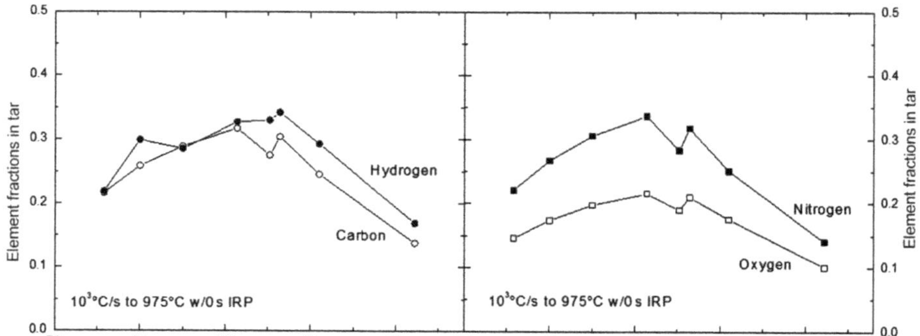

Fig. 6.18 Predicted element fractions in tar throughout devolatilization of the APCS for heating at 10^3°C/s to 975°C with no IRP at atmospheric pressure.

number of samples. Reported tar compositions are completely insensitive to rank (cf. Fig. 4.21); whereas the predictions convey appreciable sample-to-sample variability, reported tar compositions are much more erratic than those in Fig. 6.18.

Since tar shuttling determines tar composition, it is not surprising that all predicted element fractions in tar increase for progressively faster heating rates, as seen in Fig. 6.19. This tendency is weaker for oxygen than for the other three elements because oxygen is eliminated from tar precursors faster than the other elements, so tars formed latest tend to have very little oxygen at even the fastest heating rates. Since tar yields diminish for progressively higher pressures, the predicted element fractions in tar in Fig. 6.19 also diminish because, again, tar shutting determines tar compositions.

The relation between the MWDs of primary tars and their precursors in the condensed phase is illustrated in Fig. 6.20 for rapid heating to a temperature hot enough to achieve ultimate yields. In the left panel, the predicted metaplast yield passes through a maximum at about 475°C, which is only slightly hotter than the temperature at which tar production begins. Tar production continues through 800°C, beyond which additional noncondensables determine the approach to ultimate weight loss

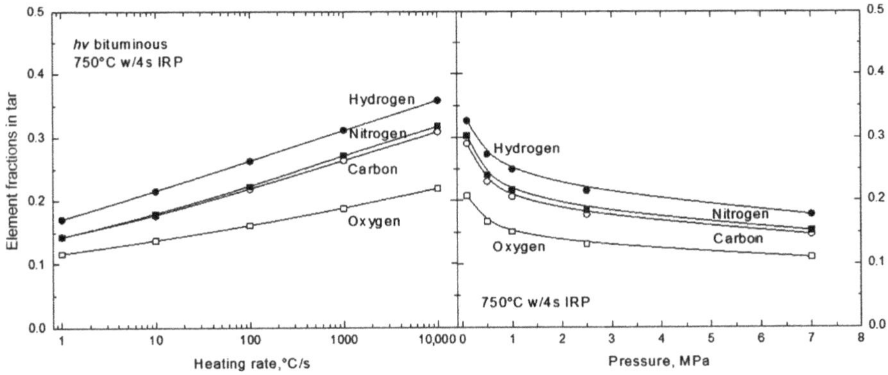

Fig. 6.19 Predicted element fractions in tar (left) after heating at different rates to 750°C with a 4 s IRP at atmospheric pressure and (right) for different pressures.

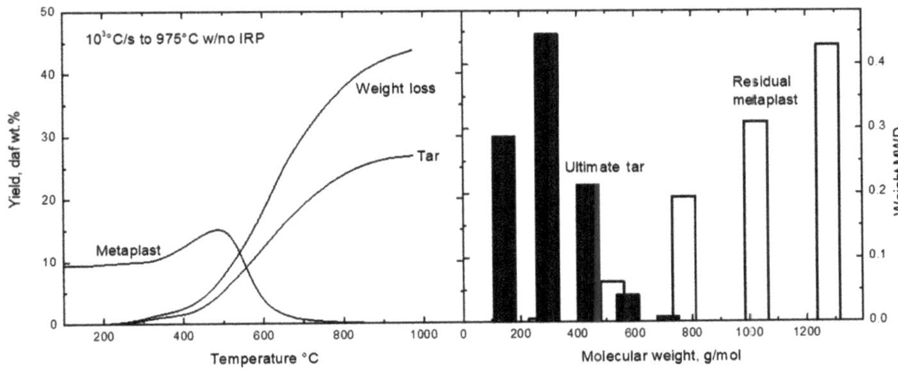

Fig. 6.20 (Left) Predicted weight loss and yields of tar and metaplast and (right) weight MWDs of ultimate tar and residual metaplast from *hv* bituminous coal for heatup at 10^3°C/s to 975°C with no IRP at atmospheric pressure.

at 1000°C. The right panel shows the weight-based MWDs of the cumulative ultimate tar sample and the residual metaplast at the end of the thermal history. The predicted tar MWD is coarsely resolved with five MW increments. It has the form of a gamma (γ)-distribution, in which the MWD abruptly rises from a minimum value of 100 g/mol and passes through a maximum below 300 g/mol into a tail that extends to weights of several hundred g/mol. The tars are skewed toward the lighter weights, compared with their precursors, simply because hydrocarbons with lower molecular weights become progressively more volatile. Yet the tar MWD extends toward very large weights because even these heavy hydrocarbons have finite vapor pressures. The metaplast MWD monotonically increases for progressively greater molecular weights, reflecting the vaporization of lighter tar precursors throughout devolatilization. This MWD appears to extend to weights that are double the weight of the largest tar component. However, this is an artifact of comparing the cumulative tar sample, which contains the lightest tar components released at the coolest temperatures, with the residual metaplast, which has been depleted of all but the heaviest components at the maximum temperature in the test. Consequently, the final incremental contribution to the cumulative tar sample would cover the same range of molecular weights as the residual metaplast in Fig. 6.20.

The predicted impact of variations in coal quality, heating rate, and pressure on the number average tar weights, M_n, are compiled in Table 6.5. Tars prepared from coals of progressively higher rank become lighter, especially in the transition from the lowest ranks through *hv* bituminous. Tar weights are essentially uniform for all low volatility coals. The predicted trend is at odds with the lone dataset on this effect (Unger and Suuberg, 1984), which showed substantially lighter tars from both a lignite and a *lv* bituminous coal compared with two *hv* bituminous samples. Unfortunately, this finding cannot be taken at face value because GPC-determinations of tar MWDs have larger uncertainties for coals of progressively lower rank, due to stronger interferences from the progressively greater levels of tar-O from low-rank coals.

Table 6.5 Predicted M_n for ultimate tars from the APCS and for different heating rates and pressures with *hv* bituminous

Coal quality		**Heating rate**		**Pressure**	
Coal-C (daf wt.%)	**M_n (g/mol)**	**q (°C/s)**	**M_n (g/mol)**	**P (MPa)**	**M_n (g/mol)**
72.5	354	1	199	0.1	261
75.0	325	10	218	0.5	237
77.5	306	100	238	1.0	228
80.7	277	1000	262	2.5	216
82.6	263	10,000	285	7.0	203
83.2	261				
85.5	252				
91.1	259				

Tar MWDs need to be accurately measured for an extensive suite of diverse coals to subject the predicted tendency to a stringent evaluation.

The predicted M_n-values for a bituminous coal for ranges of heating rate and pressure display the expected shift toward lighter tars for progressively higher pressures. They also display a pronounced shift toward heavier tars with progressively faster heating rates, due primarily to the positive temperature dependence in the saturated vapor pressure of metaplast. This tendency is much stronger than the neutral impact of heating rate variations seen in the few laboratory datasets on this effect described in Section 4.2.7.

6.2.6 Noncondensable gas yields

The yields of the major noncondensable species are determined by the compositions of peripheral groups and bridges in any coal sample. In FLASHCHAIN® both factors are coalesced into the compositions of labile bridges and char links assigned from the proximate and ultimate analyses and proton and carbon aromaticities. All elements initially in a bridge that are not incorporated into a char link during bridge conversion are released as noncondensable gases.

Distributions of the three main oxygenated gases are determined by stoichiometric coefficients based on $(O/C)_B$ (cf. Section 5.3.1). This approach is evaluated with RCFR data on the dynamics of oxygenated gas release in Fig. 6.21, and for a diverse assortment of coals from a CPP in Fig. 6.22. In Fig. 6.21, the gas yields are plotted against the extent of primary devolatilization, evaluated as the measured fractional weight loss for progressively greater transit times through the RCFR. The predicted CO_2 and H_2O yields are within the measurement uncertainty throughout primary devolatilization for all three coals. The analysis correctly predicts that CO_2 and H_2O yields approach equal proportions for coals of lower rank, and that these proportions fall to less than 1:2 for *hv* bituminous coals. Carbon monoxide is predicted and observed to be the most abundant oxygenated gas for these three coals. Since the dynamics are accurate throughout devolatilization, these cases also corroborate the premise that char link formation expels CO_2 and H_2O from the onset of primary devolatilization, and that CO production rapidly accelerates by the end of tar formation, and eventually overtakes the production of CO_2 and H_2O.

The evaluation of ultimate oxygenated gas yields in Fig. 6.22 covers the entire rank spectrum with rapid devolatilization data from a CPP, and with slow heating data from TGA tests with the APCS. The analysis clearly depicts the tendency for diminishing yields of all three species across the rank spectrum, as well as the proper proportions of these three gases. Predicted CO_2 yields are accurate to 1 daf wt.% in all but four cases in the CPP dataset, and for just over half the H_2O yields. The predicted CO yields are slightly less accurate than the H_2O yields. The same tolerance is satisfied with the CO_2 yields from the TGA dataset, which is not coincidental since, according to FLASHCHAIN®, the yields of neither CO_2 nor H_2O are affected by variations in heating rate. The discrepancies are much greater for the TGA H_2O yields, for which the measured yields are much greater than the FLASHCHAIN® predictions, but typically smaller for CPP H_2O yields. This problem is in the measurements because the

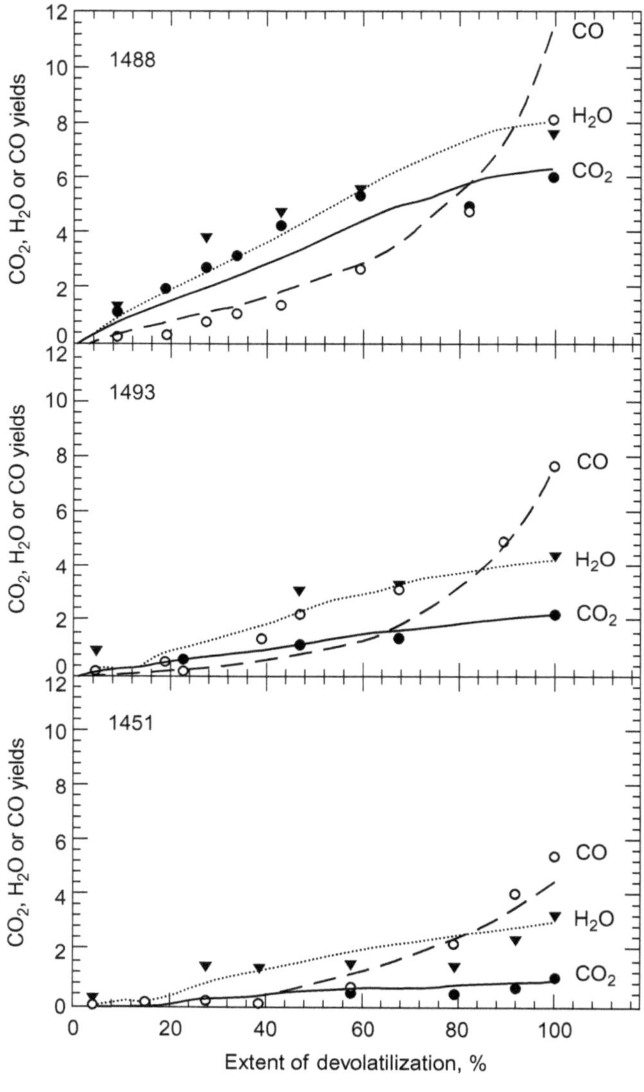

Fig. 6.21 An evaluation of predicted yields of (○) CO, (▼) H_2O, and (●) CO_2 from (top) subbituminous, (middle) *hv* bituminous, and (bottom) *hv* bituminous coals during heatup at about 10^4°C/s. The x-axis indicates the percentage of the ultimate weight loss based on the measured yields for different transit times through a RCFR.
Reproduced from Niksa S. Flashchain theory for rapid coal devolatilization kinetics. 7. Predicting the release of oxygen species from various coals. Energy Fuels 1996;10:173–87 with data from Chen JC, Niksa S. Coal devolatilization during rapid transient heating. Part 1. Primary devolatilization. Energy Fuels 1992;6:254–64 with permission from the American Chemical Society.

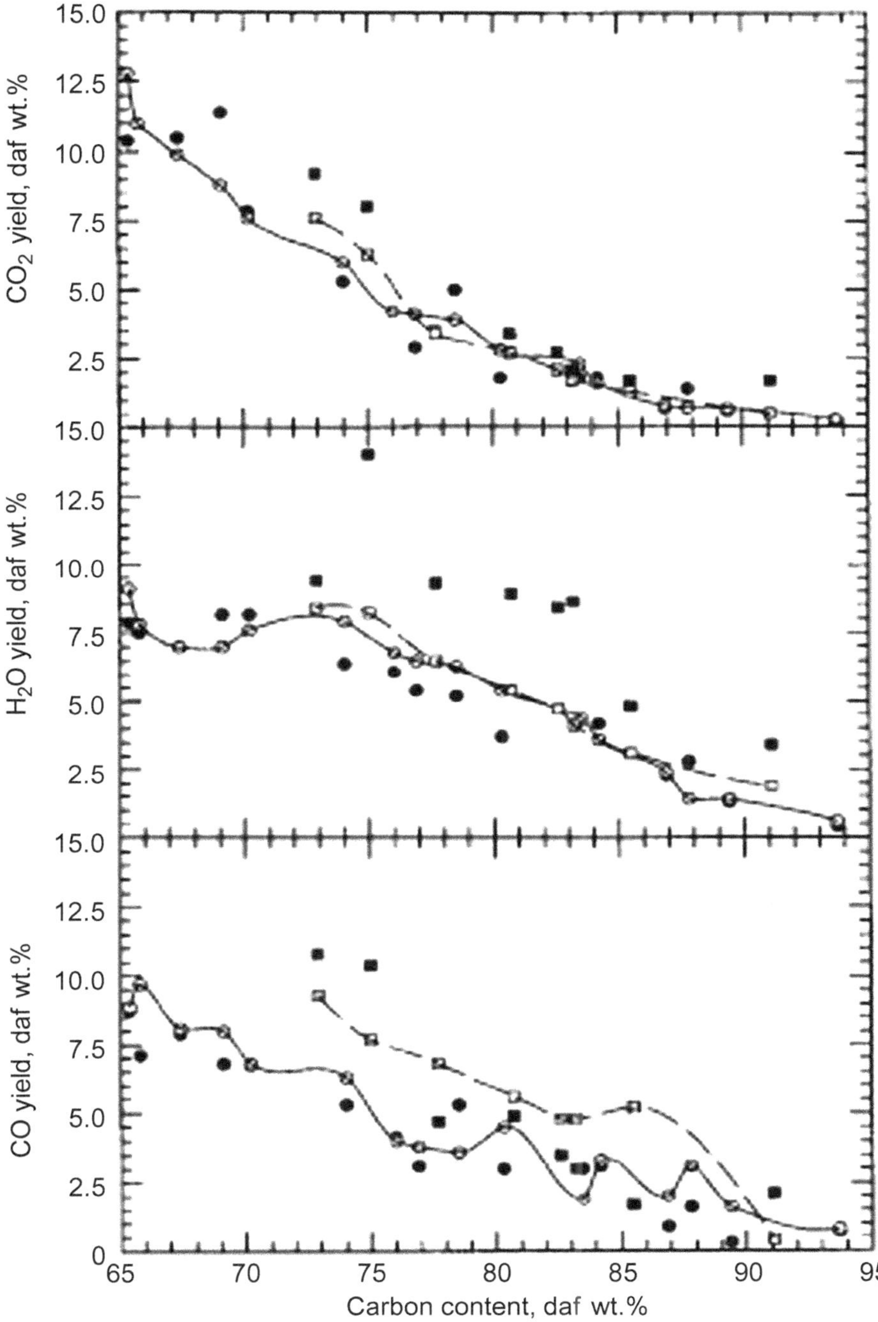

Fig. 6.22 An evaluation of predicted ultimate yields of (top) CO_2, (middle) H_2O, and (bottom) CO for (●) rapid devolatilization in a CPP at 765°C with 4s IRP (Xu and Tomita 1987) and for (■) slow heating to about 1125°C in a TGA (Solomon et al., 1990). FLASHCHAIN® predictions appear as the corresponding open symbols connected by the solid and dashed curves for CPP and TGA data, respectively.
Reproduced from Niksa S. Flashchain theory for rapid coal devolatilization kinetics. 7. Predicting the release of oxygen species from various coals. Energy Fuels 1996;10:173–87 with permission from the American Chemical Society.

TGA dataset seriously breaches the O-balances for all but one of these samples (Niksa, 1996).

Although the CO yields in the TGA data abide by the overall trend with coal quality, the quantitative discrepancies are usually more than 2 daf wt.%. The TGA CO yields are also significantly greater than the CPP CO yields for comparable coal types. The predicted CO yields also exhibit this tendency for the following two reasons: First, only the thermal severity in the TGA was sufficient to release all residual char-O as CO. Due to the lower temperature and shorter IRP in the CPP tests, some char-O remained at the end of the tests. Second, the TGA's slower heating rate would enhance its CO yields by about 1 daf wt.% over the CPP CO yields, owing to the retention of bridges and char links in the condensed phase that would be components of tar in the enhanced tar yields from the CPP tests. Both factors are responsible for the greater predicted TGA CO yields in this evaluation. FLASHCHAIN® also predicts slightly greater CO yields and uniform CO_2 and H_2O yields for elevated pressures, in accord with measured tendencies (Niksa, 1996).

In combination with extensive validation work reported previously (Niksa, 1996), the satisfactory predictions for the yields of the oxygenated gases corroborate the premise that the competing pathways for bridge scission and spontaneous char link formation also govern the production of CO_2 and H_2O from any coal type. The same rate parameters that accurately predict tar release also describe the release of CO_2 and H_2O. No additional reaction rate parameters, hypothetical ultimate yields, or conversion channels are required. Bridge conversion is also responsible for about one-quarter of CO release, the remainder being released from residual O in char links in a separate reaction from the end of tar production through the annealing stage. In contrast, the multiple DAEM channels for the oxygenated gases in FG-DVC and CPD entail dozens of adjustable parameters and multiple hypothetical ultimate yield parameters that must be specified from standardized test data for each coal sample, because their underlying premise of multiple, independent, parallel production channels is unfounded.

The predicted distributions of GHCs—the other major noncondensable gas product—are determined by $(O/C)_B$, as well as the average numbers of H and C atoms assigned to the labile bridge reactant in FLASHCHAIN® (cf. Section 5.3.4). In addition, the yield of all GHCs plus H_2 is assigned as the difference between the predicted total gas yield and the yields of the three oxygenated gases, so errors may accumulate in the predicted GHC yields. The evaluation in Fig. 6.23 covers CH_4 yields and the sum of $C_2 + C_3$ GHC yields for coals across the rank spectrum. Both GHC product yields are predicted within the measurement uncertainties for coals with more than 80% C. But there are large discrepancies for coals of lower rank in two distinct deviation patterns. The three samples with 75%–80% C gave the greatest tar yields, by far, in this sample suite, and their measured $C_2 + C_3$ GHC yields are also distinctively large. FLASHCHAIN® accurately depicts the distinctive tar yields, but overestimates both GHC yields for these coals. The second deviation pattern pertains to coals with less than 75% C. Although both measured GHC product yields are fairly uniform across this range, the predicted yields surge for coals of progressively lower rank. This deviation is hard to explain because the predicted yields of the oxygenated gases in

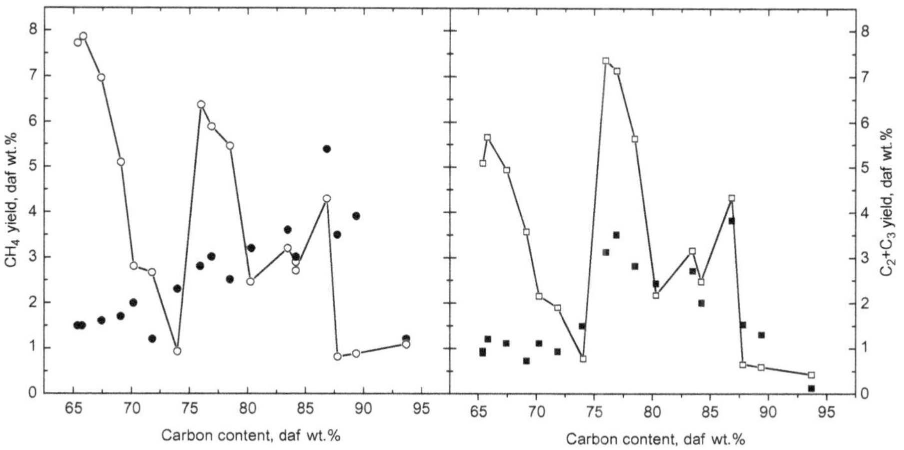

Fig. 6.23 An evaluation of predicted ultimate yields of (left) (●) CH_4 and (right) (■) $C_2 + C_3$ GHCs for rapid devolatilization in a CPP at 765°C with 4 s IRP (Xu and Tomita, 1987). FLASHCHAIN® predictions appear as the corresponding open symbols connected by the solid line segments.

Fig. 6.22 are uniformly accurate across this range, so accumulated errors are not responsible for the discrepancies in Fig. 6.23.

Oils are not resolved as a separate product lump by FLASHCHAIN®. They are incorporated into the predicted primary tar yields. For the tests in Figs. 6.22 and 6.23, measured oils yields varied from 1.5 to 3 daf wt.% for all coals except the two samples of highest rank, which gave less. The predicted H_2 yields vary from 0.2 to 0.5 daf wt.% and are within the measurement uncertainties of the measured yields.

6.2.7 Volatile nitrogen species

To interpret the release of coal-N in tar and HCN, FLASHCHAIN® allocates all coal-N to refractory aromatic nuclei, where it may be either shuttled away in tar molecules or liberated as HCN over a very broad temperature range. The rate for the HCN production channel is correlated with the O/N ratio of the whole coal, so that it decelerates for coals of progressively higher rank. The evaluation in Fig. 6.24 highlights the predominance of tar shuttling during the initial stage of rapid devolatilization. In these RCFR tests, a coal suspension was rapidly heated for progressively longer transit times until ultimate primary devolatilization behavior was recorded (Chen and Niksa, 1992). Consequently, little time was available for direct HCN release from char. The dynamics for four coals are resolved in Fig. 6.24 as a function of the fractional ultimate yield for each transit time. These predictions have tar shuttling as essentially the only means of N-release throughout most of rapid primary devolatilization, with HCN release only near the end of tar production.

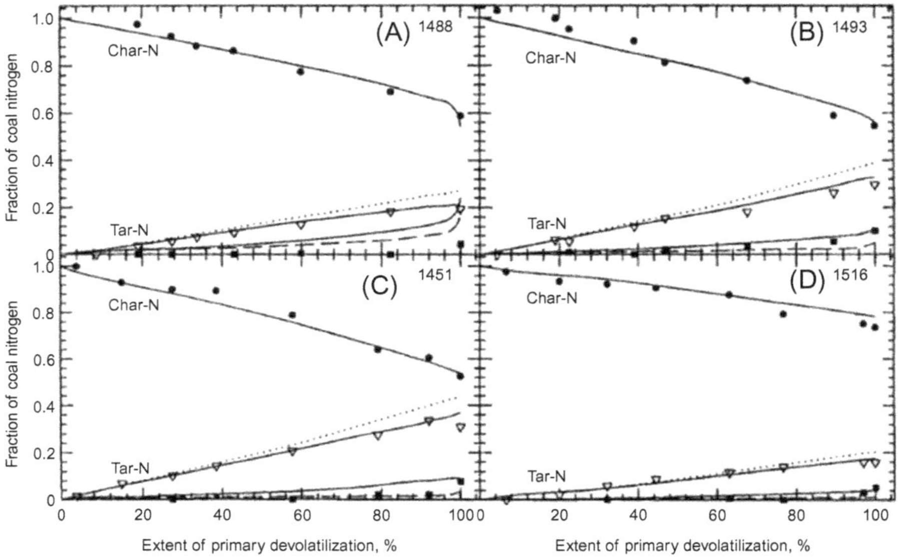

Fig. 6.24 An evaluation with measured (●) char-N, (▽) tar-N, and (■) HCN from a RCFR with subbituminous (1488), *hv* bituminous (1493), and *mv* bituminous (1516) during heating at about 10^4°C/s (Chen and Niksa, 1992). The *x*-axis indicates the percentage of the ultimate weight loss recorded for each transit time in the tests.
Reproduced from Niksa S. Flashchain theory for rapid coal devolatilization kinetics. 6. Predicting the evolution of fuel nitrogen from various coals. Energy Fuels 1995;9:467–78 with permission from the American Chemical Society.

The predominance of tar shuttling is apparent for a much broader selection of coals in Fig. 6.25, which shows the coal quality impacts on the ultimate fractional partitioning of coal-N into char, tar, and HCN + NH_3 in a RCFR at atmospheric pressure. The predicted tar-N levels are nearly the same as the fractional tar yields, whereas the gaseous N-species comprise much smaller coal-N fractions than gas yields for all but the subbituminous coal, which produced much more NH_3 than HCN. For the other coals, HCN contained 10% of coal-N or less. The predicted char-N and tar-N fractions are fairly uniform for all ranks through *hv* bituminous, which reflects the few samples of these ranks in the tests. Additional coals would expand the sample-to-sample variability in the predictions. The predicted tar-N levels fall off sharply for low volatility coals while the char-N levels surge, in accord with the data.

The evaluation in Fig. 6.26 covers the rapid devolatilization of 20 coals representing ranks from brown coals through *mv* bituminous at atmospheric pressure, where the maximum temperature of 1215°C was held for about 4 s (Kambara et al., 1995). These conditions were sufficiently severe to release major portions of coal-N as HCN and NH_3, for the coals with appreciable quaternary N. The proportions of tar-N and HCN were distorted by secondary pyrolysis in these tests, so only char-N is pertinent to this evaluation. The predicted char-N levels are within 2% of the measurements in 12 of 20 cases, and the larger discrepancies are mostly due to problems in the predicted weight loss, not in the N-release mechanisms (Niksa, 1995). FLASHCHAIN®

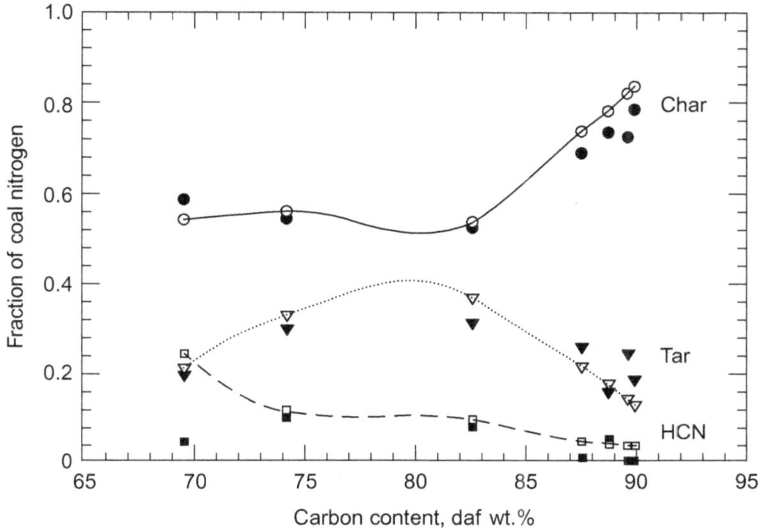

Fig. 6.25 An evaluation of predicted ultimate distributions of (○) char-N, (▽) tar-N, and (□) HCN from a RCFR with several coals during heatup at about 10^4°C/s. Predictions are open symbols connected by curves. Measured values are filled symbols, where HCN denotes HCN + NH_3 (Chen and Niksa, 1992).
Reproduced from Niksa S. Flashchain theory for rapid coal devolatilization kinetics. 6. Predicting the evolution of fuel nitrogen from various coals. Energy Fuels 1995;9:467–78 with permission from the American Chemical Society.

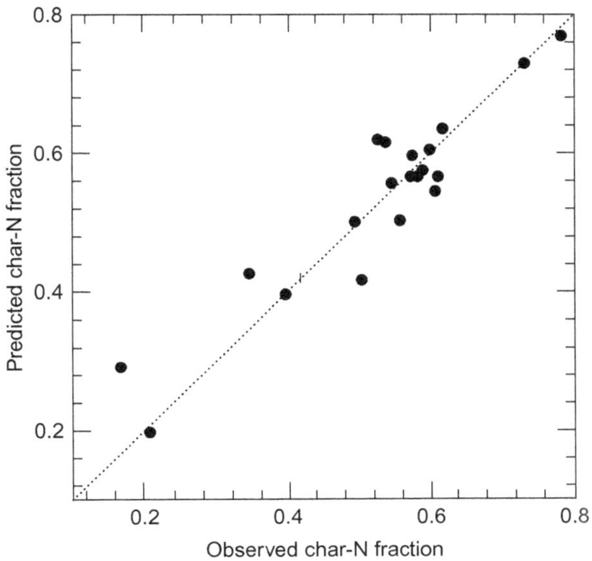

Fig. 6.26 An evaluation of predicted ultimate char-N fractions for rapid devolatilization at 1215°C for several seconds.
Reproduced from Niksa S. Flashchain theory for rapid coal devolatilization kinetics. 6. Predicting the evolution of fuel nitrogen from various coals. Energy Fuels 1995;9:467–78 with permission from the American Chemical Society.

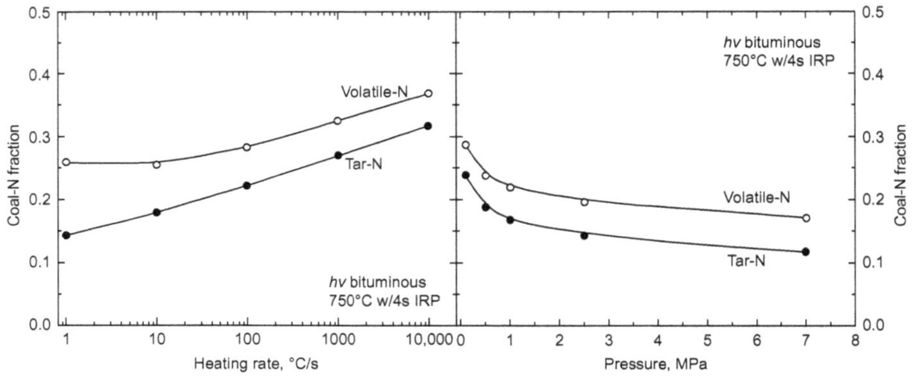

Fig. 6.27 Predicted ultimate nitrogen partitioning into volatiles and tar for various (left) heating rates and (right) pressures at 750°C with 4 s IRP.

accurately predicts slower release of residual-N in char after the end of tar production from coals of progressively higher rank.

The central role for tar shuttling is responsible for the enhanced N-release for progressively faster heating rates in Fig. 6.27, where volatile-N denotes the predicted sum of tar-N plus HCN. The incremental enhancements for each order-of-magnitude increase in the heating rate are the same for both volatile-N and tar-N, but only for heating rates faster than 100°C/s. At slower heating rates, the extended contact times allow HCN release to partially compensate for the lower levels of tar-N. But for heating rates faster than 100°C/s, the pyrrolic- and pyridinic-N retained in the condensed phase at a slower heating rate that would otherwise be shuttled away in tar with faster heating are not released as HCN on the time scale of tar production. This is consistent with the direct indication in Fig. 6.24 that tar shuttling is the only means of N-release throughout all but the latest stages of primary devolatilization at the fastest heating rates.

The predominance of tar shuttling explains why tar-N levels diminish for progressively higher pressures; in fact, the predictions in Fig. 6.27 closely match the measured tar-N levels in Fig. 4.32, albeit for a different *hv* bituminous sample. Although the measured volatile-N levels were insensitive to pressure, the predicted levels are nearly as sensitive as the tar-N levels. This deviation is due, in part, to the shorter IRP used in the predictions than in the tests (of 4 vs. 10 s).

On a molar basis, coal-N is a sparse element present in fewer than half of all aromatic nuclei and in only about one in ten ring structures, on average. So nitrogen decomposition chemistry does not play any discernible role in the bridge conversion chemistry responsible for the formation of tar precursors. Accordingly, coal-N is simply shuttled into the vapor phase as a refractory component of tar molecules. The nitrogen released by this mechanism is just a random sample of the nitrogen in the fragment population in the condensed phase, weighted by flash distillation. No additional chemical reactions or physical transformations are needed to describe the shuttling of coal-N by tar. Tar shuttling is responsible for the coal quality impact on fuel-N release, whereby volatile-N levels pass through a weak maximum for *hv* bituminous coals, then plummet for low volatility coals, which mimics the rank dependence of tar yields (cf. Fig. 6.3). So the satisfactory depiction of coal-N release for diverse coals during rapid devolatilization is determined by the accuracy of predicted tar yields. This connection explains why tar-N levels are enhanced by faster heating rates, and diminish for progressively higher pressures.

Accurate predictions for generalized thermal histories, particularly those for slower heating rates, very high temperatures, and extended IRPs, entail an additional channel for HCN production. This channel partially obscures the connection between the levels of volatile-N and tar-N given sufficient time and/or sufficiently hot temperatures to generate substantial HCN from residual char-N. The activation energy distribution assigned for this channel is broader than any assigned for bridge conversion in even the lowest rank coals, where bridge compositions are extremely heterogeneous because of the abundance of oxygen. Qualitatively, this indicates an extremely broad thermal response for HCN production that is due, in all likelihood, to multiple precursors for HCN in char-N. Evidently, these precursors become more stable as their surrounding domains of aromatic rings become more extensive, which is especially pronounced in low volatility coals. Moreover, aromatic domains in char also grow during devolatilization, so residual char-N becomes more refractory, especially in thermal histories that sustain annealing.

The most distinctive features of HCN production are revealed most clearly in the observed growth in volatile-N for faster heating rates. It is impossible to represent this tendency with an ordinary rate constant in a SFOR regardless of the parameter values. With any SFOR, volatile-N is predicted to diminish slightly as heating rates to the same ultimate temperature are increased, simply because there are fewer precursors to HCN in the char whenever tar production is enhanced by faster heating rates. Since the SFOR rate of HCN production would be proportional to the initial concentration of char-N, relatively less HCN is released for fewer precursors. In contrast, by representing HCN production with a DAEM, asymptotic yields of HCN are achieved at any sustained ultimate temperature, irrespective of heating rate. The same relative conversion is achieved for any initial precursor concentration. So with this mechanism the volatile-N, as the sum of tar-N plus HCN, increases for faster heating rates, but only from the enhancement to tar-N levels. Consequently, the tendency for more volatile-N for faster heating rates conclusively demonstrates that HCN production from char-N displays an extremely broad thermal response that requires a DAEM. The heating rate dependence also reveals that a kinetic competition apportions coal-N between the tar shuttling and HCN production channels. The scheme for coal-N release in FG-DVC incorporates hypothetical ultimate yield parameters for HCN and NH_3 which must be adjusted to depict the competition for different thermal histories, pressures, and coals. FLASHCHAIN® does not, yet it accurately predicts the partitioning between tar-N and HCN for any coal type. The only caveat is that the analysis does not distinguish HCN from NH_3, and NH_3 yields are often larger than HCN yields from low-rank coals. In such cases, HCN should be regarded as the sum of HCN plus NH_3.

6.2.8 Volatile sulfur species

Sulfur release comprises tar shuttling, char link production, peripheral group elimination, thiophene-S destruction at elevated temperatures, and pyrite decomposition. The predicted partitioning of coal-S among the major product classes in Fig. 6.28 is within

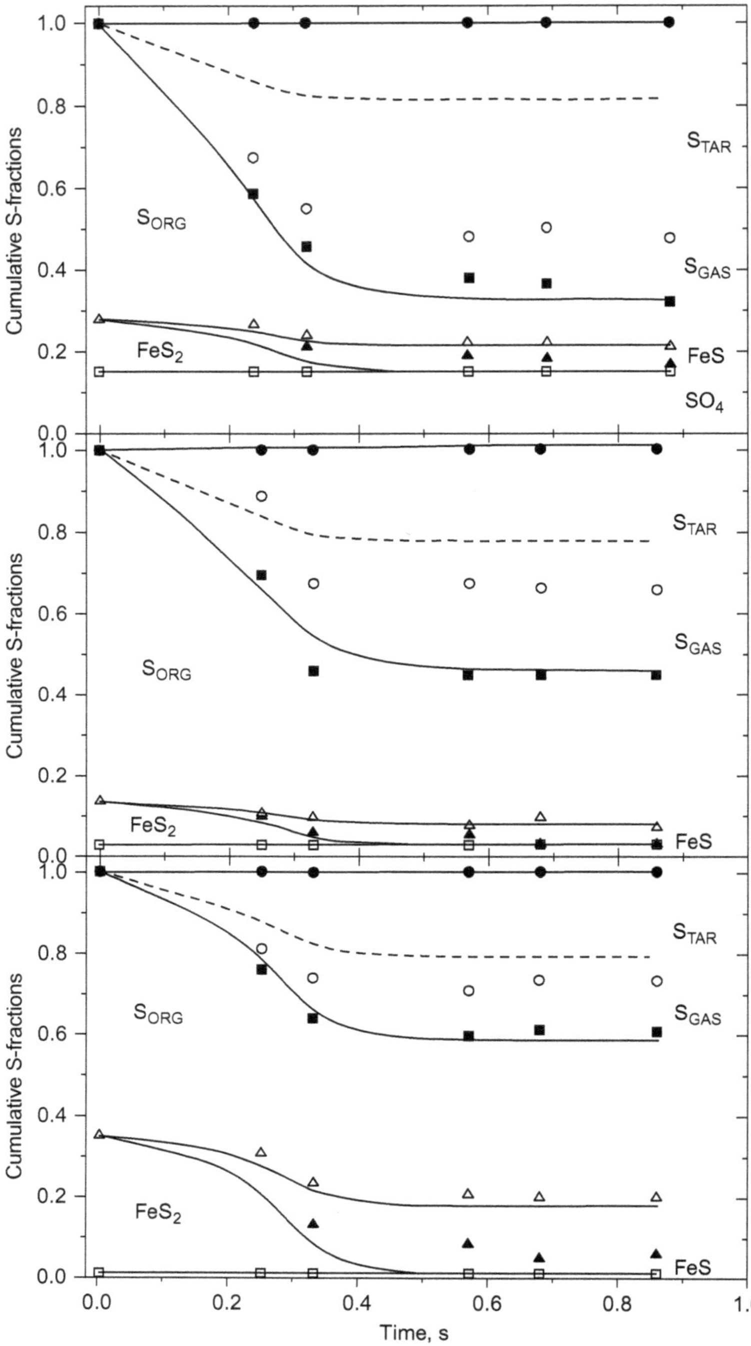

Fig. 6.28 Predicted S-product distributions from (top, middle) two *hv* bituminous and (bottom) one *mv* bituminous rapidly heated to 960°C under 0.1 MPa N_2. Cumulative contributions appear for sulfur in (□) S_{SO4}; (▲) FeS_2; (Δ) FeS; (■) S_{ORG}; (○) S_{GAS}; and (●) S_{TAR}, as fractions of coal-S. Reproduced from Niksa S. Flashchain theory for rapid coal devolatilization kinetics. 8. Modeling the release of sulfur species from various coals. Energy Fuels 2017;31:4925–38 with permission from the American Chemical Society.

the measurement uncertainties for all product classes except gas-S and tar-S. Pyrite decomposition is accurately depicted for these three coals, as is the decomposition of S_{ORG}. But the portions of tar-S and gas-S are only realistic with the *mv* bituminous, and grossly distorted with both *hv* bituminous coals. Additional datasets that resolve all the major S-species along with total and tar yields throughout primary devolatilization for at least a dozen coals are needed to resolve these discrepancies. Moreover, none of the available datasets for rapid heating conditions at elevated pressures could be qualified for model validation work, so the predicted impact of pressure on S-release remains to be validated. FG-DVC uses hypothetical ultimate yield parameters for H_2S and COS from both organic coal components and pyrite, in competition with tar shuttling. CPD does not yet describe the release of any sulfur species in tar or non-condensables. Hence, accurate predictions for the release of all volatile-S species from individual coal samples remain to be demonstrated.

6.3 Indirect modeling capabilities

The original primary goal of network depolymerization reaction mechanisms was to accurately predict the complete distribution of primary products from any coal at any operating conditions of commercial interest. In all but a handful of aspects of devolatilization behavior, this goal has been met. Beyond the original scope, network models eventually opened up additional capabilities that require accurate estimates for some of the intermediate reaction products that cannot normally be monitored directly. This section covers two such capabilities: accurate assignments for devolatilization rates and for particle swelling behavior.

6.3.1 *Nominal rates of devolatilization*

There is enormous potential for confusion whenever someone tries to represent a reaction mechanism as complex as coal devolatilization with a rudimentary reaction rate law and to estimate a nominal devolatilization rate. Indeed, the research literature is full of inconsistent and fundamentally incorrect material on what nominal rates of devolatilization are, how they may be assigned, and how they should be used.

In principle, nominal devolatilization rates can be measured with TGAs, which continuously monitor sample weight throughout a thermal history and automatically differentiate the weight histories to assign overall devolatilization rates in real time. The literature has a multitude of these rate assignments but only for heating rates that are much slower than our operating domain (cf. Fig. 5.6). None have been reported for heating rates from 100 to 10^4°C/s. In this rapid heating domain, rates can only be indirectly assigned from time-resolved product yields, which often introduce inordinate uncertainties because of poor time resolution, ambiguous thermal histories, and uncertain product yields. Although these obstacles only become worse for progressively faster heating rates, rate assignments based on model simulations are unaffected by these problems throughout our entire operating domain.

6.3.1.1 Assigning nominal rates of devolatilization

It is always possible to identify the parameters in simple, global rate laws for devolatilization that will closely mimic the predictions from more sophisticated models like FLASHCHAIN®. Niksa and Lau (1993) introduced the analyses that assign parameters in an SFOR and DAEM from FLASHCHAIN® predictions, and the commercial implementation of FLASHCHAIN®, called PC Coal Lab®, automatically fits the SFOR, DAEM, and the C2SM to FLASHCHAIN® predictions. Any rate law can be processed in the same way. Here nominal devolatilization rates are based on a SFOR because its rate parameters are frequently reported in the technical literature, and Arrhenius diagrams are routinely used to display the temperature dependence in the rate.

The parameters in any rate law can be specified from any individual devolatilization history from FLASHCHAIN®. As illustrated below, nominal rates do not extrapolate very well beyond the conditions used to assign their parameters. In other words, it is better to assign multiple parameter sets for a range of operating conditions, rather than assign some single best-fit parameter set that minimizes a least-squares error function over all conditions simultaneously. Also, the parameter assignments should be based on nonisothermal analyses because devolatilization in practical applications almost always occurs while the fuel is being heated to some ultimate temperature. So the best parameter estimates will be obtained when the total residence time in the FLASHCHAIN® simulation coincides with the duration of the devolatilization stage in the subject application as closely as possible. With rate assignments based on WMR simulations, devolatilization during the isothermal reaction period should be omitted from the analysis to follow, or used with a more comprehensive analysis (Niksa and Lau, 1993).

The single first-order reaction (SFOR) for devolatilization is

$$\frac{dV(t)}{dt} = A \exp\left(-\frac{E_a}{RT}\right)(V^\infty - V(t)) \tag{6.1}$$

where $V(t)$ is the instantaneous volatiles yield; V^∞ is the hypothetical ultimate volatiles yield; A is a pseudo-frequency factor, and E_a is an apparent activation energy. In this rate law, A, E_a, and V^∞ are adjustable parameters that vary with temperature, heating rate, pressure, and coal type. At the outset, it is important to realize that their magnitudes have no mechanistic significance whatsoever, because this simple reaction rate expression cannot possibly represent the numerous mechanisms that, in actuality, govern the kinetics of coal devolatilization. The premise that coal contains a fixed amount of precursors to volatiles, V^∞, is implicit in the SFOR yet, in actuality, ultimate devolatilization yields vary with heating rate, pressure, and coal type in ways that no global expression can depict. Similarly, A and E_a are simply numbers that can faithfully mimic devolatilization kinetics, provided that they are applied within their restricted range of applicability.

The parameters A, E_a, and V^∞ are usually assigned from laboratory test data. This analysis uses FLASHCHAIN® as a virtual coal laboratory to synthesize simulation "data" that can subsequently be analyzed for rate parameters just like test measurements would be analyzed, but with much better time resolution. We first evaluate dV/dt, $V(t)$, and V^{∞} for the operating conditions of interest from the predicted weight loss as a function of time and temperature from FLASHCHAIN®, then assign A and E_a by rearrangement of the first-order rate law, as follows:

$$\langle k \rangle = A\ \exp\left(-\frac{E_a}{RT}\right) = \frac{(V^{\infty} - V(t))}{\dfrac{dV(t)}{dt}} = \frac{(1 - X_{DV}(t))}{\dfrac{dX_{DV}(t)}{dt}} \tag{6.2}$$

where $<k>$ is the devolatilization rate constant and X_{DV} is the extent of devolatilization, $V(t)/V^{\infty}$. The expression on the far right shows that the parameter assignments for A and E_a are independent of V^{∞}; i. e., V^{∞} can be specified from measurements or any comprehensive reaction mechanism like FLASHCHAIN®, even though $X_{DV}(t)$ and dX_{DV}/dt can only be evaluated from a validated mechanism. In Eq. (6.2), the weight loss rate, dV/dt, is evaluated as the sum of the total rates of tar and gas release from FLASHCHAIN®; V^{∞} is the sum of the predicted yields of gas and tar at times long enough to achieve an asymptotic ultimate yield for the heating rate, pressure and coal sample under consideration; and $V(t)$ is the instantaneous sum of the predicted yields of gas and tar. The thermal history in the simulation specifies the sample temperature at every instant in time, so conventional Arrhenius diagrams can be prepared by plotting the logarithm of $<k>$ versus reciprocal temperature.

This same analysis can be applied to any of the product species predicted by FLASHCHAIN®, including weight loss, tar, gas, CO_2, H_2O, CO, C_1-C_3 GHCs, HCN, volatile-N, and char-N.

6.3.1.2 Comparing SFOR-based predictions to FLASHCHAIN® results

Considering the complexity of devolatilization mechanisms on the one hand, and the simplicity of the SFOR on the other, it is worthwhile to examine how accurately a SFOR can mimic FLASHCHAIN® predictions and how nominal SFOR rates will vary for different heating rates, pressures, and coal types. Numerous evaluations with data for broad ranges of reaction time and temperature are also available (Niksa, 1994c).

The impact of heating rate variations appears in Fig. 6.29 for WMR simulations of an *hv* bituminous coal heated at 100, 1000, 10^4, and 10^5°C/s to 1275°C. The predicted weight loss was used to define nominal devolatilization rates according to Eq. (6.2). Fig. 6.29 compares the FLASHCHAIN®-based and SFOR-based behavior in two ways, with an Arrhenius diagram and in terms of transient weight loss throughout the simulations. The weight loss diagram uses an x-axis with temperature instead of time because devolatilization is always complete before the ultimate reaction temperature is achieved in these simulations, and temperature is a suitable surrogate for time because the heating rates are uniform. The FLASHCHAIN® simulations of transient weight loss in the right panel of Fig. 6.28 clearly show that the onset of

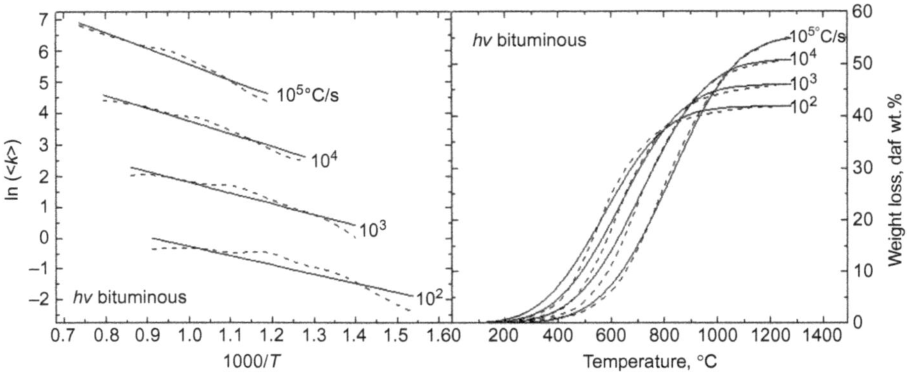

Fig. 6.29 Comparison of *(left panel, solid curves)* SFOR-based rate constants and *(right panel, solid curves)* transient weight loss with the respective *(dashed curves)* FLASHCHAIN® predictions for a bituminous coal at four heating rates to 1275°C.

devolatilization shifts to higher temperatures for faster heating rates, from ~325 to 550°C for this range of heating rates. Also, devolatilization occurs over a much broader temperature range at faster heating rates, even as the times to reach ultimate yields become shorter. The predicted ultimate yields are enhanced from 42 to 55 daf wt.% for this range of heating rates.

The FLASHCHAIN® predictions were processed according to Eq. (6.2) to assign the nominal devolatilization rate constants for an SFOR for these four heating rates. The associated Arrhenius plot appears in the left panel of Fig. 6.29. The dashed curves in the figure are the nominal rates based on Eq. (6.2) with the weight loss rate, instantaneous weight loss, and ultimate yields predicted by FLASHCHAIN® as input. The lines are least-squares fits to these curves whose slopes define $-E_a/R$ and whose *y*-intercept determines the logarithm of A in the SFOR. In the fitting procedure, the first and last several percent of the weight loss histories were discarded due to the sensitivity characteristics of Eq. (6.2). So the lengths of the curves indicate most, but not all, of the devolatilization periods. The linear fits of the SFOR to the FLASHCHAIN® simulations are not exact because the FLASHCHAIN®-based rates vary throughout devolatilization for all heating rates. Nonetheless, the fits are within useful tolerances since the correlation coefficients are between 0.91 and 0.98.

Small discrepancies in the SFOR fits are apparent in the right panel of Fig. 6.29, where weight loss transients based on the SFOR fits are compared directly with the transient weight loss from FLASHCHAIN®. The SFOR overpredicts the weight loss during the initial stages at all heating rates. However, once the mass loss reaches 20 daf wt.%, the agreement is virtually exact at all heating rates. The overpredictions at early times are universal flaws in SFOR fits for all fuel types and all operating conditions although, given their relatively small magnitude, they would not significantly degrade the performance in process simulations.

The Arrhenius plot in Fig. 6.29 clearly shows that a single set of SFOR rate parameters cannot describe devolatilization over a broad range of heating rates. In fact, nominal devolatilization rates increase in almost direct proportion to increases in the heating rates. Yet the apparent activation energies inferred from the slopes of the Arrhenius plots are reasonably similar for all heating rates, changing from 27 to 46 kJ/mol as heating rates were increased from 10^2 to 10^5°C/s. The increases in the devolatilization rates are almost entirely associated with increases in the A-factors. Here A-factors increased from 16 to 4×10^4, at an approximate rate of 15 times per order of magnitude increase in the heating rate. In light of these variations in the SFOR parameters, each set of A, E_a, and V^∞ values should only be applied over about one order of magnitude variation in the heating rate. It is also worth keeping in mind that ultimate yields are usually much more important in CFD simulations than devolatilization rate parameters, because many furnace and gasification reactor simulations do not have sufficient grid increments to resolve devolatilization over more than a few spatial increments.

Compared with the strong impact of heating rate variations on the SFOR parameters, the effects of pressure variations and coal quality are small, albeit with qualifications. The kinetic parameters, A and E_a, are virtually independent of pressure, as seen in Fig. 6.30. This figure shows the SFOR fits to FLASHCHAIN® predictions for heating an *hv* bituminous coal at 10^3°C/s to 1275°C at pressures from 0.1 to 7 MPa. The trend is for slower devolatilization for progressively greater pressures, although the quantitative variations are hardly greater than the uncertainties in the fits to the FLASHCHAIN®-based rates.

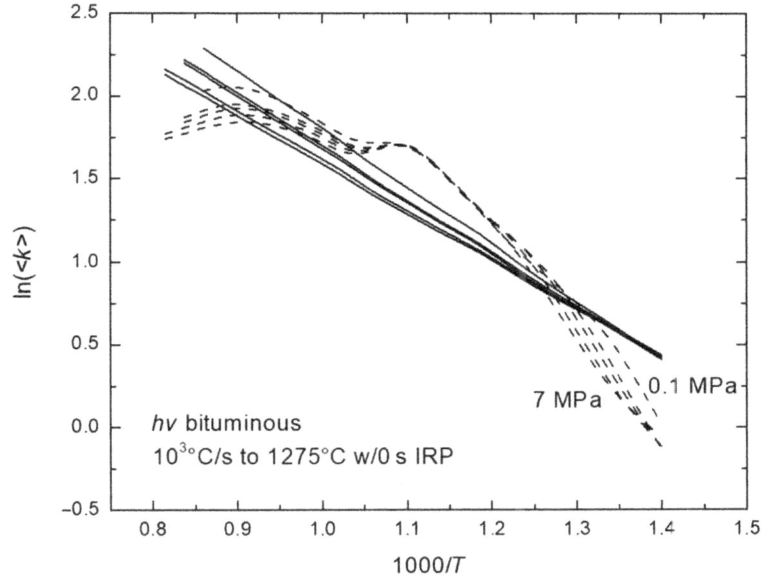

Fig. 6.30 Comparison of *(solid lines)* SFOR-based rate constants and the *(dashed curves)* FLASHCHAIN® predictions for a bituminous coal at 10^3°C/s to 1275°C at 0.1, 0.5, 1.0, 2.5, and 7 MPa.

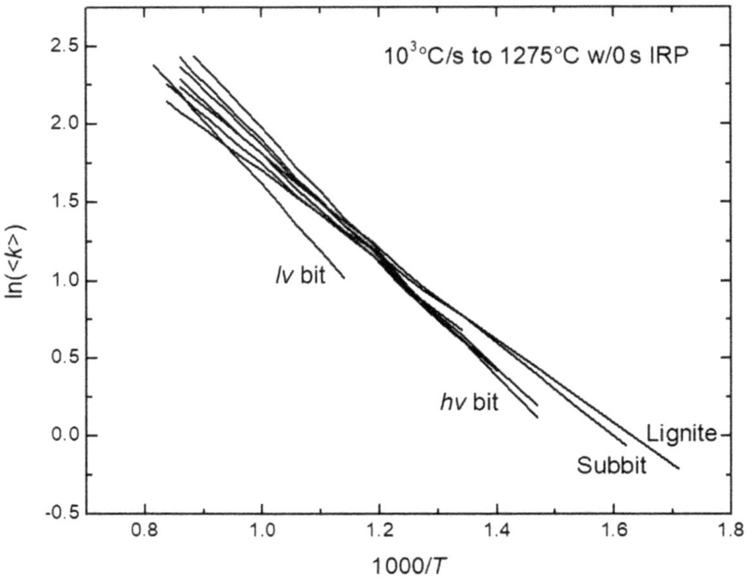

Fig. 6.31 SFOR-based rate constants for the APCS at 10^3°C/s to 1275°C at 0.1 MPa.

The fuel quality impacts are illustrated in Fig. 6.31 with the results for the APCS for heating at 10^3°C/s to 1275°C at atmospheric pressure. Rate variations for coals across the entire rank spectrum are never as substantial as those due to a single order-of-magnitude increase in heating rate. Rate variations with rank segregate into two categories. For ranks from lignite through *hv* bituminous, the rate variations are minimal, except that low-rank coals begin to decompose at substantially lower temperatures (cf. Fig. 5.10) and their devolatilization rates remain faster through most of tar production than those for *hv* bituminous samples. Nominal rates for these ranks vary by a factor of three at the onset of devolatilization, but by only forty percent during the later stages. Low volatility coals comprise the second category. They begin to devolatilize at much higher temperatures and sustain slower rates than the other ranks. Since their E_a values are comparable to those for *hv* bituminous and their rates are slowest, their *A*-factors are the lowest of all.

In contrast to the weak influence of pressure and rank on the kinetic parameters, ultimate yield parameters are acutely sensitive to pressure and rank variations. Whenever a SFOR is used to represent any aspect of devolatilization behavior, V^∞ should be updated for every heating rate, pressure and coal sample under consideration.

6.3.1.3 Practical applications and mechanistic interpretations

Since the discrepancies between the FLASHCHAIN® results and the SFOR fits are inconsequential, these rate parameter assignments provide the best means to incorporate FLASHCHAIN® predictions into CFD and other applications simulators.

Considering the programming complexity and inordinate computational burden associated with building a mechanism like FLASHCHAIN® into the applications simulator, this approach is really the only feasible means to do this. Users first run the application simulator with default estimates for devolatilization kinetics and the ultimate volatiles yield for the subject coal. The default rate parameters can be based on plots like Fig. 6.31 for the user's coals of interest, and from the data in plots like Figs. 6.2 and 6.3 for the ultimate yields. From the simulation results, they estimate the thermal histories where devolatilization occurs in the application, preferably by averaging over the population of particle trajectories within a coal injection stream. Multiple average thermal histories can be assigned for different ranges of the coal PSD, or for different sections of the coal feed stream.

The thermal history or histories are then imposed in stand-alone FLASHCHAIN® simulations with the properties of the coal(s) under consideration. The predicted weight loss history is then analyzed to assign the SFOR parameters and ultimate yield, as illustrated in the previous section. In situations where multiple thermal histories are appropriate, it may also be necessary to express the assigned kinetic parameters and/or ultimate yields as an explicit function of the size range in the coal PSD. Alternatively, the various sets of devolatilization parameters can be related to ranges of particle heating rate, as in the tabulated devolatilization process (TDP) model of Hashimoto et al. (2012a, b). Whatever the format, the assigned parameters for devolatilization are then entered into the applications simulator, as the means to incorporate the devolatilization histories from FLASHCHAIN® into the process simulation. After the process simulation has been updated, the thermal field is inspected for consistency with the original estimate. If the updated thermal history for the devolatilization zone in the application is markedly different than the original estimate, then the devolatilization kinetics are updated until the changes relax to a consistent set of devolatilization parameters.

Most of the steps in this procedure are automated in PC Coal Lab®, the commercial distribution package for FLASHCHAIN®. Users enter the standard coal properties and the operating conditions that determine the thermal history for the FLASHCHAIN® simulation. They then select the desired form of the devolatilization rate expression for the process simulation, as SFOR, C2SM, or DAEM. During a run, the program automatically assigns the rate parameters and ultimate yield for the subject devolatilization rate law. Users then incorporate these parameter assignments into their applications simulator, either through the graphical user interface, if possible, or within a user-defined function that is compiled into the application simulator at the start of a simulation. Since the parameter assignments are fully automated, users can quickly and conveniently obtain accurate devolatilization kinetics and yields for any coal across the full domain of conditions in commercial applications.

The SFOR rate assignments also carry important implications for mechanistic interpretations of primary devolatilization. In older literature, disparities among activation energies estimated for relevant chemical reaction rates and nominal devolatilization rates have been regarded as indications that mass or heat transfer mediates devolatilization mechanisms. These arguments recognize that nominal devolatilization rates have activation energies that are far lower than the bond dissociation energies of familiar organic functional groups in thermal decompositions. But roles for transport phenomena, either heat or mass, are definitively contradicted by the absence of a particle size dependence in any aspect of coal devolatilization in the operating domain of commercial interest. Moreover, the nominal devolatilization rates based on FLASHCHAIN® in this section conclusively demonstrate that heat and mass transfer limitations have nothing to do with the low apparent activation energies in nominal devolatilization rates, since the mechanisms behind the FLASHCHAIN® simulations are completely free of all heat and mass transfer resistances.

Another source of confusion has been the way that the pseudo-frequency factors in nominal rate constants are functions of heating rate and reaction time at constant temperature, especially when rate constants are used to assess consistency among different datasets. Most of the conjecture surrounding the scatter among points on Arrhenius diagrams does not properly account for the genuine variability of devolatilization rates for different test conditions, especially for variable heating rates. Fig. 6.29 clearly demonstrates that devolatilization data for diverse heating rates cannot possibly define a single set of SFOR parameters. Specifying nominal rates with FLASHCHAIN® demonstrates further that pseudo-frequency factors change in rough proportion to changes in heating rate, whereas activation energies are nearly uniform. Nominal rates should never be directly compared to assess consistency among reactors that impose different thermal histories, or even at different instants among cases that have the same thermal history.

6.3.2 Particle swelling behavior

FLASHCHAIN® has not yet been connected to any of the morphological changes occurring during devolatilization, so it does not factor into any of the results in this section. Both FG-DVC and CPD have been combined with analyses that describe the growth of bubbles in a viscous melt during devolatilization to estimate particle swelling. Swelling is important because the initial particle size and density of char particles are among the determining factors on combustion and gasification rates. It is usually estimated with swelling factor correlations, which were presented during the development of Eq. (2.9) in Chapter 2. The detailed analyses in this section aim to provide an analytical framework that automatically depicts how swelling factors change for variations in coal quality, heating rate, and pressure.

All analyses of swelling progress through three distinct stages: softening, plasticity, and resolidification. The softening stage begins with the onset of gas production, and ends when the condensed phase becomes viscous enough to seal off all pores through the external particle surface. The viscosity of the condensed phase remains above the threshold for viscous behavior throughout this stage. Being a volumetric process, devolatilization increases the internal particle porosity throughout particle softening, which reduces the density at uniform particle size. The plastic stage begins when the viscosity of the organic coal phase falls below some critical value, which signals the potential for bubble growth, coalescence, and rupture within the coal particle. Thereafter, melt viscosity passes through a deep minimum during the plastic stage, depending on coal quality and the operating conditions. Being driven by the bubble dynamics, swelling is also strongly affected by these factors. The resolidification stage begins when the melt viscosity increases beyond the threshold for viscous behavior. Trapped bubbles within the resolidified char determine the porosity which, together with the swollen particle diameter, determines the ultimate particle density. The number and dispersion of trapped bubbles determines the char morphology.

Throughout devolatilization, the magnitude of swelling, d_P/d_{P0}, passes through a maximum. The expansion is driven by the accumulation of gases within bubbles inside a particle, which raises the internal pressure to expand the bubble walls against surface tension. All or only portions of volatiles may accumulate in bubbles, depending on the heating rate and ambient pressure. Particles contract when a large central bubble or large voids rupture at the particle surface, which relaxes the internal pressure to the ambient level, and thereby allows the balance between pressure and surface tension to move the bubble boundary back toward the particle center.

The two primary connections among these physical transformations and network depolymerization mechanisms are (1) The reaction mechanisms evaluate gas production rates as the sum of release rates of all noncondensables and tar throughout devolatilization; and (2) The depolymerization/repolymerization submodels give fragment MWDs that are related to nominal viscosities for the reacting coal phase.

The different analyses for the physical transformations are categorized as either single-bubble or multibubble treatments. Solomon and Hamblen (1985) developed a single-bubble analysis that used the FG submodel to evaluate gas production rates, and treated the coal phase as a mixture of hexamer chains, although the melt viscosity was not described. Later, this group used FG-DVC to evaluate the following empirical expression for the development of fluidity during devolatilization (Solomon et al., 1992):

$$\eta = C\exp\left(-\frac{E_\eta}{RT^*}\right)\exp\left[\frac{k_E\varphi_S}{1-\dfrac{\varphi_S}{\varphi_C}}\right] \tag{6.3a}$$

where η is the viscosity of the coal phase; C, E_η, and k_E are adjustable constants; T^* is the maximum temperature used to evaluate viscosity, regardless of the actual particle temperature; and ϕ_S is the volume fraction of solids that has a critical value ϕ_C, for which the viscosity becomes infinite. During devolatilization, ϕ_S first diminishes as shorter fragments accumulate into a liquid phase, then increases while crosslinking reactions consume metaplast fragments. Although this analysis accurately interpreted plasticity data over a wide domain of conditions (Solomon et al., 1992), it was not used to explicitly estimate swelling factors.

All other swelling analyses used the expression for melt viscosity reported by Oh et al. (1989):

$$\eta = \frac{C\exp\left(-\dfrac{E_\eta}{RT^*}\right)}{[1-\varphi_M]^{\frac{-1}{3}}-1} \tag{6.3b}$$

where ϕ_M is the metaplast mass fraction in the condensed phase. With CPD, ϕ_M is evaluated as the cumulative mass of fragments in the condensed phase that are not connected to the nominally infinite Bethe lattice.

Sheng and Azevedo (2000) incorporated CPD into their single-bubble analysis. During the plastic stage, the accumulation of volatiles within bubbles is based on a balance between the volatiles production rate and diffusive escape through the outer shell of a single, central bubble. The bubble radius expands according to

$$\frac{dr_b}{dt} = \frac{r_b}{4\eta}\left(P_b - P_0 - \sigma\left[\frac{1}{r_b} - \frac{1}{r_p}\right]\right) \tag{6.4a}$$

where r_b and r_p are radii for the bubble and its outer shell; P_b and P_0 are the internal and ambient pressures; and σ is the surface tension of the melt. The bubble expands continuously throughout devolatilization until the wall strength of the shell, S_W, is exceeded, according to the following criteria:

$$\frac{1.5r_b^3(P_b - P_0)}{r_p^3 - r_b^3} - P_0 \geq S_W \tag{6.4b}$$

Upon rupture, the bubble size relaxes to some smaller steady value whose size and shell thickness satisfy the steady-state for Eq. (6.4a).

This analysis gives reasonable values for the swelling factor across broad ranges of temperature and heating rate, and predicts less swelling for progressively faster heating above 20,000°C/s. But only two *hv* bituminous samples were covered in the data evaluations (Sheng and Azevedo, 2000).

The multibubble treatments are based on the physics developed by Oh et al. (1989) which accounted for bubble growth, coalescence, and rupture. Yu et al. (2004) used CPD for devolatilization, and omitted bubble coalescence and all resistances to transport and rupture, except for diffusion through the outer shell of a cenosphere. The bubble population forms a foam whose outer shell of bubbles passes through the particle surface at the following rate:

$$\frac{dn_b}{dt} = -\left(\left[\frac{3(r_p - r_b)^2}{r_{p0}^3}\frac{dr_b}{dt}\right]\right)n_b \tag{6.5}$$

where n_b is the number of bubbles within a particle. According to this analysis, bubble growth is the only mechanism that releases volatiles from a molten particle.

Initially, n_b is a fixed value based on an estimated number of macropores of an average, uniform size. Depending on the gas production rate, n_b diminishes throughout devolatilization, and may reach the limit of a single large bubble, which depicts a cenosphere. Hence, the value of n_b at the resolidification point determines the ultimate char morphology. As seen in Fig. 6.32, the morphologies comprise dense char structures with uniform small voids, through homogeneous and heterogeneous foams, through crassispheres with only a few large voids, to cenospheres. The authors do not explain the mechanisms responsible for the morphologies that have bubbles of

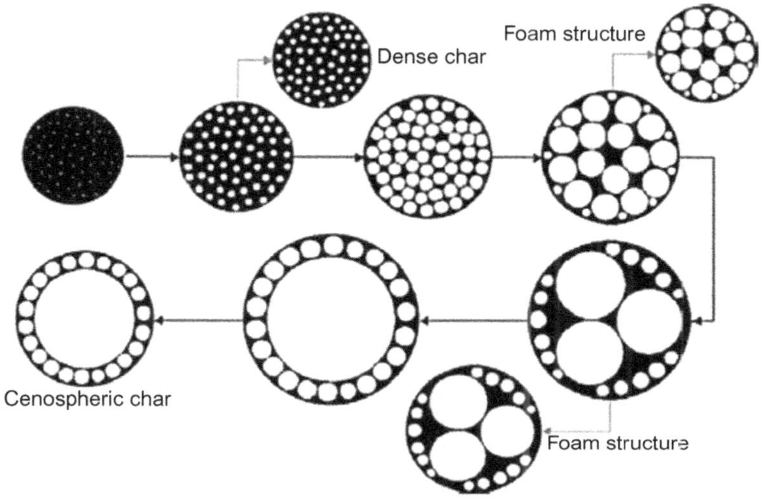

Fig. 6.32 Illustrations of various char morphologies predicted by the analysis of Yu et al. (2004).
Reproduced with permission of Elsevier.

disparate sizes, although the illustrations imply that they are associated with the diffusional resistance through the outer shell of viscous material around large voids and bubbles.

The analysis gave accurate values for the swelling ratios for five density fractions of an *hv* bituminous coal under one standardized test condition, and also gave reasonably accurate char morphologies in terms of the cenospheres, crassispheres, and fusinoid particles described in Section 4.2.6. This later prediction is at least as noteworthy as the swelling factors, because dense char morphologies make disproportionate contributions to flyash LOI from PCC furnaces. The analysis also predicts slightly lower swelling factors for progressively faster heating rates above 10^{4}°C/s, and is also the first to depict that swelling factors pass through a maximum near 1 MPa for progressively higher pressures (Yu et al., 2004).

Yang et al. (2014) developed another multibubble analysis with CPD, based on the morphology of a single large bubble surrounded by much smaller bubbles in the external shell. The prescribed morphology was intended to simplify the description of bubble coalescence. The analysis also uses Eqs. (6.3a), (6.4a), (6.5), so the predicted char morphologies may shift toward a cenosphere structure under some operating conditions. This analysis is the only one in the literature that reports internal pressures throughout devolatilization and, unfortunately, the magnitudes are preposterous. The maximum internal pressures vary from 20 to 35 MPa, which exceeds the yield strength of metallurgical coke by a few orders of magnitude. Evidently, this problem was rectified in the most recent report (Yang et al., 2015) by dramatically reducing the strength of bubble shell walls from as much as 1000 MPa in the first study to $0.1(P_0)^{0.667}$ MPa, although the basis for the connection to the ambient pressure remains unclear.

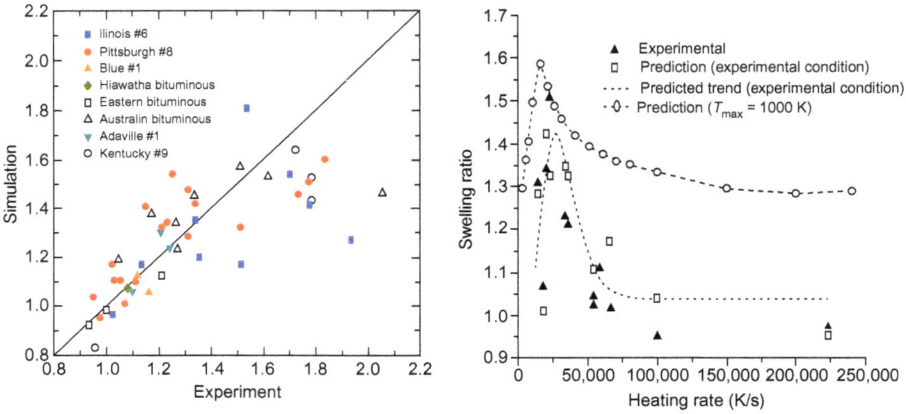

Fig. 6.33 Parity plot on predicted swelling factors for (left) lignite (Adaville #1), subbituminous (Blue #1), and *hv* bituminous coals; and (right) a broad range of heating rates. Reproduced from Yang H, Li S, Fletcher TH, Dong M. Simulation of the swelling of high-volatile bituminous coal during pyrolysis. Part 2. Influence of the maximum particle temperature. Energy Fuels 2015;29:3953–62 with permission from the American Chemical Society.

The real strength of this analysis lies in the validation work over an extremely broad domain of coal quality and operating conditions. As seen in Fig. 6.33, the predicted swelling factors were compared with data for eight samples representing ranks from lignite through *hv* bituminous, and for the complete range of heating rates faster than several degrees per second. The evaluation for heating rate variations contains one curve of predictions based on the reported test conditions, and another for a uniform ultimate temperature of 727°C to highlight the impact of heating rate alone, all other conditions uniform. This analysis also satisfies an evaluation with data for temperatures from 500 to 1200°C, and depicts the maximum in swelling factors across a broad pressure range.

At this point in the development of physical mechanisms for particle swelling, it is fair to say that the analyses explain the most important tendencies across broad ranges of heating rate, ultimate temperature, and pressure, but the coal quality impacts are not accurate enough for practical applications. One generic concern is that the predicted internal pressure needed for accurate swelling predictions are high enough to introduce inadvertent dependences on the initial particle diameter, at odds with the absence of any size dependence in the laboratory database. Note that the initial particle size is explicit in Eq. (6.5) so that, even though any network depolymerization mechanism that incorporates the FDA will predict gas production rates that are insensitive to pressure, the postulated physics of swelling introduces an independent sensitivity to the initial size. Another concern is that the required thermophysical properties are extremely difficult, if not impossible, to monitor under realistic operating conditions, which explains why different authors use grossly different values for numerous input parameters.

Until these issues are resolved, readers should also consult the engineering correlations developed from the extensive databases compiled to validate the postulated physical mechanisms (Shurtz et al., 2011, 2012).

6.4 Status summary

In this section, predictive capabilities of FLASHCHAIN® for primary coal devolatilization are evaluated separately for practical applications in process simulations and for their scientific implications.

At a minimum, a submodel for coal chemistry in a process simulator must accurately assign a devolatilization rate, the ultimate volatiles yield, and the aggregate compositions of both volatiles and char. The yields and product compositions are regarded as more important than the kinetics, because processes with p. f. grinds are often simulated with coarse grids that cover a devolatilization zone with only a few grid points anyway. For furnace simulations, the partitioning of coal-N into volatile-N and char-N is also required, and in gasifier simulations, the partitioning of coal-N and coal-S are usually required. The most complex CFD submodels resolve primary tar from aggregate volatiles, so that the relatively slow oxidation reactivity of soot can be represented. They may also isolate noncondensable fuels (CO, GHCs, H_2) and pollutant precursors (HCN, NH_3, H_2S, COS) from inert volatiles, to better estimate CO, NO_X, and S-species emissions. The practical imperative is to predict all these quantities with an absolute minimum of laboratory prerequisites, preferably with only the proximate and ultimate analyses for each coal sample.

The reported database on ultimate weight loss and tar yields clearly shows that the sample-to-sample variability must be reckoned with. For over 20 years, FLASHCHAIN® has sustained blind evaluations that demonstrate accurate predictions for ultimate and tar yields from individual coal samples. The predicted total and tar yields depict the distinctive devolatilization behavior of individual samples, given only a sample's proximate and ultimate analyses. The accuracy extends across the entire rank spectrum, and also spans the operating domain for our subject utilization technologies. In the vast majority of cases, the predicted yields are within the measurement uncertainties without any heuristic parameter adjustments.

The predicted C/H/O/N compositions for char and, by subtraction with coal compositions, for aggregate volatiles are accurate to a similar degree. The predicted coal-N partitioning is especially reliable because it has sustained many more formal evaluations than for the other elements. At this point, only two ambiguities remain: One pertains to HCN production from char-N during annealing for long contact times at very high temperatures, which can only be resolved after specialized testing with suites of diverse coals has been reported. The other pertains to the resolution of NH_3 and HCN in volatile-N from low-rank coals, which can be resolved with XPS analysis of the parent coal to assign quaternary-N levels. There are also ambiguities surrounding the retention of char-O with low volatility coals, due to the large measurement uncertainties on char-O determinations at very low oxygen levels. The partitioning of coal-S accurately depicts transformations to coal-S in the condensed phase, but does not yet accurately predict gas-S and tar-S levels for diverse coal samples. Many additional datasets that resolve tar and total yields as well as the S-species transformations are needed to rectify this situation.

The predicted ultimate yields for CO and H_2 are accurate for any coal type, but GHC yields are accurate only for ranks of *hv* bituminous and higher. Predicted GHC yields for lignites and subbituminous coals exhibit large discrepancies.

The procedure to evaluate nominal devolatilization rates from FLASHCHAIN® predictions is ideally suited to process simulations, because it circumvents laborious programming jobs and inordinate computational burdens, while it retains the rudimentary rate expressions that are already supported by the most popular applications packages. Provided that users specify sets of kinetic parameters for narrow ranges of thermal histories, particularly within one order of magnitude variation in heating rate, global rate expressions will accurately depict weight loss histories from FLASHCHAIN®. Hence, FLASHCHAIN® satisfies all essential requirements for applications in process simulators, and most of the capabilities for the most demanding simulation applications as well.

But from a scientific standpoint, there are several major opportunities for improvement. The greatest deficiencies pertain to the mechanisms that determine the molecular compositions of noncondensable products. The current scheme is based on only the assigned elemental compositions of the labile bridge reactant in a whole coal. This approach depicts the correct proportions of CO, CO_2, and H_2O for different ranks, but not much of the sample-to-sample variability. The discrepancies are greater for predicted GHC yields, even for the *hv* bituminous and low volatility coals that are reasonably well described. For low-rank coals, the predicted GHC yields are not within useful quantitative tolerances of measured values, which needs to be rectified. Perhaps a better resolution of the chemistry for spontaneous char link formation from labile bridges will bring better accuracy. One can reasonably expect generic tendencies to emerge from simulations based on realistic free-radical chain reaction mechanisms among substituted polynuclear aromatics. But it is much less clear how these tendencies can be incorporated into network depolymerization mechanisms without undermining their computational expedience, manageable numbers of rate parameters, and minimal input data requirements.

Another marker for potential improvement is the consistent underprediction of tar-O levels for coals across the rank spectrum (cf. Fig. 6.18 vs. Fig. 4.21). The resolution of this flaw may be no more involved than increasing the retention of bridge-O in char links, so that more oxygen is shuttled away as char links in tar fragments during the early stages of devolatilization. This excess oxygen would be released from fragments in the condensed phase at elevated temperatures once tar production has finished, to accurately depict the release of nearly all char-O once ultimate yields have been achieved. In fact, one implementation of this scheme was demonstrated by Su et al. (2015), and discussed in Section 5.3. Unfortunately, this work did not specifically focus on tar compositions, so this suggestion still remains to be evaluated. If the suggestion fails, then additional chemistry that directly relates oxygen functional groups to the production of tar precursors will need to be developed, so that tar precursors preferentially contain more coal-O than fragments that never disintegrate into the size range for flash distillation.

The tendencies in the tar MWDs from low-rank coals are another aspect of tar characterization that warrants a closer examination. FLASHCHAIN® predicts that tar MWDs shift toward heavier weights with coals of progressively lower rank. This tendency is at odds with the lone lab study on this issue (Unger and Suuberg, 1984), which used a single low-rank sample and tar MWDs monitored with GPC. Before the MWD shift can be addressed in the modeling, the MWD determinations need to be carefully validated for the excessive tar-O levels in these samples, especially since the legitimate concerns over the suitability of GPC for this application have not yet been definitively resolved. If, in fact, predicted tar MWDs from low-rank coals need to be shifted toward lower weights, it may be necessary to include a separate, mostly aliphatic coal component in the fragment mixture for the parent coal. Without this component, the abundance of aliphatic material in low-rank coals dramatically increases the assigned weight of labile bridges, and this excessive bridge weight is responsible for the heavier tar MWDs from low-rank coals. With a purely aliphatic component that rapidly disintegrates into GHCs but not much tar, the assigned bridge weights would be much lighter, and thereby shift the predicted tar MWDs toward lighter weights, if necessary.

Also, the predicted impact of variations in heating rate on tar MWDs for all coals is much stronger than in measured MWDs although, here too, only a few different coals were studied with a protocol that is over twenty years old. Given accurate MWD data for many more coal samples, one can then attempt to better resolve the relative kinetics for bridge conversion and peripheral group elimination in ways that eliminate the problems across a broad range of heating rates. Generally speaking, the complete lack of interest in monitoring tar MWDs during the past two decades or so has been a serious impediment to additional refinements to network depolymerization models.

Finally, FLASHCHAIN® is currently restricted to applications with heating rates faster than about 1°C/s, to ensure that secondary pyrolysis within the particles does not affect the product distribution; and also to particle sizes within the regime where mass transfer does not mediate the overall devolatilization rate. There are some utilization technologies such as coke ovens and fixed bed gasifiers that operate in the excluded domain of conditions. The secondary chemistry developed in Chapters 7 and 8 needs to be combined with detailed intraparticle transport mechanisms for volatiles escape to cover the regimes for both slow heating and large particle sizes.

References

Chen JC, Niksa S. Coal devolatilization during rapid transient heating. Part 1. Primary devolatilization. Energy Fuels 1992;6:254–64.

Hashimoto N, Kurose R, Hwang SM, Tsuji H, Shirai H. A numerical simulation of pulverized coal combustion employing a tabulated-devolatilization-process model (TDP model). Combust Flame 2012a;159:353–66.

Hashimoto N, Kurose R, Shirai H. Numerical simulation of pulverized coal jet flame employing the TDP model. Fuel 2012b;97:277–87.

Kambara S, Takarada T, Toyoshima M, Kato K. Relations between functional forms of coal nitrogen and NO_X emissions from pulverized coal combustion. Fuel 1995;74(9):1247–53.

Matsuoka A, Ma Z-X, Akiho H, Zhang Z-G, Tomita A, Fletcher TH, Wojtowicz MA, Niksa S. High-pressure coal pyrolysis in a drop tube furnace. Energy Fuels 2003;17:984–90.

Niksa S. Flashchain theory for rapid coal devolatilization kinetics. 2. Impact of operating conditions. Energy Fuels 1991;5:665–73.

Niksa S. Predicting the evolution of fuel nitrogen from various coals. Proc Combust Inst 1994a;25:537–44.

Niksa S. Flashchain theory for rapid coal devolatilization kinetics. 5. Interpreting rates of devolatilization for various coal types and operating conditions. Energy Fuels 1994c;8:671–9.

Niksa S. Flashchain theory for rapid coal devolatilization kinetics. 6. Predicting the evolution of fuel nitrogen from various coals. Energy Fuels 1995;9:467–78.

Niksa S. Flashchain theory for rapid coal devolatilization kinetics. 7. Predicting the release of oxygen species from various coals. Energy Fuels 1996;10:173–87.

Niksa S, Lau C-W. Global rates of devolatilization for various coal types. Combust Flame 1993;94:294–307.

Niksa S, Liu G-S, Hurt RH. Coal conversion submodels for design applications at elevated pressures. Part I. Devolatilization and char oxidation. Prog Energy Combust Sci 2003;29:425–77.

Oh MS, Peters WA, Howard JB. An experimental and modeling study of softening coal pyrolysis. AIChE J 1989;35:775–89.

Sheng C, Azevedo JLT. Modeling the evolution of particle morphology during coal devolatilization. Proc Combust Inst 2000;28:2225–32.

Shurtz RC, Kolste KK, Fletcher TH. Coal swelling model for high heating rate pyrolysis applications. Energy Fuels 2011;25:2163–73.

Shurtz RC, Hogge JW, Fowers KC, Sorensen GS, Fletcher TH. Coal swelling model for pressurized high particle heating rate pyrolysis applications. Energy Fuels 2012;26:3612–27.

Solomon PR, Hamblen DG. Pyrolysis. In: Schlosberg RH, editor. Chemistry of Coal Conversion. NY: Plenum; 1985. p. 121.

Solomon PR, Serio MA, Carangelo RM, Bassilakis R, Gravel D, Baillargeon M, Baudais F, Vail G. Analysis of the Argonne premium coal samples by thermogravimetric Fourier transform infrared spectroscopy. Energy Fuels 1990;4:319–33.

Solomon PR, Best PE, Yu ZZ, Charpenay S. An empirical model for coal fluidity based on a macromolecular network pyrolysis model. Energy Fuels 1992;6:143–54.

Su P, Shien J, Ling L. Depolymerization model for coal devolatilization: bridges and side chains as the reaction centers. Energy Fuels 2015;29:2162–76.

Unger PE, Suuberg EM. Molecular weight distributions of tars produced by flash pyrolysis of coal. Fuel 1984;63:606–11.

Xu W-C, Tomita A. Effect of coal type on the pyrolysis of various coals. Fuel 1987;66:627–31.

Yang H, Li S, Fletcher TH, Dong M. Simulation of the swelling of high-volatile bituminous coal during pyrolysis. Energy Fuels 2014;28:7216–26.

Yang H, Li S, Fletcher TH, Dong M. Simulation of the swelling of high-volatile bituminous coal during pyrolysis. Part 2. Influence of the maximum particle temperature. Energy Fuels 2015;29:3953–62.

Yu J, Lucas J, Wall T, Liu G-S, Sheng C. Modeling the development of char structure during the rapid heating of pulverized coal. Combust Flame 2004;136:519–32.

Further reading

Niksa S. Flashchain theory for rapid coal devolatilization kinetics. 4. Predicting ultimate yields from ultimate analyses alone. Energy Fuels 1994b;8:659–70.

Niksa S. Flashchain theory for rapid coal devolatilization kinetics. 8. Modeling the release of sulfur species from various coals. Energy Fuels 2017;31:4925–38.

Tar decomposition

7

Nomenclature

A	scaled molar concentration of aromatic nuclei in tar
A_0	molar concentration of aromatic nuclei in coal, moles/cm^3-coal
A_{GAS}	molar concentration of aromatic nuclei per unit gas volume, moles/cm^3
A_i	pseudo-frequency factor in a reaction involving species i, s^{-1}
a_k	moles of CO, CO_2, and H_2O from spontaneous charring for $k = 1, 2, 3$, respectively
A_T	pseudo-frequency factor for secondary tar destruction, s^{-1}
^1A_T	pseudo-frequency factor for primary tar production, s^{-1}
B	scaled molar concentration of labile bridges in tar
b_i	stoichiometric coefficient for product i for oil production
C	scaled molar concentration of char links in tar
$c_{O,i}$	stoichiometric coefficient for product i during oils production
$c_{R,i}$	stoichiometric coefficient for product i during addition to soot
E_{al}	threshold energy for conversion of bridges with aromatic sulfides
E_{COAL}	mass percentage of element E (= C,H,O,N,S) in coal, daf wt.%
E_i	activation energy in a reaction involving species i, kJ/mol
E_T	activation energy for secondary tar destruction, kJ/mol
^1E_T	activation energy for primary tar production, kJ/mol
E_{TAR}	mass percentage of element E (= C,H,O,N,S) in tar, daf wt.%
$^Ef_{TAR}$	tar composition as a mass fraction of the percentage of element E in coal
G	scaled molar concentration of noncondensable gases from tar decomposition
j	index for the degree of polymerization in tar molecules
J^*	maximum extent of depolymerization for primary tar molecules
k_i	rate constant for a reaction involving species i
$^1k_{TAR}$	global rate for primary tar production, s^{-1}
L_{COAL}	coal loading, g/cm^3
M_n	number-average molecular weight of tar, g/mol
O	scaled molar concentration of oils from tar decomposition
R	scaled molar concentration of soot
S	scaled molar concentration of peripheral groups in tar
S_{al}	scaled molar concentration of aliphatic sulphides
S_{ar}	scaled molar concentration of aromatic sulphides
S_{th}	scaled molar concentration of thiophene sulfur
T	Scaled molar concentration of tar molecules
t_j	tar chain with j monomer units
V^∞	ultimate yield parameter, daf wt.%
X_{T+O}	index for the extent of tar decomposition
Y_i	scaled molar yield of product i
$Y_{SOOT}(t)$	instantaneous yield of soot, daf wt.%
Y_{SOOT}^∞	hypothetical ultimate soot yield, daf wt.%

Process Chemistry of Coal Utilization. https://doi.org/10.1016/B978-0-12-818713-5.00007-1

$^{1}Y_T(t)$ instantaneous yield of primary tar, daf wt.%
$^{1}Y_T^{\infty}$ hypothetical ultimate yield of primary tar, daf wt.%
$^{2}Y_T(t)$ instantaneous yield of secondary tar, daf wt.%
ΔY_T difference between the instantaneous yields of primary and secondary tar, daf wt.%
ΔY_T^{∞} difference between the hypothetical ultimate yields of primary and secondary tar, daf wt.%

Greek symbols

η average moles of nitrogen per aromatic nucleus
ν_B scission selectivity coefficient for bridge conversion
ν_C stoichiometric coefficient for gas production during spontaneous charring
ν_{Rj} stoichiometric coefficient for soot production via addition of secondary tar
$\nu_{R,O}$ stoichiometric coefficient for soot production via addition of oils
ρ_0 bulk coal density, g/cm^3
σ_i std. dev. about the mean energy for a reaction involving component *i*, kJ/mol
Θ_C residual moles of oxygen in a char link
ξ_{th} moles of thiophene sulfur per aromatic nucleus

Subscripts

B labile bridges
C char links
N aromatic nuclei
R soot
S peripheral groups
SN soot nucleation
SC soot addition
t_j tar *j*-mer

This chapter is devoted to the decomposition of primary tar. It formally considers decomposition under only inert, nonreactive atmospheres, although decomposition rates are often so fast that the same decomposition sequence occurs under oxidizing and reducing atmospheres as well. This is especially so in pulverized coal flames, where the products of tar decomposition and secondary volatiles pyrolysis are the fuels that actually burn, rather than primary products. In contrast, tar conversion under elevated H_2 partial pressures often gives completely different product distributions, as described in Chapter 9.

Whenever primary volatiles are released into a nonreactive atmosphere at 500–600°C, primary tars spontaneously decompose into polynuclear aromatic hydrocarbons (PAH) which, under more severe conditions, coalesce into soot. Throughout this decomposition and condensation chemistry, the tar reactant eliminates its excess hydrogen as H_2 and additional GHCs; its oxygen as mostly CO; its nitrogen as HCN; and its sulfur as H_2S. At temperatures hotter than 800–900°C, GHCs and oils add to a growing soot product. These additions nearly compensate for the loss of CO from tar on a mass basis, so that soot yields are only moderately lower than primary tar yields. Tar decomposition at moderate temperatures also produces oils, which are mixtures of benzene, toluene, and xylene (BTX) and phenol, cresol, and xylenol (PCX). With many coals, tar decomposition is the predominant source of oils under rapid heating conditions.

Tar decomposition enhances the yields of nearly all the primary noncondensables, except that H_2O and CO_2 yields are hardly perturbed. While tars are being converted under inert atmospheres, C_1-C_4 primary GHCs are reformed into only CH_4 and C_2H_2, with small amounts of ethylene. This reforming chemistry is covered in Chapter 8, to sharpen this chapter's focus on how tar decomposition in isolation enhances the yields of noncondensable gases.

This chapter first presents the most important features of tar decomposition in terms of laboratory data, including impacts of both fuel quality and operating conditions. It then reviews a FLASHCHAIN®-based reaction mechanism that interprets tar decomposition behavior for any coal at any operating conditions, and surveys the performance with a validation database. The final section illustrates how the full mechanism specifies parameters in simple, global schemes that mimic the full results in application simulators, like CFD.

7.1 Commercial impacts

At temperatures $\sim>550°C$, depending on coal quality and residence time, tar decomposition is unavoidable because it does not require any gaseous reactants or catalytic solids. The temperature threshold that destabilizes tar becomes cooler for tars that contain progressively more oxygen, as do the primary tars from coals of progressively lower rank, and also as residence times are extended. Tar decomposition is often the second stage in the chemistry of commercial coal utilization technologies, and it matters in any scheme that operates above 600°C, primarily for two reasons. First, the accompanying enhancements to the noncondensable products are always appreciable enough to affect the reforming chemistry in gasifiers, especially at moderate-to-low temperatures, and also NO_X production in coal combustors. Tar decomposition often more-than-doubles the levels of CO and H_2 from primary devolatilization, which is comparable to or greater than the CO and H_2 produced by reforming the primary GHCs with steam. These reducing agents, in conjunction with the additional GHCs, also drive aerodynamic NO_X abatement within the burner belts of coal-fired furnaces, as described in Chapter 8. In addition, tar decomposition dramatically enhances HCN levels which, as the main reactant in homogeneous coal-N conversion, subsequently affect NO_X levels. Tar decomposition may also be the predominant thermal source for oils; however, even though oils are valuable chemical feedstocks, their yields must be boosted much further for commercial viability by processing under elevated H_2 pressures, as described in Chapter 9.

The second major impact of tar decomposition occurs at temperatures above 800–900°C, where PAH, oils, and the remaining GHCs coalesce into soot. Soot formation transforms a substantial volatile fuel component, tar, into a carbonaceous solid, and also sequesters highly reactive GHCs and oils into this very slow-burning solid. In pyrolysis processes that regard tar as a product or chemical feedstock for subsequent refining, sooting directly consumes tar and, if run to completion, eliminates it entirely. The impact is not much weaker in combustors and gasifiers. Even though soot agglomerates remain smaller than a micron in size, they contain no metallic elements to catalyze their subsequent conversion. So soot conversion via both combustion and gasification is necessarily much slower than tar vapor conversion by the analogous homogeneous combustion or gasification processes. In fact, soot can contribute to flyash LOI levels for CFBCs fired with anthracites and other nonreactive low volatility coals, although LOI from PCC furnaces rarely contains appreciable soot unless the furnace is badly out-of-tune. Soot also acts as a substrate for heterogeneous NO reduction by CO in near-burner flame zones, and some tar-N remains in soot, which is partially converted into NO via the same channels that convert char-N primarily into NO. Finally, soot is a major source of luminosity at the base of coal flames, and enhances radiant heat transfer rates to waterwalls in burner belts, as described elsewhere (Fletcher et al., 1997).

One does not need to understand the chemistry of tar decomposition to appreciate its impact. The datasets covered in Chapter 4 showed that primary tar yields represent about a quarter of the organics in all but low volatility coals and anthracites, and tar yields can exceed one-third for the richest subbituminous and *hv* bituminous coals at the fastest heating rates. Moreover, primary tars contain up to 40% of coal-O, 40%–50% of organic coal-S, and up to one-third of coal-H and coal-N. When a relatively rich *hv* bituminous coal is processed at moderate temperatures, tar decomposition releases an additional 8 daf wt.% CO, up to 5% oils plus a comparable amount of GHCs, and smaller amounts of H_2, HCN, and H_2S. None of these enhancements are inconsequential. At flame temperatures, this coal generates similar levels of CO, H_2, HCN, and H_2S plus an amount of soot only moderately smaller than the original primary tar yield. Clearly, tar decomposition transforms primary products into the fuel mixtures that actually burn in flames and are reformed in gasifiers and, in hot furnace gases, adds substantial amounts of soot to the fuel mix.

Tar decomposition also plays a prominent role in this author's general strategy to organize the chemistry of coal utilization. The tar decomposition mechanism developed in this chapter represents a bridge from the phenomenological reaction mechanisms for primary devolatilization to the phenomenal knowledge-base on hydrocarbon combustion chemistry developed during the last 50 years. Elementary reaction mechanisms are already validated and available to easily simulate the chemistry of mixtures of the noncondensable organics with single-ring compounds and lighter species in both combustion and gasification environments, including pollutant formation. Consequently, there is no need to resort to phenomenology or rudimentary global reaction schemes to simulate the conversion of the noncondensable fuel mixtures. In other words, the phenomenological mechanism for primary devolatilization predicts distributions of primary products, then the phenomenological mechanism for tar decomposition predicts how primary products are transformed into the noncondensable fuel mixtures that actually burn in furnaces and are reformed in gasifiers on short reaction time scales. The transformations in furnaces and gasifiers are analyzed with validated elementary reaction mechanisms in Chapter 8.

The ultimate goals are different for coal-fired combustors and entrained-flow gasifiers on the one hand, and for low and moderate temperature gasifiers on the other. For high temperature combustors and gasifiers, the main goal is to specify the yield of soot, and the distributions of noncondensable fuels and pollutant precursors that compete for the available O_2 with char. This is crucial because soot is converted on similar time scales to larger char particles, whereas finer char is converted at comparable rates to the noncondensable fuel mixtures. The reaction kinetics for tar decomposition are unimportant in these situations, because the process can be assumed to be instantaneous at these elevated operating temperatures whereby secondary products are released at the rate of primary devolatilization. In other words, tar decomposes as fast as it is released into the gas phase, so the tar decomposition rate is instantaneous. In contrast, for low and moderate temperature gasifiers, the goal is to describe the concentrations and MWDs of tar and PAH throughout the process, because these condensable species are responsible for many operational problems. Tar decomposition kinetics are crucial because they determine tar concentrations downstream of the gasifier, where operational problems tend to occur.

7.2 Laboratory prerequisites

The schematic diagram in Chapter 4 that introduced tar decomposition as an aspect of the devolatilization stage in coal conversion (cf. Fig. 4.1) delineates two channels for temperatures below and above 900°C. At moderate temperatures, the main products are PAH and oils, plus the enhanced yields of CO, GHCs, H_2, HCN, and H_2S. Characteristic times for tar decomposition at moderate temperatures are comparable to those for primary devolatilization. At elevated temperatures, the only products are soot, CO, CO_2, H_2O, CH_4, C_2H_2, H_2, HCN, and H_2S, and the transformations occur too quickly to resolve in time with measurements.

As a brief intermediate step in a complex sequence of chemical processes, tar decomposition cannot be monitored in any large-scale coal utilization system. It occurs in about ten milliseconds in furnaces and entrained flow gasifiers, which is too fast for extractive sampling. Flow patterns are too complex for sampling probes anyway. In fluidized systems, solids loadings are much too heavy to permit any access at all, and mixing patterns are too convoluted for extractive sampling along a time coordinate. So tar decomposition has only been characterized at lab scale.

In general, laboratory characterizations of tar decomposition are easier at moderate temperatures. PAH are easier to handle than primary tar, because the oxygen and nitrogen in tar substantially increase the viscosity of tar liquids, enhance carcinogenicity, and interfere with methods to monitor MWDs. The main challenge at elevated temperatures is the very short time scale for tar decomposition. There simply are no means to monitor tar decomposition on-the-fly other than extractive sampling followed by an analysis to assign conversions. Optical diagnostics provide qualitative fingerprints for tar and PAH, but no quantitative information on concentrations.

In any testing, the most important consideration is to clearly resolve tar decomposition chemistry from primary devolatilization. Two-stage reactors that isolate primary devolatilization in a first stage from secondary pyrolysis in a second stage and eliminate char conversion altogether have proven to be especially useful, particularly when primary products are kept below the temperature threshold for tar decomposition during transport to a second reactor that converts volatiles at a specified hotter temperature. Once the volatiles chemistry has been isolated in a second reactor stage, it is usually straightforward to impose specified temperatures and reaction times under a specified test pressure and known gas composition. Tests that monitor complete product distributions and are therefore able to close material and element balances are most valuable; in fact, without tight balance closures, it is impossible to distinguish intermediates from products. The tar yields must be resolved either in time or across a temperature range to assign kinetics, and supplemental tar characterizations provide valuable insights, particularly MWDs. The widest variety of coal samples is always desirable and, at a minimum, each fuel must be described by a proximate and ultimate analysis.

More formally, the following testing features are required in any dataset used to evaluate tar decomposition mechanisms:

1. *Coal properties*—Only the proximate and ultimate analyses are required. Strictly speaking, only an ultimate analysis plus the proximate volatile matter on a dry, ash-free (daf) basis is necessary.
2. *Pressure*—Usually a uniform test pressure will be specified although a pressure history can also be handled.
3. *Thermal history*—Sufficient information must be available to assign the temperature of the fuel sample throughout an entire test, as well as the thermal history for secondary chemistry in a second reactor stage. This information may comprise the thermal history of a sample support, as in WMRs, or the sample properties and ambient conditions that determine a sample's thermal history, as in EFRs. Operating conditions for a second reactor stage should be specified with thermocouple data, and reaction times or RTDs must be monitored or assigned by testing teams.
4. *Gas composition*—Entrainment gas compositions and coal loadings must be specified to assign the concentrations of volatiles during secondary volatiles conversion. Coal loadings should be comparable to those in the subject application. Reactive gases (H_2, O_2, steam) should be omitted.
5. *Distribution of primary volatiles*—Complete product distributions for primary devolatilization must be monitored at the conditions used to generate them in the tests on tar decomposition. These distributions must cover the most severe thermal processing in the tar decomposition tests, and should represent ultimate primary products. Elemental compositions for primary tars are especially valuable. All major primary noncondensables should be monitored, including CO, CO_2, H_2O, H_2, H_2S, HCN, and C_1-C_4 GHCs. Tar MWDs are also valuable.
6. *Index for secondary chemistry*—The progress of secondary chemistry must be specified quantitatively with a measured conversion index, which is usually based on the ultimate aggregate yield of all condensed aromatic products of devolatilization in the first stage reactor, which can be evaluated as the sum of primary tar plus oils. Elemental compositions for secondary tars are especially valuable. All major noncondensables should be monitored, including oils plus all primary noncondensables plus C_2H_2. Aerosol products should also be monitored, so that soot levels can be evaluated after tars and oils are separated out. Secondary tar MWDs are useful, especially around the threshold for sooting.

The assignment of thermal histories is one cumbersome requirement. The advantages of WMRs and CPPs over EFRs in this regard were covered in Section 4.1.3. WMRs, CPPs, and thermobalances almost always support the fuel in a very thin layer, usually no more than 1–2 particle diameters thick. Packed bed reactors (PBRs) of much greater sample volumes have also been used as first stage pyrolyzers, because they raise tar concentrations into a second reactor, and thereby improve the resolution of tar decomposition. But this configuration also raises two concerns. First, steep radial gradients will obscure the thermal history in the volatiles preparation stage unless very slow heating rates are imposed. Second, volatiles released at the bed inlet may react with char downstream so it may be impossible to generate primary devolatilization products with PBRs. Unless tar characteristics are monitored at the bed outlet, it is impossible to know whether secondary chemistry came into play within the packed bed. The safest operating mode with PBRs is to keep temperatures below the threshold for any tar decomposition at all.

Notwithstanding these ambiguities, one needs to specify only a nominal heating rate for the primary devolatilization stage, provided that the first stage reactor also imposes an IRP that is long enough to attain asymptotic ultimate volatiles yields at the stated reactor temperature. Much better precision would be required to resolve primary devolatilization kinetics during heat-up, but that is not necessary for any of the relevant literature datasets on secondary volatiles conversion. Our focus on relatively rapid heating rates—faster than 1°C/s—ensures that residence times for volatiles within fuel particles are too short for appreciable tar conversion before the tar reaches the free stream. In addition to the ease of assigning thermal histories, modern WMRs also have the advantage of a simple means to eliminate secondary chemistry in the first reactor stage: by a sweep gas flow. In modern WMRs, CPPs, and thermobalances, an inert flow moves across the sample support to transport volatiles away from the heated sample support and either into a holding tank or directly into a second reactor stage.

Tar decomposition is inevitable in EFRs because the process stream is necessarily hotter than the particles during particle heating, which is when devolatilization occurs. The notable exception is the RCFR described in Section 4.1.3 which heats a dense fuel suspension by radiation transfer rather than convection. This system is ideally suited to study tar decomposition because the temperature difference between the coal and entrainment gases can be regulated at will by varying the particle loading and furnace element temperature. Much cooler gas temperatures deliver pristine primary devolatilization products, whereas gases as hot as the suspension give controlled extents of tar decomposition, depending on the transit times through the reactor. Another advantage is that the relatively dense particle loadings generate high volatile species concentrations in the effluent, so that mass and element balances can be closed in individual runs.

FBRs with freeboards offer an intrinsic segregation of primary and secondary chemistry. The fluidization conditions determine the heating rate, reaction temperature, and residence time for primary devolatilization, whereas the freeboard size determines the residence times for secondary chemistry among gaseous species. Usually freeboard temperatures are controlled somewhat independent of the FBR temperature.

The most common second-stage reactors are electrically heated tubular flow reactors (TFRs) and FBR freeboards. A shock tube has also provided very useful information (Doolan et al., 1987). Reported axial temperature profiles along flow tubes always exhibit a steep parabolic form that may or may not contain an appreciable isothermal section in the middle. They are clearly not isothermal reactors, and the temperature profiles must be factored into kinetic interpretations. Chemistry is resolved in time by varying flowrates or tube sizes. Residence times have also been regulated by packing the flow tube into a PBR. Quartz is regarded as the most inert packing material, whereas char, carbons, and alumina are definitely reactive (Serio et al., 1987).

Oxygen is the most reactive gas with tar; so reactive, in fact, that flame fronts always envelop very hot regions in which tar decomposition is essentially instantaneous. So O_2 must be completely eliminated. Hydrogen is the only gas in commercial applications that disrupts the progression from tar to PAH to soot, albeit only with very high H_2 pressures. Neither steam nor CO_2 affect tar decomposition on the time scale for primary devolatilization, particularly the production of soot, except that steam catalytically reforms tars on char from coals of lowest rank (Hayashi et al., 2002). But steam reforms primary noncondensable volatiles, and CO_2 promotes dry reforming of noncondensable volatiles. So these gases may complicate the resolution of tar decomposition in measured product distributions.

In addition to these prerequisites for the regulation of operating conditions, the datasets must resolve the progress of tar decomposition. The most direct resolution is referenced to ultimate primary tar yields from devolatilization in a first reactor stage that permits no secondary chemistry whatsoever. Such pristine volatiles are most often prepared in swept WMRs or CPPs, RCFRs, or in FBRs operated no hotter than 600°C with bituminous coals; 550°C with low rank coals; or 500°C with biomass.

Reported product compositions from low temperature FBRs are in excellent agreement with those from swept WMRs (Doolan et al., 1987), and from swept CPPs (Hayashi et al., 1995). A suitable index on tar decomposition can be evaluated as the extent of reduction in the ultimate primary tar yield during conversion in a second reactor stage. Elemental tar compositions and oil yields should be monitored for several sets of secondary pyrolysis conditions. The most important gas yields to resolve are GHCs, CO, and H_2 because these species are the major intermediates in tar decomposition. Oil yields should also be resolved, especially during the initial stages of tar decomposition. Cyanide is important whenever NO_X formation is relevant. Elemental compositions of primary char are needed to close element balances for the primary product distributions. Other char characteristics are unimportant if an ultimate weight loss was attained in the first reactor stage, as is usually the case.

7.3 Laboratory database on tar decomposition

This section presents the most important attributes of tar decomposition in laboratory datasets, first, for moderate temperatures, where soot production is relatively unimportant, and then for furnace and gasifier temperatures, where soot becomes the predominant ultimate product of tar decomposition. The major organic products are characterized before pollutant precursors.

7.3.1 Major organic products of tar decomposition

The broad temperature range for tar decomposition is seen in the secondary tar yields from several coals in a FBR in Fig. 7.1. Unlike most tests described in this section, this series imposed the same temperature on the bed and freeboard, and nominal gas transit times through both sections varied from 1 to 0.5 s. (Throughout this chapter, the longest time in a range of transit times through a flow reactor corresponds to the coolest temperature, and vice versa.) The most striking feature is the breadth of the temperature range for all coal types. Primary tars continue to be released up to 600°C from *hv* bituminous coals, and up to 550°C from brown coals. The tar yields pass through maxima at 600°C for *hv* bituminous and at about 500°C for brown coals. The decomposition kinetics appear to be more variable with low rank tars, but this only reflects the small number of samples. Most important, tars are relatively refractory intermediates, decomposing over a range of 550°C in the 0.5–1 s residence times in these tests. The same range pertains even to tars with very high oxygen levels. The apparent temperature window would become cooler for progressively longer residence times.

The major products of tar decomposition appear in Fig. 7.2 as increments in the distribution of primary tar-C. This study prepared primary tars at 600°C in a FBR, then monitored their decomposition in a TFR at hotter temperatures with transit times from 1 to 0.7 s (Ledesma et al., 1998). This group used the measured C-distribution of primary products to specify the contributions for tar decomposition by subtraction. Through 800°C, tar decomposes into additional GHCs and $CO + CO_2$ in roughly equal proportions, consistent with similar surges in aggregate GHC yields during the initial

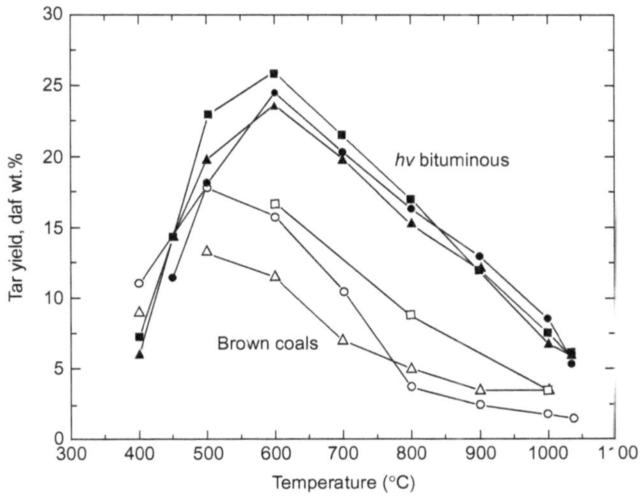

Fig. 7.1 Secondary tar yields from brown and *hv* bituminous coals versus FBR temperature. Reproduced from Li CZ, Nelson PF. Fate of aromatic rings systems during thermal cracking of tars in a fluidized-bed reactor. Energy Fuel 1996;10:1083–91 with permission from the American Chemical Society.

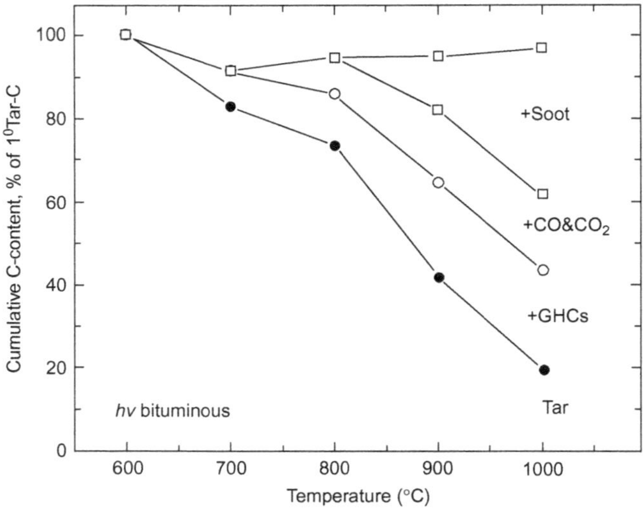

Fig. 7.2 Distribution of major products of tar decomposition with *hv* bituminous coal, as a percentage of the C in primary tar.
Reproduced from Ledesma EB, Li CZ, Nelson PF, Mackie JC. Release of HCN, NH_3, and HNCO from the thermal gas-phase cracking of coal pyrolysis tars. Energy Fuel 1998;12:536–42 with permission from the American Chemical Society.

stages of tar decomposition reported by other groups (Serio et al., 1987; Xu and Tomita, 1989; Hayashi et al., 1995; Jia et al., 2004). Above 800°C, soot becomes a major decomposition product for one-second residence times, accounting for more than one-third of the primary tar-C.

A very similar laboratory set-up was used to identify changes to individual GHC species throughout tar decomposition, except that transit times through the TFR were only 100–200 ms. As seen in Fig. 7.3, the temperature range was extended to 1100°C to compensate for the shorter residence times, which enabled this test series to cover more than 80% of primary tar decomposition. Indirect indications of soot production were reported for 900°C and hotter. The heavier GHC species pass through maxima that occur at lower temperatures for progressively heavier species. Even the CH_4 levels pass through a maximum, albeit at the hottest temperature among all GHC species; however, there is no indication that CH_4 will also vanish like the heavier GHCs. The only distinctive GHC is C_2H_2, which accumulates continuously throughout tar decomposition. All these tendencies are evident in datasets from Serio et al. (1987), Xu and Tomita (1989), Hayashi et al. (1995), and Jia et al. (2004).

Also in Fig. 7.3, the level of C_6H_6, as a surrogate for oils, increases throughout tar decomposition, which suggests that oils are a tar decomposition product. It could conceivably indicate that substituted oils are converted into unsubstituted oils during tar decomposition, except that this transformation diminishes rather than increases the aggregate oils yield, although oils yields do not diminish unless soot is formed.

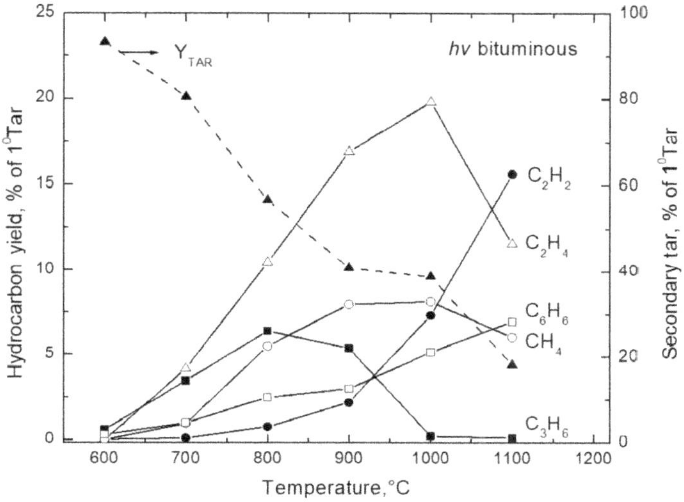

Fig. 7.3 *(Solid curves)* Distribution of major GHC species from tar decomposition with *hv* bituminous coal; and *(dashed curve)* the level of secondary tar, in wt.% of primary tar. Reproduced from Doolan KR, Mackie JC, Tyler RJ. Coal flash pyrolysis: secondary cracking of tar vapours in the range 870–2000 K. Fuel 1987;66(4):572 with permission from Elsevier.

The source of the noncondensable tar decomposition products is clearly identified as the heteroatoms in primary tar in Fig. 7.4, which shows the elemental composition of secondary tar throughout tar decomposition, along with the associated H/C ratio. Secondary tar spontaneously loses heteroatoms throughout tar decomposition, so that carbon accumulates until an ultimate H/C ratio of about 0.7 is achieved. This value is the nominal value for fully condensed, unsubstituted PAH compounds. Tar-O is eliminated more completely than the other heteroatoms, although this dataset does not resolve O-release from N- and S-release. The important implication is that primary tars do not decompose into a fixed distribution of products. Rather, they continuously eliminate their heteroatoms as noncondensable gases to ultimately form a diverse assortment of PAH species. The imperative is to identify the noncondensable products associated with the release of each heteroatom.

The only way to make these identifications is to monitor all major products of tar decomposition to close the balances on mass and the major elements in every test. This challenge was first met by Xu and Tomita (1989), who used a tubular IR quartz furnace for primary devolatilization connected to a PBR for regulated secondary pyrolysis. They compiled the complete product distributions in Table 7.1 across a broad temperature range with transit times to 14 s with one *hv* bituminous coal. Among the oxygenated gases, only CO is released during tar decomposition, because both CO_2 and H_2O yields are hardly perturbed. This important assignment is corroborated by Chen et al. (1992) and Hayashi et al. (2000, 2002), but inconsistent with Jia et al.'s finding (2004) that CO_2 yields tripled during the decomposition of tars from an *hv* bituminous coal. Carbon monoxide should tentatively be regarded as the main product

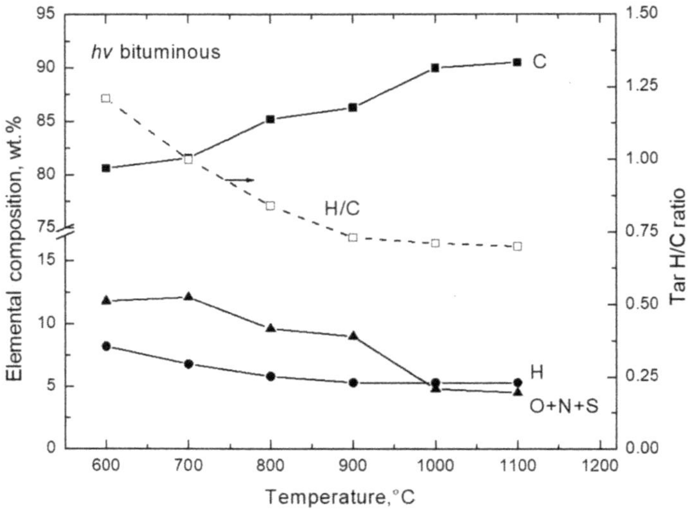

Fig. 7.4 *(Solid curves)* Elemental composition of secondary tar from *hv* bituminous coal; and *(dashed curve)* atomic *H/C* ratios.
Reproduced from Doolan KR, Mackie JC, Tyler RJ. Coal flash pyrolysis: secondary cracking of tar vapours in the range 870–2000 K. Fuel 1987;66(4):572 with permission from Elsevier.

of the release of tar-O from any coal, whereas CO_2 and H_2O levels are relatively unaffected by tar decomposition, pending additional clarification. The implication is that condensation reactions among the oxygenated functional groups on metaplast fragments that form primary H_2O and CO_2 cannot occur with these same functional groups on gaseous tar molecules (Hayashi et al., 2000). An important caveat for H_2O will be presented after this dataset.

Hydrogen yields more-than-double during tar decomposition for two reasons: (1) Hydrogen must be expelled when tar molecules are converted into PAH, because PAH is much more aromatic than primary tar; and (2) Heavier GHCs released during tar decomposition are ultimately converted into binary mixtures of CH_4 and C_2H_2, which enhances H_2 levels. Reported H_2 yields roughly tripled during tar decomposition at moderate temperatures with an *hv* bituminous coal (Jia et al., 2004).

In Table 7.1, the oils yields are fairly uniform through the initial stages of tar decomposition, but then diminish above 800°C for residence times longer than a few seconds. The reported coke yields were deposits on tube walls rather than soot recovered along with tar as an aerosol product. They remained small through 700° C, then surged at hotter temperatures, suggesting that soot production began around 800°C for the long residence times in these tests. Some other datasets show the same tendency (Chen et al., 1992), whereas others show oils production throughout tar decomposition at lower temperatures (Hayashi et al., 1995).

The caveat pertaining to whether or not the release of tar-O also produces H_2O was revealed in remarkably accurate tests that monitored primary product distributions in a

Table 7.1 Noncondensable gases from the decomposition of primary tar from an *hv* bituminous coal for ranges of temperature and residence time

T (°C)	*t* (s)	CO_2	H_2O	CO	H_2	CH_4	C_2-C_4	Oils	Coke
Temperature study									
500	7	1.7	4.0	2.1	0.4	3.5	3.1	2.9	1.2
600		1.6	4.3	2.0	0.4	3.6	3.6	3.1	1.4
700		1.4	4.1	2.5	0.5	4.6	5.1	2.8	2.3
800		1.9	4.1	2.7	0.6	5.4	4.2	2.4	3.1
900		2.0	3.6	3.4	0.9	5.7	2.1	1.8	4.9
Time study									
800	0	1.3	4.1	1.4	0.2	2.8	2.4	2.9	0
	0.2	1.4	4.1	1.7	0.3	3.9	4.8	2.9	2.4
	0.4	1.4	4.0	1.9	0.3	4.1	4.8	3.7	2.0
	2	1.4	4.0	2.6	0.4	4.8	4.4	2.6	2.2
	4	1.4	4.1	2.1	0.3	4.3	4.0	2.5	2.5
	7	1.9	4.1	2.7	0.6	5.4	4.2	2.4	3.1
	14	1.4	3.9	2.2	0.5	5.3	3.5	2.2	3.0

Reproduced from Xu W-C, Tomita A. The effects of temperature and residence time on the secondary reactions of volatiles from coal pyrolysis. Fuel Process Technol 1989;21:25–37 with permission from Elsevier.

CPP, and secondary product distributions in an EFR with a brown coal (Hayashi et al., 2000). The differences among the levels of CO + CO_2, GHCs, and H_2O from the two reactors were especially meaningful. They showed that GHCs were produced during tar decomposition, but their aggregate yield passed through a maximum at 850–900°C, coincident with the onset of sooting. The sum of CO + CO_2 yields grew continuously throughout tar decomposition, which we attribute to the elimination of tar-O mostly as CO. The H_2O yields were unaffected by tar decomposition through 800°C. At hotter temperatures in the EFR, they sharply decreased, coincident with analogous reductions in the tar level and a surge in the H_2 level, as seen in Fig. 7.5. By comparing test results before and after alkali and alkaline earth cations were removed from the coal, this group proved that H_2O was eliminated in a heterogeneous, catalytic reforming of secondary tars on char particles. This form of heterogeneous tar decomposition is probably unimportant for coals of higher rank, unless they have exceptional levels of ion-exchangeable cations. But for the lowest rank coals, heterogeneous tar decomposition is not negligible compared to homogeneous tar decomposition.

The only study to report changes in the MWDs of secondary tars throughout tar decomposition monitored secondary pyrolysis along the freeboard of a fluidized bed with an *hv* bituminous coal (Katheklakis et al., 1990). Both tar yields and M_n-values are collected in Table 7.2. The tar yields more-than-double while temperatures were increased from 400°C to 600°C, as expected, reaching a maximum value of 31 daf wt.%. The tar yields then diminish for progressively hotter temperatures, reaching ultimate values of 25% and 18% for transit times of 0.8 and 4.5 s, respectively. For the shorter transit time, the tar MWDs shift toward heavier values up to

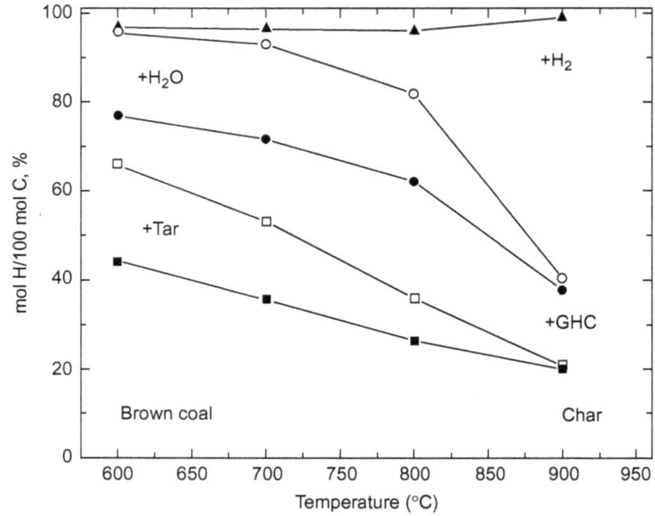

Fig. 7.5 Cumulative H-distributions for the decomposition of tar from a brown coal. Reproduced from Hayashi J-I, Takahashi H, Iwatsuki M, Essaki K, Tsutsumi A, Chiba T. Rapid conversion of tar and char from pyrolysis of a brown coal by reactions with steam in a drop-tube reactor. Fuel 2000:79:439–47 with permission from Elsevier.

Table 7.2 Yields and M_n-values for secondary tars for two transit times across a FBR freeboard

T (°C)	0.8 s		4.5 s	
	Y_{TAR}	M_n	Y_{TAR}	M_n
400	14.1	460	13.5	420
500	26.1	510	24.6	460
580	31.0	540	28.1	400
650	27.5	520	24.4	310
750	24.6	350	18.2	260

Reproduced from Katheklakis IE, Lu S-L, Bartle KD, Kandiyoti, R. Effect of freeboard residence time on the molecular mass distributions of fluidized bed pyrolysis tars. Fuel 1990;69(2):172 with permission from Elsevier.

600°C, which is the tendency expected for primary tars in isolation. For hotter temperatures and for all temperatures hotter than 500°C with the longer transit time, secondary tars become much lighter; in fact, both M_n-values for 750°C are comparable to or even lighter than the weight of a monomer unit in the subject coal. This shift is potentially misleading because the elimination of heteroatoms and GHCs from secondary tar is responsible for the shift toward lighter weights, rather than extensive depolymerization into smaller chain sizes. Unfortunately, MWD measurements like these are subject to large uncertainties, and have not yet been reported for the onset of sooting, where tar MWDs are expected to shift toward heavier values as tar molecules recombine into soot nuclei.

Transformations among the major species during the early stages of soot formation were characterized in an EFR with two brown coals (Hayashi et al., 2002). The distribution of coal-C among the tar decomposition products in Fig. 7.6 shows that, with these two brown coals, both CO and soot accumulated for progressively hotter temperatures, while levels of both tar and GHCs either diminished or passed through a weak maximum. The CO_2 yields either diminished or increased, albeit very slightly. Roughly half the primary tar was converted into soot at 800°C in the 14 s residence time. This dataset clearly shows that tars and GHCs are directly converted into soot during tar decomposition once the thermal severity exceeds a threshold of 800–900°C, depending on residence time, even with the most heavily oxygenated tars from coals of the lowest rank.

Ultimate soot yields from coals across the rank spectrum are collected in Fig. 7.7. These measurements were compiled from several studies that imposed thermal severities sufficient to convert all or nearly all secondary tar into soot (Nenniger et al., 1983; Wornat et al., 1987; Chen et al., 1992; Rigby et al., 2001; Umemoto et al., 2016). At face value, they can be regarded as the ultimate soot levels, but wide variations in the coal loadings among the tests used to generate these data are probably responsible for some of the apparent coal quality impacts in Fig. 7.7, as discussed below (cf. Fig. 7.18). Soot is an appreciable product with any coal type, provided that the processing severity is sufficient to completely convert primary tars. Ultimate soot yields are not that much lower than expected primary tar yields for these test conditions, except that soot yields from

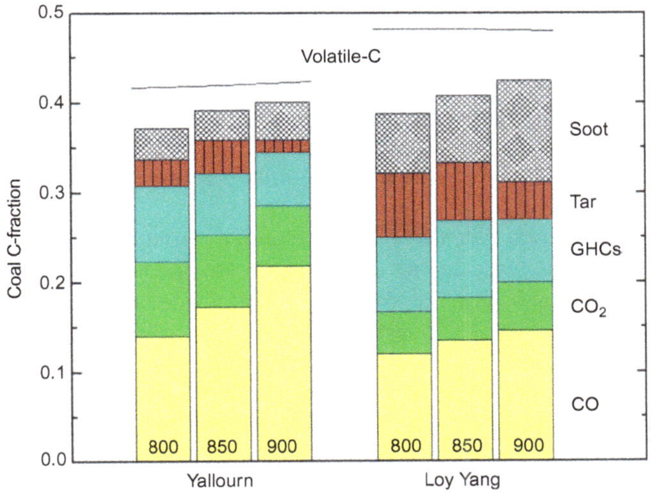

Fig. 7.6 Cumulative C-distributions for the decomposition of tars from two brown coals. Reproduced from Hayashi J-I, Iwatsuki M, Morishita K, Tsutsumi A, Li CZ, Chiba T. Roles of inherent metallic species in secondary reactions of tar and char during rapid pyrolysis of brown coals drop-tube reactor. Fuel 2002:81:1977–87 with permission from Elsevier.

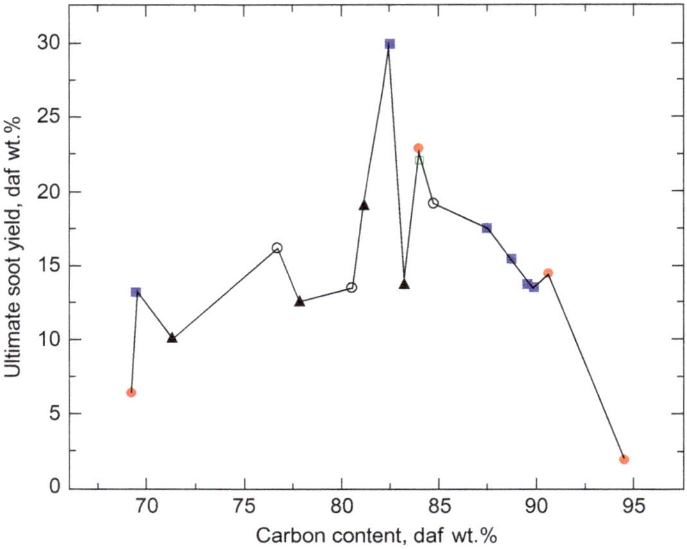

Fig. 7.7 Ultimate soot yields for conditions severe enough to decompose nearly all secondary tar.

low rank coals are appreciably lower than the expected tar yields. One reason is the abundance of oxygen in the tars from these ranks, which is expelled as noncondensables, and thereby reduces the soot mass. But variations in the coal loadings also come into play. For higher ranks, soot yields also display a similar rank dependence to primary tar yields, whereby soot yields are maximized for *hv* bituminous coals, then fall off sharply for low volatility coals and nearly vanish for anthracites. The sample-to-sample variability in the soot yields is every bit as significant as it is for primary tar yields although, again, variations in coal loadings obscure this aspect, and many more low rank and low volatility coals need to be tested to fully demonstrate the actual sample-to-sample variability.

The elemental compositions of soot formed from tar decomposition are unremarkable in some respects, but distinctive in the abundance of O, N, and S; N- and S-contents appear in Fig. 7.8. With any coal, C-contents of soot are dispersed about 92–96 wt.% while the H-contents vary from 0.75% to 1.25%. As seen in Table 7.3, soot *C*/*H* ratios increase for coals of progressively higher rank, although a nominal value of 9–10 seems to pertain to all the bituminous ranks. More data is needed to clarify this attribute. Among the heteroatoms, S-contents are most variable, which is not surprising considering that only thiophene-S can be incorporated into soot at elevated temperatures, and thiophene-S is a highly variable component of coal-S. Nitrogen contents roughly track coal-N levels, albeit at lower absolute magnitudes, because the bulk of coal-N is present in pyrollic and pyridinic forms that can be incorporated into a nascent soot phase.

The product distributions for two coals in Fig. 7.9 cover nearly complete tar decomposition, and serve to summarize all the observations in this section about the main

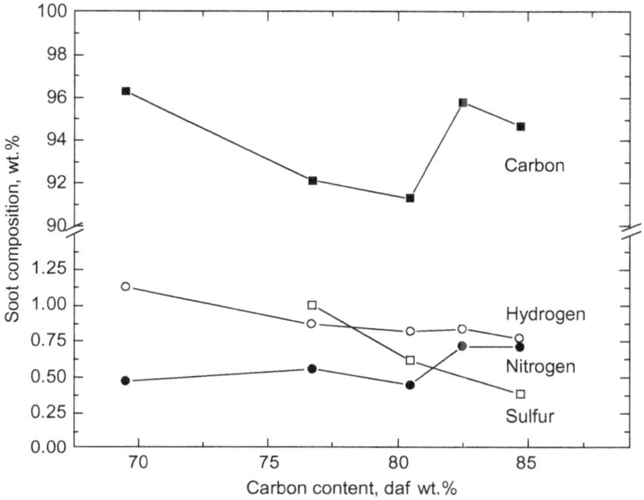

Fig. 7.8 Elemental compositions of ultimate soot samples.
Reproduced from Rigby J, Ma J, Webb BW, Fletcher TH. Transformations of coal-derived soot at elevated temperature. Energy Fuel 2001;15:52–9 with permission from the American Chemical Society.

Table 7.3 Ultimate *C/H* ratios of soot for several coals (Chen et al., 1992; Rigby et al., 2001) assembled with permission from the American Chemical Society.

Coal-C (daf wt.%)	Soot *C/H*
69.5	7.0
76.7	8.8
80.5	9.3
82.5	9.5
84.7	10.2

conversion channels throughout tar decomposition (Chen et al., 1992). These datasets are the only ones in the literature that characterize nearly complete tar decomposition along with a complete distribution of primary devolatilization products (Chen and Niksa, 1992a). These RCFR tests closed the balances on mass and *C/H/N* to within ±5 daf wt.% in individual runs, and increased the thermal severity in succeeding runs to progressively increase extents of tar decomposition. The scale along the x-axis is an index for the extent of tar decomposition, X_{T+O}, defined as unity minus the sums of secondary tar and oils scaled on the sum of primary tars and oils; an X_{T+O} of zero denotes primary products whereas 100 denotes complete conversion to soot. The RCFR was operated to minimize all secondary pyrolysis chemistry in the tests on primary devolatilization, which appear as the first bar on the left for each coal. Then the reactor was operated to gradually increase particle and gas temperatures at fixed transit times to impose specific extents of tar decomposition in succeeding runs, to resolve

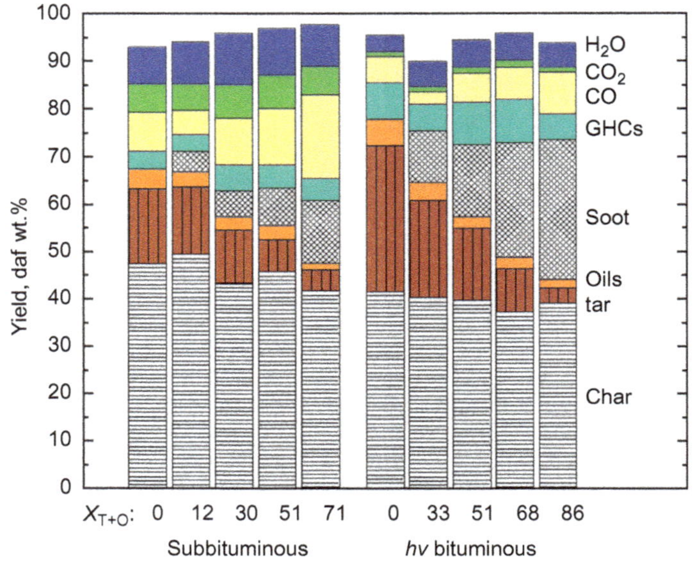

Fig. 7.9 Distributions of major products from primary devolatilization throughout tar decomposition for two coals in a RCFR.
Reproduced from Chen JC, Castagnoli C, Niksa S. Coal devolatilization during rapid transient heating. Part 2. Secondary pyrolysis. Energy Fuel 1992;6:264–71 with permission from the American Chemical Society.

product distributions for progressively greater extents of tar decomposition. Breaches in the mass balances are due to the omission of H_2, HCN, and H_2S, for the most part.

Due to the hotter temperatures in successive runs to achieve progressively greater extents of tar decomposition, char yields diminish slightly across these series due to the annealing chemistry that releases CO, H_2, H_2S, and HCN from char. The secondary tar yields diminish in consecutive runs, whereas oils yields either diminish continuously or pass through a weak maximum. Similarly, GHC yields also pass through weak maxima with both coals. Consequently, tars, oils, and GHCs all contribute to the accumulation of soot. With both coals, the sum of the yields of tar, oils, and soot are fairly uniform throughout and equal the sum of primary tar plus oils. Whereas CO levels expand continuously throughout tar decomposition, the CO_2 and H_2O yields are nearly uniform throughout.

The distributions of GHCs throughout tar decomposition in Fig. 7.10 show that C_2H_2 becomes the most abundant GHC species, by far, and that lesser amounts of CH_4 and, perhaps, C_2H_4 survive secondary pyrolysis. Acetylene is the only GHC species whose level does not pass through a maximum, suggesting that it is a major decomposition product of all heavier GHC species, and of the more saturated C_2 species. All heavier GHC species are eventually eliminated. Based on firmly established tendencies for combustion systems, acetylene addition to soot is probably mostly

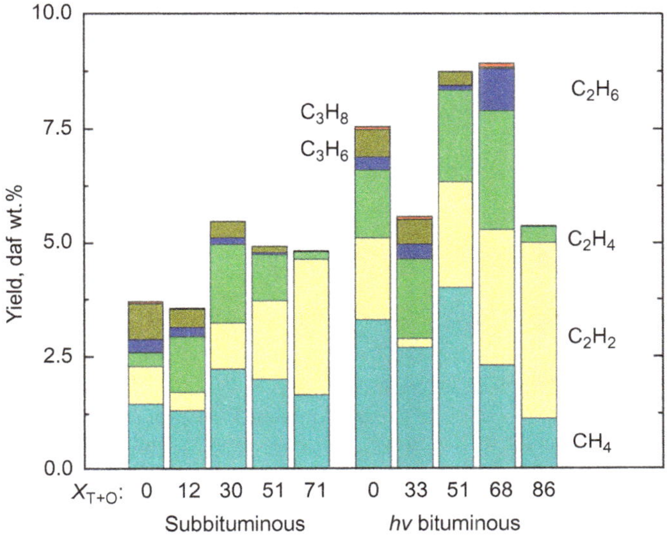

Fig. 7.10 Distributions of GHC species from primary devolatilization throughout tar decomposition for two coals in a RCFR.
Reproduced from Chen JC, Castagnoli C, Niksa S. Coal devolatilization during rapid transient heating. Part 2. Secondary pyrolysis. Energy Fuel 1992;6:264–71 with permission from the American Chemical Society.

responsible for the reduction in the ultimate aggregate GHC yield, particularly for the *hv* bituminous sample.

7.3.2 Pollutant precursors

Apparently, the release of tar-S throughout tar decomposition has not yet been monitored in any of the laboratory testing reported to date. So this section is devoted to the fate of tar-N throughout tar decomposition. The partitioning of tar-N appears in Fig 7.11 as cumulative fractions of the nitrogen in primary tar (Ledesma et al., 1998). With this *hv* bituminous coal, HNCO is the earliest N-species from tar decomposition, although its level quickly saturates through a broad maximum until it is completely eliminated at 1000°C. Both HCN and NH_3 are released simultaneously above 700°C although, again, NH_3 is an intermediate whose level saturates by 850°C until it is completely eliminated at 1000°C. In contrast, HCN levels rapidly increase for progressively hotter temperatures above 700°C and, consequently, most tar-N ultimately appears as HCN. The amounts of NH_3 and HNCO are appreciable from 700°C to 900°C, although the greatest levels coincide with breaches in the N-balance of up to 20%. The breaches could not have been caused by additional release of char-N, because the operating conditions in the first-stage FBR were uniform across this series of tests. Although the status of intermediate noncondensable

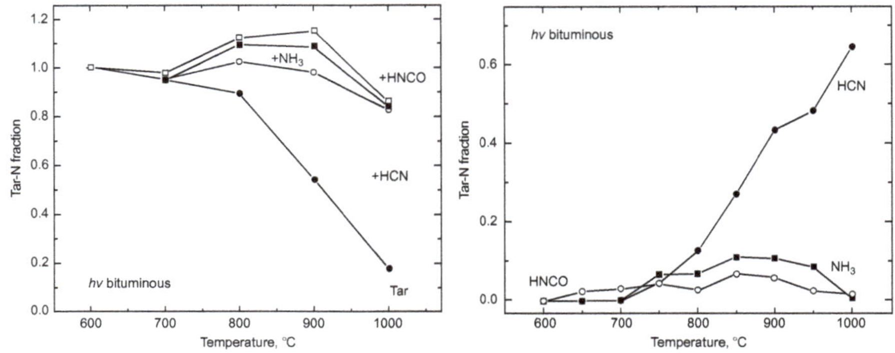

Fig. 7.11 (left) Cumulative distribution of tar-N species throughout tar decomposition with *hv* bituminous coal; and (right) release of noncondensable N-species across a temperature range. Reproduced from Ledesma EB, Li CZ, Nelson PF, Mackie JC. Release of HCN, NH_3, and HNCO from the thermal gas-phase cracking of coal pyrolysis tars. Energy Fuel 1998;12:536–42 with permission from the American Chemical Society.

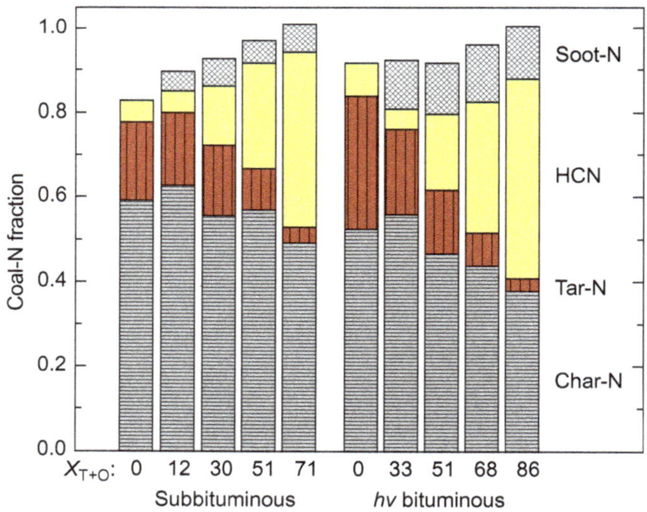

Fig. 7.12 Distributions of N-species from primary devolatilization throughout tar decomposition for two coals in a RCFR.
Reproduced from Chen JC, Castagnoli C, Niksa S. Coal devolatilization during rapid transient heating. Part 2. Secondary pyrolysis. Energy Fuel 1992;6:264–71 with permission of Elsevier.

N-species remains ambiguous, these data clearly establish that HCN is the only stable noncondensable N-species at hotter temperatures.

The N-species distributions in Fig. 7.12 expand the predominant role of HCN to hotter temperatures, and also demonstrate the incorporation of tar-N into soot. This accounting of N-partitioning includes char-N and is therefore complete; the larger breaches in the N-balances with the subbituminous reflect the omission of NH_3, consistent with its status as a low-temperature intermediate in the interpretation of Fig. 7.11. The most striking feature in Fig. 7.12 is that the sum of secondary tar-N and soot-N roughly equals the primary tar-N level throughout the first half of tar decomposition. But during the latter half, almost all tar-N is expelled as HCN, so that the contributions for soot-N stay the same. Char-N levels fall throughout due to the hotter temperatures in succeeding runs, via thermal annealing chemistry.

Ultimately, soot-N comprises 7 and 13% of coal-N with the subbituminous and *hv* bituminous, respectively, which are definitely not negligible because nitrogen within a carbonaceous matrix is much more likely than any gaseous N-species to produce NO in a flame. Soot-N also comprised 7% of coal-N with a *lv* bituminous in the same system (Chen and Niksa, 1992b). In tests at hotter temperatures in the flue gas from a flat-flame burner with three *hv* bituminous coals, soot-N comprised 4%, 6%, and 8% of coal-N (Rigby et al., 2001). Collectively, these portions of coal-N in soot with different coals are highly variable simply because the primary tar yields always exhibit substantial sample-to-sample variability, and tar shuttles coal-N into soot.

7.4 Reaction mechanisms for tar decomposition

Despite the extensive database of older laboratory studies on tar decomposition, reaction mechanisms for tar decomposition have been developed only very recently (Niksa, 2017a, b; Umemoto et al., 2017), so this chemistry has not yet been incorporated into design simulations for coal utilization technologies. Among the three network depolymerization mechanisms for primary devolatilization, FLASHCHAIN® has been expanded for tar decomposition with (Niksa, 2017b) and without sooting (Niksa, 2017a), and is the basis for this section. FG-DVC contains an option that allows primary tars to continue to lose their functional groups in the vapor phase surrounding the fuel at the same rates applied in the condensed coal phase (Serio et al., 1988), which is fundamentally incorrect. This approach interpreted reported distributions of noncondensable gas products from an entrained flow reactor (Solomon and Hamblen, 1985), where primary tars were immediately exposed to gases that were hotter than their parent particles. Both tar cracking into unsaturated GHCs and repolymerization of tars into soot were omitted from this analysis, even though the acute thermal severity in the tests would surely have converted all aromatic components of primary tars into soot. Umemoto et al. (2017) reformulated the CPD model with explicit functional groups and aromatic units with one to three rings, so that the primary devolatilization mechanism could be coupled with a 1107-step elementary reaction mechanism to describe PAH production from lighter aromatic components in the gas phase. Soot formation was omitted from the analysis, yet the bulk of the validation tests imposed acute thermal severities that completely converted tar components into soot. This approach also requires ^{13}C NMR analysis for every coal sample to specify the functional group distributions in the parent coal.

This section is based on the extensions to FLASHCHAIN® theory (Niksa, 2017a, b). Throughout the discussion, all chemistry is homogeneous and involves reactants, intermediates, and products that remain vapor species, except for soot. Heterogeneous tar decomposition on surfaces of char, inorganics, and metals is omitted.

At moderate temperatures tar decomposes into PAH, oils, GHCs, CO, H_2, HCN, and H_2S, but does not form soot. At hotter temperatures PAH, oils, and GHCs coalesce into soot while releasing their heteroatoms at additional noncondensable gases. What, if any, new concepts or phenomena are needed in mechanisms for tar decomposition above and beyond those already incorporated into FLASHCHAIN®? Since tar constitution shares numerous features in common with coal macromolecules, the tar reactant can be represented with the same four structural components as coal: Aromatic nuclei (A), labile bridges (B), peripheral groups (S), and char links (C). Since the structural components are the same, many of the reaction processes in FASHCHAIN® also factor into tar decomposition: Bridges will still be subject to both scission and spontaneous charring; peripheral groups will still be eliminated; and nuclei will remain refractory throughout, except for the release of their nitrogen as HCN and their thiophene-S (S_{th}) as H_2S. Two tar chains may recombine into a longer tar molecule, which increases the sizes of tar chains and may also release noncondensables. Since the tar reactant is in a vapor phase at the outset, flash distillation from a condensed phase into the vapor phase does not come into play. Even so, FLASHCHAIN®'s fragment statistics will also describe how bridge scission shifts the tar MWD toward lighter values; how spontaneous charring inhibits subsequent scissions; and how bimolecular recombination shifts the tar MWD toward heavier values, except that the terms for flash distillation are omitted.

The only new and distinctive processes in the mechanism for tar decomposition are the production of oils from tar monomers, nucleation of the heaviest tar chains into soot, and addition of lighter tars and oils to a nascent soot phase. Otherwise, the tar decomposition mechanism is an abridged version of FLASHCHAIN®. So readers unfamiliar with the concepts, state variables, and rate expressions in FLASHCHAIN® should review Chapter 5 before proceeding further. Of course, even with the same basic reaction mechanism, the absolute magnitudes and rank dependences of the rate parameters for tar decomposition will be different.

7.4.1 Reaction scheme

The main reactant for the mechanism is a stream of primary tar molecules which are distributed in molecular weight. Primary tars may spontaneously react as soon as they are released into an inert stream if the temperature is sufficiently hot, or they may be transported at cooler temperatures into a second, dedicated environment for their sequential decomposition. Primary tars' macromolecular structure is rendered as a mixture of chains ranging in size from a monomer to a specified degree of polymerization, $2J^*$. The degree of polymerization of the largest tar chain is twice that of its largest metaplast precursor in the condensed coal phase to accommodate bimolecular recombination. Heavier species are omitted because chain concentrations fall off sharply for progressively greater degrees of polymerization. Even though the degree of polymerization is restricted, tar weights extend to 3000–5000, depending on the parent coal constitution.

Table 7.4 Diagrams for the 11 proposed reactions.

Reaction	Diagram
Bridge scission	$t_{j-k} - t_k \rightarrow s{\cdot}t_{j-k} + s{\cdot}t_k$
Spontaneous charring	$t_{j-k} - t_k \rightarrow t_{j-k} = t_k + \nu_C G$
Bimolecular recombination	$t_j{\cdot}s + t_k \rightarrow t_j = t_k + (\nu_C/2)G$
Peripheral group elimination	$t_j{\cdot}s \rightarrow t_j + (\nu_C/2)G$
Release of char-O	C-O-C $\rightarrow$ CO + C′
Release of N in nuclei	$A\text{-}\eta N \rightarrow \eta HCN + A'$
Release of S_{th} in nuclei	$A\text{-}\xi_{th}\, S_{th} \rightarrow \xi_{th}H_2S + A'$
Oil production	$t_1{\cdot}s \rightarrow b_O O + \Sigma b_i G_i$
Soot nucleation	$t_j \rightarrow \nu_{Rj}\, R + \Sigma c_{ij}G_i;\ J^* + 1 < i < 2J^*$
Tar addition to soot	$t_j + R \rightarrow (\nu_{Rj} + 1)R + \Sigma c_{ij}G_i;\ 1 < j < J^*$
Oil addition to soot	$O + R \rightarrow (\nu_{R,O} + 1)R + \Sigma c_{i,O}G_i$

Eleven chemical processes, diagrammed in Table 7.4, represent the spontaneous disintegration of tar molecules: bridge scission; spontaneous charring; bimolecular recombination; peripheral group elimination; release of oxygen from nascent char links; release of nitrogen from nuclei; release of S_{th} from nuclei; oil production from hydrogenated tar monomers; soot nucleation from the largest tar molecules; soot addition by smaller tar molecules; and addition of oils to the soot phase. Rates of these reactions only partially determine rates of product evolution, because of the independent influence of fragment statistics.

The scission of a bridge in a tar j-mer (i.e., j-linked nuclei) generates a t_{j-k} and a t_k, with one-half of a char link and a peripheral group on each of the newly created ends (represented by "s" in Table 7.4). Since no noncondensable products form, bridge scission does not change the mass in the tar lump. Like the labile bridges in coal macromolecules, a continuous distribution of activation energies represents the thermal response of the varied bridge structures in tar chains.

Spontaneous charring converts a labile bridge between segments of a chain into a new char link, without affecting the chain size, and converts part of the bridge material into noncondensables. This process is governed by the same energy distribution and selectivity coefficient as bridge scission. Noncondensables also form by bimolecular recombination of two chain ends to form a new char link, provided that the ends of at least one of the participating chains contain a peripheral group. The diagram in Table 7.4 shows recombination involving one attached group, whereas cases with two or no peripheral groups also factor into the analysis.

Noncondensables also form by peripheral group elimination, which is represented with a SFOR. Oxygen is released from tar as CO_2, CO, H_2O, and as a substantial component of oils. The three permanent gases are expelled during spontaneous charring of bridges, then CO is released from nascent char links during the later stages via an additional distributed-energy decomposition reaction for residual char-O. The scheme introduces three stoichiometric coefficients for the three gaseous products, which are biased toward CO production with minimal amounts of CO_2 and H_2O.

The release of nitrogen from nuclei is tracked with η, the average moles of nitrogen per mole of nuclei in the total tar sample at any instant. Since nitrogen is such a sparse element on a molar basis, its decomposition chemistry cannot play an important role in the depolymerization and cross-linking channels. And since it is relegated to aromatic nuclei, it does not participate in the bridge conversion mechanisms that govern the tar MWD and the production of the major noncondensable gases. Nitrogen in nuclei decomposes into HCN via a distributed-energy reaction. Aliphatic-S (S_{al}) and aromatic sulfides (S_{ar}) are expelled as H_2S and COS via three skeletal transformations: They are directly released during spontaneous charring of bridges, and indirectly converted into peripheral groups during bridge scission which subsequently either directly decompose into noncondensable S-gases, or are eliminated during bimolecular recombination. By analogy with the release of pyridinic- and pyrrolic-N, S_{th} is expelled in a new distributed-energy reaction that produces only H_2S with the stoichiometric coefficient ξ_{th}, the average moles of S_{th} per mole of nuclei in the total tar sample.

One new reaction process in this scheme is the production of oils from tar monomers only. Oils are produced spontaneously along with O-, N-, and S-gases and olefins, which are lumped together as C_2H_4. Since BTX is not purely aromatic, oils cannot be produced from a nucleus unless it contains at least one and preferably two peripheral groups to sustain the hydrogenation. The production of oils forms b_O moles of oil per monomer, depending on the number of aromatic carbons per nucleus. This C-number is evaluated from tar properties in FLASHCHAIN®'s constitution submodel and varies from 10 to 14 across the coal rank spectrum. There are also b_1 moles of residual noncondensables—mostly GHCs—plus the O- and S-species specified earlier for peripheral group elimination, plus the HCN and H_2S released from the decomposing nucleus. If the decomposing monomer contains two peripheral groups, it will generate more oil, olefins, and additional O- and S-species, plus the usual amounts of noncondensable gases from peripheral group elimination.

Heteroatoms are continuously eliminated from both nuclei and char links without disrupting their connections to other structures in a chain. These processes alter the molecular weights of these structural components, and thereby lighten the tar MWD. Conversely, tar monomers are completely eliminated from the tar population after they disintegrate into oils and noncondensable GHCs, which shifts the tar MWD toward heavier molecular weights.

The three conversion channels for soot production in Table 7.4 are also new. A nucleation channel is resolved from the two channels that add tars and oils to the nascent soot phase, where R denotes the moles of soot units. Nucleation is spontaneous and restricted to the heaviest half of the tar MWD, whereas the lightest half may only add to existing soot particles. Similarly, oils add to existing soot particles. Like oils, the soot is assigned a nearly uniform composition for all coal types. Most of the carbon in the tar/PAH precursor is retained in the soot phase during nucleation. Hydrogen is eliminated as H_2 to increase the atomic C/H ratio of soot to 8, consistent with reported values for secondary volatiles pyrolysis and for flame-generated soot (cf. Table 7.3). All oxygen is eliminated as CO, and all three sulfur forms are eliminated as H_2S and COS. But only two-thirds of the tar-N is released as HCN, again, for consistency with experimental observations (cf. Fig. 7.12). Since the proportions of bridges, char links, and peripheral groups differ among chains of different sizes, the stoichiometric coefficients for H_2, CO, H_2S, and COS are evaluated separately for each chain size. But a single conventional Arrhenius rate constant, k_{SN}, is applied to all chain sizes, because the rate-limiting process of thermal annealing should be the same for all precursors.

Once formed, the soot phase collects additional carbon from the smaller tar molecules and oils. The smaller tar chains contain O, S, and N and therefore release H_2, CO, H_2S, COS, and HCN when they add to soot. Oils contain O but neither N nor S and release only H_2 and CO. These processes are implemented with the same rate constant k_{SC} for the primary condensation reaction of all hydrocarbons, assuming a common rate-limiting step of thermal annealing.

Rate equations were developed for the proposed reaction set with the analytical approach used for FLASHCHAIN®. Completely new rate equations describe the production of oils and noncondensables from tar monomers, and of soot and noncondensables from tar and oils (Niksa, 2017b). Here we consider only one of the new conservation equations, to show that soot formation imparts an explicit dependence on the coal loading into the process kinetics. The soot production rate accounts for the nucleation of the larger chains plus the subsequent condensation of smaller chains and oils:

$$\frac{dR}{dt} = k_{SN} \sum_{j=J^*+1}^{2J^*} \nu_{Rj} t_j + k_{SC} R \sum_{j=1}^{J^*} \nu_{Rj} t_j + \nu_{R,O} k_{SC} RO \tag{7.1a}$$

On a molar basis, ν_{Rj} simply equals j, based on the postulated connections among soot units and tar monomers. All reaction species are scaled by the moles of aromatic nuclei in coal per unit gas volume. This scaling factor is defined from

$$A_{GAS} = \frac{A_0}{\rho_0} L_{COAL} \tag{7.1b}$$

where A_{GAS} is the moles of coal nuclei per unit gas volume; A_0 is the moles of nuclei per unit coal volume; ρ_0 is the bulk coal density; and L_{COAL} is the mass loading of coal per unit gas volume in the process stream. As shown previously (Niksa and Kerstein, 1991) the ratio A_0/ρ_0 is evaluated from the molecular weights of structural components. When this scaling factor is introduced into Eq. (7.1a), it can be cancelled out of the accumulation term, dR/dt, and the first term on the right, but remains as a multiplicative prefactor on k_{SC} in both remaining terms. In other words, due to the bimolecular order postulated for bimolecular recombination and for soot addition, the associated rate constants increase in proportion to the coal loading in whatever tests or applications are under consideration. Since all testing teams in this field have apparently been unaware of this loading dependence, its impact will be demonstrated in the simulation cases in Section 7.4.3.

7.4.2 Tar constitution submodel and rate parameter assignments

Tar composition is markedly different than a parent coals' composition because primary tars are always enriched in hydrogen, and because they contain different proportions of the structural components. So all aspects of tar constitution and all the rate parameters are evaluated from the elemental composition of the whole primary tar sample predicted by FLASHCHAIN®, rather than the standard coal properties. In addition, both carbon and proton aromaticities are evaluated from a database for petroleum asphaltenes, which covers the range of *H*/*C* ratio for primary coal tars.

All stoichiometric coefficients are evaluated from tar constitution, and variations in these coefficients are primarily responsible for interpreting the large variations in the product distributions reported for diverse coal types. Since the functional group distributions, and molecular weights and elemental compositions of the structural components assigned for tars are similar to those for the parent coal, the stoichiometric coefficients are also similar. The main exceptions are the coefficients for tar-O release, which markedly skew the release of bridge-O toward CO production, at the expense of CO_2 and H_2O.

The rate parameters in the tar decomposition mechanism have been reported (Niksa, 2017b), and most are fixed for all coals. Those for S-release could not be specified because H_2S and COS release has not yet been monitored during tar decomposition. Mean activation energies are much lower for tar decomposition than for primary devolatilization, except that the energy for HCN production is substantially greater. Also, all the energy distributions for tar decomposition are much narrower than those for the same reaction processes in primary devolatilization. The mean energies for peripheral group elimination and oil production are not independent, because both noncondensables and oils can form only if a peripheral group is present on the participating tar monomer. It is the difference in the activation energies that must be adjusted to match predicted and measured product yields.

The pseudo-frequency factors for recombination and soot addition must be adjusted in rough proportion to changes in the coal loadings in the tests under consideration. That for oil production was fixed for all but a few validation cases. But the frequency factor for HCN production was adjusted across a broad range of values for individual coal samples, like the frequency factor for HCN production during primary devolatilization (Niksa, 1995).

7.4.3 Predicted behavior

The transient tar decomposition history in Fig. 7.13 illustrates the most important general features at moderate temperatures, based on an *hv* bituminous coal heated at 200° C/s to 825°C with a 3 s IRP, which is typical for coal processing in fluidized beds. In these calculations the thermal histories for the sample and surrounding gases are the same. Primary tars are released while the coal is being heated, and achieve their ultimate yield just at the end of the heating period. Tars begin to decompose when the temperature reaches 500°C; then the decomposition rate accelerates for progressively hotter temperatures; tars then continue to decompose long after the ultimate primary

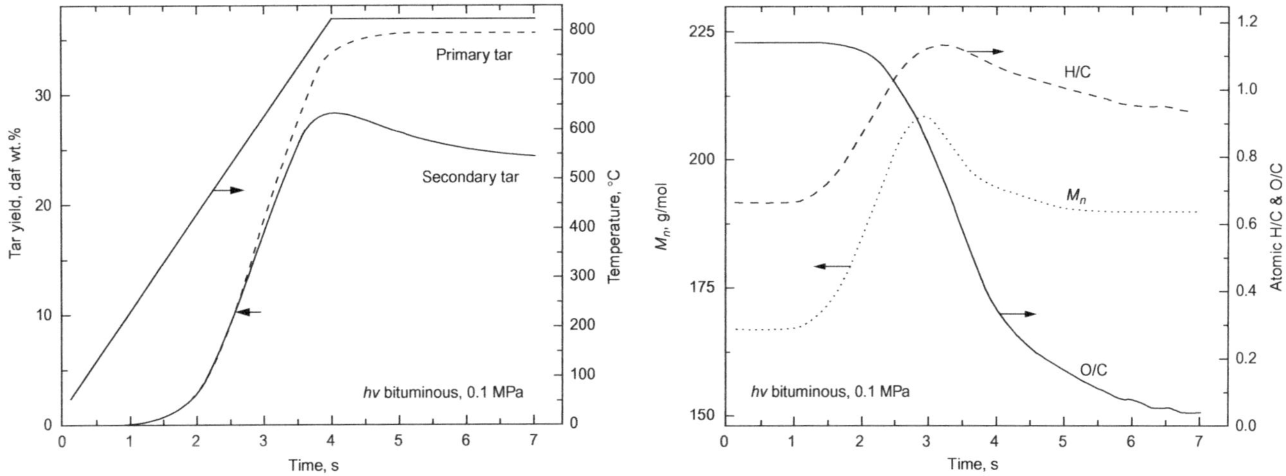

Fig. 7.13 (Left) Transient yields of primary and secondary tar and the thermal history and (right) M_n and atomic *H/C* and O/C ratios of secondary tar for 200°C/s to 825°C with a 3 s IRP at 0.1 MPa with an *hv* bituminous coal.
Reproduced from Niksa S. A reaction mechanism for tar decomposition at moderate temperatures with any coal type. Fuel 2017a;193:467–76 with permission from Elsevier.

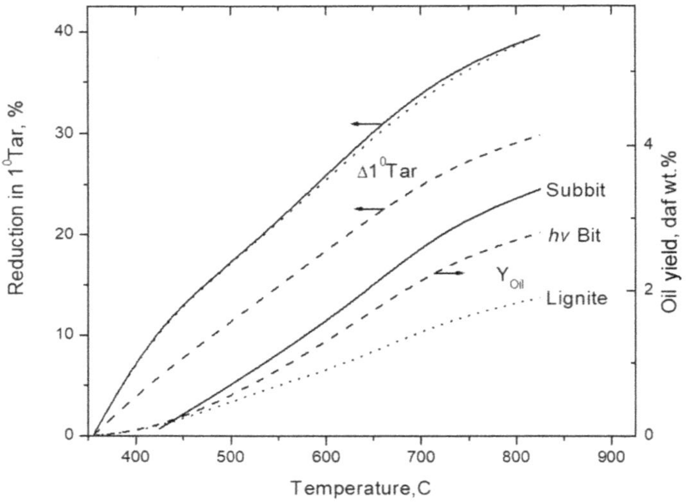

Fig. 7.14 Reduction in tar yields and oil yields versus temperature from *(dotted curve)* lignite, *(solid)* subbituminous, and *(dashed) hv* bituminous coals for 200°C/s to 825°C with a 3 s IRP at 0.1 MPa.
Reproduced from Niksa S. A reaction mechanism for tar decomposition at moderate temperatures with any coal type. Fuel 2017a;193:467–76 with permission from Elsevier.

tar yield was achieved. Consequently, the secondary tar yield passes through a maximum as primary tar production winds down. The M_n-values also pass through a maximum, which reflects the tendencies, on the one hand, for heavier primary tar at progressively hotter temperatures and a shift toward heavier tar from the conversion of tar monomers into oils and, on the other hand, for lighter tar due to the release of heteroatoms as noncondensables. Ultimately, secondary tar is lighter than primary tar (but only if no soot forms). Almost all tar-O is expelled, while the H/C ratio relaxes to a typical value for condensed PAH with no substituents.

Some of the coal quality impacts are resolved in Fig. 7.14 in terms of the percentage reductions in primary tar yields and the oils yields for a lignite, subbituminous, and *hv* bituminous coals. These are ultimate values across a broad temperature domain. The reductions in tar yields from the low rank coals are substantially greater than for the *hv* bituminous coal, primarily because their primary tars contain much more oxygen. With the subbituminous, the reduction is greater because the oil yield is maximized for this coal's elevated H-content of 6.4 daf wt.%. The lignite produced just over half as much oils as the subbituminous.

The validation cases for this mechanism in the literature cover the datasets in Figs. 7.1, 7.3, 7.4, 7.6, 7.8, and 7.12, and Tables 7.1 and 7.2 (Niksa, 2017a,b), so only a few cases are considered here. The tar decomposition history in Fig. 7.15 is shifted toward hotter temperatures by about 100°C compared to those in Fig. 7.1, which reflects the shorter residence times of only 200–100 ms for progressively hotter test temperatures. Both systems used fluidized beds of similar design, but tar

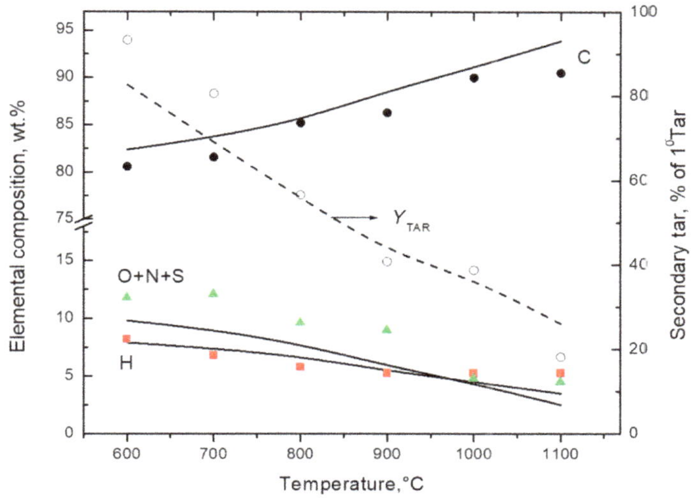

Fig. 7.15 *(○ and dashed curve)* Secondary tar levels as a percentage of primary tar (of 28.6 daf wt.%) and *(filled symbols and solid curves)* tar compositions throughout tar decomposition with subbituminous coal in a fluidized bed/TFR system (Doolan et al., 1987).
Reproduced from Niksa S Flashchain theory for rapid coal devolatilization kinetics. 9. Tar decomposition across a broad temperature domain. Energy Fuel 2017b;31:9080–93 with permission from the American Chemical Society.

decomposition in Fig. 7.15 was staged in a coupled TFR. The predicted tar decomposition history is within measurement uncertainties throughout. The measured compositions of secondary tar show that all heteroatoms and most tar-H are preferentially expelled during tar decomposition, so that tar-C increases throughout. The predicted accumulation of carbon and release of hydrogen in secondary tar are very accurate, although the sum of O + N + S is under-predicted by 2%–3% throughout. However, the predicted compositions sum to 100% whereas the measured values exceed it.

The yields of the major product lumps from tar decomposition are interpreted in Fig. 7.16. The cumulative distributions of secondary pyrolysis products are resolved into inorganic gases (IOG) (CO_2, H_2O, CO, and H_2), GHCs, oils (BTX, PCX, C_5-C_7 GHCs), and tar. The predictions shown for 400°C are for primary devolatilization at the stated conditions in the first reactor stage of this 2-stage system. No measured values were available for comparison. At the onset of secondary pyrolysis oils appear accompanied by small increases in the yields of IOGs and more substantial enhancements to the GHCs. The tar levels diminish from 20.4 daf wt.% for primary devolatilization to 7.2% at the hottest test temperature (which is less than the extents of decomposition in Figs. 7.1 and 7.15, presumably due to the much lower volatiles concentrations in these tests). These changes are greatest for secondary pyrolysis temperatures from 500°C to 700°C for the 7 s residence time in these tests.

At 900°C, the measured yields of oils and GHCs in Fig. 7.16 diminish while that for IOG continues to increase. The predicted oils and GHC yields diminish above 800°C,

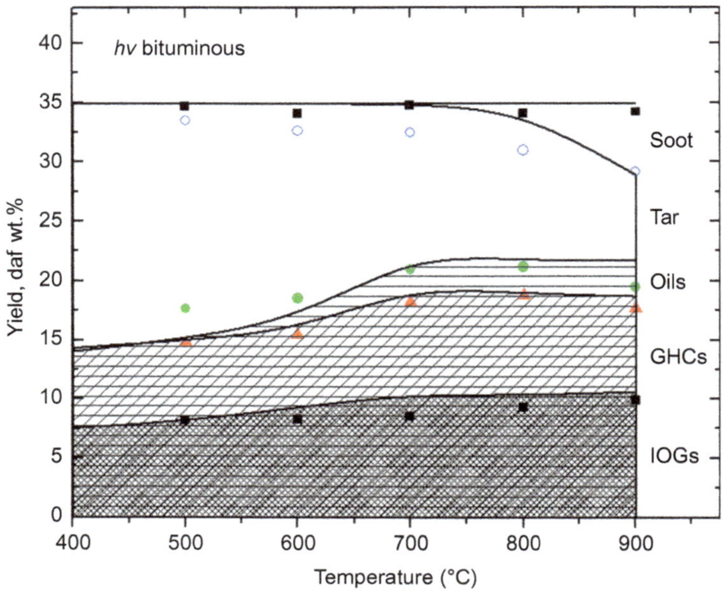

Fig. 7.16 Predicted main product lumps for secondary pyrolysis after 7 s (curves) with measured values for IOGs (■), GHCs (▲), oils (●), secondary tar (○), and total weight loss (■) from an *hv* bituminous coal (Xu and Tomita 1989).
Reproduced from Niksa S Flashchain theory for rapid coal devolatilization kinetics. 9. Tar decomposition across a broad temperature domain. Energy Fuel 2017b;31:9080–93 with permission from the American Chemical Society.

but not by as much as the measured values. The reported coke yields gradually increased from 1.2 to 4.9 daf wt.% across the entire temperature range, whereas the predicted soot yields accumulate only above 700°C for the 7 s residence time.

Complete distributions of all major products except H_2, H_2S, and HCN for different extents of tar decomposition appear in Fig. 7.17. These distributions were obtained in a RCFR that was operated in two modes. Primary products were generated with dilute coal loadings to maintain entrainment gas temperatures much cooler than the particle temperatures, and thereby minimize tar decomposition. Tar decomposition products were recorded at hotter furnace temperatures and loadings four times greater, which gave comparable thermal histories for both particles and gas. The scales along the x-axis, X_{T+O}, give the extents of tar decomposition defined previously. The values based on the measured and simulated products for each test differ primarily because of uncertainties on the thermal histories in this test series. Even so, these product distributions cover virtually the entire range of tar decomposition. In each set of five distributions, the leftmost results are the primary products, and the four distributions to the right are for progressively more severe thermal histories. Since maximum temperatures increased in the four progressively more severe tests on tar decomposition, the total volatiles yields also increased.

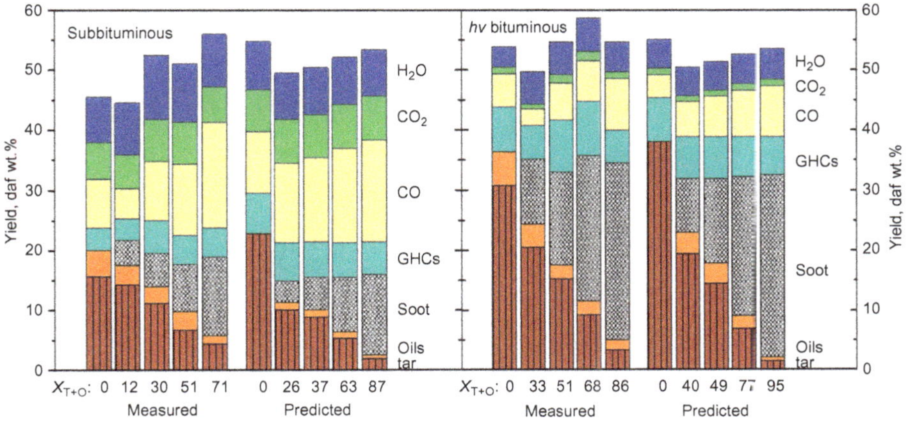

Fig. 7.17 Volatile product distributions for primary devolatilization and for progressively more severe secondary pyrolysis with (left) subbituminous and (right) *hv* bituminous coals in a RCFR (Chen et al., 1992).
Reproduced from Niksa S Flashchain theory for rapid coal devolatilization kinetics. 9. Tar decomposition across a broad temperature domain. Energy Fuel 2017b;31:9080–93 with permission from the American Chemical Society. Bottom scale shows extents of tar decomposition.

The simulated primary products are essentially the same as the measured primary distribution for the *hv* bituminous, except that the predicted primary tar yield equals the sum of the measured yields of tar plus oils, because FLASHCHAIN® does not include oils among the primary products. The simulated primary distribution for the subbituminous generally has somewhat more of each product than the measured distribution, which closes a deficit of almost 10% in the mass balance for this particular test. The following tendencies during tar decomposition are clearly apparent in both the simulated and measured distributions: Yields of oils and tar diminish continuously while soot accumulates into one of the major ultimate products; CO levels increase continuously while yields of CO_2 and H_2O are hardly perturbed; GHC yields pass through a weak maximum. The tar decomposition mechanism accurately interprets these trends, except that the maxima in GHC yields is either weaker than measured or, for the subbituminous, not present at all.

Fig. 7.18 shows the ultimate soot yields seen previously in Fig. 7.7, except that only tests that specified their coal loadings are included. The predictions for all cases were based on the kinetic parameters from the interpretation of Chen et al.'s data in Fig. 7.17, except that the frequency factors for bimolecular recombination and addition of tar and oils to soot were roughly re-scaled in proportion to the coal loadings. The maximum temperatures and residence times in all these tests were comparable. As seen in Fig. 7.18, the predicted soot yields for the five re-scaled cases are nearly exact, which is especially encouraging since the five coals span the entire rank spectrum, and all five simulations are based on the same rate parameters. The predictions for the

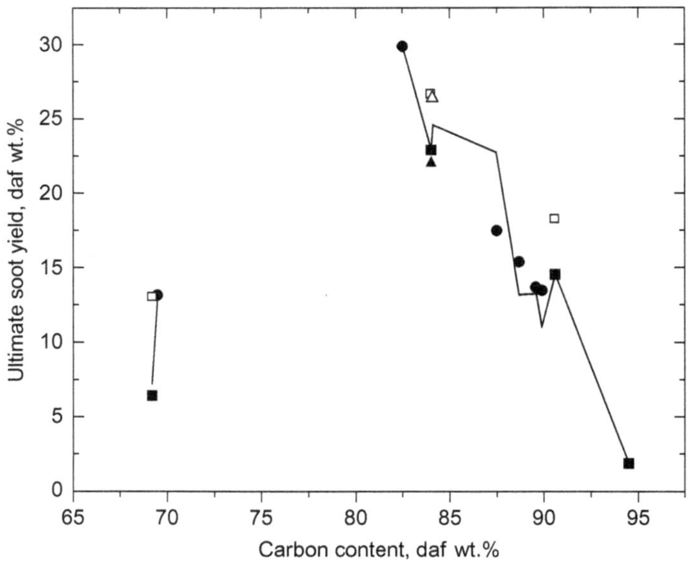

Fig. 7.18 Evaluation of reported soot yields from (●) Chen et al. (1992); (■) Nenniger et al. (1983); and (▲) Wornat et al., (1987). The line segments without data points show the model predictions. The corresponding open symbols show the predictions for the five coals tested by Nenniger et al. and Wornat et al. for the much greater coal loading used by Chen et al. (1992). Reproduced from Niksa S Flashchain theory for rapid coal devolatilization kinetics. 9. Tar decomposition across a broad temperature domain. Energy Fuel 2017b;31:9080–93 with permission from the American Chemical Society.

same *hv* bituminous (with 84 daf wt.% C) used in independent test series are essentially the same, in agreement with the measured values. The open symbols for these five cases are the corresponding predictions for the much greater loadings in the tests of Chen et al. to display the coal quality impacts for a fixed loading. Whereas the anthracite (with 94.5% C) gave essentially the same yield for the greater loading, the predicted soot yields for the other coals were much greater. That from the lignite doubled, while those from the bituminous coals increased by 15%–25%. Fig. 7.18 also contains the evaluation of soot yields for both coals in Fig. 7.17, and four low volatility coals tested in the RCFR at the same tar decomposition conditions (Chen and Niksa, 1992b). The predicted soot yields for five of these coals are within measurement uncertainties. However, for the sixth coal (at 87.5% C) the predicted soot yield is 9 daf wt.% too high. More than half this discrepancy should be attributed to the measured value, because the measured soot yield for less severe conditions was 2 wt.% greater than the 17.5% in Fig. 7.18, and there was 8 wt.% residual secondary tar remaining to be converted. Consequently, the predicted value of 22.8 daf wt.% is probably more accurate than the measured value in Fig. 7.18.

Tar decomposition entails a continuous elimination of heteroatoms as noncondensable products, as well as elimination of the monomer size class as oils and noncondensables and, ultimately, the coalescence of PAH and oils into a condensed soot phase. The interplay among these channels is complex, yet tar-N is simply shuttled by the parent nuclei among fragments in the tar MWD until it is expelled as HCN either by the HCN elimination reaction or as the nuclei are incorporated into oils and soot. In other words, none of the N-transformations affect the course of tar decomposition, per se, and N-species transformations are almost completely described by the original tar decomposition mechanism. Only one additional assumption is needed to depict tar-N transformations with the tar decomposition mechanism: One-third of tar-N is incorporated into soot.

Fig. 7.19 illustrates the accuracy of this approach for subbituminous and *hv* bituminous coals, and similar cases with low volatility coals have also been reported (Niksa, 2019b). The simulated N-species distributions are essentially the same as the measured distributions for all six coals, and accurately depict the variations in char-N from 0.4 to 0.8 for this range of coal rank. All the predicted levels of tar-N and char-N are within measurement uncertainties for primary devolatilization and all stages of tar decomposition. The predicted soot-N levels are also accurate, and also exhibit the weak maximum in the measured distributions in Fig. 7.19. The predicted HCN levels are similarly accurate, albeit for only the two most severe secondary pyrolysis conditions.

The main tendency for all coal types is the same: Most tar-N is expelled as HCN whereas soot-N is an appreciable portion of volatile-N levels for all coal ranks. In fact, the relationship between tar-N and soot-N is remarkably straightforward for any coal type: The mean fraction of tar-N converted into the ultimate soot-N in the measured distributions was 0.33 versus 0.39 in the predicted distributions. There is no appreciable variation with coal rank, although more samples need to be tested to thoroughly characterize this rank dependence. The std. dev. for the measured conversions is 0.043, which is tight considering that the primary tar yields varied from 18 to 38 daf wt.% among these tests.

From a practical standpoint, it is not necessary to predict even the primary tar-N levels to accurately estimate ultimate soot-N levels because, to good approximation, the coal-N fractions in primary tars are directly proportional to primary tar yields (Chen and Niksa, 1992a, b). So the ultimate partitioning of coal-N into HCN, soot-N, and char-N during the initial stages of p. f. firing requires two types of calculations: (1) A primary devolatilization mechanism to predict tar yields under rapid heating conditions and (2) A submechanism to predict HCN release from char throughout devolatilization up to the point of char ignition. Beyond the ignition point the bulk of char-N is released as NO at the char burning rate (Chen and Niksa, 1992b). The methodology is described further by Niksa (2019b), although the implications are particularly important. The incorporation of tar-N into soot substantially reduces the amount of coal-N amenable to aerodynamic NO_X abatement strategies for two reasons: (1) Because soot-N will only be released while the soot burns out and (2) Because tar makes up the majority of volatiles with all bituminous coals, and is still a major portion of volatiles from low-rank coals. With low volatility coals, the coal-N fractions as soot-N are relatively small in absolute terms, but are actually relatively large in relation to the volatile-N levels, because tar-N constitutes 75%–90% of

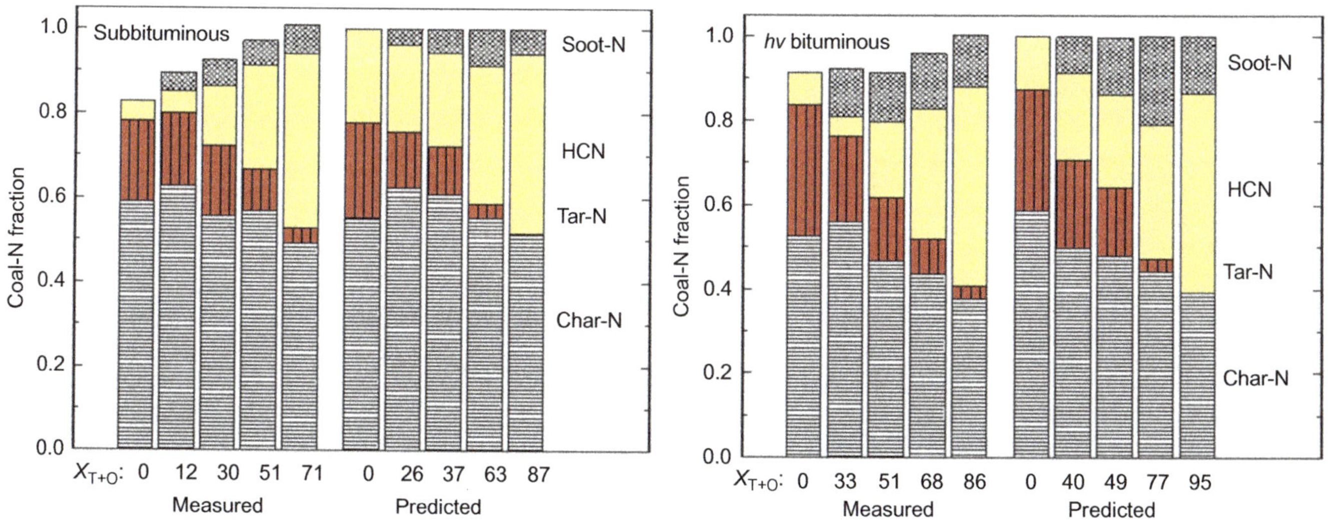

Fig. 7.19 N-species distributions for primary devolatilization and for progressively more severe secondary pyrolysis with (left) subbituminous and (right) Pit. #8 *hv* bituminous coals in a RCFR.
Reproduced from Niksa S. Predicting nitrogen release during coal tar decomposition. Proc Combust Inst 2019b;37:2765–72 with permission from Elsevier. Bottom scale shows extents of tar decomposition.

volatile-N. The tar-N contribution diminishes though the high volatility ranks but this is more-than-compensated by their relatively large tar yields. Consequently, *hv* bituminous have the greatest soot-N levels of all.

Soot-N counteracts the rapid release of coal-N into the furnace gases, where the N-species transformations can be skewed by staging toward N_2 production at the expense of NO. It preferentially generates NO as the soot burns, and thereby raises predicted NO emissions in CFD furnace simulations that resolve it from the other forms of coal-N. A scheme to incorporate soot-N conversion into CFD simulations is outlined in Section 7.5.2.

Fortunately, nearly all the essential information to accurately depict the kinetics and product distribution from tar decomposition can be deduced from the compositions and yields of primary tars. In the validation work reviewed to this point, FLASHCHAIN® specified the compositions and yields. In turn, predicted primary tar compositions were processed through the tar constitution submodel to specify the proportions and compositions of the four structural components, and all stoichiometric coefficients in this analysis. Since the vast majority of kinetic parameters are either fixed for all coals or pinned to variations in the structural components, tar constitution and the associated variations in the stoichiometric coefficients are the distinguishing factors in the distinctive behavior of individual primary tar samples. Moreover, kinetics for tar decomposition are far less sensitive to coal quality than primary devolatilization kinetics, because the activation energy distributions for the analogous reaction channels are much narrower, and therefore subject to much less variation in their thermal response. Also, mean activation energies are much lower for tar decomposition than for primary devolatilization, except that the energy for HCN production is substantially greater.

7.5 Global rates for tar decomposition and soot production

Section 6.3.1 already discussed the potential for confusion and complications when global rates are applied to mechanisms as complex as coal devolatilization. That potential may be even greater with nominal rates of tar decomposition, because SFORs and competitive reactions in parallel are fundamentally inconsistent with the dynamics of tar decomposition. The goal is to predict the yield of tar throughout its decomposition, and the accumulation of soot at elevated temperatures. To illustrate the complexity, re-visit the left panel of Fig. 7.13, which shows the predicted transient yields of primary and decomposing tar throughout a heating stage to 825°C followed by isothermal reaction for an additional 3 s.

The reduction in the tar yield during secondary pyrolysis is clearly apparent in the two yield curves because the same thermal history was imposed in the separate simulations for primary devolatilization and tar decomposition. Tar begins to decompose around 500°C at this heating rate. The decomposition rate accelerates as the tar is heated to hotter temperatures, then the decomposition reaction overtakes the production rate of primary tar, so the secondary tar yield turns over and relaxes to a saturation value of 24 daf wt.% versus 35 daf wt.% of primary tar.

Such dynamic behavior would be expected for a simple sequential decomposition. But the form of the secondary tar yield in time is not amenable to representation with a SFOR. Rate expressions such as a SFOR necessarily relax to the ultimate yield parameter, V^{∞}, without any overshoot. It is simply impossible to depict the decomposition profile in Fig. 7.13 with any of the rate laws applied to primary devolatilization. The schemes developed in the next two sections accurately describe secondary tar levels throughout tar decomposition at moderate temperatures, and tar and soot levels at elevated temperatures, respectively.

7.5.1 Nominal rates of tar decomposition at moderate temperatures

To identify a suitable rate law for tar decomposition at moderate temperatures, first notice that the difference between the yields of primary and secondary tar in Fig. 7.13 exhibits the characteristic dynamic behavior of a SFOR, initially growing at an accelerating rate before it relaxes to a saturation value. The release of noncondensables into the gas phase by primary devolatilization and secondary tar decomposition is well described by the following two reactions:

$$\frac{d^1Y_T}{dt} = {}^1R_T = {}^1k_T\left({}^1Y_T^{\infty} - {}^1Y_T(t)\right) \tag{7.2a}$$

$$\frac{d\Delta Y_T}{dt} = {}^2R_T = k_T\left(\Delta Y_T^{\infty} - \Delta Y_T(t)\right) \tag{7.2b}$$

The first rate in Eq. (7.2a) describes the release of primary tar with a conventional SFOR; and the second describes the difference between the instantaneous primary and secondary tar yields; i.e., $\Delta Y_T(t) = {}^1Y_T(t) - {}^2Y_T(t)$. Once ${}^1Y_T$ and ΔY_T are evaluated, ${}^2Y_T(t)$ is evaluated as ${}^1Y_T - \Delta Y_T$.

These global reactions contain two rate constants, ${}^1k_T$ and k_T, each in Arrhenius form, so there are two activation energies, ${}^1E_T$ and E_T; two pseudo-frequency factors, ${}^1A_T$ and A_T, and two hypothetical ultimate yield parameters, ${}^1Y_T^{\infty}$ and ΔY_T^{∞}, to assign. The parameters for the devolatilization rate of primary tar are specified with the method of Niksa and Lau (1993) from FLASHCHAIN® predictions. Then the predicted values of $\Delta Y_T(t)$ from the tar decomposition mechanism are analyzed in the same way for the second set of parameters. The fit with the sequential two-step reaction model is compared to predictions from the full tar decomposition mechanism in Fig. 7.20 for a subbituminous coal at two heating rates. The global rate laws assigned this way depict the maximum in the secondary tar yield at the end of primary tar production, and accurately describe the transient yields of secondary tar throughout both thermal histories for both coals. Of course, the assigned rate parameters for each

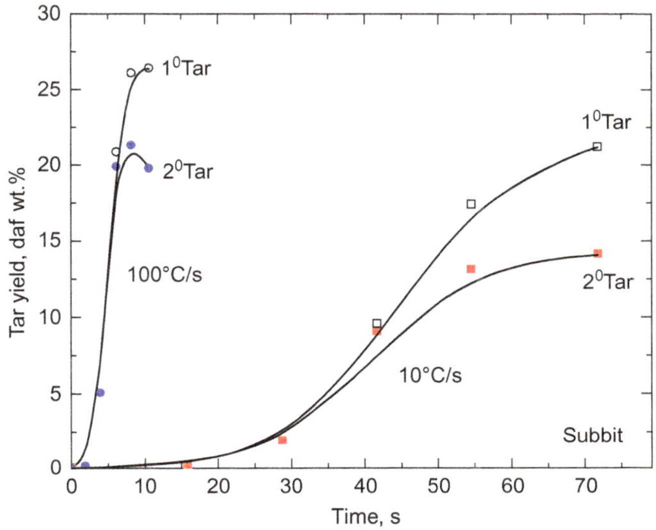

Fig. 7.20 *(Curves)* Fits of primary and secondary tar yields from Eqs. (7.2a), (7.2b) to (points) predictions from the full mechanism for subbituminous coal for 10 and 100°C/s to 825°C at 0.1 MPa.
Reproduced from Niksa S. A reaction mechanism for tar decomposition at moderate temperatures with any coal type. Fuel 2017a;193:467–76 with permission from Elsevier.

set of operating conditions are different because no global rate law could possibly describe tar decomposition over a broad domain of conditions. As seen in Table 7.5, the assigned parameters are also substantially different for subbituminous and *hv* bituminous coals and for both heating rates. This is not a limitation because rate parameters can be evaluated for each set of conditions in seconds on ordinary personal computers. The detailed mechanism also specifies the mass-based stoichiometric coefficients for all PAH, oils, and noncondensable products of tar decomposition, which are compiled in Table 7.5 for the conditions in Fig. 7.20. The units on these values are g/g primary tar.

7.5.2 Estimating ultimate soot yields at elevated temperatures

For applications at elevated temperatures, ultimate soot yields could be evaluated from the full reaction mechanism for tar decomposition, like ${}^{1}Y_{T}^{\infty}$ and ΔY_{T}^{∞} in the global analysis for moderate temperatures. But there is a much simpler method, provided that the operating conditions are severe enough to achieve the ultimate soot yield that incorporates all the aromatic components from primary devolatilization. The central premise is that a yield and elemental composition of primary tar determines the ultimate soot yield (Niksa, 2019a). Tar compositions are expressed as

Table 7.5 Assigned parameters for Eqs. (7.2a), (7.2b) for the cases in Fig. 7.20.

	Subbituminous		***hv* bituminous**	
	10°C/s	**100°C/s**	**10°C/s**	**100°C/s**
${}^{1}A_T$	1.3	1.3	1.3	1.6
${}^{1}E_T$	3.9	3.8	4.3	4.3
${}^{1}Y_T^{\infty}$	0.5	0.5	1.5	0.8
A_T	0.1	0.1	0.1	0.1
E_T	2.1	2.1	2.4	2.8
ΔY_T^{∞}	1.0	1.7	1.3	2.0
PAH	0.2	0.9	0.2	1.1
Oil	33.2	33.2	33.8	31.2
CH_4	0.054	0.043	0.043	0.029
C_2's	0.042	0.033	0.021	0.014
C_3's	0.022	0.018	0.014	0.009
CO	0.012	0.010	0.005	0.004
H_2O	0.042	0.036	0.038	0.032
CO_2	0.022	0.019	0.015	0.013

Reproduced from Niksa S. A reaction mechanism for tar decomposition at moderate temperatures with any coal type. Fuel 2017a;193:467–76 with permission from Elsevier.

products of the ultimate analysis of a parent coal on a daf basis (E_{COAL}, where E = C, H, O, N, S), and the fractional release of that element in primary tar (${}^{E}f_{TAR} = E^{1}_{TAR}Y_T/E_{COAL}$). Ultimate soot yields are estimated from the following observations:

(1) Atomic $(H/C)_{SOOT}$ ratios of soots from all coals are fixed at 0.125, consistent with measured values (cf. Table 7.3).
(2) All tar-O is expelled as CO, but the mass loss from the soot phase is nearly compensated by C_2H_2 addition, to maintain consistency with uniform sums of tar plus oils plus soot throughout secondary volatiles pyrolysis (cf. Fig. 7.9).
(3) Two-thirds of tar-N is expelled as HCN, in accord with Fig. 7.12.
(4) Tar-S is fully expelled as H_2S, pending further corroboration of the lone study that reported S in soot so far (Rigby et al., 2001).
(5) All primary GHCs are converted into CH_4, C_2H_2, and H_2, whereby all C_2-species form acetylene and C_3-species and heavier form CH_4 and C_2H_2.

Accordingly, the ultimate soot yield, Y^{∞}_{SOOT}, is evaluated from

$$Y^{\infty}_{SOOT} = {}^{1}Y_{TAR} - \frac{\frac{2\,{}^{O}f_{TAR}O_{COAL}}{16} - 27\left(\frac{2}{3}\right){}^{N}f_{TAR}N_{COAL}}{14} - \frac{34.06\,{}^{S}f_{TAR}S_{COAL}}{32.06} - {}^{TAR}Y_{H_2} \tag{7.3a}$$

where ${}^{TAR}Y_{H_2}$ is the amount of H_2 expelled from tar to achieve $(H/C)_{SOOT}$, which is evaluated from

$$
\begin{aligned}
{}^{TAR}Y_{H_2} = {}& {}^{H}f_{TAR}H_{COAL} \\
& - \left(\frac{H}{C}\right)_{SOOT}\left[\frac{{}^{C}f_{TAR}C_{COAL}}{12} + \frac{{}^{O}f_{TAR}O_{COAL}}{16} - \frac{\frac{2}{3}{}^{N}f_{TAR}N_{COAL}}{14}\right] + \frac{2{}^{O}f_{TAR}O_{COAL}}{16} \\
& - \frac{\frac{2}{3}{}^{N}f_{TAR}N_{COAL}}{14} - \frac{2{}^{S}f_{TAR}S_{COAL}}{32.06}
\end{aligned}
\tag{7.3b}
$$

where succeeding terms on the right represent (i) H in primary tar; (ii) H incorporated into soot; (iii) C added via exchange of C_2H_2 for CO; (iv) C lost with HCN; (v) H added via exchange of C_2H_2 for CO; (vi) H lost with HCN; and (vii) H lost with H_2S. Additional relations for coals with insufficient hydrogen to produce enough GHCs to fully exchange C_2H_2 for the CO that shuttles away tar-O were also needed for some coals (Niksa, 2019a).

The central premise is that yields and compositions of primary tar determine ultimate soot yields from any coal. We have already seen (in Chapter 4) that ultimate primary tar yields vary from a few weight percent to as much as 40 daf wt.%, and pass through a broad maximum through the high and medium volatile bituminous ranks. Primary tars contain slightly more carbon than their parent coals, and approximately 20% more hydrogen; oxygen contents are slightly lower, and the nitrogen contents are comparable. The C- and H-contents of primary tar increase for coals of progressively higher rank, whereas O-contents diminish until they nearly vanish with low volatility coals. Sulfur levels are usually less than half of coal-S, because only organic-S can be shuttled away with primary tars, and organic-S is often less than half of total-S. So one can expect large variations among ultimate soot yields as well.

The hypothesis that primary tar yields determine ultimate soot yields is formally evaluated in Fig. 7.21, which plots measured Y_{SOOT}^{∞} versus measured values of ${}^{1}Y_{T}$. Indeed, ultimate soot yields are directly proportional to primary tar yields for coals across the rank spectrum. The correlation coefficient is 0.89, and the std. dev. is 2.5 daf wt.%, which is consistent with the measurement uncertainty in the determinations of soot and tar yields in these validation tests. The constant of proportionality of 0.816 indicates that, on average, 18% of primary tar components are not incorporated into the soot phase by the end of secondary volatiles pyrolysis. This is a remarkably low value considering that it is smaller than the proportion of aliphatic carbon in most primary tars, and much smaller than the amounts of CO released to eliminate tar-O from the soot phase for all but the low volatility coals. Yet this reduction percentage applies equally to coals across the rank spectrum, even when as much as a quarter of the tar mass is oxygen. Such a low value clearly indicates that GHCs expelled from primary tar add to the nascent soot phase at some later stage in the process.

A parity plot for predicted versus measured ultimate soot yields also appears in Fig. 7.21. The plus-symbols in the figure show the raw predicted ultimate soot yields, which are systematically greater than the measured yields because only 10% of the primary tar yields were excluded from the predicted soot phase. The filled circles

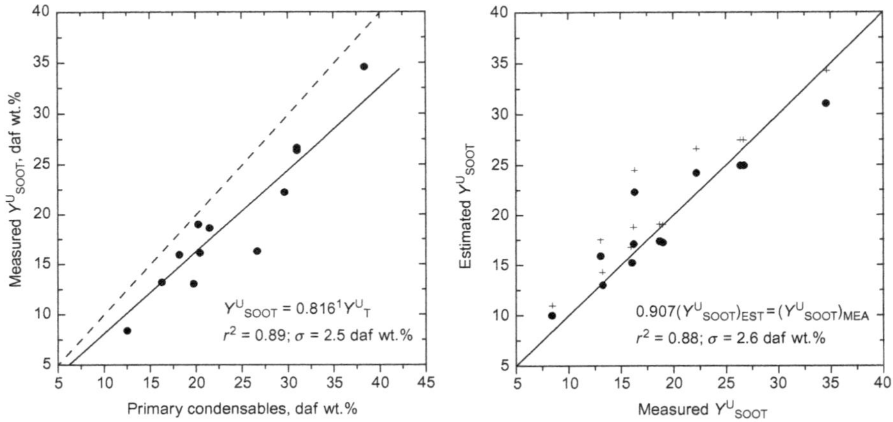

Fig. 7.21 (Left panel) Regression of measured ultimate soot yields versus measured primary tar yields; and (right) regression of predicted ultimate soot yields on measured ultimate soot yields with both (+) raw and (●) re-scaled predicted values.
Reproduced from Niksa S. Predicting ultimate soot yields from any coal. Proc Combust Inst 2019a;37:2757–64 with permission from Elsevier.

denote predicted yields that were re-scaled by the ratio of the constants of proportionality in the regressions on primary tar yields of 0.816/0.900, which is 0.907. Once re-scaled, the predicted ultimate soot yields have almost precisely the same relation to primary tar yields as the measured soot yields, since the correlation statistics in both panels in Fig. 7.21 are essentially identical.

An evaluation of the raw and adjusted predicted ultimate soot yields is plotted against the C-contents of the parent coals in Fig. 7.22 to illustrate the major tendencies with coal quality, and also the substantial sample-to-sample variability among coals of the same nominal rank. The adjusted predicted soot yields are within the measurement uncertainty in all cases except for the coal with 89.6% C, whose predicted yield is too high by several weight percent. The analysis correctly depicts greater soot yields among low-rank coals with progressively greater C-contents, and diminishing soot yields for progressively higher ranks than *hv* bituminous. Soot yields are maximized with *hv* bituminous coals, and can exceed one third of the organic coal mass under the most favorable conditions. The predicted soot yields also depict the substantial sample-to-sample variability with coals of the same nominal rank. Ultimate soot yields from Eqs. (7.3a), (7.3b) are as accurate as those from the full tar decomposition mechanism and easier to evaluate, given accurate values for the yields and elemental compositions of primary tar.

As seen in Fig. 7.22, there is currently a large gap in measured soot yields under known coal loadings for high-C lignites, all subbituminous, and low-C *hv* bituminous coals. The analysis indicates that maximum yields extend over most of the subbituminous and *hv* bituminous portions, albeit with substantial sample-to-sample variability. But the gap in Fig. 7.22 must be filled to validate these expectations.

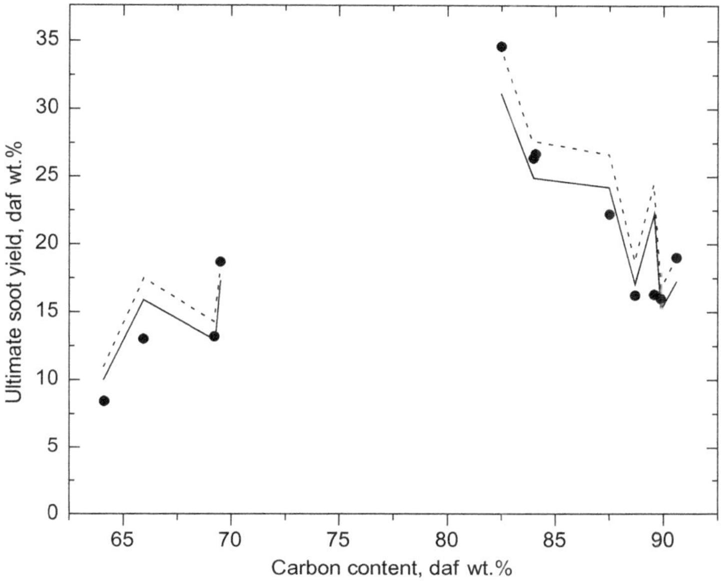

Fig. 7.22 (●) Measured and *(dotted segments)* raw and *(solid)* adjusted predicted ultimate soot yields versus C-content of the parent coal.
Reproduced from Niksa S. Predicting ultimate soot yields from any coal. Proc Combust Inst 2019a;37:2757–64 with permission from Elsevier.

Finally, it is worth noting that the ultimate soot yields in this chapter were measured or predicted for truly inert atmospheres and therefore represent the maximum values that can form from entire distributions of primary volatiles. In commercial applications, particularly in combustion technologies, sooting can be disrupted by oxidation of oils and GHCs before they are able to add to the nascent soot phase, in which case ultimate soot yields will be as much as 40% lower than values in this chapter. This situation is discussed further in Section 8.6 in connection with volatiles combustion for the very heavy suspension loadings injected into utility furnaces.

7.5.3 Nominal rates for tar and soot at elevated temperatures

Simulations of the dynamics of tar decomposition and soot production at elevated temperatures require adding only one additional SFOR for soot production to the two-step global scheme for moderate temperatures, according to

$$\frac{dY_{\mathrm{SOOT}}}{dt} = R_{\mathrm{SOOT}} = k_{\mathrm{SOOT}}\left(Y_{\mathrm{SOOT}}^{\infty} - Y_{\mathrm{SOOT}}(t)\right) \tag{7.4}$$

where $Y_{\mathrm{SOOT}}^{\infty}$ is the ultimate soot yield for the coal, heating rate, and pressure under consideration. In addition to an ultimate soot yield, the kinetic parameters in

Eqs. (7.2a), (7.2b), (7.4) must be specified to complete the global kinetic analysis. The dynamic behavior for tar decomposition in Fig. 7.23 represents the rapid heating rates imposed on dense coal suspensions near the burners in PCC furnaces and the fuel injectors in entrained-flow gasifiers. Tar decomposition resembles a simple sequential decomposition in which primary tar is converted into intermediate secondary tar, which subsequently decomposes into soot. But as noted previously, the form of the secondary tar yield in time is clearly not amenable to representation with a SFOR.

This reaction scheme contains three rate constants, ${}^1k_T$, k_T, and k_{SOOT}, each in Arrhenius form, so there are three pairs of activation energies and pseudo-frequency factors, and three hypothetical ultimate yield parameters, Y_T^∞, ΔY_T^∞, and Y_{SOOT}^∞, to assign. The parameters for the devolatilization rate of primary tar are specified with the method of Niksa and Lau (1993) from FLASHCHAIN® predictions. Then the predicted values of ΔY_T^∞ from the full mechanism for tar decomposition are specified in the same way. The ultimate soot yield is specified with Eqs. (7.3a), (7.3b), and the soot production kinetics are specified to fit the transient soot yields from the full mechanism.

The fit with the sequential three-step reaction model of Eqs. (7.2a), (7.2b), (7.4) is compared to predictions from the full tar decomposition mechanism in Fig. 7.23 for PCC conditions. The global rate laws assigned this way depict the maximum in the secondary tar yield near the end of primary tar production, and accurately describe the transient yields of secondary tar and soot throughout. Of course, the assigned rate parameters for each set of coal properties and operating conditions will be different because no global rate law could possibly describe tar decomposition over a broad domain of conditions for different coal types.

As seen in the Arrhenius diagram in Fig. 7.23, the assigned parameters for primary and (not shown) secondary tar are substantially different for different heating rates, but the soot production rate is essentially the same. The primary tar production rate increases in proportion to changes in the heating rate while the activation energy stays nearly uniform, as expected. But the soot production rate is uniform for all heating rates, even though soot is produced at hotter temperatures for progressively faster heating rates. The nominal soot production rate has a frequency factor of $2.13 \times 10^6\ s^{-1}$ and activation energy of 126.3 kJ/mol, both of which are much greater than the parameters for primary tar. The same behavior was observed in cases with diverse coals at the same heating rate, except that the variations in rates for both tar and soot production were even smaller. Hence, the analysis requires coal- and condition-specific rate parameters for primary and secondary tar production, but not for soot production. As seen in the tabulation in Fig. 7.23, ultimate yields for primary tar and, therefore, ultimate soot yields change for different operating conditions (and also for different coals). This is not a limitation because sets of rate parameters for primary and secondary tars and ultimate tar and soot yields can be evaluated from FLASHCHAIN® for each set of conditions in seconds on ordinary personal computers, given only the proximate and ultimate analyses of the parent coals. In addition to the rate parameters, the detailed mechanism also specifies the mass-based stoichiometric coefficients for all soot, PAH, oils, and noncondensable products of tar decomposition, in g/g primary tar, like those in Table 7.5.

Regarding the partitioning of tar-N into HCN and soot-N, the detailed tar decomposition mechanism certainly does not need to be built into CFD simulations. The simplest

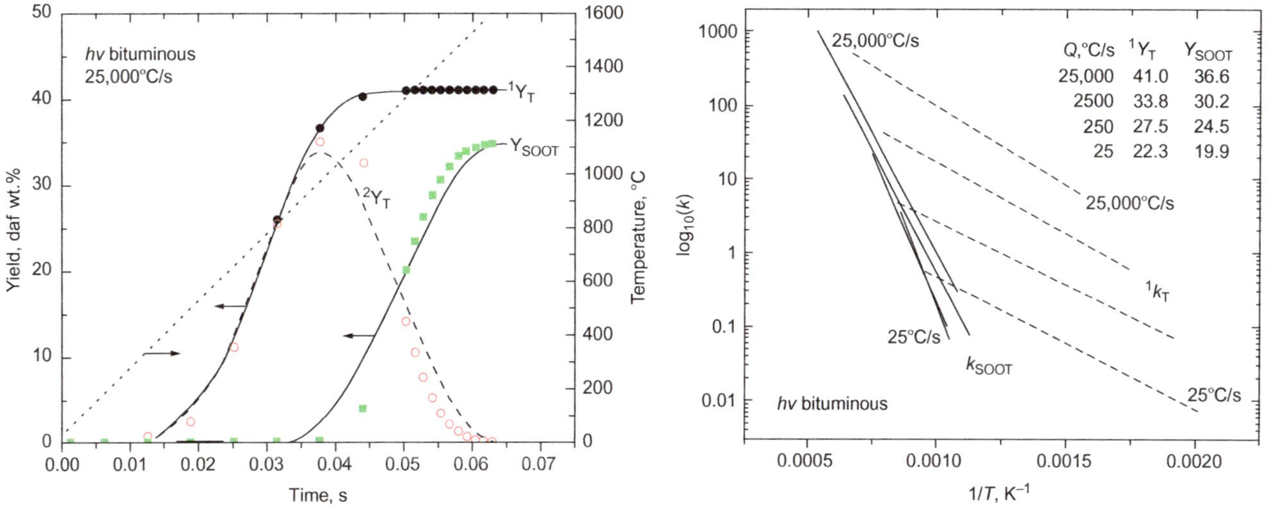

Fig. 7.23 (Left panel) Yields of primary and secondary tar and soot from *(points)* the full tar decomposition mechanism and *(curves)* the three-step global scheme for PCC conditions; and (right) production rates for *(dashed)* primary tar and *(solid)* soot for heating rates from 25 to 25,000°C/s. Lengths of the lines indicate the temperature interval for appreciable production.
Reproduced from Niksa S. Predicting ultimate soot yields from any coal. Proc Combust Inst 2019a;37:2757–64 with permission from Elsevier.

approach for furnace simulations is to first estimate the ultimate levels of soot-N, HCN, and char-N at the char ignition point as described in Section 7.4.3, and then allow these values to develop at the devolatilization rates of either tar or total weight loss. The dynamics can be better resolved with independent global rate expressions for tar production and for HCN from the decomposition of char-N. The three-step reaction sequence in this section fit to predictions from the full tar decomposition mechanism is even more accurate because it resolves the lag between tar and soot production, which becomes appreciable at the cooler temperatures of CFBC, AFBC, and PFBC.

7.5.4 Instantaneous conversion of primary volatiles into secondary products

Ultimate soot yields specified in Section 7.5.3 are the basis to calculate a complete distribution of secondary products from a distribution of ultimate primary volatiles, on the premise that tar decomposition is instantaneous. Additional specifications are only required for the noncondensable gases in secondary products. Under the assumption that moisture and CO_2 are unaffected by secondary pyrolysis, the secondary noncondensables are CH_4, C_2H_2, CO, H_2, H_2S, and HCN. The heavier GHCs are converted into CH_4 and C_2H_2, whereby all C_2-species form acetylene while C_3-species form CH_4 and C_2H_2. The supplements to the primary yields for the surviving GHCs are

$$^{2}Y_{CH_4} = \frac{16}{42}{}^{1}Y_{C_3H_6} + \frac{16}{44}{}^{1}Y_{C_3H_8} \tag{7.5a}$$

$$^{2}Y_{C_2H_2} = \frac{26}{42}{}^{1}Y_{C_3H_6} + \frac{26}{44}{}^{1}Y_{C_3H_8} + \frac{(2)26}{56}{}^{1}Y_{C_4H_8} - \frac{26}{16}{}^{O}f_{TAR}O_{COAL} \tag{7.5b}$$

In Eq. (7.5b), the final term accounts for the exchange of C_2H_2 for CO during the elimination of tar-O. The supplements to the yields of CO, H_2S, and HCN are evaluated from

$$^{2}Y_{CO} = \frac{28}{16}{}^{O}f_{TAR}O_{COAL} \tag{7.5c}$$

$$^{2}Y_{H_2S} = \frac{34.06}{32.06}{}^{S}f_{TAR}S_{COAL} \tag{7.5d}$$

$$^{2}Y_{HCN} = \frac{27}{16}{}^{N}f_{TAR}N_{COAL} \tag{7.5e}$$

And the supplemental H_2 is evaluated with

$$^{2}Y_{H_2} = \frac{2}{28}{}^{1}Y_{C_2H_4} + \frac{(2)2}{30}{}^{1}Y_{C_2H_8} + \frac{2}{44}{}^{1}Y_{C_3H_8} + \frac{(2)2}{56}{}^{1}Y_{C_4H_8} + {}^{TAR}Y_{H_2} \tag{7.5f}$$

Table 7.6 Predicted ultimate distribution of secondary products for the primary products from the fastest heating condition in Fig. 7.23, in daf wt.%.

Primary		**Secondary**	
Tar	**41.0**	**36.9**	**Soot**
CH_4	1.6	0.2	CH_4
C_2H_4	0.6	0.0	C_2H_2
C_2H_6	0.5		
C_3H_6	0.6		
C_3H_8	0.0		
H_2	1.4	4.7	H_2
CO	2.8	5.8	CO
CO_2	2.2	2.2	CO_2
H_2O	5.0	5.0	H_2O
HCN	0.3	0.9	HCN
H_2S	0.3	0.5	H_2S

where ${}^{TAR}Y_{H_2}$ is evaluated from Eq. (7.3b). The total amounts of all these secondary products are evaluated from the sums of the primary yields plus secondary supplements (as ${}^1Y_i + {}^2Y_i$).

As an illustration, Table 7.6 compiles the ultimate distributions of primary and secondary products for the fastest heating condition in Fig. 7.23, where the secondary products are based on the relations in this section. The primary tar yield is among the highest for any coal type for typical p. f. firing conditions, so this case accentuates the incorporation of hydrocarbons into soot. Indeed, all GHCs nearly vanish from the secondary products, because the supplement to the CO yield nearly matches the sum of primary GHCs. In other words, the exchange of C_2H_2 for CO from tar decomposition consumed almost all the available primary GHCs. The levels of H_2 and HCN in the secondary products were nearly triple their respective primary yields, while the levels of CO and H_2S doubled due to the elimination of heteroatoms from primary tar. Moisture and CO_2 yields were unaffected by tar decomposition. Collectively, all these supplements excluding CO account for the reduction in the soot yield compared to primary tar.

With other coal types, the relative changes to tar and soot are similar, although more GHCs usually survive tar decomposition than in this case. Also, the supplements to CO yields grow for coals of progressively lower rank, due to the abundance of oxygen in low rank coal tars. Most important, the distributions of secondary products are the ones that actually burn in pulverized coals flames. Consequently, they are the ones that should be used in CFD furnace simulations. The method in this section provides a simple yet accurate calculation scheme that converts primary products into secondary products, and is recommended whenever the distributions of primary products has either been measured or accurately predicted.

7.6 Status summary

Tar decomposition entails a continuous elimination of heteroatoms as noncondensable products, as well as elimination of the monomer size class as oils and noncondensables and, ultimately, the coalescence of tars and oils into a condensed soot phase. The proposed 11-step mechanism for tar decomposition contains seven steps with direct analogues in FLASHCHAIN®'s primary devolatilization mechanism, plus four new steps that produce oils from tar monomers, nucleate the largest tar molecules into soot, and add oils and lighter tar to a nascent soot phase. Tar decomposition combines two distinct submechanisms: One is the typical single-step conversion of a reactant into multiple products, exemplified by the production of oils and the condensation of oils and tar into soot; the other is a continuous transformation of secondary tar via release of heteroatoms whose only defined product is the ultimate PAH composition based on a limiting atomic H/C ratio. These two distinct transformations are closely coupled because the continuous elimination of peripheral groups from tar chains reduces the probability that monomers will form with peripheral groups still attached, as required for oils production. Any factors that promote spontaneous charring, such as more tar-O, slower heating rates, and hotter temperatures, suppress oils production. This coupling explains why oil yields are greatest for subbituminous and *hv* bituminous coals. As in FLASHCHAIN®, bridge conversion in tar favors scission over spontaneous charring for tars whose labile bridges contain progressively less oxygen and more hydrogen. In primary devolatilization, this premise is a cornerstone for accurate predictions of total and tar yields for individual coal samples. In the context of secondary volatiles pyrolysis, diminishing tar-O levels with coal of progressively higher rank should also be regarded as the determining rank dependence for the oil yields from tar decomposition as well.

This premise also factors into the rank-dependence of ultimate soot yields. Soot yields from low rank coals are relatively low because the abundance of coal-O suppresses primary tar yields. And since the abundance of oxygen in these primary tars can only be eliminated as noncondensables, a smaller fraction of the tar mass adds to the soot phase, compounding soot suppression from low-rank coals. With low volatility coals, primary tars contain hardly any oxygen, so most of the tar mass is converted into soot, which explains why more soot forms from low volatility coals than low rank coals, even when their primary tar yields are comparable. For anthracites, soot yields diminish even further in proportion to the reductions in primary tar. Consequently, soot yields are maximized with *hv* bituminous coals, and can approach one third of the organic coal mass under the most favorable conditions. As seen in Fig. 7.18, there is currently a large gap in measured soot yields under known coal loadings for high-C lignites, all subbituminous, and low-C *hv* bituminous coals. The FLASHCHAIN®-based mechanism indicates that maximum soot yields extend over most of the subbituminous and *hv* bituminous portions, albeit with substantial sample-to-sample variability. But the gap in Fig. 7.18 must be filled before this maximum can be analyzed in detail.

The validation work on the full mechanism also corroborates the premise that the kinetics for bimolecular recombination and soot addition must explicitly depend on coal loading and, by inference, pressure. Consequently, rate parameters for soot addition should be adjusted in proportion to variations in the coal loading when datasets from different test campaigns are analyzed. The satisfactory validations from many sources in this chapter underscore the importance of specifying a coal and/or volatiles loading for every test on tar decomposition kinetics. A precise functional form of a coal loading dependence in tar decomposition kinetics cannot currently be unraveled, simply because relatively very few testing teams reported sufficient information to specify their coal loadings. Tests with variable coal loadings, all else the same, should eventually provide the best characterization of this factor. It will be especially interesting to see the limiting coal loadings that give (i) the minimum soot yields associated with the behavior of isolated individual particles and (ii) the maximum soot yields for a specified residence time that represent complete conversion of all the aromatics in primary oils and tars into soot.

Notwithstanding the complexity of the full tar decomposition mechanism, only a handful of heuristics are required to accurately estimate ultimate soot yields from any coal at elevated temperatures. Two features are absolutely fundamental; viz., (1) Ultimate soot yields are directly proportional to primary tar yields, where the definition of primary tar must be expanded to include all condensed aromatic products of primary devolatilization and (2) In the nascent soot phase C_2H_2 is exchanged for the CO that shuttles away tar-O, so elemental compositions of primary tar are also required. Neither premise has a fundamental basis nor is in any way mechanistically plausible. Instead, they are empirical observations that potentially obscure the extensive decomposition of primary tar into PAH and noncondensables, particularly additional CO and GHCs, and the subsequent nucleation of PAH into soot and addition of PAH, oils, and C_2H_2 from GHC reforming to the soot phase. Provided that the bulk of the C_2H_2 eventually added to soot comes from primary tar rather than coal, the global transformation appears to be an exchange of C_2H_2 for CO, which is nearly neutral on a mass basis. Moreover, it makes no difference whether the lightest primary tars decompose into oils early in the process, provided that oils eventually add to the soot phase as well.

Given ultimate yields of primary tar from FLASHCHAIN® and of soot, either from the full tar decomposition mechanism or the abridged algebraic analysis, only three SFORs are needed to accurately depict the dynamics of tar decomposition, including the maximum and ultimate saturation levels of secondary tar. Whereas the kinetic parameters for primary and secondary tars display the expected variations for different heating rates and with different coals, the global rate of soot production is hardly perturbed across the full domain of heating rate and coal quality. Given a distribution of primary devolatilization products for a particular coal, the calculation scheme in Sections 7.5.3 and 7.5.4 delivers a complete distribution of secondary products, which are the products that actually burn in pulverized coal flames. Secondary products are easily incorporated into CFD furnace simulations under an assumption that tar decomposition is instantaneous, whereby they are generated at a primary devolatilization rate based on a single SFOR. Since tar is the most abundant primary product with virtually all coals, and since soot yields are comparable to primary tar yields, CFD simulations will be improved by incorporating distributions of secondary products, particularly in their resolutions of heat release rates, radiation transfer, and NO_X levels in near-burner regions.

References

Chen JC, Niksa S. Coal devolatilization during rapid transient heating. Part 1. Primary devolatilization. Energy Fuel 1992a;6:254–64.

Chen JC, Niksa S. Suppressed nitrogen evolution from coal-derived soot and low-volatility coal chars. Proc Combust Inst 1992b;24:1269–76.

Chen JC, Castagnoli C, Niksa S. Coal devolatilization during rapid transient heating. Part 2. Secondary pyrolysis. Energy Fuel 1992;6:264–71.

Doolan KR, Mackie JC, Tyler RJ. Coal flash pyrolysis: secondary cracking of tar vapours in the range 870–2000 K. Fuel 1987;66(4):572.

Fletcher TH, Ma J, Rigby JR, Brown AL, Webb BW. Soot in coal combustion systems. Prog Energy Combust Sci 1997;23:283–301.

Hayashi J-I, Amamoto S, Kusakabe K, Morooka S. Evaluation of vapor-phase reactivity of primary tar produced by flash pyrolysis of coal. Energy Fuel 1995;9:290–4.

Hayashi J-I, Takahashi H, Iwatsuki M, Essaki K, Tsutsumi A, Chiba T. Rapid conversion of tar and char from pyrolysis of a brown coal by reactions with steam in a drop-tube reactor. Fuel 2000;79:439–47.

Hayashi J-I, Iwatsuki M, Morishita K, Tsutsumi A, Li CZ, Chiba T. Roles of inherent metallic species in secondary reactions of tar and char during rapid pyrolysis of brown coals drop-tube reactor. Fuel 2002;81:1977–87.

Jia Y, Huang J, Wang Y. Effects of calcium oxide on the cracking of coal tar in the freeboard of a fluidized bed. Energy Fuel 2004;18:1625–32.

Katheklakis IE, Lu S-L, Bartle KD, Kandiyoti R. Effect of freeboard residence time on the molecular mass distributions of fluidized bed pyrolysis tars. Fuel 1990;69(2):172.

Ledesma EB, Li CZ, Nelson PF, Mackie JC. Release of HCN, NH_3, and HNCO from the thermal gas-phase cracking of coal pyrolysis tars. Energy Fuel 1998;12:536–42.

Nenniger RD, Howard JB, Sarofim AF. Sooting potential of coals. Proc 1983 int conf coal sci. Pittsburgh, PA: IEA; 1983. p. 521–4.

Niksa S. Flashchain theory for rapid coal devolatilization kinetics. 6. Predicting the evolution of fuel nitrogen from various coals. Energy Fuel 1995;9:467–78.

Niksa S. A reaction mechanism for tar decomposition at moderate temperatures with any coal type. Fuel 2017a;193:467–76.

Niksa S. Flashchain theory for rapid coal devolatilization kinetics. 9. Tar decomposition across a broad temperature domain. Energy Fuel 2017b;31:9080–93.

Niksa S. Predicting ultimate soot yields from any coal. Proc Combust Inst 2019a;37:2757–64.

Niksa S. Predicting nitrogen release during coal tar decomposition. Proc Combust Inst 2019b;37:2765–72.

Niksa S, Kerstein AR. Flashchain theory for rapid coal devolatilization kinetics. 1. Formulation. Energy Fuel 1991;5:647–65.

Niksa S, Lau C-W. Global rates of devolatilization for various coal types. Combust Flame 1993;94:294–307.

Rigby J, Ma J, Webb BW, Fletcher TH. Transformations of coal-derived soot at elevated temperature. Energy Fuel 2001;15:52–9.

Serio MA, Peters WA, Howard JB. Kinetics of vapor phase secondary reactions of prompt coal pyrolysis tars. Ind Eng Chem Res 1987;26:1831–8.

Serio MA, Hamblen DG, Markham JR, Solomon PR. Kinetics of volatile product evolution in coal pyrolysis: experiment and theory. Energy Fuel 1988;1:138–52.

Solomon PR, Hamblen DG. Pyrolysis. In: Schlosberg RH, editor. Chemistry of coal conversion. New York: Plenum Press; 1985. p. 121–251.

Umemoto S, Kajitani S, Hara S, Kawase M. Proposal of a new soot quantification method and investigation of soot formation behavior in coal gasification. Fuel 2016;167:280–7.

Umemoto S, Kajitani S, Miura K, Watanabe H, Kawase M. Extension of the chemical percolation devolatilization model for predicting formation of tar compounds as soot precursor in coal gasification. Fuel Process Technol 2017;159:256–65.

Wornat MJ, Sarofim AF, Longwell JP. Changes in the degree of substitution of polycyclic aromatic compounds from pyrolysis of a high-volatile bituminous coal. Energy Fuel 1987;431–7.

Xu W-C, Tomita A. The effects of temperature and residence time on the secondary reactions of volatiles from coal pyrolysis. Fuel Process Technol 1989;21:25–37.

Further reading

Li CZ, Nelson PF. Fate of aromatic rings systems during thermal cracking of tars in a fluidized-bed reactor. Energy Fuel 1996;10:1083–91.

Volatiles reforming and volatiles combustion

8

This chapter presents the impact, characteristics, and analyses for volatiles combustion and volatiles reforming, which are collectively called volatiles conversion. Chapter 7 considered one essential portion of volatiles reforming, tar decomposition under inert atmospheres, and Chapter 9 analyzes tar hydroconversion under elevated H_2 pressures. This chapter factors in simultaneous conversion of noncondensable fuel components and pollutant precursors, with and without O_2, and also conversion of char and soot.

The conversion rates of noncondensable volatiles are relatively fast compared to those for primary devolatilization and tar decomposition. In particular, GHCs are the fastest burning volatile components of all, and also the fastest to be reformed under reducing conditions. Consequently, it is possible to identify reaction systems such as entrained flow gasification where the compositions of the noncondensable species are in local thermochemical equilibrium even while tar and char are continuously converted on relatively long time-scales. However, elevated temperatures and appreciable contact times are needed to equilibrate product gas compositions. At moderate temperatures, gas compositions usually do not reach equilibrium compositions even though the volatiles conversion kinetics are relatively fast compared to reaction rates in the condensed coal phase.

Similar to the previous chapter, the presentation divides the behavior into two domains, one where the kinetics are so fast that they do not need to be resolved, and the other where finite-rate kinetics are essential. Where flames are present, local gas temperatures are hot enough that burning rates of volatiles are often based on the mixing rates of secondary and tertiary air streams into a coal-laden primary stream, without any finite-rate kinetics. Product distributions from such flames generally relax to the compositions for thermochemical equilibrium, albeit at the flame temperature. But such an approach ignores one of the most important aspects of volatiles combustion: that it does not occur in isolation under most commercial operating conditions. Instead, volatiles compete for the available O_2 with char and, perhaps, soot, and this competition can only be staged with finite-rate kinetics for all the various fuel components. Moreover, at the cooler temperatures associated with lower O_2 levels in some technologies, volatiles burning rates may be mediated by the chemical kinetics, so that the product compositions may not achieve equilibrium compositions in the available reaction time. For many aspects of combustion and gasification, finite-rate kinetics for volatiles conversion are essential: Important examples include fuel-N conversion in coal flames; the performance of aerodynamic NO_X abatement schemes (low-NO_X burners, staging, selective non-catalytic reduction (SNCR), reburning) in PCC furnaces, CFBCs, AFBCs, and PFBCs; and syngas compositions from all forms of gasification in fluidized beds and fixed beds, especially when GHCs are appreciable

Process Chemistry of Coal Utilization. https://doi.org/10.1016/B978-0-12-818713-5.00008-3

products. Note that temperature does not necessarily delineate the need for kinetics; this categorization also depends on the quantities of greatest interest in an analysis, and also on the specifications on the coal feed that affect char burning rates.

In previous chapters, the characteristics of devolatilization and tar decomposition were presented, first, in terms of extensive laboratory results and, then, in quantitative interpretations with phenomenological reaction mechanisms. This organizational format is retained for volatiles conversion, although relatively few datasets actually isolate this chemical stage during coal processing; rather, measured product gas compositions reflect simultaneous tar decomposition and volatiles conversion, often in partial overlap with primary devolatilization, soot conversion, and/or char conversion.

Fortunately, an assortment of validated elementary chemical reaction mechanisms is available for the gas mixtures generated by coal processing. We can be confident that the elementary reaction mechanisms developed by the combustion research community for gaseous hydrocarbon mixtures are suitable for coal-based applications as well, even if specialized heterogeneous chemistry on soot, char, and inorganic flyash components may be needed to address certain aspects of reacting coal suspensions. The strategy behind this presentation uses validated elementary reaction mechanisms to accurately interpret the laboratory database, and also as virtual laboratories to illustrate the major stages in volatiles combustion and reforming, as well as the coal quality impacts.

One can legitimately ask, "If the knowledge base on realistic coal conversion products has already been synthesized into elementary reaction mechanisms, can volatiles conversion be reduced to one sufficiently robust mechanism that can accurately simulate the entire domain of coal utilization technologies for any coal type?" Unfortunately, the answer is "No," because such a mechanism would comprise several thousand individual chemical reactions, and carry an inordinate computational burden. It is much more efficient to associate specific mechanisms with portions of the operating domain, and with particular groups of reactants. Also, it can be more efficient to first use a generic hydrocarbon conversion mechanism to describe the chemistry of the major volatiles species, and then run additional calculation passes with specialized mechanisms to describe the formation of pollutants and other minor species.

This chapter first surveys commercial impacts of volatiles conversion, then identifies two limiting scenarios for the mixing of coal volatiles with reactant gases that frame the time scales and stages of volatiles conversion. Then various datasets illustrate the essential features of volatiles combustion, including NO_X production and steam reforming. Then the presentation shifts to quantitative analyses of volatiles combustion both around individual coal particles and also in dense coal suspensions. Instead of interpreting the entire lab-scale database, a collection of chemical reaction mechanisms is first validated with the most discriminating datasets, and then extrapolated to commercial operating conditions for p.f. firing, entrained flow gasification, and the first stage of CFBC. These simulations reveal how furnace and gasifier operating conditions affect volatiles conversion, along with the coal quality impacts.

8.1 Commercial impacts

Volatiles conversion is always important in any commercial coal utilization technology because O_2 is almost always present in the coal feedstream, and operating temperatures are always hotter than ignition temperatures for GHCs. Even without O_2, volatiles conversion is unavoidable because it proceeds spontaneously without any gaseous reactants or catalytic solids. Unfortunately, volatiles conversion cannot be monitored in any large-scale coal utilization system. There are no means to monitor conversion of multicomponent gaseous fuel mixtures on-the-fly other than extractive sampling followed by an analysis to assign conversions. Gaseous volatiles are converted in tens milliseconds in furnaces and entrained flow gasifiers, which is too fast for extractive sampling. Moreover, the flow patterns are too complex for sampling probes anyway. Optical diagnostics cannot contend with the particulate loadings, vibrations, and intense background radiation. In gasifiers, the elevated operating pressures make the monitoring even more complicated. In fluidized systems, solids loadings are much too heavy to permit any access at all, and mixing patterns are too convoluted for extractive sampling along a time coordinate.

Notwithstanding the measurement uncertainties, we know that volatiles conversion generates most of the heat release in near-burner flame zones (NBFZs), which is a determining factor in the ignition and stability of coal flames, both in the burners and injectors in PCC furnaces and in the fuel injectors in entrained flow gasifiers. Since GHCs are the fastest burning fuel components of all, their burning rates factor into the time scale for flame ignition.

At temperatures $>\sim 1000°C$, CO becomes the predominant product of char oxidation. Since char burns throughout most, if not all, of a furnace, CO is released continuously along the furnace elevation. Chemistry in the gas phase converts this CO into CO_2, usually at rates sufficient to diminish CO levels at the furnace exit to a few hundred ppm or less. But when CO is produced near the quench layers along furnace waterwalls, and when burner mismanagement is responsible for plumes of flue gas with very low O_2 concentrations, volatiles conversion kinetics can become too slow to fully oxidize CO in the available furnace transit time. Under the worst of such circumstances, CO emissions can grow to tens of thousands of ppm.

All forms of aerodynamic NO_X abatement manage air mixing rates into a primary coal stream to provide sufficient residence times under sub-stoichiometric conditions to skew the conversion of volatile-N toward N_2 and away from NO. This conversion occurs among gaseous radicals and intermediate species, and the requisite contact times are determined by the reaction kinetics. Since the residual nitrogen in char and soot is primarily expelled as NO, volatiles conversion kinetics in conjunction with stream mixing rates determine NO_X concentrations at the exit of a furnace into the gas cleaning system. Similarly in SNCR, a nitrogen compound such as ammonia or urea is injected into the post-flame gases to generate amine radicals that can reduce NO from NBFZs into N_2 in the upper furnace elevations. And in reburning, GHCs or a source of GHCs like biomass is injected into the post-flame gases to generate small hydrocarbon radicals that also reduce NO from NBFZs. In both schemes, volatiles conversion chemistry in conjunction with the agent mixing rate determines how much NO is present at the furnace exit.

Volatiles conversion chemistry also converts H_2S, the primary volatile-S species from most coals, into SO_2, and subsequently converts a small portion of SO_2 into SO_3 upstream of the furnace exit. This chemistry is important, because SO_3 promotes corrosion and forms sulfates that can poison SCR catalysts and generate visible plumes in gas cleaning systems. In somewhat similar ways, acid gases such as HCl and SO_2 react with vapors of alkali and alkaline earth metals to form compounds that promote flyash deposits on heat transfer tubes, poison SCR catalysts, and corrode gas cleaning components.

Volatiles conversion in steam is an essential aspect of the initial stages of coal gasification in some, but not all, gasifiers. One key distinction is the proportion of coal and O_2 or air in the primary coal feedstream. If this stream contains sufficient O_2 to burn out the volatiles, then steam reforming is irrelevant because its rate is orders of magnitude slower than volatiles combustion. Another distinction based on steam mixing pertains only to gasifiers that do not sustain volatiles combustion. Among the many different steam injection configurations used commercially, some introduce water or steam into the stream that entrains coal suspensions into the reactor, and others rapidly mix steam into a dry coal feedstream near the coal injectors. The limiting case for steam mixing is when coal is fed as a slurry at very high suspension loadings. In any of these situations, steam is well mixed with coal during or immediately after the suspension is heated to the onset temperature for primary devolatilization. So the steam will participate as a reactant in volatiles conversion in these injector configurations and, under some operating conditions, disrupt the decomposition of tar into soot. Conversely, some steam injectors delay mixing until the volatiles are spontaneously converted by secondary volatiles pyrolysis, including tar decomposition into PAH and, if the temperature is hot enough, into soot. In this situation, steam does not disrupt the normal progression of tar decomposition for inert atmospheres. But the intermediate noncondensable products of volatiles pyrolysis are spontaneously reformed nonetheless.

Regardless of the gasifier configuration, noncondensable intermediates and products are continuously reformed in the gas phase downstream of the coal injectors as steam and CO_2 are converted into CO, H_2, and pollutants by the gasification of soot and char. In entrained flow gasifiers rapid reforming keeps the gas composition in thermochemical equilibrium even while the carbon in soot and char is gasified. But at cooler temperatures, chemical kinetics determine product gas compositions, especially when GHCs are present.

No legitimate analysis of any coal utilization technology may omit volatiles conversion.

8.2 Determining factors for volatiles conversion

The operating domain for volatiles conversion is spanned by coal quality, temperature, pressure, and the proportions of coal and oxidizer into a system. The latter proportions will be gauged by the stoichiometric air ratio, SR, which is the ratio of the air-to-fuel flowrates into the system normalized by the value for stoichiometric conversion (cf. Eq. 1.1). Since coal is entrained in suspension into most utilization technologies, SR would normally be evaluated as the primary air flowrate divided by the coal flowrate, normalized by the ratio of flowrates for stoichiometric combustion. Although this value clearly adheres to the definition of SR, it has virtually no bearing on the proportions of fuel and oxidizer during volatiles conversion, for several reasons.

First of all, this SR value represents conversion of all combustibles generated by the coal, including soot and char. But char and soot are often converted on much longer time scales than the noncondensable fuels, so the effective SR for volatiles conversion will be much greater than the nominal, whole coal value. If an SR incorporates a primary air flowrate, then an SR value based on the whole coal feedrate indicates a less oxidizing atmosphere for volatiles conversion than the actual reaction system.

Various mixing phenomena also obscure the atmosphere for volatiles conversion. All noncondensable fuels emanate from individual particles in a coal suspension. For some period after the onset of devolatilization, these fuel compounds must be segregated from any oxidizer in the form of clouds attached to their parent particles, even in the primary coal/air stream. Oxygen penetrates the clouds via diffusion and convective transport, which eventually brings the mixture compositions into their combustible range. Meanwhile, devolatilization continues to spew additional gaseous fuels into the mixture, while the clouds are detached from their parent particles by viscous drag forces. Once the clouds are detached, they are mixed into turbulent eddies and become subject to the movement of larger flow structures in the reacting suspension. Secondary and tertiary air streams are entrained into the suspension on longer time scales, increasing the O_2 concentration in the volatiles mixture.

Without appreciable entrainment of external air streams, volatiles concentrations throughout the suspension remain as spatially uniform as the dispersion of the local coal loading, because particles are the only sources of volatiles in the system. In heavily laden, momentum dominated coal jets, like the ones emanating from the injectors in T-fired furnaces and entrained flow gasifiers, particle dispersion is minimal in the near-field, so volatiles concentrations will be relatively uniform across the jet, and diminish down the jet axis as volatiles burn with the primary air. Conversely, with rapid entrainment of the external air streams, large turbulent flow structures both aggregate and disperse the particles, so that the suspension disintegrates into a multitude of clusters that produce a broad distribution of volatiles concentrations. Heavily swirled, dual-register burners epitomize this situation.

In conjunction with these essential mixing phenomena, the amounts and compositions of gaseous fuels from different coals can be grossly different. More gaseous fuels of greater calorific value are released from coals of progressively higher rank through mv bituminous, due to the relatively large yields of the diluents H_2O and CO_2 from low rank coals. Although the devolatilization rates of all but the highest rank coals are comparable, variations in the yields of gaseous fuels will generate combustible mixtures with lower SR values with coals of progressively higher rank through mv bituminous, under the same mixing conditions. Consequently, coal quality and mixing phenomena are independent determining factors on SR values for volatiles conversion.

So what is an appropriate value of SR to characterize volatiles conversion? The answer depends on both coal quality and mixing phenomena across multiple length scales. Most of this chapter is devoted to illustrations of the coal quality impacts, which can be set aside for the time being. As sketched in Fig. 8.1, the impact of mixing will be illustrated with two limiting cases delineated by the relative rates of chemical conversion vs. O_2 transport and mixing. For relatively fast conversion and slow mixing at the scale of particles, the volatiles clouds ignite and burn in less time than the time scales for viscous drag and convective mixing. Consequently, all gaseous volatiles and soot are consumed in the immediate vicinity of individual coal particles throughout the suspension, so that volatiles conversion occurs at the rates of primary devolatilization and tar decomposition. Certainly, this limit can only be achieved in primary coal streams in which, by definition, the coal particles are premixed with air, and the suspension loading is light. At face value, the appropriate SR-value should be based on the primary air flowrate and a fuel flowrate associated with the ultimate yield of only the gaseous fuels and soot. But this is not the relevant SR value, because whenever segregated volumes of fuel and air burn at their interface, burning rates are determined by the transport rates of fuel and O_2 in a so-called "diffusion flame" configuration. As is well known, the reactants are consumed in stoichiometric proportions, for which the SR is unity. The actual volatiles conversion in a coal suspension skews toward this limit for light coal loadings and elevated temperatures, O_2 levels, and pressures.

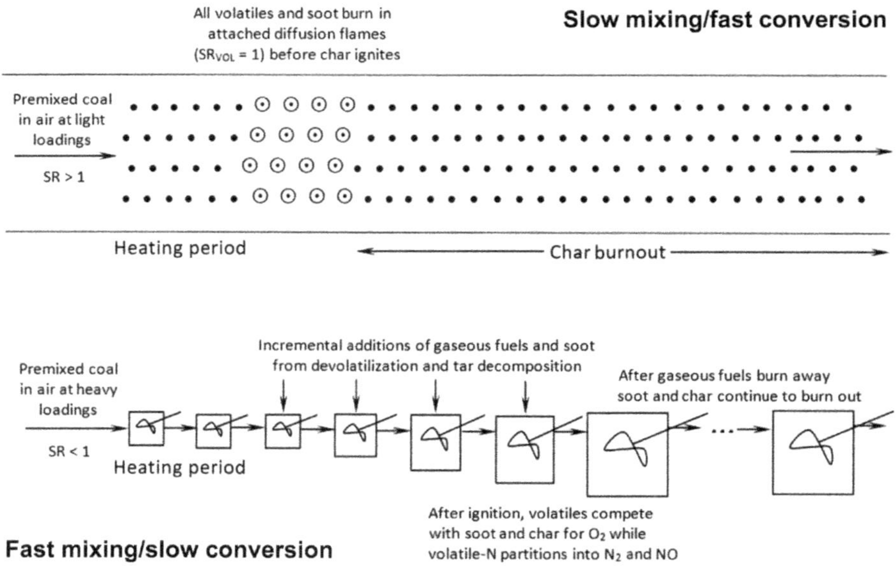

Fig. 8.1 Limiting cases for volatiles conversion determined by (top) slow mixing of volatiles and fast reaction kinetics and (bottom) instantaneous mixing of volatiles and slow reaction kinetics.

In the opposite limit, volatiles mix in shorter times than the chemical reaction time scale, so that nascent volatiles clouds are drawn throughout the entrainment stream before the fuel compounds are converted. Moreover, each increment of the ultimate volatiles yield is fully dispersed into the mixture even after the onset of volatiles conversion. As seen in Fig. 8.1, this limit will be represented as a long series of continuously stirred tank reactors (CSTRs), each of which stages the chemistry during a brief increment in the nominal residence time along the suspension. By definition, feedstreams into each CSTR instantaneously mix into the bulk flow prior to conversion. To represent volatiles conversion, discrete injections of gaseous volatiles and soot into each CSTR are based on the incremental yields of gaseous fuels and soot released by primary devolatilization and tar decomposition for the time interval associated with the mean transit time through each CSTR. If, for example, during the interval from 35 to 40 ms after injection, the coal suspension released 5 daf wt% CH_4 by devolatilization and another 1.7% by tar decomposition, then the flowrate associated with 6.7% CH_4 would be injected into the CSTR in the series that covers 35 to 40 ms after injection, along with any other released fuels and diluents. In general, these injections could also contain increments of the secondary and tertiary air streams whenever their entrainment rates are relatively fast compared to the chemical conversion time scale. But secondary and tertiary air injections were omitted from Fig. 8.1 to sharpen the focus on volatiles conversion.

The other key feature of this limit is that char competes for O_2 with gaseous fuels and soot in each CSTR. In the slow mixing limit, devolatilization/tar decomposition occur simultaneously with fast volatiles conversion; then once all gaseous fuels and soot have been consumed, O_2 reaches the char surface and the char ignites and burns out. So volatiles conversion occurs in series with char burnout, for which volatiles burning rates are governed by kinetics for devolatilization and tar decomposition, and are independent of the kinetics for char burnout. Conversely, in the fast mixing limit, each CSTR stages a competition for O_2 among gaseous fuels, soot, and char. The instantaneous O_2 concentrations are governed by the oxidation kinetics of all three fuel types, which couples the rates of volatiles conversion to the oxidation kinetics of both char and soot. This coupling accentuates the coal quality impacts on volatiles conversion, because char oxidation kinetics are particularly sensitive to coal quality (Niksa et al., 2003).

In this completely mixed limit, each CSTR in the series converts volatiles under a different SR value. Prior to the onset of primary devolatilization, the SR values are infinite, by definition, because there are no fuel compounds in the gas phase. As more volatiles and, later, more tar decomposition products are added to succeeding CSTRs, their SR values rapidly decay while some, but not all, of the gaseous fuel compounds are converted during the transit time for each CSTR. Depending on the coal rank, more or less char and, perhaps, some soot are converted simultaneously. Without entrainment of secondary and tertiary air, the ultimate SR value is that based on the ultimate yields of gaseous fuels plus the portions of char and soot that burned simultaneously with the gaseous fuels. The ultimate SR value will be smaller for progressively greater extents of burnout of soot and char.

Note that nearly all the gaseous fuels are actually converted under SR values that are greater, and sometimes much greater, than the ultimate, nominal value, even when the nominal value excludes soot and char conversion. This is because the CSTR series explicitly incorporates the time coordinate for volatiles conversion, as with streamlines through the flow in a Lagrangian reference frame that have been averaged transverse to the flow direction in each component's concentration. The number of CSTRs in the series is deliberately chosen to give relatively short residence times in each CSTR, compared to the total transit time along the coal suspension. So the series resolves all stages of volatiles conversion, beginning with particle heat-up under an infinite SR, and progressing through incremental conversion under progressively lower SR values, including sub-stoichiometric conditions for some coals and coal loadings. Note also that as more CSTRs are added to the series to improve the time resolution, the series approaches the limiting behavior of a plug-flow reactor, which sustains no axial dispersion along the flow direction and complete dispersion through the flow cross section.

It is somewhat reassuring that the well-mixed limit is reasonably consistent with the relatively uniform volatiles concentration fields across momentum dominated coal jets. But this approach clearly does not account for the large flow structures and gross inhomogeneities in volatiles concentrations along swirled coal flames. Such complications are incompatible with our focus on the coal quality impacts for volatiles conversion. Readers interested in these more complex flowfields should follow the research on large-eddy simulations of coal flames underway in Japan, China, and Germany at the time of this writing.

There is yet another generally important aspect of coal flames to consider. In suspension-fired systems, the behavior of individual particles gives way to clouds in which numerous fuel compounds—char, tar, soot, and noncondensable fuels—compete for the available O_2 while volatiles are continuously reformed, partially oxidized, and ultimately burned out. This competition is especially important for the broad PSDs and fine grinds of the coal streams fired in any commercial system, because the small sizes of some char particles compensate for the relatively slow burning rates of char compared to volatiles burning rates. In fact, in many coal burner designs, volatiles conversion overlaps with soot oxidation and substantial levels of char conversion. These situations are not beyond the reach of simulations with reactor networks, which will be demonstrated in the later sections on volatiles conversion in p. f. flames and entrained flow gasifiers. More complex situations involving simultaneous conversion of all fuel components in coal furnaces and gasifiers are covered in the second volume in this series.

8.3 Measured volatiles conversion behavior

This section presents the most important attributes of volatiles conversion in terms of laboratory datasets, first, for volatiles combustion and then for volatiles reforming. The presentation exclusively relies on laboratory-scale testing systems, where operating conditions can be varied independently, specified at will, and regulated within close tolerances, and transverse temperature gradients can be essentially eliminated. Pilot-scale combustors and gasifiers are deliberately avoided because two-phase mixing phenomena inevitably introduce sizable gradients in temperature throughout the flowfields, which complicates quantitative interpretations. The major organic products are characterized before pollutant precursors.

8.3.1 Laboratory prerequisites

Under the most favorable circumstances, one database on volatiles conversion would cover a complete range of SR from zero, which corresponds to secondary volatiles pyrolysis, through values greater than unity, where O_2 is in large excess. Another complementary database on volatiles reforming would span broad ranges of steam concentrations for broad ranges of temperature and pressure. Both databases would represent the complete coal rank spectrum over a wide range of coal loadings, to encompass both of the limiting scenarios for mixing described in Section 8.2. These circumstances have not yet been realized, although the situation for oxidizing atmospheres is considerably more developed than for volatiles reforming. The main reason is that O_2 rapidly eliminates all GHCs and oils from the mixtures, and accelerates the conversion of tar into soot by raising stream temperatures. So product recovery and analysis is straightforward, even when N-species conversion into N_2 and NO is the primary objective. However, volatiles reforming does not rapidly eliminate tars, oils, and GHCs so product analysis is much more complicated.

In any testing, the strongest imperative is to clearly resolve the conversions of GHCs, CO, H_2, tar, oils, soot, and char, since these fuel components burn at markedly different rates. Volatiles combustion tests are much easier to interpret when coal is premixed with an oxidizing carrier stream, so that mixing does not introduce inordinate transverse spatial gradients. Test series covering a range of inlet O_2 concentrations are especially illuminating because each succeeding test achieves progressively greater conversions. Under the best circumstances, conversions for each gaseous fuel, char, and soot are recorded from ignition through complete burnout. Tests in a diffusion flame configuration are well suited for work on N-species transformations. In reforming tests with added steam and CO_2, the same guidelines on premixing are applicable although more difficult to implement in practice at the highest steam levels of interest. Conversely, it is easier to minimize extents of char conversion because char gasification is much slower than char oxidation.

More formally, the following testing features are required in any dataset used to evaluate volatiles conversion mechanisms:

1. *Coal properties*—Only the proximate and ultimate analyses are required. Strictly speaking, only the ultimate analysis plus the proximate volatile matter on a dry, ash-free (daf) basis is necessary. Coal PSDs should be confined to no more than two standard mesh sizes to manage the associated distributions in particle temperature and extents of primary devolatilization and char conversion.
2. *Pressure*—Usually a uniform test pressure will be specified although a pressure history can also be handled.

3. *Thermal history*—Sufficient information must be available to assign the temperature of the fuel sample throughout an entire test, as well as the thermal history of the carrier gases in which volatiles are converted. This information may comprise the thermal history of a sample support, as in WMRs, or the sample properties and ambient conditions that determine a sample's thermal history, as in EFRs. Operating conditions for a second reactor stage should be specified with thermocouple data, and reaction times or RTDs must be monitored or assigned by testing teams.
4. *Coal loadings and particle dispersion*—Coal loadings at the inlet to the testing system must be specified, and the flowfield should be analyzed for significant particle dispersion. Coal loadings should overlap with the loadings in the target commercial applications. Simple axisymmetric flows are preferable but nonetheless sustain large gas velocity gradients in laminar flows and particle aggregation along tube walls in turbulent flows.
5. *Gas composition*—Carrier gas compositions must be specified to assign the concentrations of volatiles during volatiles conversion. Reactive gas concentrations (O_2, steam, CO_2) should span the relevant technological domain of interest.
6. *Distribution of primary volatiles*—Complete product distributions for primary devolatilization must be monitored at the conditions used to generate them in the tests on volatiles conversion, supplemented with tests with known extents of tar decomposition. At a minimum, preliminary tests without any reactive gases should be run. These product distributions should cover the most severe thermal processing in the volatiles conversion tests, and should represent ultimate products for complete primary devolatilization and tar decomposition.
7. *Incorporation of reactive gases*—The only means to evaluate the incorporation of a reactive gas into the product distributions is to monitor sufficient species to close balances on mass and the major organic coal elements. All major noncondensables should be monitored, including CO, CO_2, H_2O, H_2, H_2S, SO_2, HCN, NH_3, N_2, NO, C_1–C_4 GHCs, and oils, along with elemental compositions of tar and char.
8. *Index for volatiles conversion*—The progress of volatiles conversion must be specified quantitatively with multiple conversion indices for the different fuel components (GHCs, tar, oils, CO, H_2, soot, char). Elemental compositions for secondary tars, soot, and char are especially valuable. Aerosol products should also be recovered, so that soot levels can be evaluated after tars and oils are separated out.

Coal loadings and particle dispersion are particularly important in volatiles combustion testing because, as we shall soon see, volatiles burn in clouds attached to individual particles whenever coal is fired in dilute suspension, as in the vast majority of EFR tests in the literature. Unfortunately, it is simply impossible to resolve volatiles conversion in this testing configuration, for reasons given shortly. Moreover, NO_X production efficiencies evaluated from the behavior in dilute suspension grossly overestimate the efficiencies for realistic suspension loadings. If at all possible, suspension loadings in the target application should be imposed in lab tests that characterize volatiles conversion and, especially, NO_X production.

For tests on primary devolatilization and tar decomposition, WMRs and CPPs hold distinct advantages over EFRs in the regulation of thermal histories (cf. Section 4.1.3). These advantages disappear for tests on volatiles conversion unless a WMR or CPP are used to generate a volatiles stream that is converted in a second reactor stage. But even this latter arrangement is problematic because the concentrations of volatiles from the first stage will vary throughout devolatilization, which makes it impossible to assign SR values into the second stage throughout a test. One resolution is to accumulate all primary volatiles into a batch before it is sent through a second reactor, although it is hard to avoid deposition of heavier components and soot on the walls of a containment vessel. Another resolution is to use continuous feed systems such as fluidized beds, RCFRs, and EFRs to generate a volatiles stream for a second stage.

The most common second-stage reactors are combustion bombs and shock tubes. A combustion bomb can be used to determine the overall burning rates of combustible volatiles mixtures as functions of the composition of gaseous fuel mixtures, SR, and pressure. This method has been successfully applied to mixtures of whole non-condensable volatiles from different coals (Marlow et al., 1992, 1993). Shock tube studies provide comparable information, with the advantage that both temperature and pressure are strictly uniform and well-known throughout a test (Doolan et al., 1987).

Tests in a single reactor must be conducted with entrained coal suspensions. In conventional EFRs, the convective mixing phenomena at the particle injector introduce large uncertainties into the assignments of particle thermal histories (cf. Section 4.1.3), and these uncertainties are compounded by ambiguous particle dispersion from those same mixing phenomena. To alleviate these issues, coal loadings are almost always kept far below loadings in commercial systems, which carry additional complications explained subsequently.

RCFRs minimize both of the mixing ambiguities in EFRs because there is no preheated gas stream that mixes with the coal feedstream at the injector, and they also can be operated at much more realistic coal loadings (Niksa and Cho, 1998). Another distinct advantage is that the reactor operates as a burner at any O_2 level, including none at all, because the heat transfer and chemical conversion are driven by the intense thermal radiation from an oven instead of the heat release from a coal flame. And relatively dense particle loadings generate high volatile species concentrations in the effluent, so that mass and element balances can be closed in individual runs. Complete product distributions have been reported for broad ranges of SR and coal quality at both atmospheric (Niksa and Cho, 1996, 1998) and elevated pressures (Liu and Niksa, 2005; Eckstrom et al., 2014). Still, thermal histories of both particles and gas along RCFRs can only be assigned by calculations. Tolerances on the particle thermal histories are not stringent because only a nominal heating rate is required for the primary devolatilization stage, provided that the reactor also has residence times long enough to attain asymptotic ultimate volatiles yields at the stated reactor temperature. Gas temperature histories are subject to tighter tolerances because the gas stream sustains the chemistry of interest. Unfortunately, the calculations to assign gas thermal histories are complicated by numerous heat transfer mechanisms and boundary conditions.

Flat-flame burners are operated in two modes to characterize volatiles conversion, a single-particle limit and a one-dimensional coal flame. In the single-particle limit, a flat flame of an external gaseous fuel mixture is stabilized on a honeycomb to generate a hot stream with a specified O_2 level. Coal particles are injected in very dilute suspension onto the burner centerline, where they heat-up, ignite, and burn along the reactor elevation as isolated, individual particles. Such tests have characterized heat-up times and the duration of volatiles combustion for ranges of ambient conditions with different coal types (McLean et al., 1981; Seeker et al., 1981). To generate larger 1D coal flames, particles are injected across wire mesh stabilizers to form a much denser suspension that heats-up and ignites across a narrow increment of the furnace axis (Altenkirch et al., 1979). Such flames are especially well-suited to studies of NO_X production in post-flame gases, which has been characterized across broad ranges of SR and coal quality. But they can only resolve chemistry downstream of the region for volatiles conversion.

8.3.2 Measured characteristics of volatiles combustion

This section presents the most important attributes of volatiles conversion with O_2 in laboratory datasets, first, in the limit of individual coal particles with attached diffusion flames to clarify the time scales for volatiles conversion and the physical configuration. Then volatiles combustion is characterized across the NBFZs of coal flames for the very heavy coal loadings into PCC's and broad ranges of O_2 partial pressure and coal quality. A final section relates NO_X production to the initial stages of volatiles combustion.

8.3.2.1 Timescales for volatiles conversion in flames

In the 1980s, imaging and high-speed cinematography were used to monitor individual coal particles as they burned under uniform O_2 levels. As seen in Fig. 8.2, these images reveal mantels of soot luminosity surrounding individual particles. In time, these "clouds" grow to several particle diameters as they accelerate ahead of their parent particles, which are subject to stronger gravitational forces and drag. Initially the clouds are spherical, but their solids are ultimately drawn into filaments in the upstream wakes of their parent particles. This shape suggests that nascent soot spherules follow the streamlines around a sphere in Stokes flow until they agglomerate into filaments where the streamlines converge upstream. The shadowgraph image in Fig. 8.2 shows soot clouds both attached to and ahead of their parent particles.

Similar images were reported for nine coals (Seeker et al., 1981; McLean et al., 1981). All the hv bituminous coals generated soot clouds, provided that sizes were relatively large. However, 40 μm particles of an hv bituminous coal did not generate soot clouds, even though 80 μm particles of the same coal did (Seeker et al., 1981). Neither two lignites nor an anthracite generated soot clouds (Seeker et al., 1981; McLean et al., 1981). Since soot production is nucleated by large tar molecules and sustained by addition of secondary tar, oils, and GHCs, the absence of soot clouds with both very low and very high rank coals reflects their relatively low primary tar yields, as well as the abundance of oxygen in tars from coals of lowest rank. The absence of soot clouds with small particles of hv bituminous coal may be an indirect indication that, for the accelerated O_2 penetration rates associated with smaller particles, oxidation of tar precursors in the boundary layer can disrupt tar decomposition into soot during secondary volatiles conversion.

The time interval from the onset of soot luminosity through expansion of the cloud until the extent of the luminous region returns to the char particle diameter can be

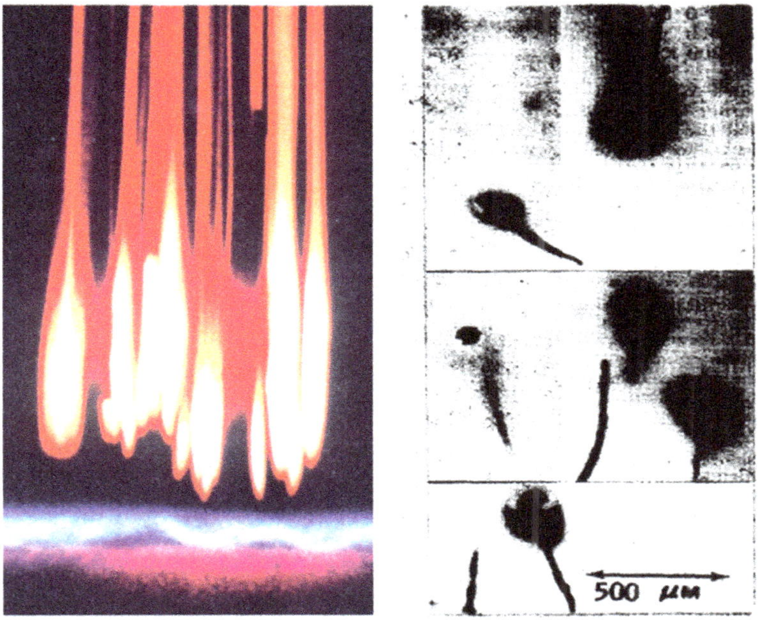

Fig. 8.2 (Left) Superimposed images of individual hv bituminous coal particles in oxidizing post-flame gases; and (right) shadowgraph images of particles in the luminous zone several milliseconds after ignition.
Reproduced from McLean WJ, Hardesty DR, Pohl JH. Direct observations of devolatilizing pulverized coal particles in a combustion environment. Proc Combust Inst 1981;18:1239–47, with permission from Elsevier.

regarded as an approximate duration of volatiles conversion. With 100 μm particles of hv bituminous coal, this period was about 5 ms (McLean et al., 1981). In general, volatiles conversion takes only several milliseconds when particles are heated at 10^4°C/s or faster. This time is considerably longer than the time to traverse laminar flames of typical gaseous hydrocarbon mixtures, because volatiles are converted in series with primary devolatilization. Consequently, the primary devolatilization rate sets the time scale for volatiles conversion, rather than the burning rate of volatiles, per se. Provided that the coal is burned as isolated, individual particles, volatiles conversion kinetics are superfluous because individual particles sustain diffusion flames whose burning rates are governed by reactant transport rates into the flame. So the images in Fig. 8.2 are valuable not only as sources of characteristic conversion times, but also because they clearly reveal the limitations of nearly all testing configurations for volatiles conversion: It is impossible to resolve volatiles conversion in time in any test with dilute coal suspensions and an abundance of O_2 because this process (i) is confined to a boundary layer around individual coal particles; (ii) is too fast for extractive sampling; and (iii) is obscured by soot luminosity.

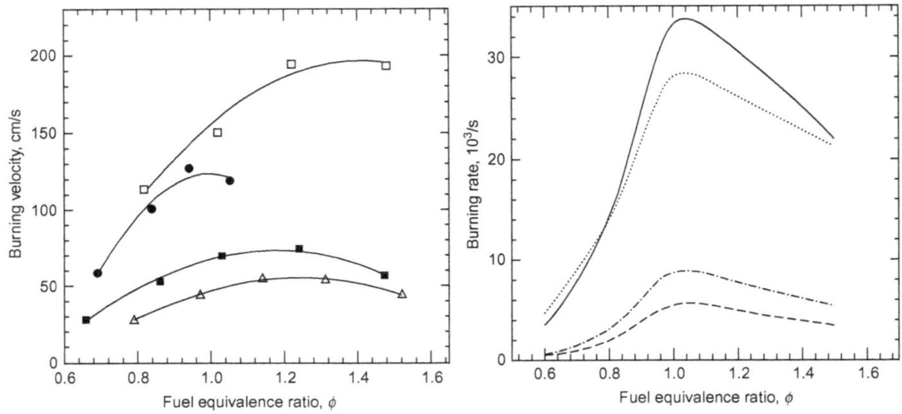

Fig. 8.3 (Left) Laminar burning velocities of noncondensable fuel mixtures for nearly complete secondary volatiles pyrolysis at atmospheric pressure and gas temperatures from 160 to 250°C from subbituminous (Δ), hvC bituminous (▪), lv bituminous (●), and hvA bituminous (□) coals. (Right) Burning rates estimated from the burning velocities for, in ascending order, subbituminous *(dashed curve)*, hvC bituminous *(dot-dash curve)*, lv bituminous *(dotted curve)*, and hvA bituminous *(solid curve)* coals (Marlow et al., 1993).

These issues notwithstanding, volatiles burning rates can be monitored directly in two stage reactors. The laminar burning velocities in Fig. 8.3 were obtained with an RCFR coupled to a combustion bomb (Marlow et al., 1993). Whole volatiles were prepared with specified extents of secondary volatiles pyrolysis in the RCFR, and stored in a holding tank to remove secondary tar, soot, and char fines. The noncondensable fuel mixtures were then mixed with O_2 under specified dilution in the combustion bomb, and burned to provide the information needed to assign laminar burning velocities. As seen in Fig. 8.3, for nearly complete tar decomposition into soot, burning velocities increase by a factor of three with coals of progressively higher rank from subbituminous through hv bituminous. The main reasons for the coal quality impacts are a tradeoff of H_2 for CO in the fuel mixtures from coals of progressively higher rank, and because H_2 burns much faster than CO. Burning velocities also increased by a factor of 3 as extents of tar decomposition in the RCFR were increased from 20% to almost 100%, because the much greater yields of H_2 and greater CO yields compensated for the elimination of GHCs (Marlow et al., 1992, 1993).

The nominal burning rates estimated from the scaling law for burning velocities also appear in Fig. 8.3 (Marlow et al., 1993). The values for different coal types span nearly an order of magnitude. Most important, the characteristic times associated with these rates are in the sub-millisecond range, diminishing from 200 μs with subbituminous to 30 μs with hvA bituminous coal. Noncondensable mixtures of secondary volatiles pyrolysis products burn at rates comparable to mixtures of conventional light hydrocarbons. But we have no assurance from these results that such rapid burning actually materializes in pulverized coal flames.

8.3.2.2 Volatiles conversion in coal flames

The only means to resolve the stages of volatiles conversion demonstrated so far is to use O_2 depletion as a "quench" on volatiles conversion chemistry. When coal at a uniform suspension loading is entrained by an oxidizing carrier gas into a 1D flow reactor, increasing the inlet O_2 level from zero through the level for stoichiometric combustion progressively moves the reaction system from secondary volatiles pyrolysis, through oxidative pyrolysis, through conversion of each of the gaseous fuels from fastest through slowest, through soot oxidation, and through char oxidation. Of course, these stages may overlap. But the value in such a test series is that it resolves the conversion of all the major fuel components, which can, in turn, be directly related to fuel-N transformations. However, these benefits can only be realized if this hypothetical flow reactor can actually operate with inlet O_2 levels as low as none. This is problematic for any conventional burner design that uses heat feedback to stabilize the coal flame. But the strong radiant field in a RCFR overcomes this difficulty.

The datasets in Fig. 8.4 were recorded with uniform loadings of 0.15 g-coal/g-carrier with subbituminous, hv bituminous, and lv bituminous coals. For these three

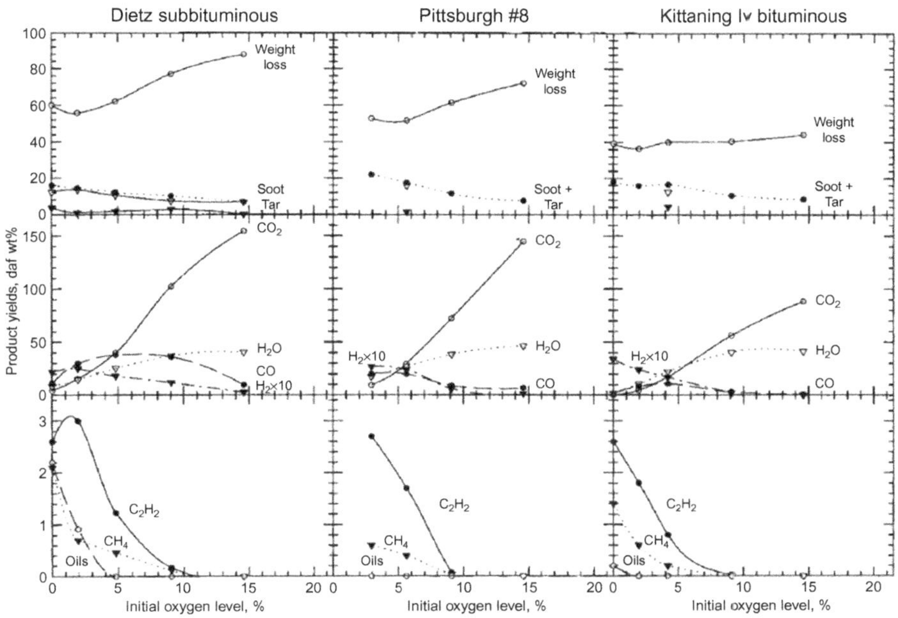

Fig. 8.4 In descending order down each column, these figures show the distributions of the major solid products, the major gaseous products, and GHCs from the oxidative pyrolysis and combustion in a 1425°C RCFR at the indicated inlet O_2 levels. Across the top row, the sum of the soot and tar yields (●) as well as the resolved yields of soot (▽) and tar (▼) appear with the weight loss (Niksa and Cho, 1998).

coals, the respective inlet O_2 levels that would theoretically consume all volatiles-derived products, including soot, were 7.9%, 9.4%, and 8.8%, respectively. The gas transit times were fixed at about 150 ms in all tests. Slip velocities for the 90 μm particles in these tests would move a soot cloud about 30 diameters away from its parent particle in the available transit time after ignition (Niksa and Cho, 1996). So envelope flames could not impose the environment for volatiles conversion in these tests. It is more likely that the gas phase was fairly well mixed. At least 85% of the inlet O_2 was consumed in every test, which was not 100% because some of the O_2 was unreacted in an annular coflow that minimized particle dispersion out of the central coal jet.

Changes in the yields of the condensed phase products clearly resolve the sequence of fuel combustion in this multicomponent reaction system. As the O_2 level was increased from 3% to 14.6% with the hv bituminous, the weight loss increased from 53 to 72 daf wt%, due entirely to char burnout. The yields of aerosol (as soot plus tar) fell from 21% to 4%, indicating either that oxidation more effectively disrupted soot formation during tar decomposition at progressively greater O_2 levels or that soot oxidation was not complete in the available transit time at even the highest O_2 level. Tar was not recovered (with any coal) at even the lowest O_2 level, because the thermal severity in these tests was sufficient to achieve ultimate soot yields with no O_2 in the carrier gas. Oils, CH_4, and C_2H_2 were the only appreciable GHCs. Among these fuels, oils were the fastest burning species, whereas CH_4 and C_2H_2 were consumed on the same time scale. The GHCs burn out faster than CO and H_2, and much faster than soot and char, as expected. Notwithstanding, even GHCs and H_2 persist while soot and char are burning away. GHCs persist for inlet O_2 levels that are roughly double the O_2 requirement for conversion of the noncondensable fuel components. The CO_2 yields increase in direct proportion to inlet O_2 levels. When O_2 was increased from 3% to 6%, 70% of the CO_2 increment came from soot oxidation (or disruption), along with contributions of 15% each for C_2H_2 and char and 2% for CH_4. So char combustion was important even before the GHCs were burned away. The yields of H_2 and CO diminished sharply as O_2 was increased from 6% to 10%. Both of these O_2 levels are much greater than the O_2 needed to burn out the noncondensable fuels only. Hydrogen was eliminated by the greatest O_2 levels, but CO persisted. Moisture yields appear to approach a saturation limit midway through the test series, which only reflects the very low H-contents in soot and char that remained to form H_2O.

These same tendencies are apparent in the products with the subbituminous coal. The only significant difference is that this char burns faster, as expected, and more effectively competes for the available O_2. Moreover, CO yields for intermediate O_2 levels are much greater than with both other bituminous coals. Only a small portion of the enhancement is due to more CO from primary devolatilization and tar decomposition; rather, the abundance of CO reflects an overlap in the partial oxidation of GHCs, soot, and char. As with the hv bituminous, the lightest GHCs persisted with nearly double the O_2 needed to burn out the noncondensable fuels if neither soot nor char was oxidized simultaneously.

As the O_2 level was increased from zero to 14.6%, the weight loss from the low volatility coal increased by only 2 wt%, indicating negligible char oxidation, and the soot yields diminished from 19 to 8 wt%. This soot did not burn as long as GHCs were present. In contrast with both other coals, this one exhibits sequential conversion of its volatile fuel compounds. GHCs, CO, and H_2 burned first, and were nearly eliminated with the theoretical O_2 level for combustion of noncondensable fuels only. But these fuels were only partially oxidized, since CO passed through its maximum yield with the O_2 level for combustion of noncondensable fuels. With 10% O_2, all gaseous fuels had burned away, and soot burned but char did not. At the greatest O_2 level in the series, soot burned out while the char finally ignited.

The product distributions in Fig. 8.4 were combined with predicted primary products from FLASHCHAIN® to determine carbon burnout indices for GHCs only, soot, all volatiles-derived products including soot, and char (Niksa and Cho, 1998). As seen in Fig. 8.5, GHCs from all coals are completely burned out, albeit with inlet O_2 levels about double the theoretical level for burnout of the gaseous fuels. Both GHCs and soot from all coals burned with the same sensitivity to inlet O_2. Although GHCs always burned out, soot burnout always saturated at about 70%, which probably reflects deactivation at the hotter temperatures associated with higher O_2 levels. The only coal quality impact is in char burnout, which falls off for chars of

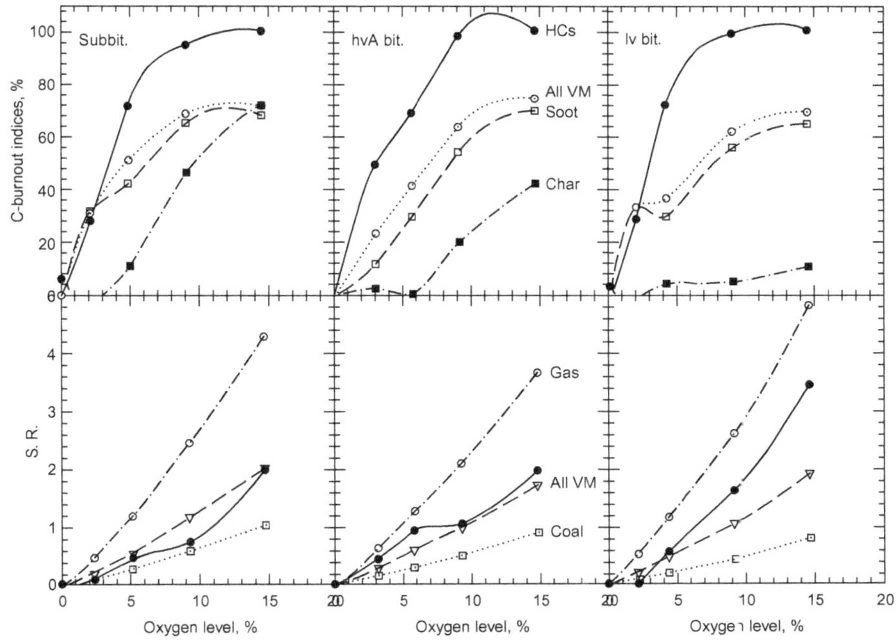

Fig. 8.5 Along the top row, these figures show the burnout indices for GHCs (labeled as HCs) (●), all volatiles-derived products, including soot (○), soot only (□), and char (▪), based on the product distributions in Fig. 8.4. The bottom row shows SR values adjusted for soot and char oxidation (● and *solid curves*) compared to values for gaseous fuels only (○), all volatiles (▽), and the whole coal (□) (Niksa and Cho, 1998).

progressively higher rank. Subbituminous char burnout matches soot burnout at the highest inlet O_2, whereas the burnout of the low volatility char hardly escapes the baseline.

So what is the most meaningful SR value to apply to the gas phase in premixed coal flames? Fig. 8.5 shows one attempt to answer this important question. The dashed lines are references based on the measured conversions of gaseous fuels, all volatiles including soot, and the whole coal. The highest inlet O_2 level reflects stoichiometric combustion of whole coal with all coal types. These values are compared to a customized SR assignment that factors out the O_2 consumed by soot and char oxidation for each inlet O_2 level (Niksa and Cho, 1998). According to this analysis, the gas phase with subbituminous coal remains reducing for all but the highest inlet O_2 level, for which it becomes strongly oxidizing due to deactivation of soot oxidation. The gas phase with both bituminous coals is reducing only up to the O_2 level for burnout of the gaseous fuels only. It then becomes much more oxidizing throughout all succeeding stages.

This study carries two important implications: First, the conversion of tar into soot undermines the expectation that gaseous fuel components burn on sub-millisecond time scales. Sooting sequesters GHCs that would otherwise be among the fastest burning fuel components. Soot yields are substantial with all coal types, and soot is among the slower burning components because it deactivates during combustion. Moreover, both CH_4 and C_2H_2 burn much slower than their burning rates in premixed laminar flames because the available O_2 is apportioned among gaseous fuels, soot, and, perhaps, char, depending on the coal rank. The second major implication is that char oxidation is often fast enough to effectively compete for the available O_2. The relative impact of char oxidation is strongly rank dependent because char burning rates diminish for coals of progressively higher rank, whereas the burning rates of gases and soot are very similar for all coal types.

Major species have also been resolved in time along 1D coal flames on a Meeker-type burner with coal loadings from 0.15 to 0.25 g-coal/g-carrier gas, which correspond to nominal SR values of 0.5 to 0.3 (Altenkirch et al., 1979; Peck et al., 1984, 1991). Since inlet O_2 levels were 23%, the volatiles conversion stage was complete in less than 10 ms. Even though concentration profiles were monitored through 70–80 ms in residence time, all GHCs and O_2 were burned out before they could be retrieved with sampling probes immediately above the burner. The slate of reported N-species closed N-balances within useful tolerances, so these datasets are especially well-suited to track N-species transformations in post-flame gases, as presented in Section 8.3.2.3. But they did not resolve conversion of the major noncondensble fuels during volatiles conversion.

Volatiles combustion was also resolved at elevated pressures in tests with a RCFR (Eckstrom et al., 2014; Liu and Niksa, 2005). The burnout indices for GHCs, soot, and char (defined like those in Fig. 8.5 for atmospheric pressure) appear in Fig. 8.6. Complete product distributions have also been reported (Liu and Niksa, 2005). Despite the similarities in the burnout profiles, the tests at pressures from 1 to 3 MPa were different in many important respects from tests at 0.1 MPa in Figs. 8.4 and 8.5: The flow was upward so gas transit times were shorter than solids transit times; the latter varied from

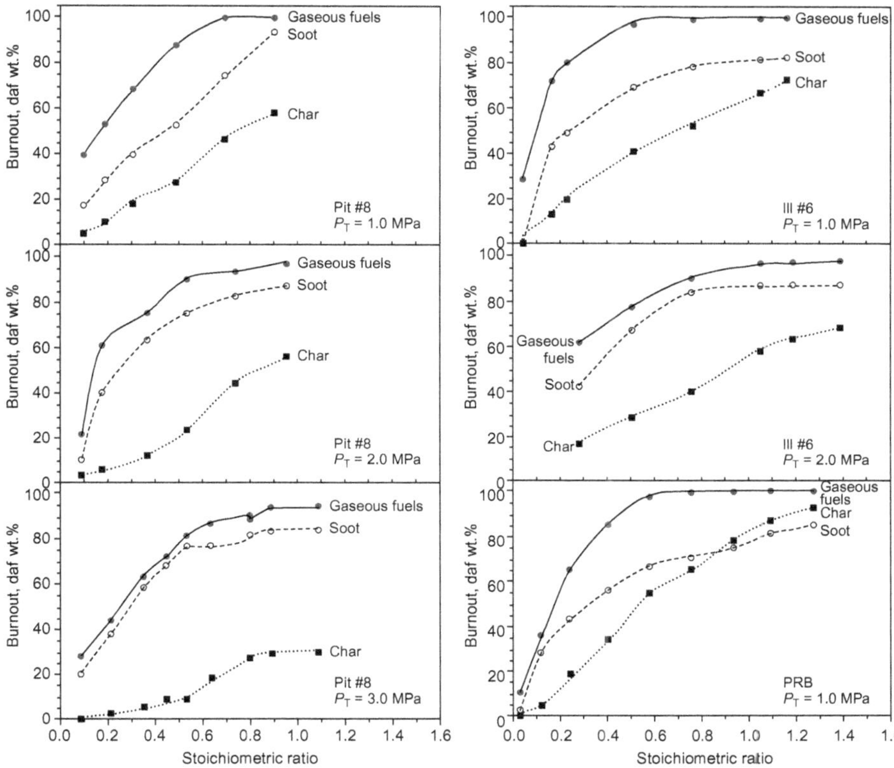

Fig. 8.6 Along the top row, these figures show the burnout indices for GHCs (●), soot only (□), and char (■), based on detailed product distributions (Liu and Niksa, 2005). Burnout indices for hvA bituminous coal for 1–3 MPa appear in descending order down the left column. Indices for hvC bituminous coal for 1 and 2 MPa appear in the upper right column, and those for subbituminous for 1 MPa appear in the lower left panel.

200 to 100 ms for progressively greater O_2 levels, vs. 150 ms at 0.1 MPa. The flow was turbulent instead of laminar; consequently, particles accumulated in an annulus near the tube wall, rather than in a central core flow (Liu and Niksa, 2005). Gas temperatures diminished for progressively higher pressures, and were subject to steep gradients from the tube wall into the flow. To manage particle agglomeration, coal loadings were reduced to 0.05 g-coal/g-carrier gas at 1 MPa, and to 0.017 at 3 MPa. Owing to these differences, these flames had 2D conical structures emanating from the tube wall that may or may not have closed on the centerline by the reactor outlet, depending on the SR and pressure in the test. In cases where the flame core remained open, CO and H_2 were present in the effluent.

All these complications notwithstanding, the product distributions in the effluent still reflect the relative conversion rates of the different gaseous fuels, soot, and char. The overlap among the conversion of the various fuels was probably attenuated by radial gradients in O_2 concentration and inhomogeneous particle dispersion. But by using O_2 depletion as the quench, these tests still progressed through the stages of fuel conversion in a systematic manner.

The C-burnout levels for the different fuels in Fig. 8.6 clearly resolve the competition for O_2 in the oxidation of GHCs, soot, and char. For these three diverse coal samples at pressures from 1 to 3 MPa, the progression in the burnout indices is very similar to that for atmospheric pressure. Compared to the behavior at 0.1 MPa in Fig. 8.5, GHCs require more O_2 for complete burnout at elevated pressures, whereas soot burnout is stronger at any particular SR value, and saturates to a greater burnout at the maximum SR condition. Soot's saturation burnout reflects weaker deactivation at the cooler gas temperatures in the tests at elevated pressure. Char burnout is stronger for all three coals at 1 MPa but falls off for the higher pressures due to the tendency for cooler gas temperatures which, in turn, reduce char particle temperatures. The same coal quality impacts between the subbituminous and hv bituminous coals seen at 0.1 MPa in Fig. 8.5 are apparent at 1 MPa: Burnout as a function of SR is roughly the same for GHCs and soot, whereas char burnout is considerably stronger for the subbituminous, as expected.

The tendencies for progressively higher pressures from 1 to 3 MPa are slightly stronger GHC burnout; stronger soot burnout; and much weaker char burnout. These tendencies are driven by much weaker char burnout, which makes more O_2 available to oxidize GHCs and soot at progressively higher pressures. All these tendencies have been quantitatively interpreted with CFD simulations (Liu and Niksa, 2005). Since the indices are presented as functions of SR, the tendencies do not indicate anything about the relative reaction rates, because no time coordinate is involved. But the similarities do suggest that the competition for O_2 by GHCs, soot, and char is not affected very much by pressure variations, which suggests similar dependences on pressure in the burning rates for the different fuel components.

In summary, the volatiles that burn in the luminous zones of coal flames are secondary pyrolysis products, including soot. Rates of volatiles combustion in flames are strongly mediated by the competition for the available O_2 among gaseous fuels, soot, and char. Consequently, volatiles burning rates in coal flames are at least an order of magnitude slower than in gaseous fuel mixtures, even though the volatiles fuel mixtures in isolation burn as fast as many gaseous hydrocarbon mixtures. Gaseous fuels are the fastest burning component in flames, but GHCs persist while substantial amounts of soot and char burn away. Burnout as a function of SR of the noncondensable fuels and soots from different coals are very similar, whereas char oxidation competes less effectively for O_2 with coals of progressively higher rank. Elevated pressure does not directly affect the relative burnout of the different fuel components, although it very likely perturbs local conditions such as gas temperatures.

8.3.2.3 Volatile nitrogen conversion

The limiting extents of particle dispersion—from isolated individual particles in an excess of O_2 to dense suspensions carried by well mixed gases under reducing conditions—clearly affects both the time scales for volatiles combustion, and its

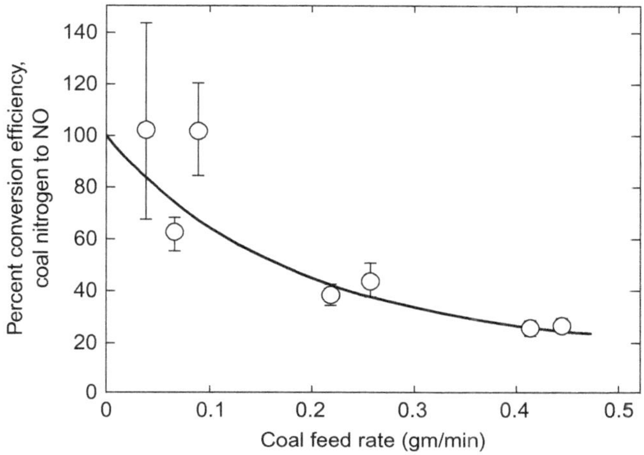

Fig. 8.7 NO_X conversion efficiency for 40 μm hv bituminous coal vs. coal feedrate into a flat flame burner.
Reproduced from Kramlich JC, Seeker WR, Samuelson GS. Observations of chemical effects accompanying pulverized coal thermal decomposition. Fuel 1988;67:1182–89, with permission from Elsevier.

progression from ignition through the complete elimination of all gaseous fuels. These limiting scenarios affect the production of NO_X from coal-N even more profoundly. The potential range of NO_X production efficiencies is clearly illustrated in Fig. 8.7, which shows that NO_X conversion efficiency grows for progressively slower coal feedrates into a flat flame burner. In this burner all coal was fed onto the centerline of a flat flame of CH_4/air. Consequently, the coal loading on the centerline diminishes for progressively slower feedrates, until it reaches the limit of isolated particles. In the single-particle limit, the NO_X production efficiency goes to completion. As the suspension loading is increased, the efficiency diminishes until it saturates at approximately 25%. With larger particles of the same coal, the limit for individual particles was 65% (Kramlich et al., 1988). In a similar burner configuration Haussmann and Kruger (1991) reported limits for isolated particles of 100% for subbituminous and 55%–60% for hv bituminous. These different magnitudes could not be attributed to NO reduction by GHCs, but appeared to be associated with soot. Tests in which NO was fed separately with coal reduced NO to N_2 only in the presence of soot and in proportion to estimated soot yields (Haussmann and Kruger, 1991).

But these quantitative differences are beside the point in any commercial context, because tests in the single-particle limit exclude the NO reduction chemistry that forms the basis for aerodynamic NO_X abatement strategies. In other words, only tests with realistic suspension loadings pertain to the chemistry in low-NO_X burners and staged burner belts in commercial furnaces. This chemistry arises in two stages, one that requires GHCs to limit NO production during the earliest stages of volatiles combustion, and a slower second stage that requires CO and H_2 to reduce NO in the post-flame gases. Both stages are characterized in the rest of this section.

Early volatile-N conversion is inextricably coupled to the oxidation of the gaseous fuels, because the oxyhydroxyl radical pool of H-atoms, OH, and O-atoms is established and sustained by hydrocarbon oxidation, and these radical species also drive N-species transformations. The N-species distributions from three diverse coals in Fig. 8.8 are especially illuminating, because they were recorded during the same tests that generated the distributions of major products in Fig. 8.4 (Niksa and Cho, 1996). With an hv bituminous coal, HCN persists as long as GHCs are present, which significantly affects volatile-N conversion. As the O_2 level was increased from 2% to 14.6%, char-N diminishes while char is consumed by combustion. It is no coincidence that the reduction in char-N is the same as the char burnout percentage. The N-content of soot also falls continuously, starting from the soot-N level for complete tar decomposition into soot. Changes in soot-N also align closely with the mass loss due to soot burnout. In principle, additional HCN could have been released from both char and soot due to the hotter temperatures in the tests with progressively more O_2. But, apparently, the HCN release kinetics are too slow to supplement the release of char-N and soot-N via oxidation.

Cyanide levels are inversely proportional to inlet O_2 until HCN is eliminated with 11% O_2. Ammonia is never appreciable, even under the most reducing operating conditions, and this hv bituminous coal does not even generate NH_3 during secondary volatiles pyrolysis. Estimates for the N_2 levels that close the N-balances vary from 9% with 6% O_2 to 53% with 14.6%. The most interesting yields in Fig. 8.8 are the NO yields. Even after a substantial portion of char-N has been converted in 10% O_2, hardly any NO appears in the products. As O_2 levels were increased from 3% to 10%, soot-N and char-N diminished by more than 8% each, while HCN was halved. Yet the NO percentage grew by only 2.5% over this range. In contrast, the NO production efficiency surges for progressively higher O_2 levels. For the increment from

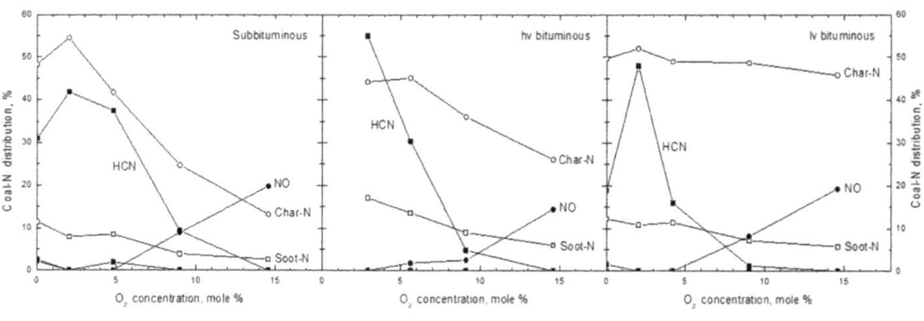

Fig. 8.8 Coal-N distributions for (left-to-right panels) subbituminous, hv bituminous, and lv bituminous coals as (■) HCN, (●) NO, (○) char-N, and (□) tar+soot-N.
Reproduced from Niksa S, Cho S. Conversion of fuel-nitrogen in the primary zones of pulverized coal flames. Energy Fuel 1996;10:463–73, with permission from the American Chemical Society.

10% to 14.6% O_2, two-thirds of the increments in HCN, soot-N, and char-N were converted into NO. Overall, however, only 14% of coal-N was converted into NO with 14.6% O_2, even while two-thirds of coal-N was converted into benign N_2.

The surge in NO production coincides with the inlet O_2 level at which the adjusted SR value for the gas phase in Fig. 8.5 flips from reducing to oxidizing atmospheres, which is certainly no coincidence. Moreover, NO production efficiencies remain very low as long as GHCs are present. But with more than 10% O_2, no GHCs are available to reduce NO into N_2, so the available N-gases are converted into and persist as NO. Evidently, the CO and H_2 that remain after all GHCs have been converted under more oxidizing atmospheres are not effective NO reductants on the short time scales imposed in these tests.

The distributions of GHCs (in Fig. 8.4) and N-species (in Fig. 8.8) with the subbituminous and low volatility coals reported by Niksa and Cho (1998) reaffirm the crucial interplay among GHCs and NO production attenuated by much greater extents of char-N conversion with the subbituminous and by much lower conversion with the lv bituminous. With the subbituminous, NH_3 is an appreciable intermediate with less than 5% O_2, which is the level for stoichiometric conversion of the gaseous fuels. With the lv bituminous, char-N levels are almost invariant, reflecting the negligible extents of char burnout with this coal. Cyanide levels pass through a weak maximum with low O_2, perhaps due to the rapid conversion of cyanogen (C_2N_2) via oxidative pyrolysis (Niksa and Cho, 1996).

With reference to the volatile-N transformations with subbituminous and hv bituminous coals, one could reasonably expect NO production to increase in proportion to the char-N conversion once all GHCs have been consumed. But with the lv bituminous, the maximum NO production is the same as that from the subbituminous, even though the respective char burnouts were 2 vs. 30 wt%. Since the different fuel components burnout sequentially with the lv bituminous, GHC concentrations are too low to completely skew the conversion of HCN toward N_2, and hydrocarbon reductants are not available to reduce the NO formed from soot-N and char-N.

The impact of hydrocarbons on NO production is summarized in more conventional terms in Fig. 8.9, which shows the reactor effluent concentrations of HCN, NH_3, and NO on a dry, O_2-free basis for the three coals. Total fixed nitrogen (TFN) is the sum of these three concentrations. With little inlet O_2, TFN is determined by HCN with both bituminous coals, and by HCN plus NH_3 with the subbituminous. TFN decays for progressively higher O_2 levels due to N_2 production, and then passes through a rank-dependent minimum. The minimum TFN is only 100 ppm for the hv bituminous for the O_2 level that eliminates all GHCs. The minimum TFN-values for both other coals occur at the same O_2 level, but are two to three times greater. Also, the minimum values are uncorrelated with the coal N-contents, which rise from 0.9 to 1.4 to 1.6 daf wt% in order of increasing rank. Beyond the minimum values, NO comprises all of TFN.

The second stage of NO_X production begins at the minimum in TFN, in post-flame gases with no GHCs so that H_2 and CO are the NO reductants. It is illustrated in Figs. 8.10 and 8.11 with the major products and N-species along 1D coal flames with subbituminous and hv bituminous coals run with loadings of 0.15 and 0.25 g-coal/g-carrier gas (Altenkirch et al., 1979; Peck et al., 1984, 1991). The respective

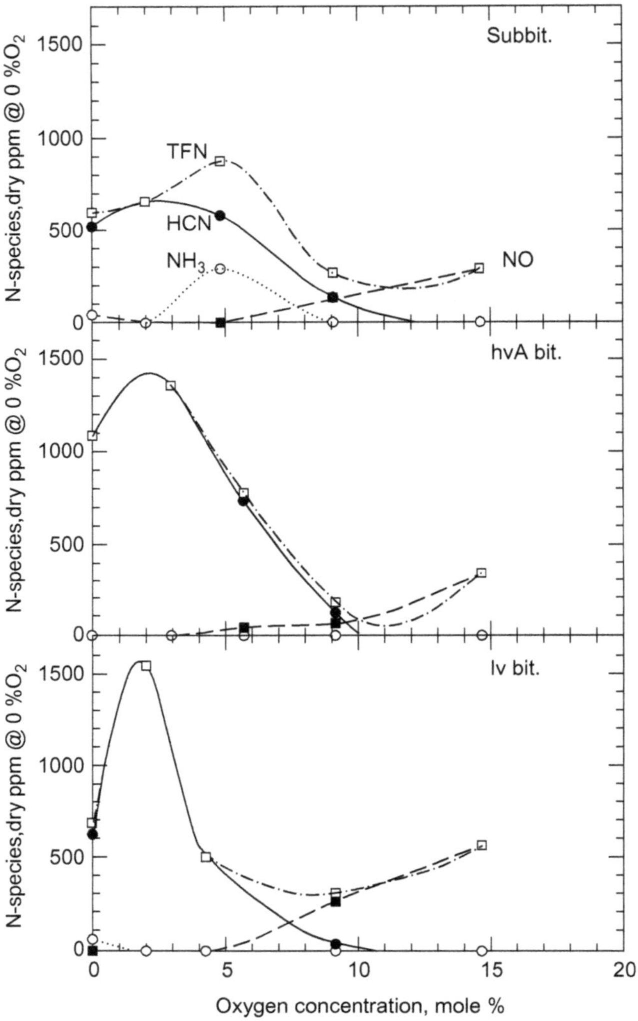

Fig. 8.9 Concentrations of HCN (●), NH_3 (○), NO (■), and their sum (□, TFN) on a dry, O_2-free basis for, in descending order, subbituminous and hv and lv bituminous coals. Reproduced from Niksa S, Cho S. Conversion of fuel-nitrogen in the primary zones of pulverized coal flames. Energy Fuel 1996;10:463–73, with permission from the American Chemical Society.

coal-based SR values are 0.6 and 0.5 for the low loading, and 0.38 and 0.33 for the high loading. The compositions of O_2, CO, CO_2, and H_2 in Ar and N-species are resolved in time in Figs. 8.10 and 8.11 for both loadings with both coals. With the low loading, O_2 is eliminated within 20 ms with the subbituminous, and in half that time with the hv bituminous, due to the lower levels of CO_2 and H_2O diluents in the secondary

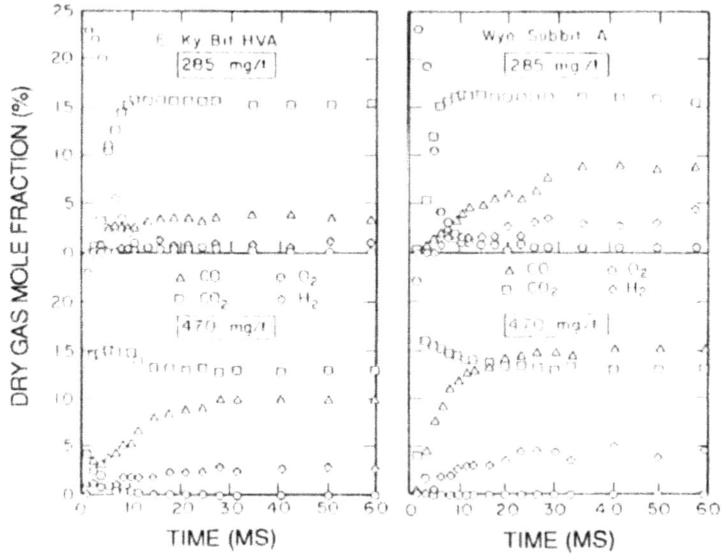

Fig. 8.10 Concentrations of H_2 (◇), CO (△), CO_2 (□), and O_2 (○) for (left) an hv bituminous and (right) a subbituminous at (upper) low and (lower) high loadings.
Reproduced from Peck RE, Glarborg, P, Johnsson JE. Kinetic modeling of fuel-nitrogen conversion in one-dimensional, pulverized coal flames. Combust Sci Technol 1991;76:81–109, with permission from Taylor and Francis Group.

pyrolysis products from the bituminous. The CO_2, CO, and H_2 levels saturate on the same time scales, except that CO and H_2 take much longer with the subbituminous, presumably due to water gas shifting because the H_2 and CO levels increase in tandem after 20 ms. The bituminous generates mixtures with hardly any H_2 and less than 5% CO, whereas the subbituminous makes about 4% H_2 and more-than-double the CO. With the high loading, O_2 was eliminated in 5 ms with both coals. The CO and H_2 concentrations continue to develop on much longer time scales to ultimate values that are more-than-double the levels in the leaner flame conditions.

The coal-N distributions in Fig. 8.11 close N-balances to within ±10% across these flames. With the low loading, coal-N is almost entirely released from both coal chars. This volatile-N is converted into NO, N_2, and a small amount of NH_3 with the subbituminous, and into only NO and N_2 with the hv bituminous. NO is produced only as long as O_2 is present. Thereafter, NO diminishes gradually across the post-flame with the subbituminous but persists at the same level with the bituminous. With the high loading, the conversions of coal-N to volatile-N are 80% and 70% with the subbituminous and hv bituminous, respectively. The levels of HCN, NO, and N_2 are substantial with both coals, although only subbituminous generates NH_3. The NO and N_2 levels clearly mimic the dynamic features in the CO and H_2 histories.

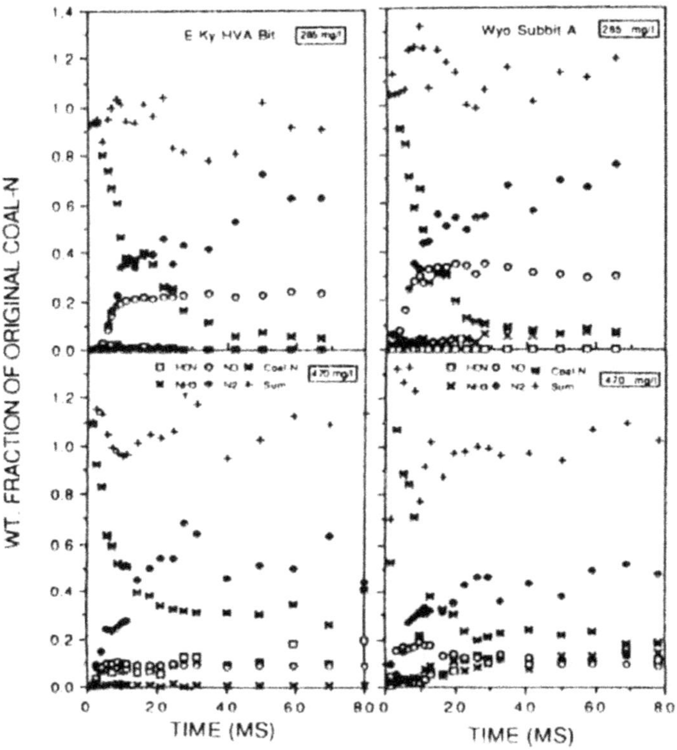

Fig. 8.11 Concentrations of char-N (*), HCN (□), NH_3 (x), NO (○), N_2 (△), and their sum (+) for (left) an hv bituminous and (right) a subbituminous at (upper) low and (lower) high loadings. Reproduced from Peck RE, Glarborg, P, Johnsson JE. Kinetic modeling of fuel-nitrogen conversion in one-dimensional, pulverized coal flames. Combust Sci Technol 1991;76:81–109, with permission from Taylor and Francis Group.

The concentration histories with subbituminous relax to just over 10% each for HCN and NH_3 and to 50% for N_2, and to 10% HCN and 60% N_2 with the hv bituminous. After accounting for char-N conversion, the respective NO production efficiencies are 10% with both coals.

In the post-flame, NO is produced only as long as O_2 is present, whereas N_2 is produced along the entire post-flame, and becomes the exclusive product of this chemistry after O_2 vanishes. Both HCN and NH_3 are completely eliminated from the post-flame with SR values of 0.5 and greater. But both species remain in the effluent under more reducing conditions, at least during the first 100 ms of conversion.

During the initial stages of coal combustion under realistic coal loadings, coal-N is transformed in two distinct stages. First, volatile-N and soot-N are converted into HCN, N_2, and with low rank coals, small amounts of NH_3. Char-N may also participate if the oxidation reactivity is sufficient to effectively compete for the available

O_2. Essentially no NO is produced by this channel. For progressively greater SR values, this channel remains the dominant one until all GHCs and HCN are consumed. For even leaner conditions, nearly all the incremental N-species released from soot-N and char-N become NO. The only coal quality impact in the first stage is on the char burning rates. As burning rates accelerate for coals of progressively lower rank, the surge in NO production occurs with leaner SR values.

In the second stage and on longer time scales, NO produced in the first stage may be reduced into N_2, perhaps with residual HCN and, with low rank coals, NH_3. Coals that generate more H_2 and CO in their post-flame gases give lower ultimate NO production efficiencies, although the predominant factor is the SR value. In practice, optimal production efficiencies occur when both HCN and NH_3 are eliminated from the effluent, and only a small amount of coal-N is converted to NO. This situation arises in under 100 ms with coal-based SR values in the vicinity of 0.6. Flames with greater SR values will give more NO and those with lower SR values may give less NO but will also emit HCN and, perhaps, NH_3.

8.3.3 Measured characteristics of volatiles steam reforming

The behavior in this section is pertinent to gasifiers that do not sustain volatiles combustion and quickly mix steam into the coal feedstream before spontaneous volatiles pyrolysis converts tar into soot. If steam is mixed too slowly to participate in volatiles conversion, then the tar decomposition mechanism developed in Chapter 7 will be applicable, and the reforming mechanisms for noncondensable gases in this chapter still apply.

For volatiles combustion, particle dispersion largely determines whether volatiles are converted in clouds on individual coal particles or in reasonably well-mixed entrainment gases. But these limits do not pertain to volatiles reforming with steam, because the reforming rates are too slow and there is no strong exotherm to affect volatiles compositions in the immediate vicinity of the parent particles. Consequently, volatiles are reformed only in reasonably well-mixed gases. Unless noted otherwise, all the chemistry in this section occurs in the gas phase only.

Volatiles steam reforming happens in two stages. Initially, steam reacts with primary tar, secondary PAH, and oils. At moderate temperatures the steam reforming converts condensable liquids into CO, CH_4, and other light GHCs. At elevated temperatures it can diminish or completely disrupt tar decomposition into soot. Steam also reacts with primary and secondary GHCs to form additional CO and H_2, so that only the lightest GHCs remain.

The second stage is the continuous reforming of the noncondensable species as steam and CO_2 are converted into CO and H_2 by gasification of soot and char. The water gas shift reaction (WGSR) is the predominant process, although other processes can also affect CO_2 and CH_4 concentrations. The distribution of volatile-N species shifts from HCN toward NH_3 as the atmosphere becomes more reducing, whereas H_2S predominates with small amounts of carbonyl sulfide (COS) and carbon disulfide (CS_2) under some operating conditions.

8.3.3.1 Steam reforming of aromatic gases

Without any reactive gases, oils and tars generate soot provided that the temperatures are hotter than about 850°C. With O_2, soot production is almost inevitable because the heat release drives temperatures well above the threshold for sooting, where soot production rates are relatively very fast. Conversely, under high H_2 pressures, soot production is completely disrupted (as explained in Chapter 9) because the bulk of the tars are hydroconverted into oils and GHCs. So the crucial question surrounding the steam reforming of oils and tar is, "Does steam disrupt soot production?" Unfortunately, the answer is definitive only for coals of lowest rank.

Soot production was definitely disrupted by steam reforming of the tars from two Australian brown coals, as seen in Fig. 8.12. These C-distributions were recorded in EFR tests with and without 0.06 MPa steam at 900°C (Hayashi et al., 2000, 2002) in a transit time of 14 s. The two brown coals were tested in both whole and demineralized forms. Without steam, both the whole and demineralized coals produce more soot for progressively hotter temperatures, as expected for spontaneous tar decomposition. With steam, soot production from the whole coals diminishes for hotter temperatures, and almost disappears at 900°C, while the yields of H_2, CO, and CO_2 grow markedly. The heaviest tar molecules were eliminated faster than lighter tars and oils. With the demineralized coals, soot production again diminishes for progressively hotter temperatures, but appreciable soot was recovered at all temperatures.

Since the measurement uncertainties in these tests are among the lowest in the literature, Hayashi et al. could demonstrate that the incremental changes in the yields of CO, CO_2, and H_2O were quantitatively consistent with steam gasification of oils and tar via $C + H_2O \rightarrow CO + H_2$ and $C + 2H_2O \rightarrow CO_2 + H_2$. Furthermore, by comparing the soot yields without steam for the whole and demineralized coals, they showed that primary moisture from devolatilization alone disrupted soot production during secondary pyrolysis.

Hayashi et al. (2002) concluded that oils and tars were reformed on alkali and alkaline earth metals in char, not in the gas phase. A follow-on study with Loy Yang brown coal clarified the role of char by running tests with and without char particles in the gas stream (Masek et al., 2007). With char, steam reforming nearly eliminated heavy PAH and soot and converted some heavy PAH into oils. Without char, reforming reduced soot production but did not significantly affect heavy PAH and oils.

Steam reforming of tars and oils does disrupt soot production with brown coals. The primary mechanism is through adsorption of heavy PAH onto alkali and alkaline earth metals on char. Reforming of the heaviest PAH in the gas phase also appears to disrupt soot nucleation. Unfortunately, these findings cannot be generalized to other operating conditions and coal types, for several reasons. PAH adsorption on char will become relatively slower at progressively hotter temperatures while tar decomposition into soot accelerates. The residence times in the tests were much longer than those in many gasifier designs. Coals of higher rank do not contain the active metals in their ion exchangeable forms, and also give much lower primary moisture yields. So for the vast majority of coals, the impact of steam reforming of tars and oils is still unclear.

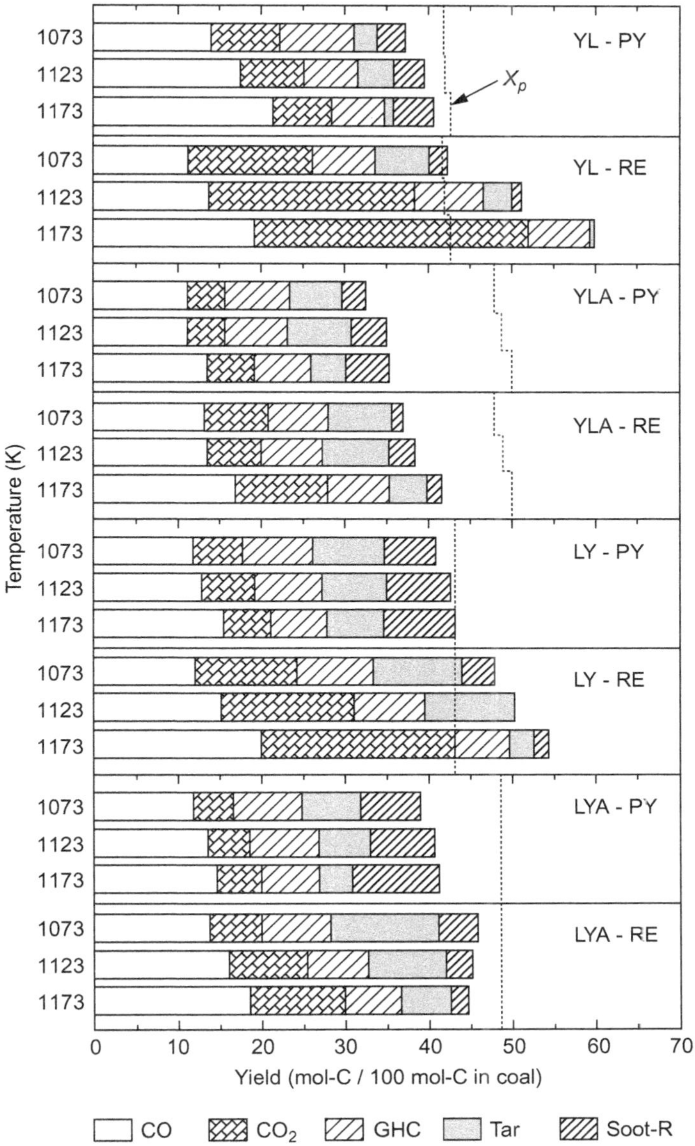

Fig. 8.12 Carbon distributions from whole (YL, LY) and demineralized (YLA, LYA) brown coals with (RE) and without (PY) 0.06 MPa steam. *Dashed lines* (X_P) indicate total carbon loss from the whole coals in a reactor configuration without any secondary chemistry. Reproduced from Hayashi J-I, Iwatsuki M, Morishita K, Tsutsumi A, Li CZ, Chiba T. Roles of inherent metallic species in secondary reactions of tar and char during rapid pyrolysis of brown coals drop-tube reactor. Fuel 2002;81:1977–87, with permission from Elsevier.

Even at moderate temperatures, it may inhibit soot production by interfering with nucleation, but it seems unlikely to convert major portions of PAH into non-condensable gases. At high temperatures, its impact may be inconsequential.

The database on tar steam reforming with other coals is too thin to clarify these ambiguities. As part of a larger testing program on CH_4/coal co-firing for chemical feedstocks, Lim et al. (2013) ran two RCFR tests with a subbituminous coal in a 25/75 mixture of steam in Ar at 3 MPa. No secondary tars were recovered in either test, and the soot yields were 1.7 daf wt% at 1300°C and 1.0% at 1400°C. However, the gas transit times were 3 s, which is sufficient for appreciable soot gasification at such elevated steam pressures and temperatures. So the time resolution was insufficient to verify any disruption of soot production.

Of course, a multitude of EFR tests on steam gasification with all coals types are available. Apparently, none resolve the conversion of volatiles during the initial stage, particularly the disruption of sooting by steam reforming, for any coal type except brown coals. Only tests that resolve all major gases and oils, tar, and soot, and also close mass and elemental balances in individual runs can characterize the coal quality impacts. Time resolution in increments of 100 ms is needed to resolve the dynamics, and to isolate the impact of reforming from soot gasification.

In addition to disrupting soot formation, steam reforming can also affect GHC levels and compositions. But in the C-distributions in Fig. 8.12, the only noticeable change is the slight consumption of GHCs whenever soot yields are increasing, which resembles the trend for progressively hotter temperatures. Steam reforming, per se, appears to have little, if any, impact with coals in both their whole and demineralized forms. So the temperature threshold for homogeneous GHC reforming is hotter than 900°C.

8.3.3.2 Equilibration of gaseous intermediates and products

During any coal gasification process, carbon and hydrogen are added continuously to the gas phase via the gasification of soot and char. Gas compositions continuously respond by reforming according to the WGSR and a methanation process, according to the following global processes:

$$CO + H_2O \leftrightharpoons CO_2 + H_2$$

$$CO + 3H_2O \leftrightharpoons CH_4 + H_2O$$

Other processes can come into play, depending on the temperature and pressure. Whether the gasification agent is O_2, H_2O, or CO_2, the WGSR will reform the gas composition toward equilibrium for the proportions of C, O, and H in the gas phase which, in turn, are determined by the extents of conversion for volatiles and char and the amounts of C, O, and H introduced with coal and the gasification agents. This latter contribution is often gauged by the O/C stoichiometry, where unity (or 100%) corresponds to complete conversion of the coal-C into CO.

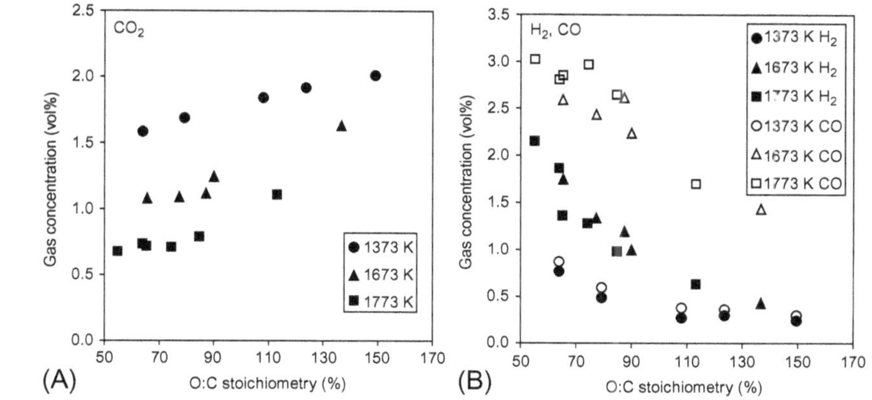

Fig. 8.13 Compositions of (A) CO_2 and (B) CO and H_2 from hv bituminous coal at 2 MPa in 3 s for a range of O/C. In the right panel, H_2 and CO appear in filled and open symbols, respectively.
Reproduced from Harris DJ, Roberts DG, Henderson DG. Gasification behavior of Australian coals at high temperature and pressure. Fuel 2006;85:134–42, with permission from Elsevier.

These factors are isolated in the measured partial gas compositions in Figs. 8.13 and 8.14 for gasification of an hv bituminous coal at 2 MPa in 3 s. The levels of CO_2 increase and those of CO and H_2 diminish for progressively greater O/C, as expected. Conversely, CO_2 diminishes while both CO and H_2 grow for progressively hotter temperatures. Methane levels normally vanish toward the low end of the temperature range in this study, although they are not shown in these figures. Fig. 8.14

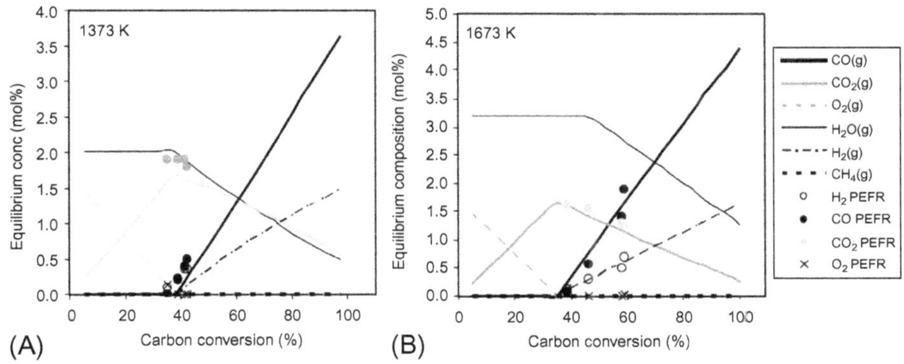

Fig. 8.14 Gas compositions from hv bituminous coal at 2 MPa in 3 s with an O/C of unity for a range of C-conversion at (A) 1100 and (B) 1400°C.
Reproduced from Harris DJ, Roberts DG, Henderson DG. Gasification behavior of Australian coals at high temperature and pressure. Fuel 2006;85:134–42, with permission from Elsevier.

shows how gas compositions change as coal-C is added to the gas phase by char gasification. At both temperatures, H_2O levels are uniform through 40%–50% conversion before they fall off in proportion to higher conversions; CO_2 levels pass through a maximum at 36% conversion. The predominance of the WGSR is evident in the H_2 and CO levels, which are inverted images of the H_2O levels. Similarly, O_2 and CO_2 have an inverse relationship through 36% conversion, where O_2 vanishes. Since both oxidation and reforming affect CO_2 levels directly, CO_2 is the only species that passes through a maximum. Temperature variations perturb the magnitudes of the species concentrations, but do not change the functional forms of their variations with the C-conversions.

Generally speaking, finite-rate kinetics for chemistry in the gas phase are not needed to predict outlet gas compositions from entrained flow gasifiers, because the compositions for thermochemical equilibrium are as accurate as they need to be. But accurate equilibrium calculations also require accurate estimates for C-conversions. Char gasification rates diminish for coals of progressively higher rank but with very substantial sample-to-sample variability (Liu and Niksa, 2004). So if the temperature, pressure, and residence times in a gasifier are kept the same, then the conversions at the outlet with different coals will vary. In turn, this variation changes the gas compositions as illustrated in Fig. 8.14. C-conversions for diverse assortments of coals cannot yet be accurately predicted from first principles. At a minimum, laboratory support is required to calibrate the initial intrinsic gasification reactivity of every sample, although large databases of conversions at full-scale are used by gasifier OEMs. In principle, coal quality indirectly affects outlet gas compositions in so far as different coal compositions give different O/C and H/C ratios in the gas phase. These variations would change exit gas compositions, except that feedrates of air or O_2 and moisture are adjusted to maintain fixed elemental ratios into the gasifier.

Below 1000–1100°C, depending on transit times and pressures, equilibrium compositions deviate from measured compositions, especially when GHCs are present. In general, equilibrium GHC levels are too low for even moderately high temperatures. Under these conditions, outlet gas compositions can only be calculated with simulations across the gasifier exit section either with elementary reaction mechanisms or some approximate scheme that can be implemented in CFD simulations. At even lower temperatures, it may be necessary to simulate the entire gasification process to accurately estimate the exit syngas compositions.

In principle, the release of coal-N and coal-S species into gases and their subsequent transformations would be as important for gasification as they are for combustion. But gas cleaning units are indispensable in gasification systems, and efficiently eliminate most of these pollutants for regulatory compliance. This probably explains why datasets that resolve the dynamics of volatile-N and volatile-S species are very difficult to find. Volatile-N species comprise NH_3, HCN, and less N_2 with virtually no NO, and over 90% of volatile-S will be H_2S with a balance of COS and minor amounts of CS_2. However, it is impossible to assess the coal quality impacts on these speciations without a large database of measured gas compositions that include these species.

8.4 Stoichiometry and thermochemistry for volatiles conversion

Large coal flames are mostly stabilized by the heat release from volatiles combustion. The fastest burning noncondensable fuels are partially consumed within burner belts along with the smallest char particles for all but coals of highest rank. Soot burnout is slower than volatiles combustion for all coal types. Contributions from char combustion are clearly rank dependent, but those from noncondensables and soot seem relatively insensitive to coal quality, albeit based on only the behavior of the three coal samples in Fig. 8.5. This section re-visits this issue with estimates for the stoichiometric air requirements for volatiles combustion and reforming, and for the enthalpy requirements of these processes.

This section also introduces a suite of eight coal samples to clarify the coal quality impacts in the calculation results. The suite comprises a brown coal (BrC), two subbituminous (subB1, subB2), three hv bituminous (hvC, hvA1, hvA2), a mv bituminous (mvB), and an anthracite. The two subbituminous and three hv bituminous were selected to span the range of volatility for these nominal ranks, primarily with low and high H-contents, and also with low and high O-contents for the hv bituminous samples. Two product distributions from FLASHCHAIN® are compiled in Table 8.1 for primary devolatilization and ultimate secondary pyrolysis, including complete tar decomposition into soot. These distributions reflect a nominal heating rate of 50,000°C/s, based on injection of 60 μm particles into gases at 1100°C with a radiation source temperature of 1300°C at atmospheric pressure. Primary devolatilization was complete within 10–15 ms, as expected. As explained in detail in Chapters 4 and 7, the weight loss is similar with progressively higher ranks through hv bituminous, whereas primary tar yields pass through a maximum for subbituminous and hv bituminous ranks. Secondary pyrolysis products have much more CO and H_2 but many fewer GHCs and, of course, abundant soot.

Table 8.1 also contains two stoichiometric air requirements for volatiles combustion, ν_{AIR}: one for primary products including tar, and one for noncondensable fuels from secondary pyrolysis, excluding soot. The air requirements for soot combustion appear separately (${}^{SOOT}\nu_{AIR}$). These requirements are expressed in g-air/g-daf coal, which explicitly factors in the fuel species yields, but excludes variations in the coals' moisture and ash contents. Similarly, the heats of combustion resolve values for whole primary volatiles, noncondensable secondary volatiles, and soot. But the analogous values for volatiles reforming are based on only the noncondensable gases, on the premise that soot gasification is too slow to keep pace with the gas phase chemistry.

Since the fuels that actually burn in coal flames are secondary pyrolysis products, Fig. 8.15 plots the stoichiometric air requirements and heats of combustion for the secondary noncondensable fuels and soot. The air requirements for gas combustion are modest through hv bituminous, and fall off sharply for higher ranks due to the reduction in yields. The sample-to-sample variability is apparent but also modest. In contrast, the air requirements for soot burnout are much more variable because they track the variations in the primary tar yields. The coal quality impacts are largely

Table 8.1 Primary and secondary products, stoichiometric air requirements, and enthalpy changes for volatiles conversion.

	Brown coal		SubB1		SubB2		hvC		hvA1		hvA2		mvB		Ant	
	1^0	2^0	1^0	2^0	1^0	2^0	1^0	2^0	1^0	2^0	1^0	2^0	1^0	2^0	1^0	2^0
W^∞	65.9	65.9	50.1	50.1	55.7	55.6	58.8	58.9	49.8	49.8	52.0	51.9	35.3	35.6	10.0	10.0
Tar	25.4	0.0	27.7	0.0	34.9	0.0	38.5	0.0	35.1	0.0	37.6	0.0	24.9	0.0	6.3	0.0
Soot	0.0	22.8	0.0	24.9	0.0	31.2	0.0	34.6	0.0	31.6	0.0	34.1	0.0	23.1	0.0	6.0
H_2	0.1	3.3	0.8	2.7	1.0	3.9	0.5	3.8	1.1	3.5	1.8	4.7	0.8	1.9	0.0	0.1
CH_4	6.4	0.6	0.9	0.1	1.4	0.2	3.4	0.5	2.2	0.3	1.4	0.2	1.9	0.1	0.0	0.0
C_2H_2	0.0	0.1	0.0	0.0	0.0	0.0	0.0	0.0	0.0	0.0	0.0	0.0	0.0	0.8	0.0	0.0
C2H4	2.0	0.0	0.4	0.0	0.6	0.0	1.5	0.0	0.9	0.0	0.5	0.0	0.4	0.0	0.0	0.0
C_2H_6	0.7	0.0	0.1	0.0	0.2	0.0	0.4	0.0	0.2	0.0	0.4	0.0	0.3	0.0	0.0	0.0
C_3H_6	1.6	0.0	0.3	0.0	0.5	0.0	1.2	0.0	0.7	0.0	0.5	0.0	0.4	0.0	0.0	0.0
C_3H_8	0.0	0.0	0.0	0.0	0.0	0.0	0.0	0.0	0.0	0.0	0.0	0.0	0.0	0.0	0.0	0.0
CO	11.1	20.3	5.1	6.8	4.3	6.9	3.0	8.8	2.4	5.9	2.0	4.6	0.8	2.6	1.3	1.3
CO_2	10.6	10.6	6.4	6.4	5.0	5.0	3.4	3.4	1.9	1.9	2.4	2.4	1.4	1.4	0.5	0.5
H_2O	7.5	7.5	8.0	8.0	7.2	7.2	6.2	6.2	4.6	4.6	5.0	5.0	3.9	3.9	1.6	1.6
HCN	0.3	0.5	0.3	1.0	0.3	0.8	0.2	0.9	0.2	1.0	0.2	0.6	0.2	1.0	0.1	0.2
H_2S	0.2	0.2	0.2	0.2	0.3	0.4	0.5	0.7	0.6	1.0	0.2	0.4	0.3	0.6	0.2	0.2
Combustion																
ν_{AIR}, g/g-daf	4.5	1.8	3.8	1.2	5.0	1.6	5.7	1.7	5.4	1.5	5.6	1.8	3.7	1.0	0.7	0.1
${}^{Soot}\nu_{AIR}$, g/g-daf		2.7		2.9		3.6		4.0		3.7		4.0		2.7		0.7
ΔH_{COMB}, kJ/g-daf	−13.4	−6.5	−10.8	−4.3	−14.3	−5.8	−16.1	−6.1	−15.1	−5.3	−16.1	−6.5	−10.2	−3.3	−1.9	−3.5
ΔH_{SOOT}, kJ/g-daf		−8.6		−6.6		−6.9		−5.9		−4.2		−5.2		−3.1		−1.0
Reforming																
ν_{AIR}, g/g-daf	1.4	0.1	1.3	0.1	1.7	0.1	2.0	0.1	2.0	0.1	1.9	0.1	1.5	0.1	0.3	0.0
ΔH_{REFR}, kJ/g-daf	−0.6	1.4	−0.5	1.3	−1.4	0.9	−2.3	0.5	−2.4	0.2	−2.3	0.3	−1.7	0.1	−0.3	0.2

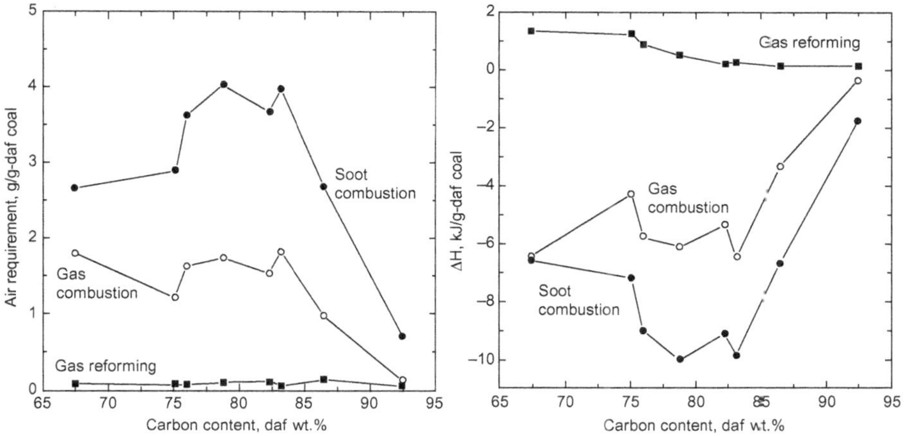

Fig. 8.15 (Left) Stoichiometric air requirements for combustion of (●) noncondensable fuels and (○) soot from secondary volatiles pyrolysis; and (right) the analogous heats of combustion. Corresponding values for reforming of the noncondensables appear as (■).

determined by the air requirements for soot, so that the requirements abruptly surge for hv bituminous coals, and diminish for both higher and lower ranks, albeit not by much for one of the subbituminous samples. Since this coal suite covers such a broad range of rank, it seems likely that the air requirements for secondary pyrolysis products from hv bituminous coals are at least 15%–20% larger than for the other ranks. For ranks through hv bituminous, the total air requirements for volatiles combustion are 50%–60% of the total air into a furnace.

The heats of combustion for noncondensable fuels exhibit the same tendency as the air requirements, albeit with stronger sample-to-sample variability due primarily to the different contributions from CO and H_2 which, in turn, reflect variations in the coal-O and -H contents. The heats of combustion for soot are essentially an inverted image of the air requirements, because both quantities directly reflect the variations in the primary tar yields. Consequently, the total heat release for volatiles combustion, including soot, is at least 20%–25% greater for subbituminous and hv bituminous coals than for other ranks. And the less volatile subbituminous (subB1) and hv bituminous (hvA1) samples have lower air requirements and weaker heat release than the other samples of the respective ranks.

For reforming of the noncondensable volatiles only, the air requirements are only about 10% of the daf-coal feed, and the enthalpy change is weakly endothermic with low rank coals, but relaxes toward thermal neutrality for coals of progressively higher rank.

Air requirements and heats of combustion for volatiles shift to lower values for elevated pressures, primarily reflecting the lower soot yields, and the sample-to-sample variability remains the same. Air requirements for the noncondensable fuels are essentially the same at 1 MPa as those in Fig. 8.15 for 0.1 MPa, but those for soot combustion at 1 MPa are lower by about 1 g/g-daf coal. The heat of

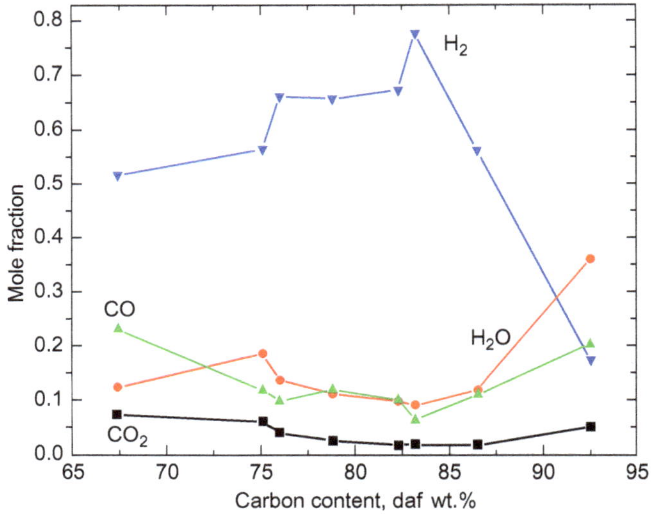

Fig. 8.16 Product gas compositions for stoichiometric reforming in air of noncondensable volatiles at atmospheric pressure.

combustion for noncondensables is stronger by 0.5 kJ/g-daf coal, whereas that for soot is weaker by 2 kJ/g-daf coal.

The product gas compositions for stoichiometric volatiles combustion are very similar for all coals except anthracite. The mole fractions of CO_2 and H_2O vary from 14% to 17% and 10% to 13%, respectively. With anthracite, the CO_2 level is nearly double, while steam levels are halved. The gas compositions for stoichiometric reforming of the noncondensable fuels are much more variable because so little air is required for stoichiometric reforming. Product gas compositions for atmospheric pressure appear in Fig. 8.16. Hydrogen levels grow from 50% to 75% for high volatility coals, then plummet for the two coals of highest rank. Steam levels are between 10% and 20% for all coals except anthracite, which has almost 40%. CO levels are between 10% and 25%, whereas CO_2 is always less than 8%. Volatiles reforming does not supply sufficient CO_2 and steam to gasify soot and char. But it is an important source of H_2, which inhibits steam gasification, and may produce CH_4 via char hydrogasification under some operating conditions. At 1 MPa, the gas compositions are very similar, except that the H_2 mole fraction is lower by 0.05 for all coal types.

8.5 Analysis of volatiles conversion around isolated particles

The burning behavior of isolated coal particles is not of much commercial interest because the chemistry responsible for aerodynamic NO_X abatement only occurs in whole coal suspensions (cf. Section 8.3.2.3). But it is worthwhile to survey the studies of volatiles combustion around individual particles to clarify the conditions that allow O_2 to penetrate a burning volatiles cloud to the char surface during devolatilization, especially for different coal types. Analyses of this reaction system were widely developed during the 1980s and 1990s, first, to support the interpretations of pyrometry signals during particle ignition (Grosshandler, 1984; Choi and Kruger, 1985); second, to delineate the operating conditions for homogeneous vs. heterogeneous ignition mechanisms (Gururajan et al., 1988, 1990); third, to specify the time scales for volatiles escape and combustion (Musarra et al., 1986; Lau and Niksa, 1992, 1993); and, fourth, to clarify aspects of the particle size dependence in N-species chemistry (Kramlich et al., 1988). This section draws heavily from Lau and Niksa's analyses, because of their coverage of the coal quality impacts. Although numerous analyses of volatiles combustion around an isolated particle are available, none were found on volatiles reforming because the chemistry is too slow to be confined to an attached cloud.

The reaction system is sketched in Fig. 8.17. It consists of the spherical boundary layer surrounding an individual coal particle throughout primary devolatilization and volatiles combustion. Volatiles were analyzed in two pairs of two components, tar and noncondensable fuels (Lau and Niksa, 1992) and soot and noncondensable fuels (Lau and Niksa, 1993). These cases represent the limits for frozen and instantaneous secondary pyrolysis, respectively. Two additional scenarios were framed by allowing soot to either burn in the same flame sheet as noncondensable fuels or penetrate the gas flame sheet and burn in the free stream. In all cases, the species burning rates were

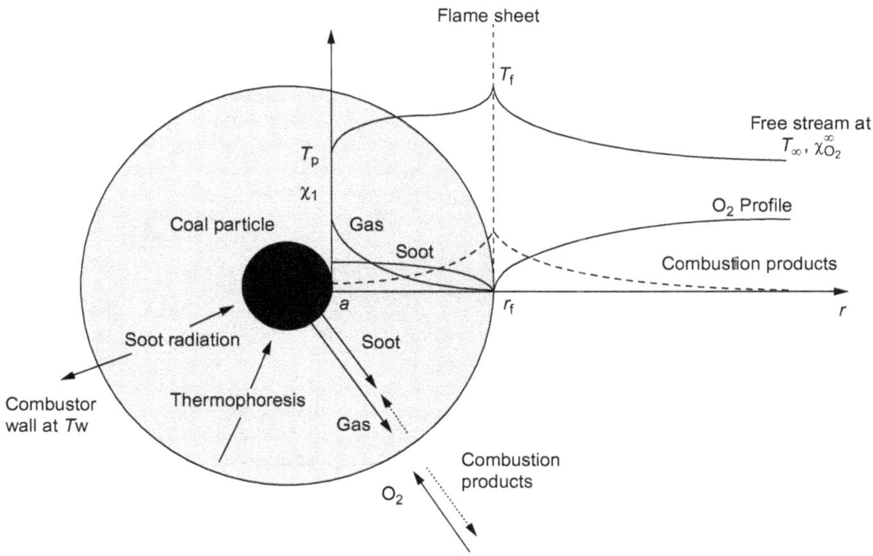

Fig. 8.17 Reaction system for the analysis of a cloud of burning volatiles around an individual coal particle (Lau and Niksa, 1993).

evaluated under diffusion control, and the flame sheet sustained stoichiometric volatiles combustion. The flame was initially attached to the particle external surface at the onset of devolatilization; then lifted off into the free stream while the devolatilization rate accelerated; then attained its maximum standoff from the particle surface with the maximum devolatilization rate; and regressed back to the char surface when devolatilization was exhausted. At that point the char particle ignited, and the heat release rapidly drove the particle temperature to a quasi-steady value governed by the gas and wall temperatures and the O_2 level.

The thermal histories for a 70 μm particle of an hv bituminous coal in 8% O_2 in N_2 at 1225°C appear in Fig. 8.18 for three scenarios on secondary pyrolysis and soot combustion. The hottest flame temperature is for frozen secondary pyrolysis because this enthalpy balance has much weaker energy feedback to the particle and weaker radiation losses, because there is no soot. The intermediate case is for instantaneous secondary pyrolysis and soot combustion in the volatiles flame sheet, and gives a maximum flame temperature over 2000°C. The coolest history is for fast secondary pyrolysis but no soot oxidation, which is unrealistic at such temperatures.

Maximum flame temperatures and flame standoffs for four diverse coals are compiled in Table 8.2. The results show that both characteristics grow for coals with progressively greater soot yields. The opposite tendency is apparent in the predicted sizes for the onset of heterogeneous ignition. That size diminishes for coals with greater soot yields because their volatiles have greater stoichiometric O_2 requirements and therefore require the faster transport rates associated with smaller particles to satisfy

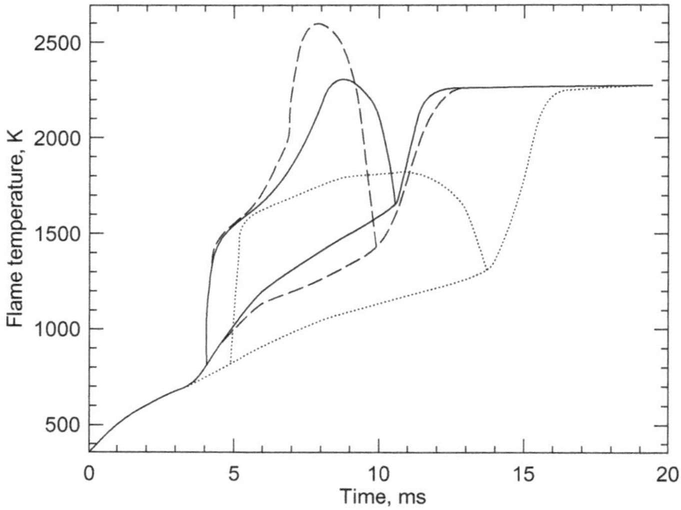

Fig. 8.18 Thermal histories of a coal particle and its attached volatiles flame for *(dashed curve)* frozen volatiles pyrolysis, *(solid)* fast secondary pyrolysis and soot oxidation, and *(dotted)* fast secondary pyrolysis and frozen soot oxidation (Lau and Niksa, 1993).

Table 8.2 Predicted combustion characteristics for individual particles of four coal types (Lau and Niksa, 1993).

	Lignite	hvC Bit	hvA Bit	lv Bit
Max T_{FLAME} (°C)	1855	1915	2050	1885
Max R_{FLAME}, r_0	5.4	6.0	7.0	5.8
Heterogeneous ignition size (μm)	18	16	8	28
Energy feedback (%)	61	62	56	64
Energy losses (%)	33	32	33	30

these requirements. The energy fed back to particles varies by less than 10% among these four very different coals, and that lost to surroundings is also insensitive to coal quality.

The flame duration in Fig. 8.18 is 6 ms, which is consistent with the estimates for the images in Fig. 8.2. These estimates were also in agreement with measured values for broad ranges of O_2 concentrations for different coal types (Lau and Niksa, 1993). This is not surprising because, according to this analysis, flame durations are determined by the predicted primary devolatilization behavior.

These types of analyses rank-order the various heat and mass transport mechanisms that determine the macroscopic characteristics of volatiles combustion, and reveal how the characteristics respond to different operating conditions. But, as seen in Table 8.2, the sample-to-sample variability is not large except for the size at heterogeneous ignition, even among coals of very different rank. The most important finding is that O_2 is excluded from the particle by the stoichiometric requirement for volatiles combustion, in conjunction with the counteracting transport mechanisms. Once the O_2 transport rate becomes fast enough to deliver the stoichiometric O_2 requirement to the particle surface against the outward volatiles flux, the flames remain attached to the particle.

8.6 Analysis of volatiles conversion in dense suspensions

Measurements to diagnose burning dense coal suspensions are impossible now and for the foreseeable future, because of the heavy particle loadings, very fast velocities, short time scales, and severe thermal histories. Consequently, the only pertinent information comes from CFD. But CFD concentration fields omit reaction mechanisms for volatiles conversion. Usually no finite-rate volatiles combustion chemistry of any kind is included in CFD chemistry submodels, because this chemistry is confined to thin layers that cannot be resolved on the coarse computational grids of current furnace simulators. Instead of using even global oxidation rate expressions, volatiles combustion submodels determine the equilibrium compositions of the combustion products from the elemental compositions of volatiles with Gibbs free energy minimization, and set the overall burning rates equal to turbulent mixing rates of secondary air streams into fuel-rich, primary coal jets.

Such analyses do not distinguish primary and secondary products, so they omit soot. Soot is often the slowest burning fuel component in the system and sequesters major portions of what began as GHCs, which have the fastest burning rates of all fuel components. So this omission distorts the competition for O_2 among the fuel components in the analysis. The equilibrium analysis for volatiles combustion also explicitly decouples conversion of char from gaseous volatiles. As seen throughout this chapter, this is an important omission from the mechanisms responsible for fuel-N conversion and NO production in furnaces, and also for the development of nascent syngas in gasifiers.

The set of reaction mechanisms in this section's simulations rectifies these shortcomings and may, in principle, provide the clearest view of the process chemistry during the earliest stages of p.f. firing under commercial conditions. But it is worth emphasizing that there are no means to stringently evaluate these findings with measurements under comparable operating conditions.

In Fig. 8.1, the limiting scenario for coal suspensions with realistic particle loadings and fast mixing among gaseous fuels, soot, char, and carrier gases is represented by a CSTR-series. This idealization reveals how coal quality and different operating conditions affect volatiles conversion during the initial stages of combustion and reforming. First the simulations for dense suspensions are validated with the RCFR datasets in Sections 8.3.2.2 and 8.3.2.3, then simulation results characterize commercial firing conditions in utility furnaces and in entrained flow gasifiers with dry and slurry coal feeds. Various applications of the same approach with various utilization technologies at pilot and commercial scale are presented in the second volume in this series.

8.6.1 Volatiles combustion in a RCFR

The RCFR system used to compile the datasets in Figs. 8.4–8.6, 8.8, and 8.9 is especially well-suited for an analysis with CSTR networks because the suspension loadings were realistic yet the inlet conditions were straightforward, and macroscopic mixing phenomena were minimal, according to CFD simulations of this furnace (Niksa and Liu, 2002). Coal was entrained in Ar with specified O_2 onto the centerline of a heated cylindrical tube to form a premixed suspension with 0.15 g-coal/g-carrier. The suspension was surrounded by an annular sheath flow with the same O_2 level and inlet velocity to minimize particle dispersion off the centerline. As suspensions moved along the flow tube, they were heated at rates exceeding 10^4°C/s to the onset temperature for primary devolatilization and released their volatiles, which then decomposed further in the gas phase, mixed with O_2, and burned. At any particular operating condition, O_2 depletion eventually 'quenched' the chemistry at an intermediate stage determined by the proportions of coal and O_2 at the inlet. Inlet O_2 levels were progressively increased in successive cases to move the process chemistry through oxidative volatiles pyrolysis, volatiles combustion, char oxidation, and soot combustion, although as noted earlier, the oxidation of gaseous fuels, char, and soot is rarely sequential, so that volatiles combustion is rarely isolated from the combustion of char and, perhaps, soot.

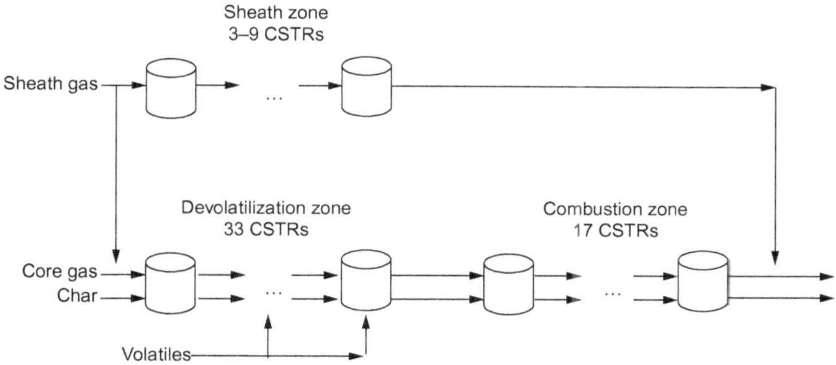

Fig. 8.19 CSTR network for the core and sheath flows in RCFR tests at atmospheric pressure with different O_2 levels (Niksa and Liu, 2002).

As explained in more detail elsewhere (Niksa and Liu, 2002; Niksa, 2019a), the equivalent reactor network was developed directly from CFD results, and appears in Fig. 8.19. First, a number of CSTRs in series was assigned to the core flow by fitting the analytical RTD expression for a series of CSTRs (Froment and Bischcoff, 1979) to the CFD-based RTDs for each test. Next, the averaged gas and tube wall temperatures as functions of transit time through the core were approximated as discrete isothermal values for each CSTR in the series. The average gas and wall temperature histories were incorporated into an enthalpy balance that determined the thermal histories for the particles in suspension. These thermal histories were then used with FLASHCHAIN® to determine the complete distributions of primary devolatilization products. Then the FLASHCHAIN®-based tar decomposition mechanism determined the incremental additions of noncondensables and soot for secondary volatiles pyrolysis. However, preliminary simulations with finite-rate tar decomposition showed that GHCs would be partially oxidized before they could add to the nascent soot phase. Accordingly, tar decomposition was evaluated at the primary devolatilization rate, and soot yields were reduced by 40% to retain all GHCs in the gas phase, as explained elsewhere (Niksa, 2019a). Volatiles combustion was simulated with an elementary reaction mechanism for homogeneous combustion and N-species conversion developed by Glarborg et al. (1998), which contains 444 elementary reactions among 66 species. All kinetic and thermodynamic parameters were assigned independently by the developers, so there are no adjustable parameters in the submodel for gas phase chemistry. The only chemistry included for soot is its oxidation, whose kinetics were based on the closed-form rate expression developed by Nagle and Strickland-Constable (Park and Appleton, 1973).

Char burning rates are determined by several mechanisms, including thermal annealing and, with low rank chars, ash encapsulation and a transition toward chemical kinetic control during the latest stages of burnout. These aspects were represented with the extended version of the Carbon Burnout Kinetics (CBK/E) model (Niksa et al., 2003). CBK/E depicts the impact of variations in gas temperature, O_2 level, and char particle size within useful quantitative tolerances. However, it is not yet possible to specify the initial char reactivity within useful tolerances from standard coal properties. This value can only be specified with a one-point calibration for each fuel; e.g., by adjusting the reactivity to match measured burnout in an EFR test, or to match measured LOI in a pilot-scale coal flame. The submodel for char-N conversion is subject to a similar calibration requirement compounded by its simplistic mechanistic premise: that a fixed fraction of char-N is converted into NO at the overall burning rate throughout all stages of char oxidation for all O_2 levels. This fraction was fixed at 0.1 for the subbituminous, hv bituminous, and lv bituminous coals in this test series. In these simulations, all the parameters in the reaction mechanisms for all stages of coal combustion are specified from the proximate and ultimate analyses of the coals, except for two. The initial char oxidation reactivity is specified by calibration with a single burnout measurement for the greatest inlet O_2 level, and the fixed fraction of char-N converted into NO has the same value for all coals and operating conditions (Niksa, 2019a).

The predicted major- and N-species for subbituminous coal are compared with measured values in Fig. 8.20; similar comparisons are available for hv and lv bituminous coals (Niksa, 2019a). This coal had the greatest extent of burnout, by far, as expected from its low rank. Whereas the char burnout approached 80% under the most

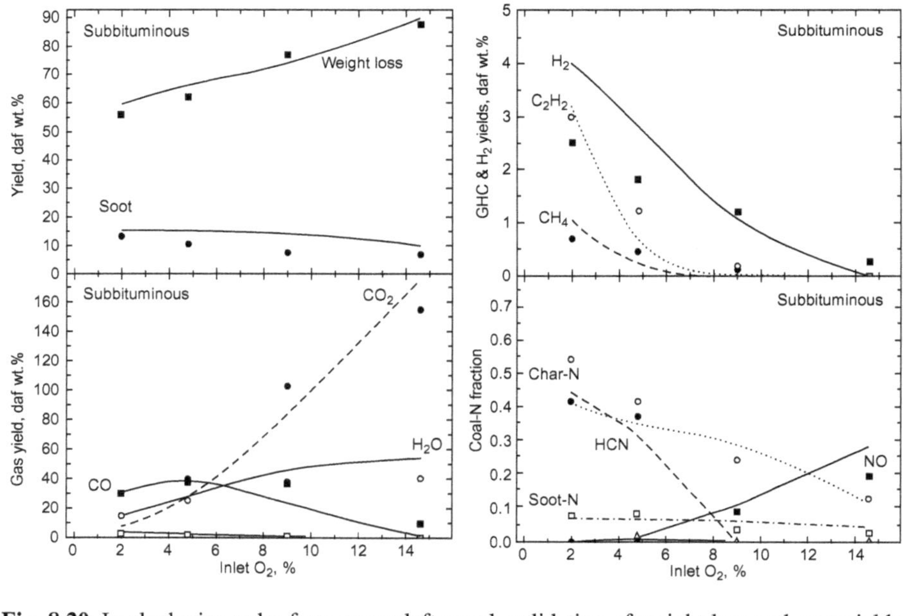

Fig. 8.20 In clockwise order from upper left panel, validation of weight loss and soot yields; yields of (● and dashed curve) CH_4, (○ and dotted curve) C_2H_2, (dot-dashed curve) C_2H_4, and (■ and solid curve) H_2; coal-N partitioning (excluding N_2) as (○ and *dotted curve*) char-N, (□ and *dot-dashed curve*) soot-N, (● and *dashed curve*) HCN, and (■ and *solid curve*) NO; and major gas intermediates and products with subbituminous coal (Niksa, 2019a).

oxidizing conditions, the maximum extent of soot burnout was 38%. Predicted yields of char and soot are within measurement uncertainties except for the soot yields at intermediate O_2 levels. Predicted yields of CO, CO_2, and H_2O are similarly accurate, and correctly indicate a CO yield of 42 wt% with 4.6% O_2, which is the greatest among the three coals. The predicted H_2 yields are accurate for oxidizing conditions although they overestimate the levels under reducing conditions, but not by inordinate amounts. The yields of both CH_4 and C_2H_2 are accurate across the full O_2 domain, and correctly vanish for all oxidizing conditions. The complete N-species distributions are within measurement uncertainties across the entire O_2 domain, except for the char-N level with 2% O_2 and the NO efficiency with 14.6% O_2. They correctly show proportionate reductions in char-N and soot-N with the respective burnout levels, and accurately depict the abrupt transition toward NO production under oxidizing conditions. The agreement in the validations with the other two coals is almost as satisfactory (Niksa, 2019a).

8.6.2 Volatiles combustion under commercial P.F. firing conditions

Under the premise that the RCFR validation tests span the chemical reaction environment immediately beyond burner nozzles in utility furnaces, the same reaction mechanisms were used to simulate the chemical transformations along a single coal jet from an injector in a 530 MW T-fired utility furnace (Niksa, 2019b). CFD simulations based on detailed engineering specifications determined the operating conditions along the coal jet, including coal mass loading, thermal histories, and transit times. A reactor network developed from the CFD results staged the chemistry in this environment with actual utility grinds for the same three diverse coal types in the RCFR tests.

The reaction system is one of the primary coal jets into the furnace throughout the period for primary devolatilization, tar decomposition, volatiles combustion, and partial oxidation of char and soot, where the total transit time was 340 ms. The jet consists of pulverized coal entrained in air at mass loadings approaching 0.5 kg-coal/kg-air. Once the coal suspension leaves the injector nozzle, it heats primarily by radiation from the coal flame, and from the heat release from fuels released from the coal. Since coals jets in commercial furnaces are momentum-dominated, entrainment of surrounding air into the primary jet was negligible during this period, and there was also no impingement with other coal jets or any wall element.

Since the subject flowfield is a premixed primary jet with no entrainment of auxiliary air, its equivalent reactor network is simply a linear CSTR series with 12 reactors for the initial nonreactive stage and 21 reactors for the radiant heating section. A series with 33 CSTRs closely approximates plug flow. The mean thermal histories for gas and radiant source temperatures from the CFD results were approximated as discrete isothermal values for each CSTR in the series. Then the same set of validated reaction mechanisms from Section 8.6.1 was used to simulate the dense burning suspension without any parameter adjustments. The same grind was used with each coal and was specified with a Rosin-Rammler PSD whose mean size was 50.8 μm. In the simulations the PSD was resolved into 25 discrete size increments. In addition to the heavy commercial loading, lighter coal loadings of 0.15, 0.25, and 0.35 kg-coal/kg-air were run with all coals to better resolve the trends in the chemistry.

The predicted flame structure in Fig. 8.21 for an hv bituminous with 0.35 kg-coal/kg-air conveys three distinct conversion time scales. The fastest burning fuels – C_2H_2, CH_4, and H_2 – appear to be converted from 140 to 200 ms. However, only about three-fourths of these fuels were released from the coal during this period, since volatiles continued to be released through 270 ms. So GHCs continue to burn away or are reformed into CO and H_2 through most of the total transit time. The second major stage from 180 to 210 ms oxidizes away CO; converts HCN into NO via an N_2O intermediate; and burns out some char, while almost all O_2 is consumed. The bulk of the HCN forms N_2 rather than NO although this case is so reducing that one third of the maximum HCN concentration remains as HCN in the effluent. The third and slowest conversion period extends from 210 to 275 ms, when soot is converted under very low O_2 concentrations and the last portion of volatiles is added to the entrainment stream to sustain further NO reduction. The CO product released from volatiles reforming and soot combustion promotes water gas shifting, which elevates the H_2 concentration. In turn, H_2 and CO convert most of the NO back into HCN along the post-flame region. With this particular coal and coal loading, char oxidation coincides with the oxidation of all gaseous fuels and N-species conversion, whereas soot oxidation in conjunction with volatiles reforming promotes water gas shifting chemistry, CO accumulation, and major extents of NO reduction along the post-flame region.

For this particular coal loading, the post-flame is sufficiently reducing to re-generate CH_4 and C_2H_2 in tandem with NO reduction. Consequently, volatiles combustion does not completely consume even the most reactive gaseous fuels. However, the simulations did not account for addition of C_2H_2 to soot, which persists at a high level throughout the post-flame, so the C_2H_2 concentration in Fig. 8.21 is ambiguous. Also, C_2H_2 and soot are predicted to co-exist with other coals and loadings throughout this chapter, and all such cases are subject to the same qualification.

The tendencies for progressively heavier coal loadings mimic those for progressively lower O_2 levels, and the quantitative values are also quite similar, as seen in Fig. 8.22. The mass loss from char and soot, the yields of GHCs and major gas species, and the coal-N distributions are mirror images of those for progressively greater inlet O_2 levels in Figs. 8.4, 8.8, and 8.20. Since the inlet O_2 level for the p.f. firing simulations was 21%, variations in the coal loading shift the nominal stoichiometric ratio, as do variations in inlet O_2 level with uniform coal loading. Accordingly, the weight loss increases for progressively lighter loadings while soot levels decrease, because more O_2 is available to burn away the carbonaceous fuels. Hydrogen, GHCs, and CO remain in the effluent only for the two heaviest loadings. These levels of the reducing agents stifle but do not completely suppress NO production. But once the reducing agents are eliminated, NO production surges for the two lightest loadings. The partitioning of HCN into NO instead of N_2 in the absence of reducing agents is one factor behind the surge in NO production, and the conversion of char-N and soot-N also contribute.

With lighter loadings the effluent gas contains less volatiles, but the dynamics of volatiles combustion are otherwise the same with all loadings. Char ignition is also the

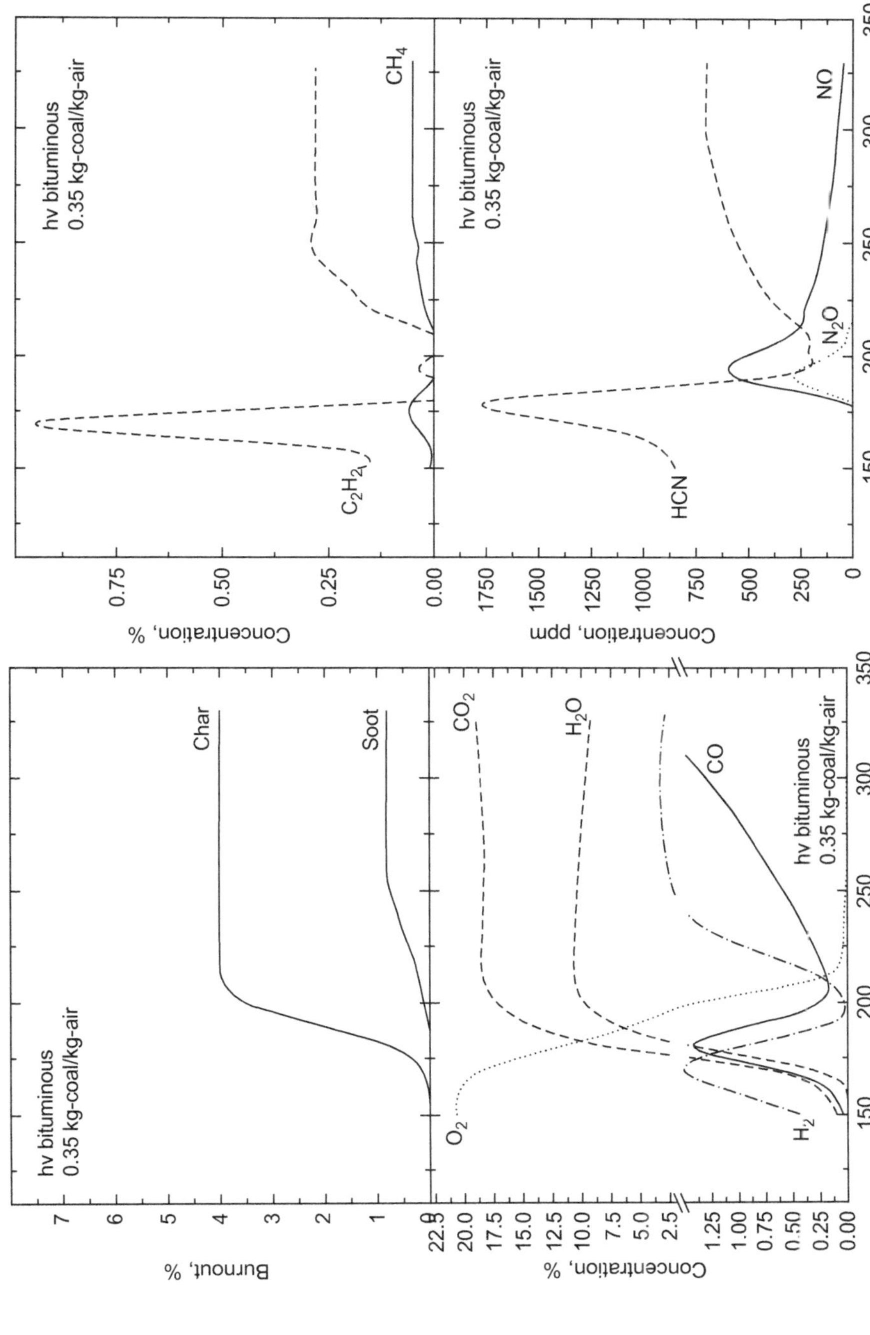

Fig. 8.21 Predicted flame structure with hv bituminous at 0.35 kg-coal/kg-air as, in clockwise order from upper left panel, extents of burnout of char and soot; the molar concentrations of CH_4 and C_2H_2; and HCN, N_2O, and NO; and O_2 and major gas intermediates (H_2, CO) and products (CO_2, H_2O) (Niksa, 2019b).

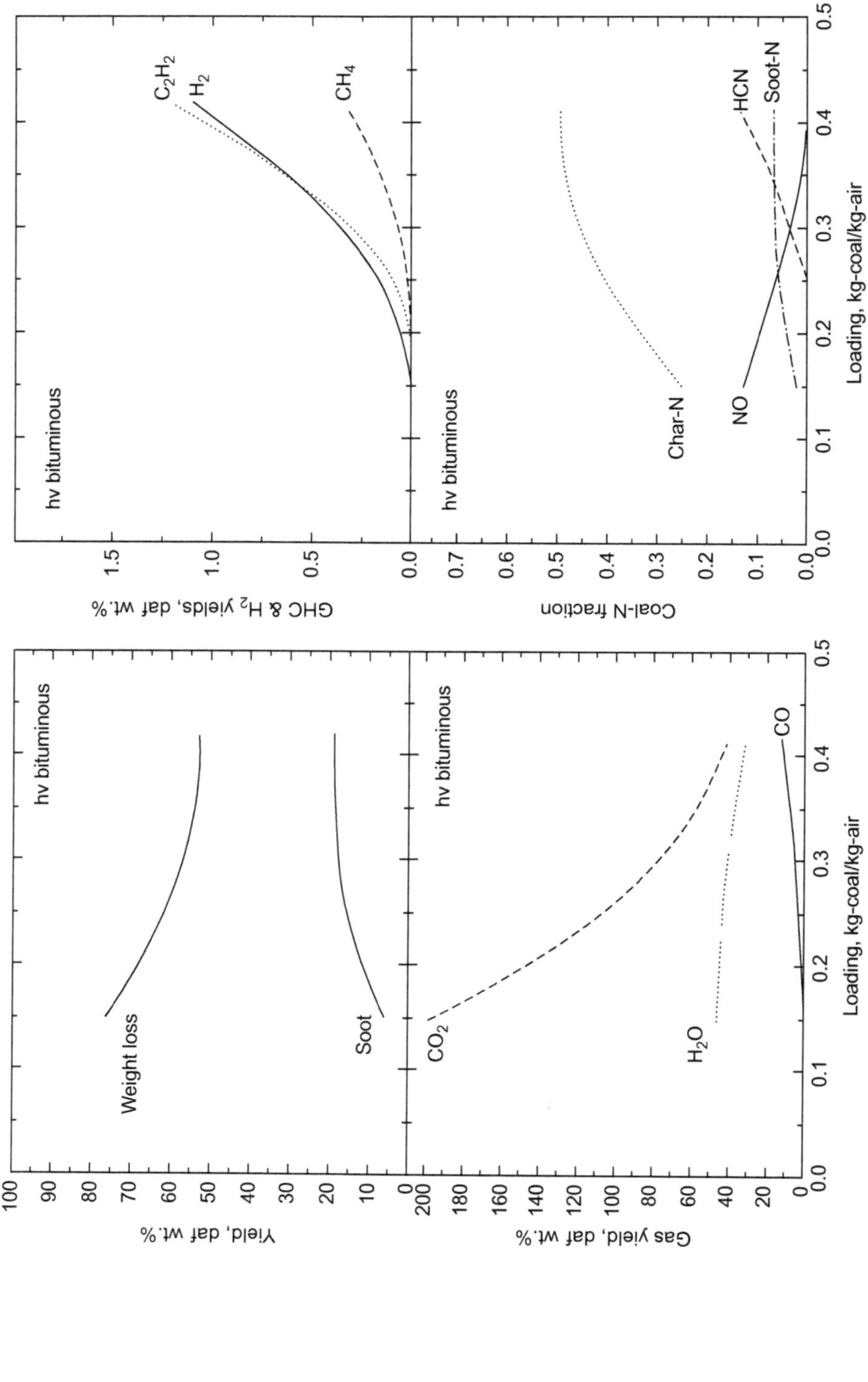

Fig. 8.22 In clockwise order from upper left panel, weight loss and soot yields; yields of *(dashed curve)* CH_4, *(dotted curve)* C_2H_2, and *(solid curve)* H_2; coal-N partitioning (excluding N_2) as *(dotted curve)* char-N, *(dot-dashed curve)* soot-N, *(dashed curve)* HCN, and *(solid curve)* NO; and major gas intermediates and products with hv bituminous coal for the full range of coal loading, where 0.42 is the commercial loading

same and, with the hv bituminous coal, char oxidation always precedes soot oxidation. Loading variations affect coal flame structure, but in only one critical aspect: O_2 is eliminated faster with progressively heavier loadings. With 0.15 kg-coal/kg-air, at least half of char and soot burn out yet O_2 persists in the effluent. But with any of the heavier loadings O_2 is eventually eliminated at shorter transit times with progressively heavier loadings.

This shift in O_2 consumption is responsible for the much greater effluent NO concentrations with the two lighter loadings in Fig. 8.23. The destruction of HCN extends over longer transit times with progressively heavier loadings, and is even reversed with the two heaviest loadings so that HCN appears in the effluent. Even so, the maximum NO concentrations are similar with all loadings and occur at about 180–190 ms. The distinctive behavior happens downstream in the post-flame region, when nearly all NO is reduced away with both heavier loadings, but the maximum NO levels are hardly perturbed with both lighter loadings. The persistence of O_2 with both lighter loadings suppresses NO reduction along the post-flame region, because H_2 and CO reductants cannot coexist with O_2 at post-flame temperatures even while both char and soot burn out along the entire post-flame with lighter loadings. Conversely, at the heavier loadings the elimination of O_2 allows the gaseous reductants to accumulate along the post-flame and drive the effluent NO concentration well below its maximum, even when extents of char and soot burnout are negligible.

As seen in Fig. 8.24, extents of subbituminous char burnout are always substantial, increasing from 17% with the heaviest loading to 82% with the lightest loading. However, soot burnout is only appreciable with the lightest loading, albeit at only one-third of char burnout. Extensive char burnout coincides with volatiles combustion, which

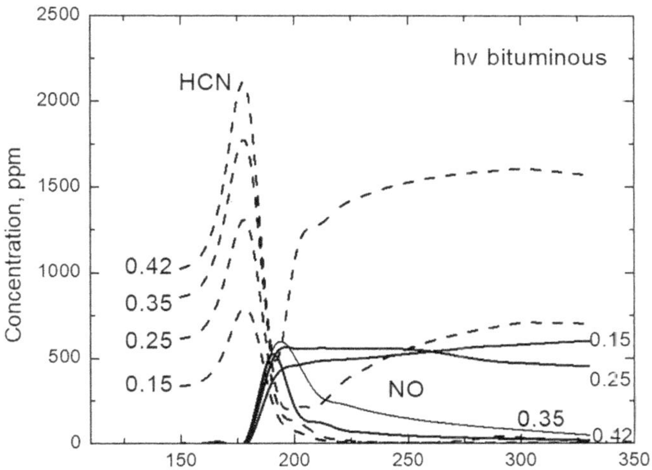

Fig. 8.23 Predicted concentration histories of *(dashed curves)* HCN and *(solid curves)* NO with the hv bituminous at four loadings, where 0.42 is the commercial loading.

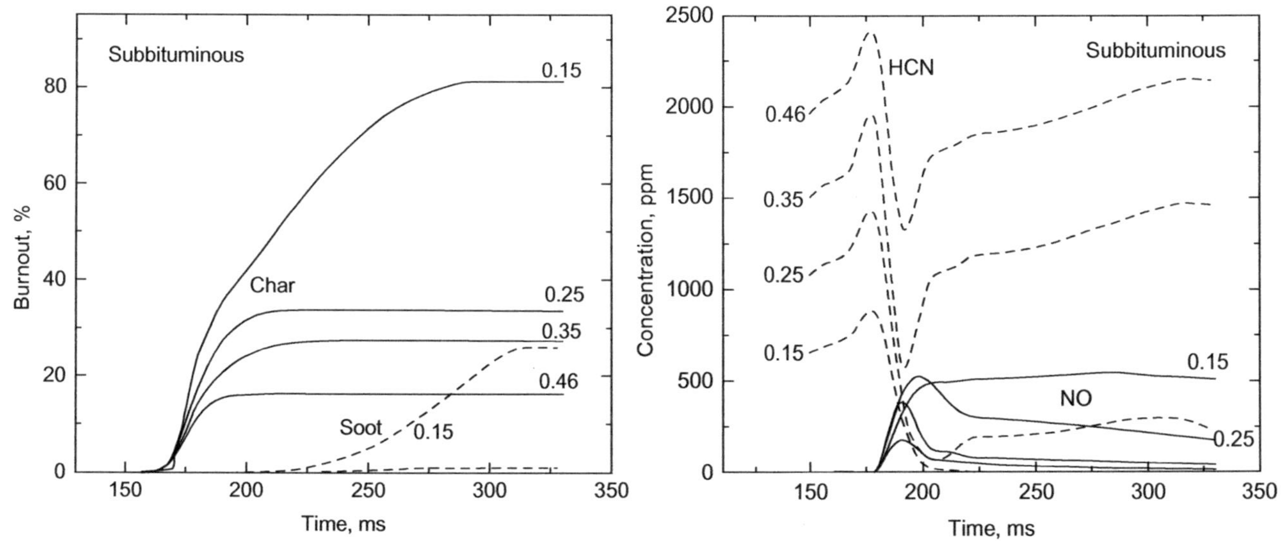

Fig. 8.24 Predicted histories with subbituminous at four loadings, where 0.46 is the commercial loading, of (left panel) extents of burnout of *(solid curves)* char and *(dashed curves)* soot; and concentrations of (right panel) *(dashed curves)* HCN and *(solid curves)* NO.

consumes O_2 even before all the gaseous fuels have burned away. Consequently, as seen in Fig. 8.24, NO reduction remains strong throughout the post-flame with the three heavier loadings, and effluent NO levels become much lower than the maximum levels at the end of volatiles combustion and nearly vanish with the two heaviest loadings. Consequently, the subbituminous generates much lower near-burner NO than the hv bituminous for two reasons: First, the maximum NO levels at the end of volatiles combustion are lower because the coal-N is lower and also because HCN destruction skews toward N_2 with progressively lower O_2 concentrations; and, second, NO reduction along the post-flame is stronger due to the greater H_2 and CO concentrations generated with greater extents of char burnout. With this coal, only the lightest loading gives an effluent NO level comparable to the maximum at the end of volatiles combustion because persistent O_2 suppresses NO reduction.

With the lv bituminous, char and soot ignite at comparable transit times, and soot burns much faster than this particular char, as seen in Fig. 8.25. More than half the soot burns out with the lighter loadings, but char burnout never exceeds 20%. Most important, the slow heterogeneous burning rates are responsible for sequential volatiles combustion and soot burnout, so that NO production is unaffected by heterogeneous combustion with both intermediate loadings. With the heaviest loading, NO diminishes from its maximum at the end of volatiles combustion, albeit only by one-half. With the lightest loading, the maximum NO concentration is the lowest of all only because the accumulation of HCN is at a minimum.

Based on these connections among the sequence of conversion of the different fuel components and NO production, the histories for H_2, CO, and O_2 concentrations with the subbituminous and lv bituminous coals in Fig. 8.26 illustrate how coal quality affects near-burner NO production. In the most favorable situation, as with the subbituminous, char burns fast enough to accelerate O_2 consumption during volatiles combustion. Rapid O_2 consumption biases the partitioning of HCN toward N_2, away from NO toward the later stages of volatiles combustion, which diminishes the maximum NO concentration. It also allows H_2 and CO reductants to accumulate along the post-flame region. In turn, these species reduce NO concentrations to well below their maximum levels at the end of volatiles combustion. In the least favorable situation, as with the lv bituminous, O_2 concentrations remain elevated throughout volatiles combustion, which skews HCN decomposition toward NO production, and thereby raises maximum NO concentrations. NO concentrations will remain elevated unless the post-flame can sustain H_2 and CO reductants. With the two lightest loadings in Fig. 8.26, O_2 persists in the effluent so that reductants are present only with the heaviest loading. But even with the heaviest loading, no reductants emerge in the post-flame region until 250 ms, so the extent of NO reduction along the post-flame is relatively small (cf. Fig. 8.25).

8.6.3 Volatiles combustion in entrained flow gasifiers

The reaction mechanisms validated in Section 8.6.1 were also used to simulate the earliest chemical transformations along the coal jets in gasifier designs licensed by Shell and General Electric Power Systems (GEPS). Operating conditions along these

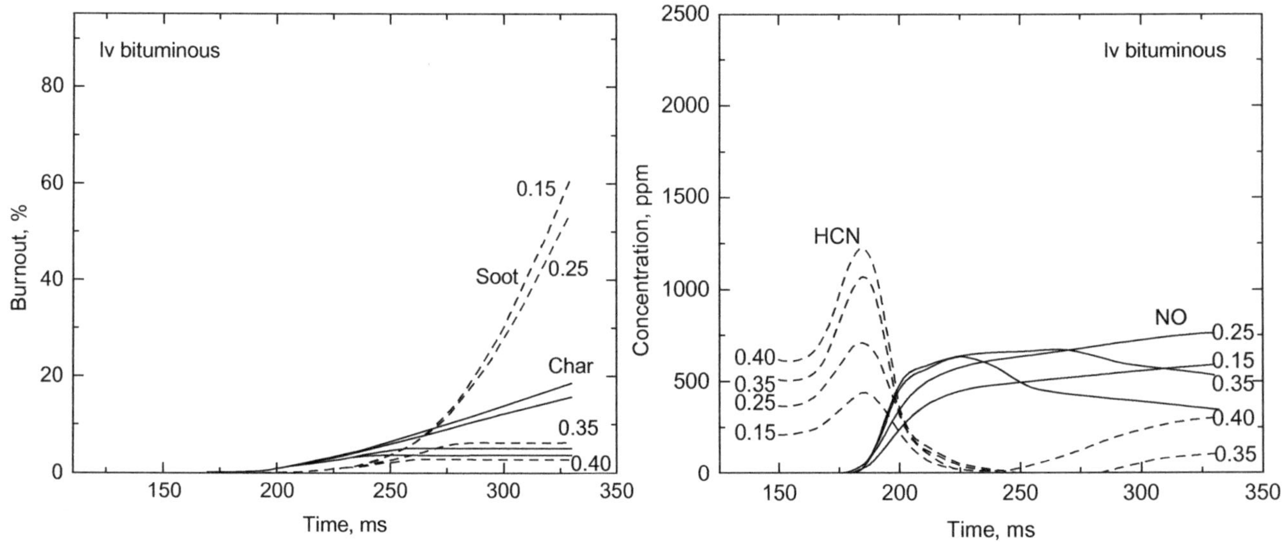

Fig. 8.25 Predicted histories with lv bituminous at four loadings, where 0.40 is the commercial loading, of (left panel) extents of burnout of *(solid curves)* char and *(dashed curves)* soot; and concentrations of (right panel) *(dashed curves)* HCN and *(solid curves)* NO.

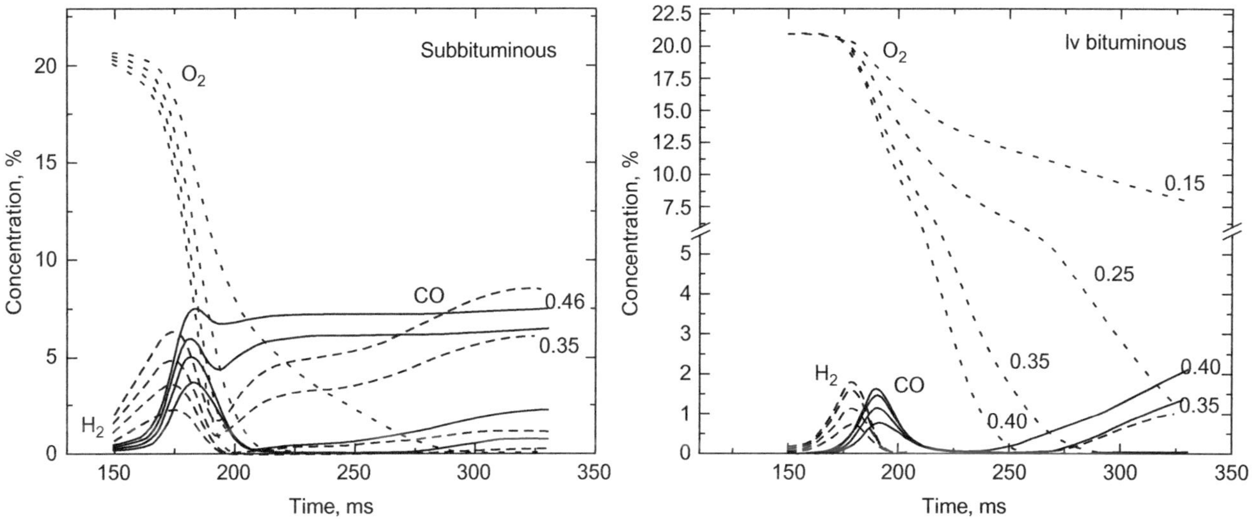

Fig. 8.26 Predicted concentration histories of *(solid curves)* CO, *(dashed curves)* H_2, and *(dotted curves)* O_2 with four loadings of (left panel) subbituminous and (right panel) lv bituminous coals.

coal jets, including coal mass loadings, moisture and O_2 flowrates, thermal histories, and transit times, were roughly characterized with CFD simulations in the literature. A network of CSTRs again stages the chemistry in this environment with actual grinds for dry feed and slurry systems for three diverse coal types (Niksa, 2019c). This section focuses on the chemical transformations that affect product distributions and carbon conversion up to the point of complete O_2 consumption through the first 400–500 ms of transit time.

The reaction system is one of the primary coal jets into a gasifier throughout the period for primary devolatilization, tar decomposition, volatiles combustion, and partial oxidation of char and soot. The jet consists of pulverized coal entrained in gas or water at mass loadings approaching 1.0 kg-coal/kg-gas and 2.0 kg-coal/kg-water. In Shell gasifiers, coal in very fine grinds is entrained in N_2 or CO_2 and co-fed with O_2 through four nozzles in an injection plane near the base of the reactor. The nozzles are offset from the reactor radius, so that the horizontal coal jets impinge into a rectangle that rotates the flow upward in a helical pattern, like the flowfield in T-fired utility furnaces. In GEPS gasifiers, coal-water slurry is fed downward through a cylindrical annulus into the reactor vessel. The coal jet is co-fed with a swirled O_2 stream that mixes into the coal jet after the water evaporates. In both designs, once the coal suspension leaves the injector nozzle, it heats primarily by radiation from the coal flame, and from the heat release from fuels expelled from the coal. Since coal jets in commercial gasifiers are momentum-dominated, entrainment of surrounding syngas into the primary jets is deemed to be minimal during this period.

Unfortunately, recent CFD simulations of Shell gasifiers do not give consistent operating conditions in the fuel injection plane. The calculations in this section were guided by the CFD results of Cao et al. (2018), which have coal velocities that never exceed 10 m/s, so that transit times within the horizontal injection plane were a few hundred milliseconds, which was sufficient to consume most of the O_2. Despite the nominally simpler flowfield in GEPS gasifiers, the operating conditions are even more uncertain than those for Shell gasifiers. Since the O_2 co-flow is swirled, the primary coal jet can be resolved into an internal recirculation zone (IRZ) immediately downstream of the nozzle, surrounded by the expanding coal jet which, in turn, is surrounded by an external recirculation zone (ERZ) (Monaghan and Ghoniem, 2012). Ultimately, these flow regions coalesce into plug flow into the reactor outlet. The main source of ambiguity is that the swirl intensity determines the strengths of the recirculation zones which, in turn, govern the entrainment of hot syngas into the coal jet. A recent detailed parametric study of the swirl number gave combustion zones that extended only 0.5 m from the nozzle for strong swirl to nearly the entire reactor for weak swirl (Bi et al., 2015). Since these ambiguities remain unresolved, the calculations in this section used the nominally 1D flowfield for a coal jet under weakly swirled conditions and therefore characterize the impact of coal grind and coal quality rather than how entrainment of syngas under strong swirl affects volatiles conversion. Readers interested in how stronger syngas entrainment would shift the behavior should consult the parametric study of coal loading in Figs. 8.22–8.26, since the impact of stronger entrainment resembles that for lighter coal loadings.

Estimated mean thermal histories for gases, the radiant source temperature, and coal were based on the exit gas temperatures for various test cases in Shell and GEPS gasifiers (Bockelie et al., 2003). The maximum values for both gas and radiation source were assumed to match the hottest exit gas temperatures (of 2400°C for Shell and 2100°C for GEPS). The approaches to the maximum values are consistent with those from CFD of the coal injectors in the T-fired furnace in Section 8.6.2, except that the time scale was extended to account for the greater pressures and coal loadings in the gasifiers. The thermal history for the GEPS reactor omits the preliminary stage for water evaporation, which extends the cool isothermal soaking period but does not affect the chemistry under consideration. The coal suspension heats at about 2000°C/s during this stage but remains below 300°C. Consequently, this stage is regarded as nonreactive, and simply adds a 200 ms lag to the onset of chemistry in the jet.

After the lag, the gases are heated to their maximum temperatures in less than 200 ms at approximately 10^4°C/s primarily by radiation feedback from the macro-scale coal flame. Both the coal and entrainment gas heat at similar rates in the Shell gasifier, while gas temperatures are as much as 500°C cooler than coal temperatures because only the particles absorb the intense thermal radiation. Coal heating rates are slightly slower than gas heating rates in the GEPS gasifier, so that gases become hotter by the end of the heating period. Mean coal thermal histories throughout primary devolatilization and tar decomposition were evaluated from a particle energy balance that incorporated the estimated thermal histories for gases and the radiation source, and the mean particle diameters.

Since the subject flowfields are premixed primary jets with no entrainment of syngas via recirculation, their equivalent reactor networks are simply linear CSTR series with 40 reactors for Shell and 45 reactors for GEPS, where about half the CSTRs covered the initial, nonreactive period. The subbituminous and hv bituminous coals from Sections 8.6.1 and 8.6.2 are used in this study as well; a lignite was substituted for the lv bituminous because low rank coals are preferred in IGCC coal gasification for their relatively fast gasification reactivities. The most variable operating conditions for both gasifiers are the O_2/coal and steam/coal ratios, because these ratios are the means to operate the gasifiers to produce essentially uniform syngas compositions from a broad range of fuel quality. With Shell gasifiers, the O_2/coal ratio is generally increased for coals of progressively greater rank to maintain H/C ratios from 0.80 to 1.00 and O/C ratios from 0.90 to 1.15. GEPS gasifiers are fed with coal slurry that has 65 wt% coal, so steam/coal ratios are between 0.50 and 0.70. The O_2/coal ratios are generally increased for coals of progressively higher rank although, as for Shell gasifiers, the variation in this parameter for coals with C-contents over 77% is minimal. For all but the lowest rank coals, the H/C varied from 1.85 to 2.10 and O/C varied only from 1.54 to 1.65 (Niksa, 2019c).

The same grind was used with each coal and was specified with Rosin-Rammler PSDs to give a mean size of 44.4 μm for Shell conditions. The GEPS gasifier fires a grind much coarser and broader than standard utility grind with a mean size of 205 μm, as needed to stabilize the slurry viscosity. Reactor pressures were the same with all coals at 2.3 MPa for Shell and 4.1 MPa for GEPS.

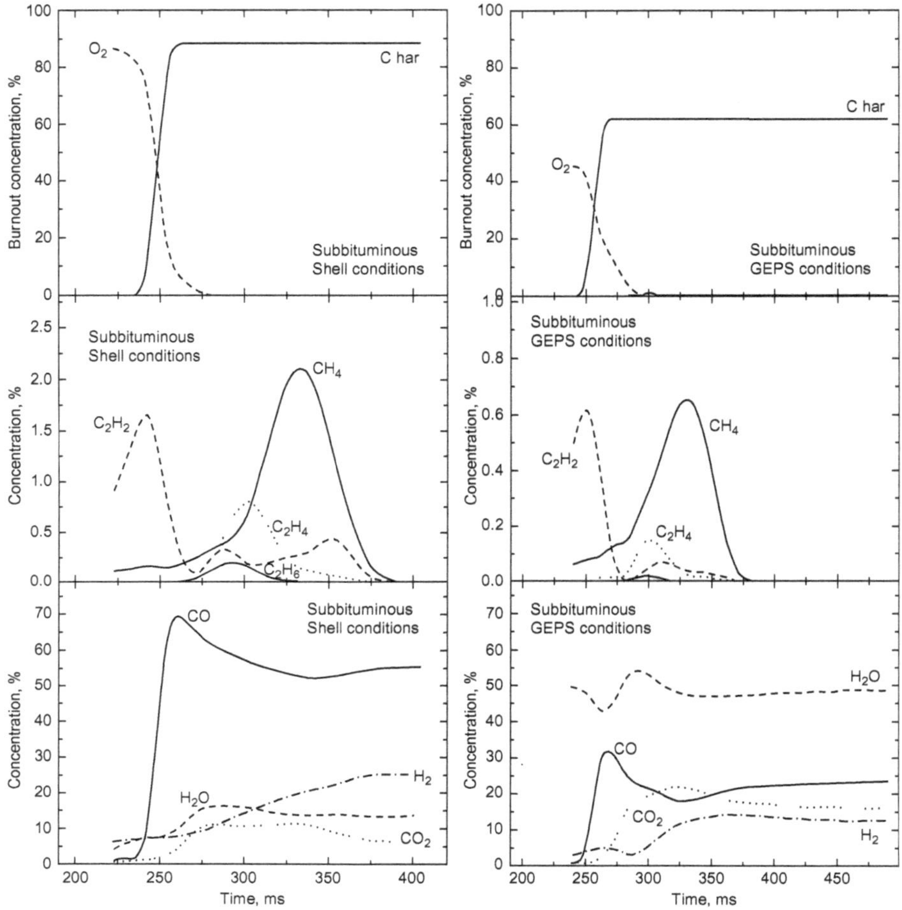

Fig. 8.27 In descending order, predicted flame structure for (left) Shell and (right) GEPS conditions with a subbituminous including (top) *(solid curve)* extents of char burnout and *(dashed curve)* O_2 concentration; (middle) the concentrations of *(solid curve)* CH_4, *(dashed curve)* C_2H_2, *(dotted curve)* C_2H_4, and *(lower solid curve)* C_2H_6; and (bottom) concentrations of *(dot-dashed curve)* H_2, *(solid curve)* CO, *(dotted curve)* CO_2, and *(dashed curve)* H_2O (Niksa, 2019c). Note change of scales between Shell and GEPS conditions on the time axes and on molar concentration in the middle panels.

The flame structure in Fig. 8.27 for Shell conditions with subbituminous conveys two distinct conversion time scales. During the first partial oxidation stage from 225 to 275 ms, most of the GHCs and char consume all the available O_2 to generate mostly CO with small amounts of CO_2 and H_2O from homogeneous oxidation. Soot burnout was negligible. Due to the fast oxidation reactivity of this char and its fine grind, just under 90% of the char burns out upon ignition. During the second gas reforming stage from 275 to 350 ms, two additional C_2 GHCs – C_2H_4 and C_2H_6 – appear as intermediate species, while C_2H_2 and CH_4 pass through their maximum concentrations before they are eliminated by 400 ms. Carbon dioxide, CO, H_2, and H_2O change to reflect the elimination of GHCs and water-gas shifting. Although the predicted effluent levels of CO and H_2 in Fig. 8.27 are lower than ultimate syngas concentrations, these values are expected to relax toward accurate values upon addition of CO and H_2 by char gasification, and the associated water-gas shifting throughout the gas quench cycle.

The flame structure with the same coal under GEPS conditions also appears in Fig. 8.27, where the partial oxidation stage extends from 250 to 290 ms and the reforming stage continues through 375 ms. The extent of char burnout is much lower than for Shell conditions due to the much coarser grind and, to a lesser extent, the lower ultimate temperatures for gases and the radiant source, and the slightly lower O_2 partial pressures for GEPS. Soot oxidation was again negligible. Since char oxidation is disadvantaged in the O_2 competition with volatiles, it releases relatively little CO so the gas phase remains relatively oxidizing as it generates twice as much CO_2 as Shell conditions. This more oxidizing environment also significantly alters the course of the reforming stage. Methane, C_2H_2, and C_2H_4 appear early in this stage, albeit at less than one-third the levels under Shell conditions. But all GHCs are partially oxidized into CO and H_2 by 385 ms. Consequently, the effluent from the NBFZ contains only the four major syngas components. It is interesting that neither volatiles combustion nor a major portion of char oxidation perturb the moisture level from its initial value, despite major shifts in the other syngas concentrations. Of course, both the CO and H_2 concentrations will more-than-double after the large amount of residual char is gasified by steam, and the gas composition is shifted during the quench cycle.

The major gas components with the lignite and the hv bituminous coal under Shell conditions appear in Fig. 8.28. The flame structure with the lignite is very similar to the subbituminous coal's (Niksa, 2019c): the extent of char burnout is only slightly greater; all GHC concentrations are 25% lower; and the nascent syngas contains slightly less CO and H_2 and slightly more CO_2 and H_2O. In contrast, the hv bituminous burnout (not shown) is only slightly greater than 60%, whereas the ultimate GHC concentrations are nearly identical to the subbituminous coal's. The lower extent of char burnout is responsible for the substantially lower concentrations of CO and H_2 in Fig. 8.28, and for the greater CO_2 and H_2O levels. Since more residual char leaves the NBFZ with the hv bituminous, the CO and H_2 concentrations in the ultimate syngas will be much greater than those in Fig. 8.28 once nearly all char has been gasified.

Under GEPS conditions, the coal quality impacts are weaker than under Shell conditions because the differences among extents of char burnout are smaller. With the lignite, burnout is only slightly greater than with the subbituminous; GHC levels are 25% lower, as for Shell conditions; and ultimate CO_2 concentrations are slightly greater, whereas CO and H_2 levels are slightly lower. Gas concentrations with the hv bituminous in Fig. 8.29 show more significant differences. The maximum GHC levels are one-third lower. Char burnout is 25% lower than with the subbituminous,

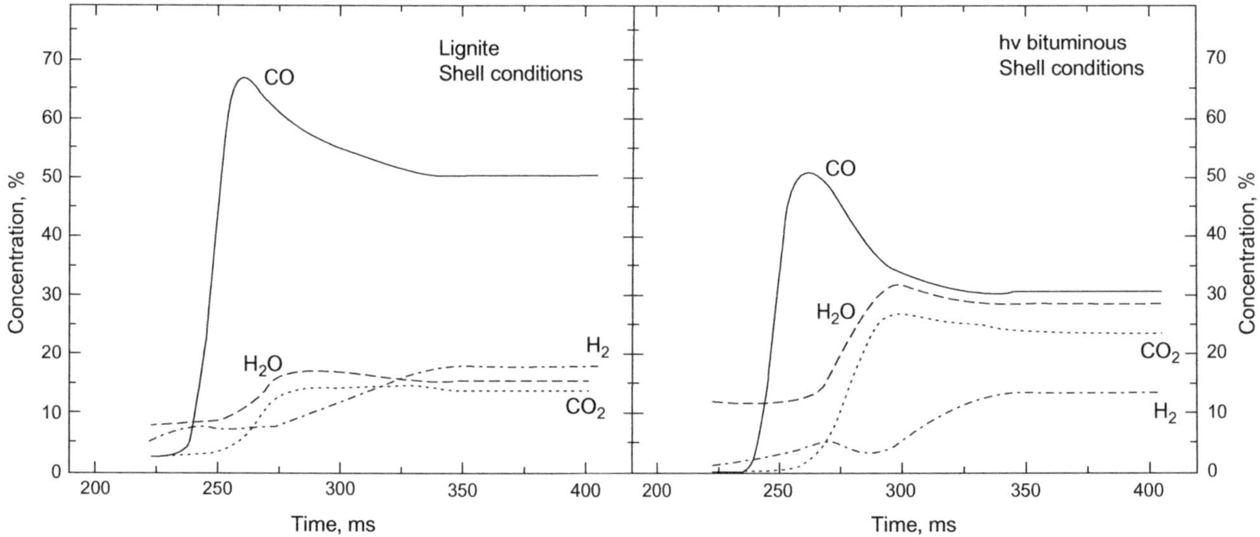

Fig. 8.28 Predicted concentrations of *(dot-dashed curves)* H_2, *(solid curves)* CO, *(dotted curves)* CO_2, and *(dashed curves)* H_2O for Shell conditions with (left) a lignite and (right) an hv bituminous.

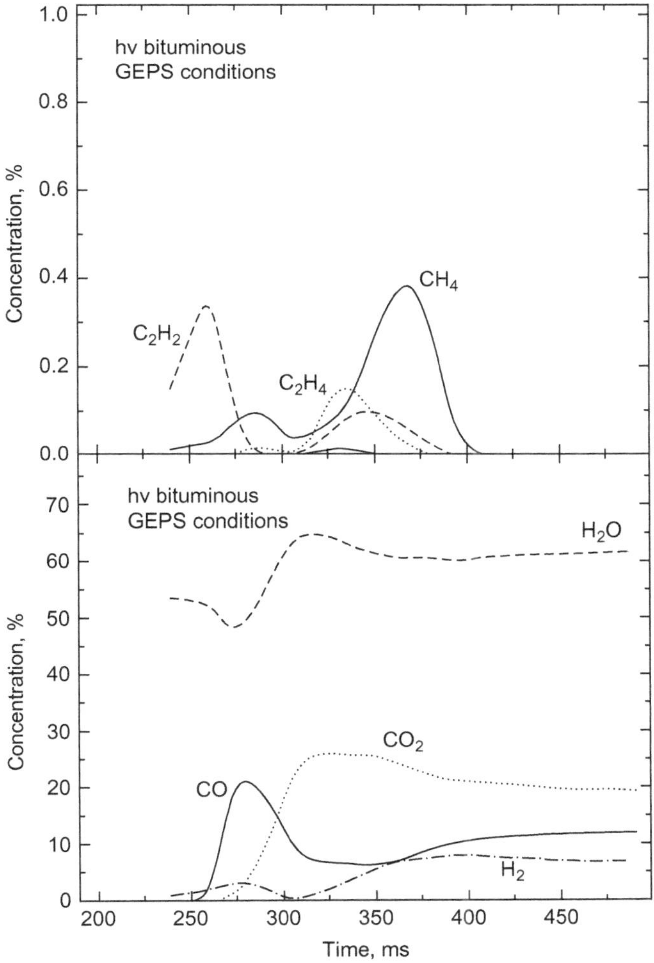

Fig. 8.29 Predicted concentrations of (top) *(solid curve)* CH_4, *(dashed curve)* C_2H_2, *(dotted curve)* C_2H_4, and *(lower solid curve)* C_2H_6; and (bottom) *(dot-dashed curves)* H_2, *(solid curves)* CO, *(dotted curves)* CO_2, and *(dashed curves)* H_2O for GEPS conditions with hv bituminous coal.

and the H_2O concentration is 25% greater than in Fig. 8.28. There is also double the CO_2 in the nascent syngas, but only half-as-much CO and H_2 as with the subbituminous. As for Shell conditions, the coal quality impacts on nascent syngas compositions from GEPS NBFZs are determined by char oxidation reactivities, in that progressively faster reactivities produce more reducing gas conditions.

Soot oxidation with all coals under both sets of operating conditions was negligible, and none of the GHCs from any coal persisted into either of the NBFZ exits.

8.6.4 Similitude for volatiles combustion at lab and commercial scales

The nearly identical tendencies for progressively greater O_2 levels in the RCFR validation database and for progressively lighter loadings in the NBFZ simulations are especially reassuring. Obviously, both variations affect the flame stoichiometry, so the similarities are not unexpected. But changes in the stoichiometry are not the only essential consideration. The thermal histories of coal particles have long been recognized as essential specifications to relate any test to commercial processing.

As seen in Fig. 8.30, it is equally important to impose gas temperature histories in lab tests like those in the commercial system if the focus is on volatiles combustion and N-species conversion. This figure shows the mean thermal histories for primary gases and coal evaluated from CFD simulations of both the RCFR and the commercial coal jet in a T-fired furnace. Mean temperature values were evaluated across the primary stream transverse to the flow direction. In both systems, the coal suspension was primarily heated by radiation from sources at nominal temperatures of 1400°C in the RCFR and of 1500–1700°C in the commercial furnace. In the RCFR, the mean gas temperature seriously lags the particle temperature through the first half of the transit time because the gas is transparent to radiation, and this stream is heated primarily by conduction off the flow tube wall and by conduction off the particle suspension. Consequently, the particle thermal histories are as much as 500°C hotter than the core gas during early times, but gas temperatures increase dramatically upon ignition and exceed the particle temperatures by the outlet (for this particular case).

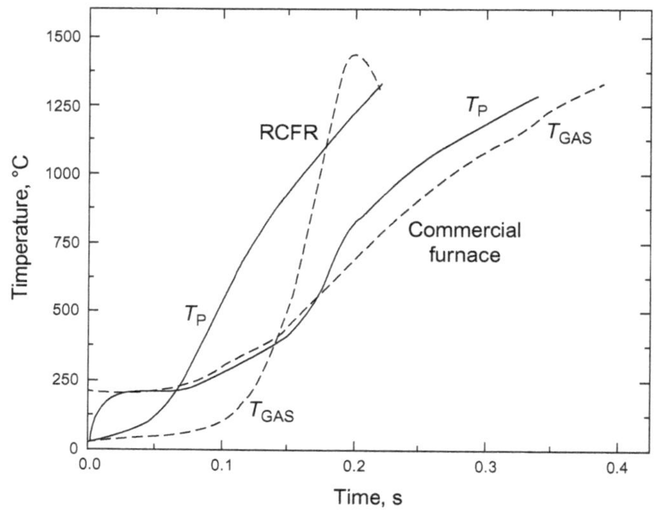

Fig. 8.30 Mean thermal histories for *(solid curves)* the coal suspension and *(dashed curves)* carrier gas from CFD simulation of the RCFR and the primary injection stream in a commercial furnace.

In the commercial coal jet, both temperatures increase at two distinct heating rates in two stages. The first heating stage lasts for 130 ms, while the primary jet received energy mostly from auxiliary air streams that were preheated to 335°C. The coal suspension is heated at about 2000°C/s during this stage but remained below 300°C. Consequently, this stage is regarded as nonreactive, and simply adds a 130 ms lag to the onset of chemistry in the jet. The second stage covers the remaining 210 ms, while the primary stream heats at approximately 10^4°C/s primarily by radiation from the macro-scale coal flame. Although the literature contains numerous estimates for coal heating rates in utility furnaces as fast as 10^6°C/s, the bulk of the coal in this primary stream does not heat nearly so fast. Indeed, momentum-dominated jets with such heavy mass loadings cannot possibly heat as fast as individual 51 μm particles. Both the coal and entrainment gas heat at similar rates, while gas temperatures are 100°C cooler than coal temperatures because only the particles absorb the intense thermal radiation.

Tests on volatiles combustion should resolve the behavior in time while both the coal suspension and its entrainment gas are being rapidly heated to flame temperatures. Primary coal jets heat up with relatively small temperature differences between the suspension and carrier gas, because the heavy particle loading closely couples both thermal histories. Consequently, the intense thermal radiation and the heavy loading in the RCFR impose heating rates and thermal histories for both particles and the entrainment gas that are very similar to the ones from the CFD simulations of the primary coal jet in the commercial furnace.

It remains to be demonstrated that any reactor in which coal is heated by convective mixing with a preheated gas stream can impose realistic thermal histories on both coal and carrier gas. The flat-flame burner developed by Peck and co-workers (Altenkirch et al., 1979; Peck et al., 1984) injected relevant coal loadings into a coal-fired Meeker burner. But to stabilize these flames, inlet O_2 levels were 23%, the post-flame gas temperatures exceeded 1700°C, and the particle sizes were even finer than a p.f. grind, so volatiles conversion was complete in less than 10 ms. Consequently, the initial stages of conversion of GHCs, H_2, and HCN could not be resolved in time with sampling probes. Even though concentration profiles were monitored through 70–80 ms in residence time, all GHCs and O_2 were burned away before they could be retrieved with sampling probes immediately above the burner. However, the slate of reported N-species for longer sampling times closed N-balances within useful tolerances, and these datasets are especially well-suited to track N-species transformations in the post-flame region. As seen in Figs. 8.10 and 8.11, these datasets for two different coals at two different loadings exhibit the same tendencies along the post-flame region as in RCFR tests and in the primary jet simulations in this chapter.

8.6.5 Volatiles reforming in CFBC dense bottom beds

One is hard-pressed to find a technology that does not co-feed coal with air or O_2, which implies that the analyses in the previous two sections illustrate many of the most important features of volatiles conversion under commercial conditions. The dense bottom beds in CFBCs are a notable exception. Counterintuitively, these beds sustain highly reducing conditions at moderately hot temperatures with minimal oxidation. Bottom beds are fluidized by an air stream from below at extremely high suspension loadings, and fed from above by either entrained coal or a coal slurry. The bed gas partitions into a fast-moving bubble phase with hardly any particles and an emulsion phase that has roughly equal volume fractions of gas and solids.

The disparate gas flowrates through bubbles and emulsion carry an important implication for volatiles conversion in bottom beds, as explained elsewhere in more detail (Niksa et al., 2017). Since all particles are confined to the emulsion, volatiles are necessarily released into the emulsion gases, which raises the question: Can the flowrate of emulsion gases accommodate the flowrate of volatiles associated with actual coal feedrates? Surprisingly, the gas flowrate through the emulsion is too slow to accommodate all the volatiles for coals of even the lowest volatility. With hv bituminous coals, most of the volatiles must spontaneously enter the bubble phase, along with all the primary air from the air distributor. With low volatility coals, including anthracites, most of the volatiles remain in the emulsion, but all O_2 from the primary air is still excluded from the emulsion at the inlet air distributor plate. Oxygen only enters the emulsion via gas exchange between the bubble and emulsion phases, which is relatively slow in the fast-bubble hydrodynamic regime. Consequently, only a few percent of volatiles are oxidized in typical emulsion transit times if burning rates are limited by gas exchange rates (Niksa et al., 2017).

This scenario sustains mixtures of volatiles fuel components at temperatures approaching 900°C for several seconds in bottom bed emulsions. To evaluate the impact of volatiles reforming, the calculations in this section were based on gas velocities of 4–5 cm/s for minimum fluidization; bubble volume fractions of 0.3; bottom gas voidages of 0.6; and a bottom bed height of 1 m, for which the emulsion gas transit time is about 7 s. Volatiles reforming chemistry was simulated for a bed operating temperature of 870°C with a coal grind whose mean size was 2 mm in both dry and slurry feeds. Conversion of both char and soot were omitted because there is no O_2 and gasification by steam and CO_2 is too slow at atmospheric pressure.

The distributions of major gases in the emulsion effluent are compiled in Table 8.3 in daf wt% for comparison with the levels in both primary and secondary volatiles. Due to the hot temperature and extended contact time, the secondary volatiles sustained complete tar decomposition into soot, so no oils appeared in the effluent. With all three coals, the effluent for a dry coal feed contains slightly less CO, H_2, CO_2, and H_2O than secondary volatiles, and also slightly less CH_4. The main transformation is the conversion of C_2H_2 into C_2H_4 and C_2H_6. With slurry feeds, the moisture concentration approached 60% and, with steam reforming, the bed effluent became even more similar to secondary pyrolysis products, again, with reforming of C_2H_2 into C_2H_4 and C_2H_6. None of these changes would affect combustion above the dense bottom beds in so-called splash zones. Moreover, the majority of volatiles must be exchanged into bubbles with all but coals of lowest volatility, where they mix with air and ignite once the bubble gas has been heated above the ignition temperature.

Table 8.3 Gaseous fuels in primary (1^0) and secondary (2^0) volatiles and in effluents from bottom bed emulsions with dry and slurry coal feeds, in daf wt%.

	Subbituminous				hv bituminous				lv bituminous			
	1^0	2^0	Dry	Wet	1^0	2^0	Dry	Wet	1^0	2^0	Dry	Wet
CH_4	2.7	2.8	2.4	2.6	3.1	3.4	2.9	3.1	1.9	4.0	3.0	3.5
C_2H_2	0.0	1.5	0.2	0.3	0.0	2.9	0.2	0.5	0.0	1.4	0.2	0.5
C_2H_4	1.2	0.0	1.0	1.0	1.3	0.0	1.7	2.0	0.7	0.2	0.9	1.0
C_2H_6	0.36	0	0.09	0.06	0.65	0	0.21	0.11	0.12	0	0.08	0.03
C_3H_8	0.96	0	0	0	1.15	0	0	0	0.57	0	0	0
CO	5.1	8.8	7.2	7.8	2.3	4.2	3.3	3.7	0.8	1.8	1.4	1.6
H_2	0.8	2.0	1.5	1.7	1.1	2.2	1.7	1.9	1.0	1.3	1.0	1.2
CO_2	6.3	6.7	5.6	6.4	1.9	2.2	1.7	2.1	1.2	1.4	1.1	1.2
H_2O	8.3	8.5	6.9	65.0	4.7	4.7	3.4	63.2	2.1	2.2	1.4	60.8

8.6.6 Summary of volatiles combustion under commercial operating conditions

Operating conditions in the T-fired utility furnace have coal loadings from 0.40 to 0.5 kg-coal/kg-gas that diminish for progressively higher coal rank. With such heavy loadings, chemistry in the primary coal suspension first transforms the full slate of primary volatiles into the mixtures of GHCs, H_2, CO, HCN, soot, and char that actually burn. Once ignited, the gaseous mixtures compete for the available O_2 with char but not soot. However, char burnout with even the most reactive char in this study remained under 20% with these heavy loadings. So volatiles combustion with commercial loadings was weakly mediated by char oxidation with the subbituminous, and essentially isolated with the hv and lv bituminous coals.

The most striking feature of volatiles combustion under commercial loadings is that it alters the species concentrations but does not necessarily add or eliminate any components. Table 8.4 shows the yields of GHCs, H_2, CO, HCN, and NO in both volatiles and the effluent from the primary coal jet. Both compositions are given in daf wt% to facilitate the comparison with volatiles yields. Oxygen was absent in the effluent with all coals. With subbituminous and hv bituminous coals, all the original volatiles components appear in the effluent at substantial levels, without any additional species because NO is negligible. With the lv bituminous, GHCs were completely eliminated and NO is appreciable. The effluents contain no more than 25% of the GHCs in volatiles and 40% of the H_2, but have two to three times more CO. Up to half the HCN remains to be converted beyond the primary jet. With most coals, volatiles combustion with commercial loadings is a relatively weak reforming process. Its greatest impact is on volatile-N conversion, for which at least half of gaseous NO precursors are converted into N_2. But even so, the ultimate fate of most coal-N is determined beyond the primary coal jet, because char-N levels in the effluent remain substantial.

However, with progressively lighter loadings, volatiles combustion progresses much further toward complete oxidation, and transforms most coal-N. In fact, with 0.25 kg-coal/kg-air, volatiles combustion eliminated all GHCs and HCN and produced

Table 8.4 Gaseous fuels and N-species in volatiles and primary stream effluent with commercial coal loadings, in daf wt%.

	GHCs		H_2		CO		HCN		NO	
	Vol[a]	Eff[b]	Vol	Eff	Vol	Eff	Vol	Eff	Vol	Eff
Subbit	9.1	1.9	4.2	1.6	9.4	23.6	1.0	0.51	0	0
hv bit	5.6	1.6	4.3	1.1	5.7	12.1	1.1	0.39	0	0
lv bit	2.1	0.0	3.6	0.5	2.1	6.4	1.0	0.07	0	0.09

[a]Species in volatiles.
[b]Species in the primary stream effluent.

substantial NO with all coals, while appreciable amounts of char burned away with all but the low volatility coal. By varying the coal loadings in the simulations, all else the same, these results clearly reveal the connections among near-burner NO production and the sequence of conversion of the various fuel components, including char. At face value, these connections should be restricted to a single flowfield archetype, because the primary coal jet in isolation represents a limiting scenario for minimum mixing of a primary stream with auxiliary air streams. But many alternative burner designs rapidly disperse primary streams into coflowing air streams. Whenever the streams mix before the onset of volatiles combustion, the reaction system is shifted toward the behavior for lighter initial coal loadings. From this standpoint, the general tendencies for variations in loading provide guidance for even much more complex mixing configurations.

Disparate extents of char burnout are the distinguishing factor that interprets the much different gas compositions from the NBFZs in Shell and GEPS gasifiers, provided that the char reactivity is sufficient to effectively compete for the available O_2 with GHCs and other volatile fuel components; i.e., with all but coals of highest rank. Since O_2 pressures and maximum reactor temperatures are comparable, the coal grind is the only disparate operating condition affecting char burnout in these NBFZs. Moisture levels are also much greater in GEPS gasifiers, but moisture acts as a diluent rather than as a reactant in this region. Consequently, the much finer coal grind in Shell gasifiers competes much more effectively for O_2 with GHCs and other volatile fuel components. All but the smallest sizes reached or exceeded 90% burnout with low rank coals under Shell conditions; but the largest quarter of the GEPS PSD could not achieve a fully ignited rapid-burn state under comparable conditions, because of its inordinate thermal capacitance. This feature explains most of the difference between extents of burnout in both gasifiers with the same coal.

Char burnout affects syngas compositions from NBFZs by making the gas atmosphere more reducing for progressively greater extents of burnout. This tendency is apparent in greater levels of CO and H_2 and lesser amounts of CO_2 and H_2O for progressively greater burnout. It explains the different ultimate compositions of nascent syngas from Shell and GEPS NBFZs with the same coals, and also explains the variations for coals of progressively greater rank in either of the gasifier's NBFZ.

The simulations also showed that GHCs are eliminated from NBFZ effluents under both Shell and GEPS conditions with all coals. This circumscribes the finding that volatiles combustion under commercial loadings in the T-fired furnace alters the species concentrations but does not necessarily add or eliminate any components of secondary volatiles. Effluents from gasifier NBFZs contain only the four major syngas components, along with minor levels of N- and S-species. Notwithstanding, the estimated gas compositions in these simulations are incompatible with the instantaneous combustion and equilibrium gas compositions within NBFZs that are usually imposed in CFD gasifier simulations, because they are strongly influenced by simultaneous and very large extents of char burnout.

References

Altenkirch RA, Peck RE, Chen SL. Disappearance of nitric oxide and cyanide in one-dimensional coal dust/oxidizer flames. Combust Sci Technol 1979;20:49–58.

Bi D, Guan Q, Xuan W, Zhang J. Numerical simulation of a GSP gasifier under different swirl angles. Fuel 2015;155:155–63.

Bockelie MJ, Denison MK, Chen Z, Senior CL, Sarofim AF. Proc. Pittsburgh coal conf., Pittsburgh, PA. 2003.

Cao Z, Li T, Zhang Q, Zhou H, Song C, Fengqi Y. Systems modeling, simulation, and analysis for robust operations and improved design of entrained-flow pulverized coal gasifiers. Energy 2018;148:941–64.

Choi S, Kruger CH. Modeling coal particle behavior under simultaneous devolatilization and combustion. Combust Flame 1985;61:131–44.

Doolan KR, Mackie JC, Tyler RJ. Coal flash pyrolysis: secondary cracking of tar vapours in the range 870–2000 K. Fuel 1987;66(4):572–8.

Eckstrom DJ, Hirschon AS, Malhotra R, Niksa S. High pressure coal combustion: characterization of the near-burner flame zone. J Sustain Energy Engr 2014;2:192–222.

Glarborg P, Alquita MU, Dam-Johansen K, Miller JA. Kinetic modeling of hydrocarbon/nitric oxide interactions in a flow reactor. Combust Flame 1998;115:1–27.

Froment GF, Bischoff KB. Chemical reaction analysis and design. New York: Wiley; 1979.

Grosshandler WL. The effect of soot on pyrometric measurements of coal particle temperature. Combust Flame 1984;55:59–72.

Gururajan VS, Wall TF, Truelove JS. The combustion of evolved volatile matter in the vicinity of a coal particle—an evaluation of the diffusion limited model. Combust Flame 1988;72:1–12.

Gururajan VS, Wall TF, Gupta RP, Truelove JS. Mechanisms for the ignition of pulverized coal particles. Combust Flame 1990;81:119–32.

Haussmann GJ, Kruger CH. Evolution and reaction of coal fuel nitrogen during rapid oxidative pyrolysis and combustion. Proc Combust Inst 1991;23:1265–71.

Hayashi J-I, Takahashi H, Iwatsuki M, Essaki K, Tsutsumi A, Chiba T. Rapid conversion of tar and char from pyrolysis of a brown coal by reactions with steam in a drop-tube reactor. Fuel 2000;79:439–47.

Hayashi J-I, Iwatsuki M, Morishita K, Tsutsumi A, Li CZ, Chiba T. Roles of inherent metallic species in secondary reactions of tar and char during rapid pyrolysis of brown coals drop-tube reactor. Fuel 2002;81:1977–87.

Kramlich JC, Seeker WR, Samuelson GS. Observations of chemical effects accompanying pulverized coal thermal decomposition. Fuel 1988;67:1182–9.

Lau C-W, Niksa S. The combustion of individual particle of various coal types. Combust Flame 1992;90:45–70.

Lau C-W, Niksa S. The impact of soot on the combustion characteristics of coal particles of various types. Combust Flame 1993;95:1–21.

Lim J-P, Steele D, del Rio Diaz-Jara D, Eckstrom DJ, Wilson RB, Niksa S, Malhotra R. A zero-CO_2 emitting process for transportation fuels from coal and natural gas sources. J Sustainable Energy Eng 2013;1:202–19.

Liu G-S, Niksa S. Coal conversion submodels for design applications at elevated pressures. Part II. Char gasification. Prog Energy Combust Sci 2004;30(6):697–717.

Liu G-S, Niksa S. Pulverized coal flame structures at elevated pressures. Part 1. Detailed operating conditions. Fuel 2005;84:1563–74.

Marlow D, Niksa S, Kruger CH. Secondary pyrolysis and combustion of coal volatiles. Proc Combust Inst 1992;24:1251–8.
Marlow D, Cho S, Niksa S, Kruger CH. Combustion of the noncondensable volatiles from various coals. Fuel Process Technol 1993;34:229–37.
Masek O, Sonoyama N, Ohtsubo E, Hosokai S, Li C-Z, Chiba T, Hayashi J-I. Examination of catalytic roles of inherent mineral species in steam reforming of nascent volatiles from the rapid pyrolysis of a brown coal. Fuel Process Technol 2007;88:179–85.
McLean WJ, Hardesty DR, Pohl JH. Direct observations of devolatilizing pulverized coal particles in a combustion environment. Proc Combust Inst 1981;18:1239–47.
Monaghan RFD, Ghoniem AF. Simulation of a commercial-scale entrained flow gasifier using a dynamic reduced order model. Energy Fuel 2012;26:1089–106.
Musarra SP, Fletcher TH, Niksa S, Dwyer HA. Heat and mass transfer in the vicinity of a devolatilizing coal particle. Combust Sci Technol 1986;45:289.
Niksa S. Simulating volatiles conversion in dense burning coal suspensions. Part 1. Validation of reaction mechanisms. Fuel 2019a;252:821–31.
Niksa S. Simulating volatiles conversion in dense burning coal suspensions. Part 2. Extrapolations to commercial p. f. firing conditions. Fuel 2019b;252:832–40.
Niksa S. Simulating volatiles conversion in dense burning coal suspensions. Part 3. Extrapolations to entrained flow gasification conditions. Fuel 2019c;252:841–7.
Niksa S, Cho S. Conversion of fuel-nitrogen in the primary zones of pulverized coal flames. Energy Fuel 1996;10:463–73.
Niksa S, Cho S. Assigning meaningful stoichiometric ratios for pulverized coal flames. Proc Combust Inst 1998;27:2905–13.
Niksa S, Liu G-S. Incorporating detailed reaction mechanisms into simulations of coal-nitrogen conversion in p.f. flames. Fuel 2002;81:2371–85.
Niksa S, Liu G-S, Hurt RH. Coal conversion submodels for design applications at elevated pressures. Part I. Devolatilization and char oxidation. Prog Energy Combust Sci 2003;29:425–77.
Niksa S, Sakurai Y, Fujiwara N. Predicting the conversion efficiencies of any coal type in CFBCs. Energy Fuel 2017;31:4507–19.
Park C, Appleton JP. Shock-tube measurements of soot oxidation rates. Combust Flame 1973;20:369–79.
Peck RE, Midkiff KC, Altenkirch RA. The evolution of nitrogen from pulverized subbituminous coal burnt in a one-dimensional flame. Proc Combust Inst 1984;20:1373–80.
Peck RE, Glarborg P, Johnsson JE. Kinetic modeling of fuel-nitrogen conversion in one-dimensional, pulverized coal flames. Combust Sci Technol 1991;76:81–109.
Seeker WR, Samuelsen GS, Heap MP, Trolinger JD. The thermal decomposition of pulverized coal particles. Proc Combust Inst 1981;18:1213–26.

Further reading

Harris DJ, Roberts DG, Henderson DG. Gasification behavior of Australian coals at high temperature and pressure. Fuel 2006;85:134–42.
Pedersen LS, Glarborg P, et al. A reduced reaction scheme for volatile nitrogen conversion in coal combustion. Combust Sci Technol 1998;131:193–223.

Hydropyrolysis and hydrogasification

9

Nomenclature

A_k	pseudo-frequency factor for reaction k, s^{-1}
A_{TAR}	mass of aromatic nuclei in primary tar, daf wt%
B	scaled labile bridges in tar as a molar concentration in the gas phase
B^*	scaled hydrogenated bridges in tar as a molar concentration in the gas phase
b_i	stoichiometric coefficient for product i for oil production via monomer decomposition weighted for the proportions of tar monomers with either one or two labile peripheral groups
b_i^*	stoichiometric coefficient for product i for oil production via monomer hydrogenation weighted for the proportions of tar monomers with either one or two hydrogenated peripheral groups
C	scaled char links in tar as a molar concentration in the gas phase
C_{H2}	molar concentration of ambient H_2
c_{ij}	stoichiometric coefficient for gaseous product i during nucleation and/or addition of tj to soot
$c_{i,O}$	stoichiometric coefficient for product i during oils addition to soot
E	fragment ends that lost labile peripheral groups as a molar concentration in the gas phase
E_k	activation energy for reaction k, kJ/mol
G	scaled molar concentration of noncondensable gases from decomposition of labile bridges and peripheral groups
G^*	scaled molar concentration of noncondensable gases from decomposition of hydrogenated peripheral groups
I	fragment ends that lost hydrated peripheral groups as a molar concentration in the gas phase
j	index for the degree of polymerization in tar molecules
J^*	maximum extent of depolymerization for primary tar molecules
k_{HG}	rate constant for char hydrogasification, moles/s-atm$^{0.5}$
k_k	rate constant for reaction process k = HY, HC, HG, or MH, $atm^{-1}\ s^{-1}$
M_n	number-average molecular weight of tar, g/mol
n_k	reaction order for H_2 in reaction k
^eN_i	number of atoms of element e in structural component i
O	scaled molar concentration of oils from both monomer hydrogenation and monomer decomposition
p_H^e	probability that a fragment end contains a hydrogenated peripheral group
p_{H2}	partial pressure of H_2, atm
R	scaled molar concentration of soot
R_{HG}	char hydrogasification rate, mol/s
$R_{O,MH}$	rate of oils production from monomer hydrogenation, mol/s
S	scaled labile peripheral groups in tar as a molar concentration in the gas phase

Process Chemistry of Coal Utilization. https://doi.org/10.1016/B978-0-12-818713-5.00009-5

S^*	scaled hydrogenated peripheral groups in tar as a molar concentration in the gas phase
T	scaled molar concentration of tar molecules
t	time, s
t_j	tar fragment with j linked aromatic nuclei
V^{∞}	ultimate yield parameter, daf wt%
X_{HG}	extent of char hydrogasification, %
Y_i	scaled molar yield of product i
$Y_{OILS}(t)$	instantaneous yield of oils, daf wt%
Y_{OILS}^{∞}	hypothetical ultimate oils yield, daf wt%
${}^1Y_T(t)$	instantaneous yield of primary tar, daf wt%
${}^1Y_T^{\infty}$	hypothetical ultimate yield of primary tar, daf wt%
ΔY_T	difference between the instantaneous yields of primary and secondary tar, daf wt%
ΔY_T^{∞}	hypothetical ultimate difference between the yields of primary and secondary tar, daf wt%

Greek symbols

η	average moles of nitrogen per aromatic nucleus
ν_B	scission selectivity coefficient for labile bridge conversion
ν_{G^*}	twice the stoichiometric coefficient for gas production during elimination of hydrogenated peripheral groups
ν_{HY}	stoichiometric requirement for H_2 in labile bridge hydrogenation
$\nu_{K,MH}$	stoichiometric coefficient for species $K = O$, H_2O, H_2S, NH_3 from monomer hydrogenation
ν_{MH}	stoichiometric requirement for H_2 in monomer hydrogenation
ν_O	stoichiometric coefficient for oil production from monomer decomposition
${}^S\nu_{OG^*}$	stoichiometric coefficient for gas production from monomer decomposition for S=1 or 2 hydrogenated peripheral groups on the tar monomer
ν_{Rj}	stoichiometric coefficient for soot for nucleation and addition of tj into soot
$\nu_{R,O}$	stoichiometric coefficient for soot in addition of oils to soot
ξ_{th}	moles of thiophene sulfur per aromatic nucleus
σ_k	std. dev. about the mean activation energy for reaction k

Subscripts

B	labile bridges
B*	hydrogenated bridges
C	char links
FC	pertains to quantities from FLASHCHAIN® for primary hydropyrolysis
HC	hydrocracking of labile and hydrogenated bridges and char links
HY	hydrogenation of labile bridges and peripheral groups
MH	monomer hydrogenation
N	aromatic nuclei
R	soot

S	labile peripheral groups
S*	hydrogenated peripheral groups
SN	soot nucleation
SC	soot addition
TD	monomer decomposition
tj	tar j-mer

The terms "hydropyrolysis" and "hydrogasification," respectively, denote primary devolatilization and char conversion under elevated H_2 pressures. Under the most favorable operating conditions, hydropyrolysis substantially increases primary tar yields and ultimate volatiles yields, especially with low rank and hv bituminous coals. However, for the rapid heating rates in most entrained-coal technologies, these enhancements do not materialize. Even so, hydropyrolysis is an effective means to hydroconvert primary tars into chemical feedstocks on the relatively short time scales for chemistry in the gas phase. On much longer time scales, hydrogasification converts char into CH_4, albeit at rates that are much slower than char gasification in steam and CO_2. Overall, hydropyrolysis and hydrogasification can convert most of a coal into BTX and CH_4 without catalysts at moderate temperatures, although H_2 pressures as high as 15 MPa are required.

The chemistry driven by elevated H_2 pressures can substantially alter each of the four main stages of coal conversion: (1) primary devolatilization is directly affected by hydrogenation of bridges and peripheral groups in the condensed coal phase, which enhances the pool of tar precursors; (2) tar hydrogenation dramatically skews tar decomposition toward the production of oils and GHCs at the expense of PAH and soot; (3) volatiles reforming generates additional GHCs due to the predominance of H-atoms in the radical pool, and is kinetically accelerated at elevated pressures; and (4) char hydrogasification further enhances the levels of GHCs on the longest time scales of all. Since the impacts on each stage can be substantial, hydropyrolysis is treated as a separate topic to emphasize the expansions of the underlying reaction mechanisms covered in previous chapters. Readers should already be familiar with the material in Chapters 4–8 before proceeding further.

This chapter presents reaction mechanisms that accurately describe how H_2 affects each of the four conversion stages, along with sufficient test results to cover variations in the main operating conditions and the coal quality impacts. The presentation begins with the commercial implications and the prerequisites for detailed characterizations in laboratory tests. Then measured total and tar yields are surveyed across a broad domain of operating conditions, without regard for the distributions of non-condensable gas products. This focus on only the coarsest aspects of coal conversion isolates hydropyrolysis and hydrogasification from tar hydrogenation and volatiles reforming, since the latter two stages do not affect coal conversion. Then distributions of all major hydropyrolysis products are presented, first, to guide and validate a tar hydrogenation mechanism and, ultimately, to assess the impact of volatiles reforming chemistry in conjunction with the mechanisms for all other stages. After the detailed mechanisms have been developed, accurate quasi-global reaction schemes are developed for deployment in CFD simulations.

9.1 Commercial impacts

Hydropyrolysis is a process rooted in petroleum hydrotreatment. Coal liquefaction most closely mimics hydrotreatment, because it is conducted in a slurry that contains catalyst and a liquid solvent donor. As a simpler processing option with a similar basis, hydropyrolysis exposes coal to very high H_2 pressures to hydrogenate and hydrocrack primary tars into chemical feedstocks on the relatively short time scales for chemistry in the gas phase. This hydroconversion may also extend into the condensed coal phase throughout devolatilization, as evidenced by much greater tar yields than for primary devolatilization alone under some operating conditions.

Throughout the 1980s, coal hydropyrolysis technologies were aggressively developed to produce chemical feedstocks from coal as a technological response to the OPEC oil shocks. Most schemes used fluidized beds in series to manage H_2 pressures as high as 10 MPa at temperatures of 700–800°C. Alternative schemes dubbed "flash hydropyrolysis" processed fine coal grinds in entrained flow to impose very fast heating rates, including conventional entrained flow gasifier designs as well as re-configured rocket nozzles. But the mechanistic analysis in this chapter exposes "flash hydropyrolysis" as the oxymoron that it is, because coal components cannot be hydrogenated on the short time scales imposed by very fast heating rates.

The interest in hydropyrolysis technology waned long ago, and the prospects for any re-emergence are bleak as long as world hydrocarbon liquid supplies meet the demands of transportation sectors. Notwithstanding, hydropyrolysis and hydrogasification chemistry are directly relevant to coal gasification at moderate temperatures, particularly when GHCs in the synthesis gas are generated to boost calorific values, and when coal is co-fired with natural gas. Such processes often sustain H_2 pressures approaching 1 MPa, which is high enough, in principle, to enhance primary tar yields, hydroconvert tar into oils and GHCs, and convert char into CH_4 via hydrogasification.

Even though nominal gasification rates are greater for steam and CO_2 than for H_2, the huge domain of syngas compositions among different gasification technologies includes situations where char hydrogasification consumes appreciable portions of char and supplements syngas heating values through CH_4 production. Unfortunately, hydropyrolysis is extremely difficult to monitor in any large-scale coal utilization system because extremely high system pressures complicate all forms of sampling; flow patterns are too complex for sampling probes anyway; and optical diagnostics are unavailable.

9.2 Stages of hydropyrolysis and hydrogasification

Hydropyrolysis is the first stage in the chemistry of coal processing under elevated H_2 pressures. We know that some coal components are hydrogenated by H_2 during primary devolatilization because the yields of both tar and gas are enhanced under elevated H_2 pressures, and the enhancements are much greater under slow heating conditions than during rapid coal devolatilization (Eklund and Wanzl, 1981).

Tar enhancements become greater for slower heating rates, which is the inverse of the heating rate dependence for primary devolatilization under inert atmospheres. This inverted heating rate dependence could conceivably connect to the extensive rearrangements associated with the catalytic hydrotreatment of petroleum resids, which converts condensed aromatics into naphthene rings that decompose into GHCs, especially longer aliphatics and olefins. Only very slow heating provides sufficient reaction time for such complex chemistry. For rapid heating conditions, two much less extensive roles for H_2 are more plausible: (1) direct hydrogenation of labile bridges and peripheral groups in coal without any transformations of aromatic nuclei; and (2) suppression of recombinations among the ends of mobile fragments in the condensed phase that would otherwise introduce refractory char links into the nascent char phase, and thereby diminish the pool of tar precursors. These reactions utilize H_2 within the condensed coal phase, either as a dissolved species in the coal melt during the plastic stage of bituminous coal decomposition or as adsorbed species on the pore surfaces of solid reacting particles of coals of low or very high rank. In either case, the level of H_2 within the condensed phase is directly related to the H_2 pressure within the particle, which is related to the ambient H_2 pressure through the transport of H_2 against the outward flow of volatiles.

Once in the condensed phase, dissolved H_2 may react with the coal components in various ways, including rupture of hydroaromatics, saturation of olefins into aliphatics, and conversion of carboxylic acids into alcohols. The conversion of carboxylic acids is especially important, because carboxylic acids are crosslinking agents that skew bridge conversions toward the spontaneous generation of refractory char links away from scission of longer fragments into smaller tar precursors. On a phenomenological level, bridge hydrogenation shifts bridge decomposition chemistry away from spontaneous condensation into char links toward scission.

Hydropyrolysis introduces two new phenomena into the mechanisms for primary devolatilization: (1) a gaseous reactant in primary devolatilization chemistry and (2) a role for mass transport to deliver this reactant into the condensed phase during primary devolatilization, albeit not necessarily a rate-limiting one.

Tar components are hydrogenated like components in the condensed coal phase and also hydrocracked, but without any transport resistances or phase partitioning of H_2 from a surrounding atmosphere. The hydrogenations shift bridge conversion toward scission; incorporate H_2 from the ambient gas into bridges and peripheral groups; suppress recombination of tar fragments; and convert tar monomers into oils. Oils are mixtures of BTX and PCX. Hydrocracking breaks labile and hydrogenated bridges and char links among tar nuclei, which shift the tar MWD toward lighter fragments and promote oils production. All these effects stimulate the decomposition of the lightest tar molecules into oils and GHCs, and undermine the progression toward PAH and soot. Tar hydroconversion skews the distribution of major products toward oils, GHCs, and moisture at the expense of PAH, soot, and the oxides of carbon.

Volatiles reforming under elevated H_2 pressures generates a predominance of H-atoms in the oxyhydroxyl radical pool that shifts the GHC distribution toward CH_4 and C_2H_6, and reforms HCN into NH_3. Water-gas shifting, the Boudouard reaction (of CO into CO_2), and the hydrogenation of carbon into CH_4 act in concert to enhance the production of steam and CH_4 at the expense of both oxides of carbon.

Once hydropyrolysis, tar hydroconversion, and volatiles reforming have finished, H_2 directly converts the carbon in char into CH_4. Hydrogasification is exothermic, and the gasification exotherm perturbs the char temperature throughout conversion, albeit weakly. But many of the complexities in the gasification mechanisms for steam and CO_2 are absent because hydrogasification is relatively very slow, particularly at the moderate temperatures of commercial interest. Both film diffusion and intraparticle transport of H_2 through the internal pore system are relatively fast for all but the largest particle sizes; i.e., the effectiveness factor approaches unity and, consequently, neither transport rate ever becomes rate limiting. Any description of how particle size and density vary through a hydrogasification history usually relaxes to a size-independent limit due to the slow hydrogasification kinetics; i.e., the particles are converted at uniform size and variable density, consistent with complete penetration of the internal pore system and full accessibility to the internal surface area. Similarly, ash encapsulation is also inconsequential because the hydrogasification kinetics are slow.

There are two potential obstacles to a reaction-controlled analysis of hydrogasification. First, the fact that hydrogasification is much slower than gasification by steam and CO_2 has discouraged lab characterization of hydrogasification, so the available kinetic information is confined to decades-old tests that emphasized conversion over the dynamics. The second complication pertains to thermal annealing. As yet, there is no direct evidence that thermal annealing affects char hydrogasification kinetics, which simply reflects the absence of any tests on this aspect. Later in this chapter evidence is presented to assert that the hydrogasification surface chemistry is probably not coupled with the surface chemistry for gasification by steam and CO_2, and that there are probably no common adsorbed intermediates or primary reaction sites. So it is unlikely that the same thermal annealing kinetics for conventional gasification pertain to hydrogasification. But it remains to be determined if and when annealing affects the hydrogasification reactivity. In addition, the char hydrogasification rate varies by much less than an order of magnitude for coals across the rank spectrum (Niksa, 2018a), which is a much smaller range than the one for rates of char gasification in steam or CO_2.

Under elevated H_2 pressures, pyrite decomposes to release more of its sulfur than under inert atmospheres, ultimately forming elemental Fe plus the theoretical maximum amount of H_2S. Most H_2S forms during hydropyrolysis, although residual amounts are also released throughout hydrogasification.

9.3 Laboratory prerequisites

Extents of hydropyrolysis and hydrogasification cannot be resolved as distinct contributions in any large scale coal utilization system. One reason is elevated H_2 pressures require the tightest conceivable pressure vessels because leaks can form invisible torches and carry enormous blast potential. Another is that the contributions of hydropyrolysis to primary devolatilization and of hydrogasification to char conversion are almost always relatively minor. In fluidized systems, solids loadings are much too heavy to permit any access at all, and mixing patterns are too convoluted for extractive sampling along a time coordinate. So this chemistry has only been characterized at lab scale.

Laboratory characterizations are also complicated by the elevated H_2 pressures, particularly in the continuous feeding of a coal stream into a pressurized flow field. Buoyancy can distort reactant transit times in downflow systems because it strengthens for progressively higher pressures. Since the temperatures of commercial interest are moderate, neither materials selection nor time resolution are special challenges. Distributions of the products of primary hydropyrolysis are just as complex as primary devolatilization products. But subsequent stages of conversion radically simplify these distributions by progressively eliminating tars, PAH, and soot. Ultimately, only the lightest GHCs, oils, and moisture comprise the major volatile products.

The greatest challenge is to manage the different reaction time scales to focus a test series on a particular stage of conversion. Indeed, the literature is full of datasets devoted to hydropyrolysis which, upon closer examination, are found to be completely unaffected by any hydrogenation chemistry within the condensed coal phase. Two characteristics of primary devolatilization under inert gases critically constrain the operating domain for hydropyrolysis. First, primary devolatilization is spontaneous, so there are no means to retard its progress. Second, primary devolatilization rates increase in rough proportion to the heating rate and, consequently, devolatilization time scales become shorter for progressively faster heating rates. The crucial implication is that the hydrogenation of coal components underlying hydropyrolysis can only occur before those same coal components are spontaneously converted by the reaction channels of primary devolatilization. Consequently, testing with slow heating rates provides the only means to ensure that coal components are hydrogenated before they spontaneously decompose. Rapid devolatilization at heating rates of 10^3°C/s and faster does not provide enough time for appreciable bridge hydrogenation so tar yields are not appreciably enhanced under the H_2 pressures associated with entrained-flow coal gasification technology. Conversely, tar yields are enhanced to greater extents under progressively slower heating rates and for progressively higher H_2 pressures.

Another challenge is to resolve primary hydropyrolysis products from those affected by tar hydrogenation, because elevated H_2 pressures accelerate the tar hydroconversion rates. All test records should include both solids contact times and gas transit times from the sample to the quench point. The best information in current literature on primary products is based on an indirect approach that monitors the major products at the same pressure under H_2 and an inert gas. The products monitored under inert atmospheres are primary devolatilization products if the temperatures and contact times for gases in the tests were managed to minimize extents of tar decomposition. These distributions are similar to primary hydropyrolysis products, although hydropyrolysis products may contain more tar and additional GHCs. Both primary tars and noncondensable gases from hydropyrolysis incorporate some of the ambient H_2, and these additions can be substantial portions of the coal-H levels. In principle, tests under H_2 can nearly eliminate tar hydroconversion by keeping the ambient gas cool and by minimizing hot zone contact times. However, in practice it is nearly impossible to establish that none of the primary tars have been hydrogenated, even under 0.1 MPa H_2 and temperatures as cool as 600°C, because oils and GHCs are both the main tar hydroconversion products and significant primary products as well. Tight closures on C– and O-balances are the most stringent means to evaluate reported hydropyrolysis distributions, and should be a primary goal for all testing campaigns. But even these are insufficient to definitively establish that tar hydroconversion was negligible.

Similarly, volatiles reforming can only be managed by minimizing the temperatures and contact times for gases in the tests to minimize its extent. Even when C– and O-balances are complete, it is impossible to establish that reported product distributions were not affected by volatiles reforming.

The final conceptual challenge is to clearly resolve the ultimate yield for hydropyrolysis from the contributions due to hydrogasification. Surely, the nominal time scale for hydropyrolysis is much shorter than that for hydrogasification. But the nominal time scale covers tar production, rather than the later stages of hydropyrolysis, when CO, CH_4, HCN, H_2S, and H_2 are released at much slower rates. The release of these primary products does seem to overlap with the initial stages of hydrogasification in most datasets, particularly those taken at 900–1000°C and hotter. Instead of an asymptotic approach to a plateau in the weight loss across a broad temperature range, one sees instead a shift toward a smaller slope in the weight loss curve due to the contributions from hydrogasification. Under such circumstances, an ultimate hydropyrolysis yield is not that clear. However, it can be definitively resolved by recording when nearly all the additional weight loss equals the increase in the CH_4 yield. Also, complementary tests that impose the same thermal history under an inert atmosphere resolve the primary devolatilization rate, which is a good approximation to the hydropyrolysis rate because hydrogenation does not affect the nominal time scale for tar production. For heating rates of 10^3°C/s or faster, hydrogenation will be negligible, so the ultimate yields for primary devolatilization and hydropyrolysis are essentially the same, and extents of char hydrogasification are apparent in the weight loss recorded under H_2 (cf. Fig. 9.4).

There are several prerequisites for tests that provide data that are suitable for model validation work on hydropyrolysis and hydrogasification. Tests on hydropyrolysis should impose heating rates from 1 to 100°C/s. Tests with heating rates of 10^3°C/s and faster should only be used to clearly resolve extents of char hydrogasification because rapid heating eliminates the hydrogenation of coal components during devolatilization. The range of test temperatures should include tests hot enough to achieve ultimate primary volatiles yields under both H_2 and an inert gas at every test pressure. Either tests across a broad temperature range or multiple runs for several time increments at a uniform heating rate and temperature are needed to reveal the reaction kinetics. All tests should be at a uniform pressure although total and H_2 pressures need not be the same. Whether coals are entrained in suspension or tested in batch mode, as in fluidized beds and WMRs, the samples must be classified into narrow size fractions of no more than two standard sieve sizes. Datasets that cover a suite of coal samples and uniform test conditions are the most valuable.

More formally, the following features are required of a dataset to be used to evaluate mechanisms for hydropyrolysis, tar hydroconversion, and hydrogasification:

(1) *Coal properties*—Proximate and ultimate analyses are absolutely essential for every coal sample. Additional information from specialized analytical testing is not strictly required but may be informative. A mean particle size or PSD is essential.
(2) *Pressure*—Usually a uniform test pressure will be specified although a pressure history can also be analyzed. The H_2 pressure must be known but does not have to equal total pressure.
(3) *Thermal history*—Sufficient information must be available to assign the temperature of the sample as a function of time throughout an entire test. This requirement may entail direct monitoring of sample temperatures, entrainment gas velocity fields, and/or particle transit times or CFD simulations. In addition, an independent thermal history must also be assigned for the gas phase either by direct measurements of temperature and contact time or with CFD simulations.
(4) *Primary devolatilization behavior*—For the many reasons discussed above, researchers are strongly encouraged to conduct parallel tests series under both H_2 and an inert gas. Primary devolatilization behavior as a reference condition will quantitatively clarify the impacts of hydropyrolysis, tar hydroconversion, and volatiles reforming, and enable unambiguous assignments for extents of hydrogasification.
(5) *Impact of secondary chemistry*—Whenever volatiles are released into a flow that is hotter than the parent coal particle, volatiles will be transformed by secondary chemistry, and elevated H_2 pressures accelerate tar hydroconversion into oils and GHCs. The extent of this transformation should be monitored as the decay in tar yields, preferably from a complete primary tar yield. Tar should be clearly distinguished from oils in the resolution of liquid products. Transit times and temperatures of gaseous products from their point of release to a quench location must be specified. At moderate temperatures, tens of seconds of gas transit time may be required to monitor an ultimate product distribution. Product distributions should be monitored under both H_2 and inert gas for the same thermal history to clarify the sources of noncondensable gases.
(6) *Relevant aspects of hydropyrolysis behavior*—Total weight loss and a tar yield should always be monitored, whereas the best datasets monitor all major products and their elemental compositions so that balances on C/H/O/N/S can be closed in individual tests, although such resolution is a formidable challenge. H-balances are difficult to close because tar, gases, and char incorporate ambient H_2 at significant portions of coal-H levels. But H-contents in oils, tar, and gases should always be monitored to track extents of hydrogenation. The main hydropyrolysis products are char, tar, oils, GHCs, CO, CO_2, H_2O, H_2, NH_3, and H_2S. Oils, CH_4, C_2H_6, NH_3, and H_2S are the ultimate products of tar hydroconversion and should always be monitored. Elemental compositions of tar and char are especially valuable. Changes in char size and bulk density are valuable. Tar MWDs were highly instrumental in advancing comprehensive reaction mechanisms but are almost never monitored any longer.
(7) *Relevant aspects of hydrogasification behavior*—Incremental mass loss must be monitored throughout char hydrogasification. Simultaneous CH_4 yields are especially valuable. Changes to particle density and size are valuable, although other aspects of char morphology are usually not very informative.

WMRs are the laboratory workhorses on coal conversions for hydropyrolysis and hydrogasification because they are easily contained within a cool vessel to test at elevated H_2 pressures, and the entire system can be operated remotely behind blast mats or other protection. CPPs share these advantages. Numerous advantages for both systems were reviewed in Chapter 4, including sweep gases that move primary products away from the hot sample support for recovery before they can be hydrogenated. Buoyancy circulation at elevated H_2 pressures has undermined electronic control strategies to impose prescribed thermal histories at the highest pressures of interest (Gibbins et al., 1991), so flow control devices are usually necessary. Also, WMRs do not regulate the conditions for controlled tar hydrogenation and volatiles reforming, unless they are connected to a TFR in a second stage.

Pioneering work on hydropyrolysis at the US Bureau of Mines utilized PBRs that are still used today in process development work. Large beds packed with tens of grams of coal which are slowly heated in tubular furnaces are not pertinent to our operating domain and our focus on the process kinetics. But versions of the hot rod reactor (HRR), which process less than a gram of sample and are heated resistively at rates as fast as 50°C/s, can be operated to meet our objectives. Gibbins et al. (1991) demonstrated that total conversions from HRRs are lower than WMR yields under some, but not all, directly comparable conditions, primarily due to heterogeneous deposition of tar onto bed particles and to hindered penetration of H_2 throughout the entire coal bed. The latter factor is pronounced with caking coals. Both concerns have been satisfactorily resolved by operating with gas velocities that explicitly demonstrate no impact of mass transfer limitations on yields, and by processing mixtures of coal with sand or quartz to maintain a permeable bed (Bolton et al., 1987). Gas flowrates must be increased at progressively higher pressures to impose the same sweep gas velocity in all tests. Since PBRs cannot be quenched to resolve conversions in time and coal contact times extend to tens of minutes, they inevitably prepare product distributions that are affected by primary hydropyrolysis, tar hydroconversion, volatiles reforming, and char hydrogasification. Accordingly, they are well-suited to identify the operating conditions that maximize a particular product lump, such as oils or CH_4.

Being continuous throughput systems, EFRs provide the most relevant data for flash hydropyrolysis technologies but they are also the most difficult to operate at elevated H_2 pressures. Many EFR designs have been based on the pioneering system developed at Brookhaven National Laboratory by Steinberg and co-workers (Fallon et al., 1980), in which pressurized H_2 is preheated and injected into a coal jet at the inlet to a heated tubular reactor, and the process stream is quenched with a water spray at the end of the isothermal section. To shorten the reactor length, newer EFRs are operated as free-fall reactors (FFRs), into which coal is fed under gravity into a tubular furnace section. By adjusting the H_2 flowrate and/or the length of the isothermal furnace section, these reactors impose coal contact times up to 2 s, and gas transit times beyond 40 s. EFRs and FFRs delivered the best time resolution to date on tar hydroconversion, which identified the optimal operating conditions to maximize either oils or CH_4 yields.

The datasets in this chapter will demonstrate that many of the most important characteristics of primary hydropyrolysis are rooted in tar production. See Chapter 4 for guidance on recovering and handling this important product lump. It is the responsibility of any experimentalist to demonstrate that yields reported as tar yields actually comprise the complete distribution of all organic products that condense at room temperature, excluding water. In most hydropyrolysis and gasification test systems, char is collected at 200–400°C to avoid tar condensation on the particles, and tar is recovered in two collectors in series, one cooled with ice or dry ice and the other cooled by liquid N_2. Sometimes "tar" and "oils" denote the products from the respective collectors, but often the usages are much more ambiguous. In this book, "oils" exclusively denotes mixtures of BTX and PCX, whereas "tar" denotes all forms of heavier condensable organics and PAH beginning with naphthalene.

9.4 Coal conversion during hydropyrolysis and hydrogasification

This section presents the most important attributes of coal conversion during hydropyrolysis and hydrogasification in laboratory datasets. It covers the yields of char, tar, and gas in aggregate, and illustrates the impacts of all the major operating conditions. But detailed distributions of the major product species are relegated to Section 9.5, because product distributions are radically altered by tar hydroconversion chemistry, which will be analyzed with a separate reaction mechanism. Curves in the figures are simulation results from the mechanism developed in Section 9.4.2.

9.4.1 Laboratory database on coal conversion during hydropyrolysis and hydrogasification

It is uncommon for a single dataset to capture the most distinctive features of a complex reaction process. But the data in Fig. 9.1 are like a Rosetta Stone for hydropyrolysis and hydrogasification, because they represent negligible to very substantial yield enhancements from hydropyrolysis against a backdrop of negligible to substantial extents of hydrogasification. The various test series swept through four heating rates to 700°C with 10 s IRP under five H_2 pressures with a hv bituminous coal (Guell and Kandiyoti, 1993). The total test time diminished from 685 to 77.5 to 16.8 to 10.7 s while heating rates were varied from 1 to 10^3°C/s. The weight loss grows for progressively faster heating rates at the lowest pressure; is nearly neutral at the next two intermediate pressures; and then diminishes for progressively faster heating rates at both the highest pressures. Tar yields are enhanced by faster heating at both lower pressures; insensitive to heating rate at 1 and 2 MPa; and diminish for progressively faster heating at the highest pressure. So for progressively greater pressures, the variations in tar yields due to faster heating change from yield enhancements to negligible changes to yield suppression. The available data (in Section 4.2.3) shows a shift from tar yield enhancements due to faster heating at atmospheric pressure to negligible changes at elevated pressures for primary devolatilization under inert atmospheres (due to hindered flash distillation of tar precursors). Elevated H_2 pressures invert the heating rate dependence and this dataset, more than any other, clearly demonstrates that.

The inverted heating rate dependence at the highest H_2 pressures indicates that the hydrogenation of coal components can be more forceful than the suppression of tar release by hindered flash distillation, and thereby enhance the tar yields. Most

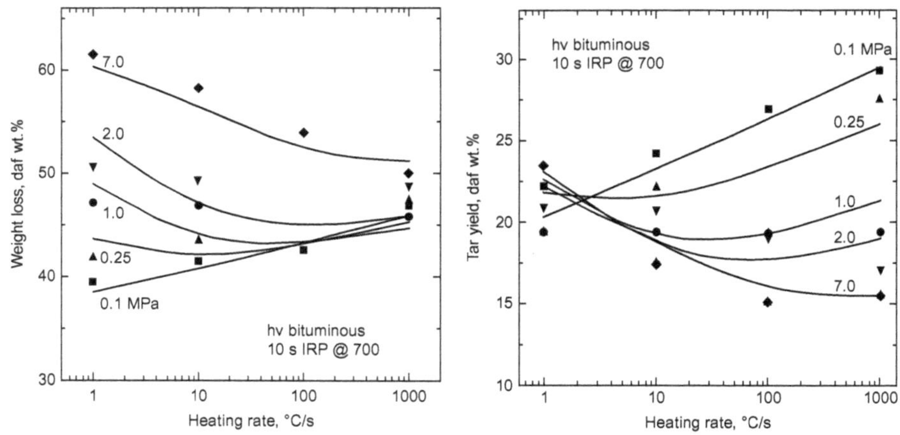

Fig. 9.1 (Left) Weight loss and (right) tar yields for various heating rates to 700°C with 10 s IRP under (■) 0.1 MPa He and H_2 pressures of (▲) 0.25, (●) 1, (▼) 2, and (◆) 7 MPa. Reproduced from Niksa S. Flashchain theory for rapid coal devolatilization kinetics. 10. Extents of conversion for hydropyrolysis and hydrogasification of any coal. Energy Fuels 2018a;32:384–95, with permission from the American Chemical Society.

importantly, hydrogenation wins this competition only at the slowest heating rates, because hydrogenation is slow chemistry. Only tests with the slowest heating rates provide enough time for hydrogenation to appreciably enhance the tar yields before the coal components are converted by spontaneous devolatilization channels. At the fastest heating rates, there is too little time for hydrogenation to enhance tars yields before the spontaneous channels take over, so tar yields diminish monotonically for progressively greater pressures, as they would under inert atmospheres.

The data in Fig. 9.2 illustrate the joint impact of variations in temperature and H_2 pressure for extended IRPs (Heyd, 1982). The left panel covers extended IRPs at three temperatures at 3.4 MPa H_2, and the right panel covers extended IRPs for three pressures at 750°C. The heating rate was 10^3°C/s in all tests. The most striking feature is that weight loss continues to increase at all temperatures and pressures, even though IRPs were extended to almost 200 s in half the series. The only exception is the apparent asymptotic weight loss at 900°C after 40 s IRP, although it is implausible that char hydrogasification would stop at some intermediate extent of conversion. All other test series gave weight loss that increased for progressively longer IRPs, as expected. This same laboratory reported datasets showing that nominal devolatilization rates are insensitive to pressure and, most important, that primary devolatilization is finished in only 1 s IRP at 750°C (cf. Fig. 4.10). Consequently, all the weight loss after 1 s IRP at 750 and 900°C in Fig. 9.2 should be attributed to char hydrogasification. At 530°C, devolatilization is not finished in the available reaction times; in fact, the quantitative interpretation (in Section 9.4.2) attributes all the conversion through 90 s IRP to devolatilization alone, suggesting an onset temperature of roughly 600°C for hydrogasification. These datasets also show that the time scale for hydrogasification is at least

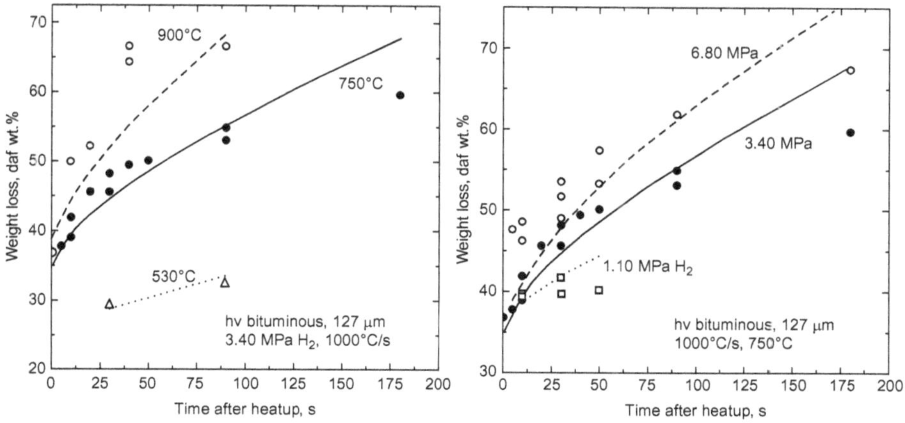

Fig. 9.2 Weight loss from hv bituminous coal after heating at 10^3°C/s (left) to (○ and *dotted curve*) 530, (● and *solid curve*) 750, and (△ and *dashed curve*) 900°C under 3.40 MPa H_2; and (right) to 750°C under (□ and *dotted curve*) 1.10, (● and *solid curve*) 3.40, and (○ and *dashed curve*) 6.80 MPa H_2.
Reproduced from Niksa S. Flashchain theory for rapid coal devolatilization kinetics. 10. Extents of conversion for hydropyrolysis and hydrogasification of any coal. Energy Fuels 2018a;32:384–95, with permission from the American Chemical Society.

two orders of magnitude longer than the time for primary devolatilization. Hydropyrolysis and hydrogasification must be resolved as distinct conversion mechanisms because they act on grossly different time scales.

The case in Fig. 9.3 also resolves independent contributions from hydropyrolysis and hydrogasification because the tar yields can only be affected by hydropyrolysis, whereas weight loss includes greater contributions from hydrogasification for progressively higher H_2 pressures. Due to the rapid heating rate in these tests (Guell and Kandiyoti, 1993), the tar yields exhibit the continuous decay toward a saturation limit for progressively greater pressures seen for primary devolatilization under inert atmospheres. The predicted enhancements due to hydrogenation are <1 daf wt% at the highest H_2 pressure. In contrast, the enhancements to weight loss due to hydrogasification are roughly as large as the absolute tar yields at the highest H_2 pressures. Weight loss passes through a minimum near 1 MPa, then increases in direct proportion to the H_2 pressure. The minimum reflects the acute sensitivity to pressure in the tar yields for the lowest test pressures and the positive reaction order for H_2 pressure in the hydrogasification kinetics. Higher H_2 pressures give greater extents of char hydrogasification, which counteracts the lower weight loss associated with lower tar yields for progressively greater pressures.

The temperature dependence in the char hydrogasification kinetics is apparent in Fig. 9.4, which compares the weight loss under H_2 and He at 6.9 MPa for the same thermal histories (Anthony et al., 1975). Extents of char hydrogasification are simply the difference between the two curves defined by the data points, because the heating rate was fast enough that tar and total volatiles yields were hardly perturbed under H_2.

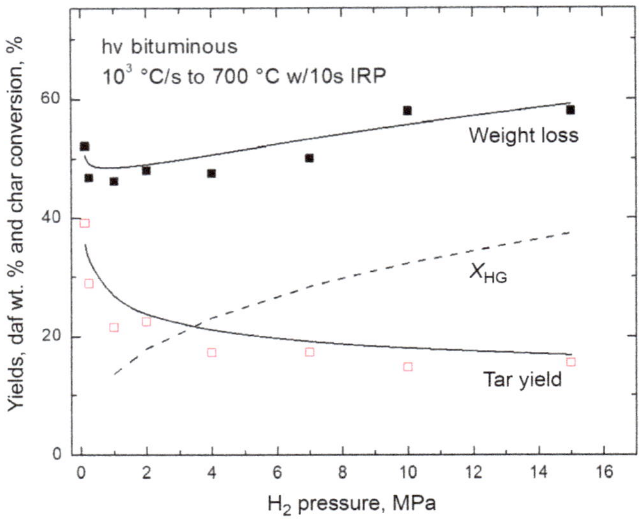

Fig. 9.3 Weight loss and tar yields from hv bituminous coal for heating at 10^3°C/s to 700°C with 10 s IRP vs. H_2 pressure. The predicted extents of char hydrogasification appear as the dashed curve.
Reproduced from Niksa S. Flashchain theory for rapid coal devolatilization kinetics. 10. Extents of conversion for hydropyrolysis and hydrogasification of any coal. Energy Fuels 2018a;32:384–95, with permission from the American Chemical Society.

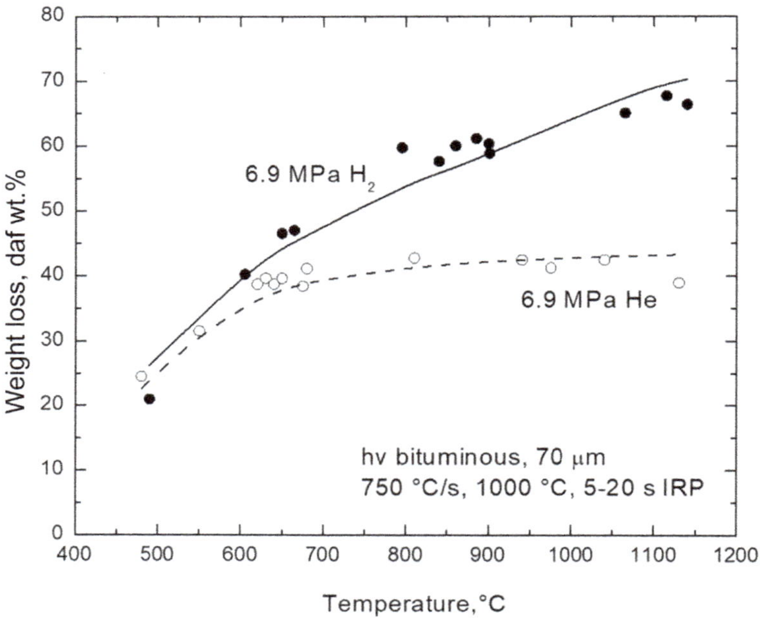

Fig. 9.4 Weight loss from hv bituminous coal for heating at 750°C/s to various temperatures with 5–20 s IRP under (● and *solid curve*) 6.9 MPa H_2 and (○ and *dashed curve*) 6.9 MPa He (Anthony et al., 1975).

They increased for progressively hotter temperatures above 600°C, suggesting that hydrogasification begins around 600°C for contact times of several seconds. This threshold value is compatible with the negligible extents of hydrogasification predicted for the 530°C case in Fig. 9.2. Extents of hydrogasification continuously increase through the hottest temperatures in the test series, albeit at a diminishing rate. Without the measured ultimate weight loss for primary devolatilization under He for comparison, it would be impossible to accurately estimate the ultimate hydropyrolysis yield from only data under an elevated H_2 pressure.

The temperature dependences in the weight loss from a brown coal (LY) and an hv bituminous (MN) are compared in Fig. 9.5 for rapid heating with a 2 s IRP under 7 MPa H_2 (Strugnell and Patrick, 1996). For both coals, the weight loss increases sharply to 650°C, then increases in smaller increments for progressively hotter temperatures. Neither coal exhibited an approach to asymptotic ultimate yields over this temperature range. In fact, the predicted extents of hydrogasfication are reasonably close for both coals and, consequently, nearly all the greater weight loss from the brown coal reflects its excessive volatiles yield, rather than a faster hydrogasification rate. This is the first indication that the very strong coal quality impact on gasification with steam and CO_2 does not pertain to hydrogasification.

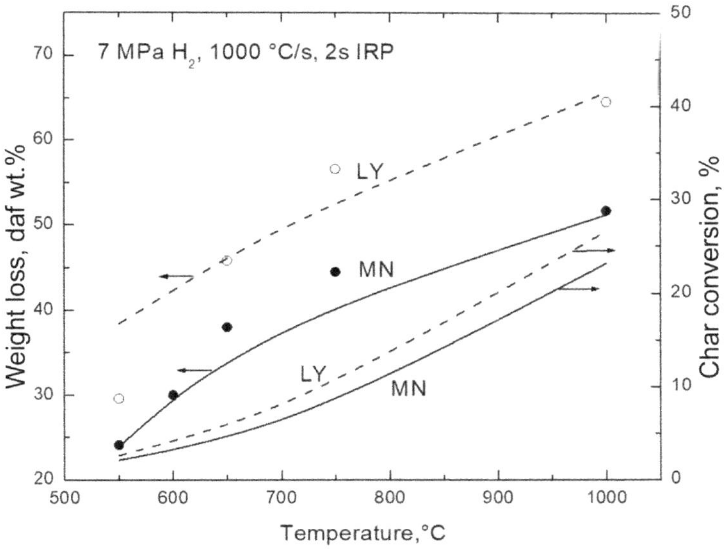

Fig. 9.5 Weight loss vs. temperature from a (LY) brown coal and (MN) hv bituminous for heating at 10^3°C/s to various temperatures with 5 s IRP under 7.0 MPa H_2. Predicted extents of char hydrogasification appear on the right ordinate.

Reproduced from Niksa S. Flashchain theory for rapid coal devolatilization kinetics. 10. Extents of conversion for hydropyrolysis and hydrogasification of any coal. Energy Fuels 2018a;32:384–95, with permission from the American Chemical Society.

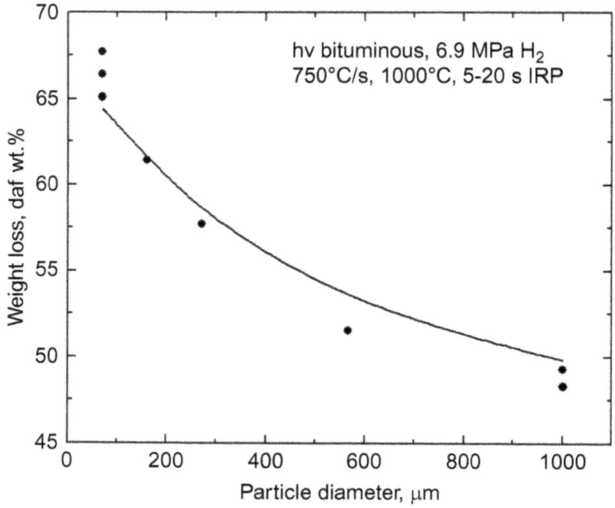

Fig. 9.6 Weight loss from different size cuts of hv bituminous under 6.90 MPa H_2 for heating at 750°C/s to 1000°C with IRPs from 5 to 20 s.
Reproduced from Niksa S. Flashchain theory for rapid coal devolatilization kinetics. 10. Extents of conversion for hydropyrolysis and hydrogasification of any coal. Energy Fuels 2018a;32:384–95, with permission from the American Chemical Society.

Primary devolatilization under inert gases is unaffected by variations in particle size provided the size is smaller than a few millimeters. In contrast, hydrogasification yields diminish for progressively larger sizes (Anthony et al., 1975), as seen in Fig. 9.6. Since all these yields are greater than the ultimate volatiles yields for these conditions (Seen in Fig. 9.4 for 6.9 MPa He), this dataset indicates that extents of char hydrogasification diminish for progressively larger particle sizes.

The impact of coal quality on hydrogasification kinetics is apparent in Fig. 9.7 as the difference between the yields for pyrolysis under He and hydrogasification under H_2 for the same pressure and thermal history (Strugnell and Patrick, 1995). The two results for both the extremes in coal rank show much faster hydrogasification of the lignite than the lv bituminous sample. But the results for the samples with the four intermediate rank indices in Fig. 9.7 have extents of hydrogasification from 50 to 65 daf wt% even though they represent ranks from subbituminous through hv bituminous. The bulk of these data suggest that hydrogasification does not become slower for coals of progressively higher rank, like the rates of gasification by steam and CO_2. Any monotonic rank dependence for char hydrogasification will be hard to discern when many more samples are considered in the quantitative interpretation of coal quality impacts in Section 9.4.2 (cf. Fig. 9.8).

In summary, hydropyrolysis enhances tar yields, but only if heating rates are much slower than 10^3°C/s and H_2 pressures are >1 MPa (cf. Fig. 9.1). The vast majority of the literature database exhibits minimal, if any, enhancements to tar yields, because

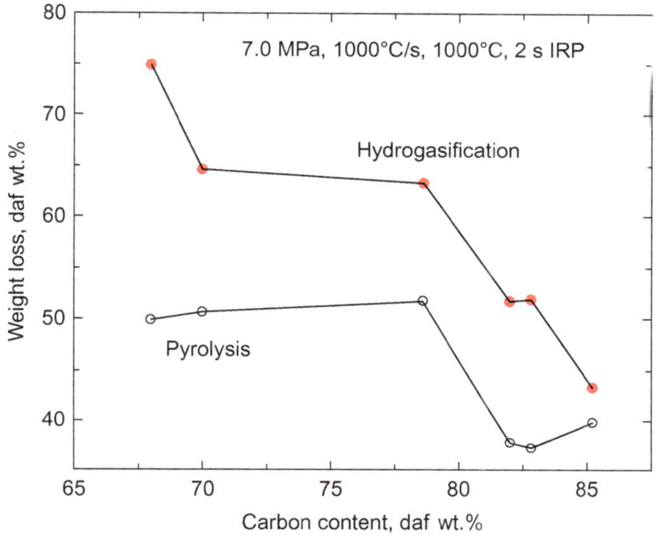

Fig. 9.7 Weight loss from various coals under 7.0 MPa of both (○) He and (●) H_2 after heating at 10^3°C/s to 1000°C with 2 s IRP.
Reproduced from Strugnell B, Patrick JW. Hydropyrolysis yields in relation to coal properties. Fuel 1995;74:481–86, with permission from Elsevier.

heating rates around 10^3°C/s were widely used to make the connection to flash hydropyrolysis technology development. In contrast, all the major tendencies in char hydrogasification kinetics across a broad domain of operating conditions are evident. Hydropyrolysis and hydrogasification must be resolved as distinct conversion mechanisms because the time scale for hydrogasification is approximately two orders of magnitude longer than times for hydropyrolysis. Extents of hydrogasification increase for progressively higher H_2 pressures and temperatures and longer contact times. They also diminish for progressively larger particle sizes.

Across a broad range of H_2 pressure, weight loss often passes through a minimum near 1 MPa then increases in direct proportion to the H_2 pressure. The minimum reflects the acute sensitivity to pressure in the tar yields for the lowest test pressures and the positive reaction order for H_2 pressure in the hydrogasification kinetics. Higher H_2 pressures give greater extents of char hydrogasification, which counteracts the lower tar yields for progressively greater pressures with rapid heating. Hydrogasification rates become appreciable around 600°C for contact times of several seconds, then extents of hydrogasification continuously increase for progressively hotter temperatures. Without data on the ultimate weight loss for primary devolatilization under an inert gas for comparison, it would be impossible to accurately estimate the ultimate hydropyrolysis yield from only data taken under an elevated H_2 pressure. Hydrogasification does not become slower for coals of progressively higher rank, like the rates of gasification by steam and CO_2.

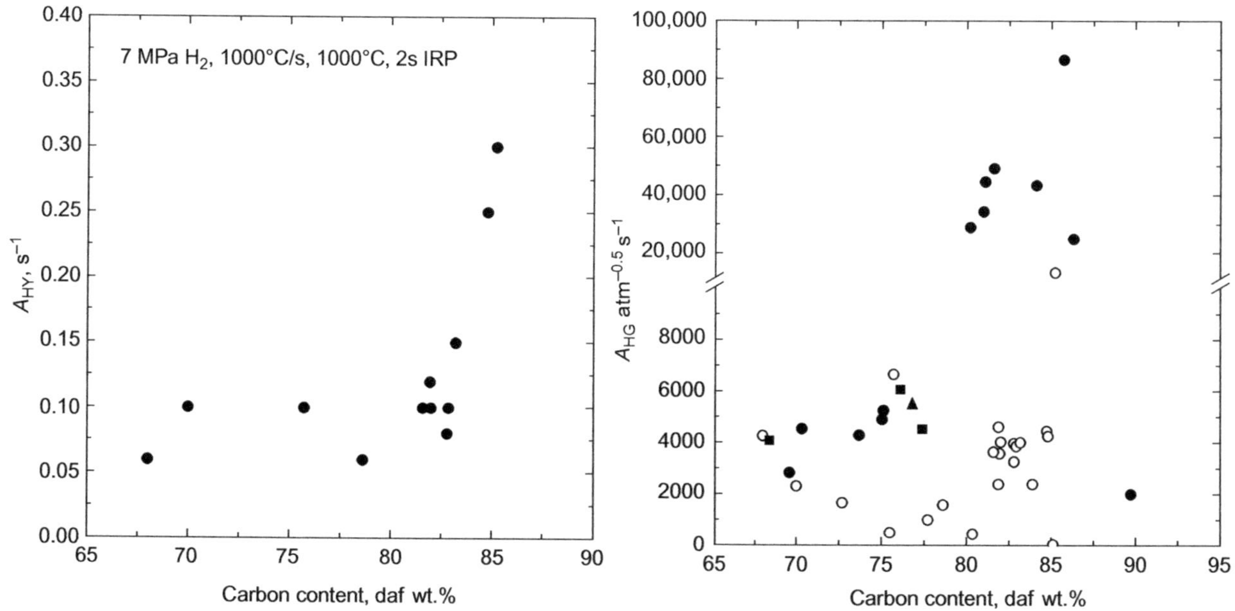

Fig. 9.8 Assigned pseudo-frequency factors for (left) bridge hydrogenation for 15 diverse coals heated at 10^3°C/s to 1000°C with 2 s IRP under 7 MPa H_2 and (right) char hydrogasification for 39 coals from Sections 9.4 and 9.5.
Reproduced from Niksa S. Flashchain theory for rapid coal devolatilization kinetics. 10. Extents of conversion for hydropyrolysis and hydrogasification of any coal. Energy Fuels 2018a;32:384–95, with permission from the American Chemical Society.

9.4.2 Reaction mechanisms to predict coal conversion during hydropyrolysis and hydrogasification with any coal

The earliest mechanisms proposed for hydropyrolysis were based on classical devolatilization theory (Howard, 1981; Suuberg, 1985), whose essential scheme is secondary redeposition of volatiles into residues which remain in the char on a time scale set by the transport mechanism for volatiles escape. Hydrogen disrupts redeposition by stabilizing the escaping volatiles, and also enhances conversion via the hydrogasification reaction. Conversely, the longer transit times through larger particles are purported to lower yields by promoting redeposition. Volatiles escape mechanisms such as either continuum or Knudsen diffusion, continuum diffusion of liquids through a melt, bulk flow through macropores, and bubble rupture and growth in a viscous melt have been analyzed. Although classical devolatilization theory is definitively contradicted by the absence of a particle size dependence in the yields for primary devolatilization under inert gases, total yields under elevated H_2 pressures do diminish for progressively larger sizes and, consequently, classical theory remains viable for hydropyrolysis. However, the mechanism developed in this section accurately interprets this size dependence without resorting to intraparticle redeposition of volatiles. Moreover, classical theory has no means to describe the behavior of different coals other than to re-specify the kinetic parameters.

Among the network depolymerization mechanisms, both FLASHCHAIN® (Niksa, 2011) and CPD (Guan et al., 2015) were expanded for hydropyrolysis applications. The base reaction mechanisms in FLASHCHAIN® were supplemented with finite-rate bridge hydrogenations that shift the selectivity of succeeding bridge conversions toward scission and away from spontaneous condensations into char links, which expands the population of tar precursors; and by suppressing bimolecular recombinations between fragments that have hydrogenated ends. This scheme is developed further throughout this section. This analysis accurately depicts the joint impacts of heating rate and H_2 partial pressure on the total and tar yields, and accurately interpreted a database of 32 coals representing the entire rank spectrum (Niksa, 2011; Niksa, 2018a). Similarly, Guan et al. (2015) substituted a bridge hydrogenation reaction for the scission reaction in CPD, and postulated that the selectivity ratio between hydrogenation and crosslinking was proportional to the ambient H_2 pressure. They also modified correlations for the distribution of noncondensable products to skew the predicted products toward CH_4, H_2O, and other light GHCs. The analysis accurately interpreted the aggregate yields of tar and gas and selected noncondensables from 15 coals subject to the same thermal history and H_2 pressure, as well as datasets covering broad ranges of temperature and H_2 pressures. However, this analysis does not include the char hydrogasification reaction and, consequently, the interpretations of the same datasets as the FLASHCHAIN® validations are fundamentally different. Problems associated with the omission of hydrogasification would become apparent in interpretations of the joint impacts of heating rate and H_2 pressure variations, which Guan et al. did not evaluate.

This section surveys the expansion of FLASHCHAIN® for hydropyrolysis and hydrogasification applications, and explains the major trends in the database on coal conversion. Both a formal mathematical development and additional evaluations are available (Niksa, 2018a,b). The data validations cover the yields of char, tar, and gas in aggregate, but detailed distributions of the major product species are relegated to Section 9.5, because product distributions are radically altered by tar hydrogenation chemistry, which entails a separate reaction mechanism.

9.4.2.1 Hydrogenation of bridges and peripheral groups

The postulated hydrogenation reactions for bridges and peripheral groups utilize H_2 within the condensed coal phase whose level is directly related to the H_2 pressure within the particle which, in turn, connects to the ambient H_2 pressure through the transport of H_2 against the outward flow of volatiles. This transport has been analyzed in detail (Niksa, 2018a). Once in the condensed phase, dissolved H_2 may react with the components of bridges and peripheral groups in various ways, including conversion of carboxylic acids into alcohols which, on a phenomenological level, shifts bridge decomposition chemistry away from spontaneous condensation into char links toward scission. Bridge hydrogenation is based on:

$$B + \nu_{HY} H_2 \rightarrow B^* \tag{9.1}$$

where B and B* are the molar concentrations of the original and hydrogenated labile bridges, respectively; and ν_{HY} is the stoichiometric coefficient for complete hydrogenation. An analogous reaction describes the hydrogenation of peripheral groups as one-half of a labile bridge. The rates of these reactions are evaluated with a first-order dependence on the H_2 pressure and a distributed-energy rate constant that introduces A_{HP}, E_{HP}, and σ_{HP} (Niksa, 2018a). The stoichiometric coefficient is evaluated from the premise that carbon in bridges is hydrogenated into a mixture of CH_4 and C_2H_4 in proportions that maintain an H/C ratio of three. All oxygen becomes H_2O and all sulfur becomes H_2S. Hence, hydrogenation leaves only short methylene hydrocarbon chains capped by methyl groups and alcohol functional groups in the hydrogenated bridge, which is the theoretical maximum extent possible.

The mean H-number in bridges throughout hydropyrolysis, ${}^{H}N_B$, is evaluated as a weighted average of the concentrations of original and hydrogenated bridges, as

$${}^{H}N_{B} = \left(\frac{B^*}{B_0}\right)\left[{}^{H}N_{B}^{0} + 2\nu_{HP}\right] + \left(\frac{B_0 - B*}{B_0}\right)^{H} N_{B}^{0} \tag{9.2}$$

where B_0 is the bridge composition without hydrogenation from FLASHCHAIN®. The value of ${}^{H}N_B$ determines how hydrogenation affects the scission selectivity during bridge conversion. In FLASHCHAIN®, this selectivity is expressed as a stoichiometric coefficient, ν_B, that gives the probability that a bridge conversion results in scission rather than spontaneous condensation into a char link. Based largely on the crosslinking character of carboxylic acids, this coefficient is evaluated as a function of the atomic O/H ratio of bridges (Niksa, 1994). As the O/H ratio falls below about 0.2, bridges mostly consist of methylene chains whose dominant conversion channel is scission and, consequently, the scission selectivity coefficient surges toward greater values. Given the H-number in bridges throughout hydropyrolysis from Eq. (9.2), the value of ν_B is updated simply by evaluating the ratio ${}^{O}N_B{}^{0}/{}^{H}N_B$, and applying the functional relation for ν_B in the coal constitution submodel. Hydrogenation increases ${}^{H}N_B$, and thereby diminishes ${}^{O}N_B{}^{0}/{}^{H}N_B$, which shifts the bridge conversion selectivity toward scission at the expense of spontaneous charring. This is how bridge hydrogenation suppresses spontaneous charring in FLASHCHAIN®.

Once a bridge has been hydrogenated, it may also interfere with another charring process, bimolecular recombination of mobile fragments. If a hydrogenated bridge later undergoes scission, its remnants will remain attached to the newly formed ends of the two new fragments. A fragment with hydrogenated ends is thereby capped by stable functionalities and unable to produce a char link in combination with another fragment end. According to FLASHCHAIN®, without hydrogenation the ends of mobile fragments can recombine regardless of whether or not the end contains the remnants of broken bridges. Since hydrogenation eliminates unsaturated components from bridges, and since such components readily condense further into aromatics, and since aromatics are essential to the refractory character of char links, hydrogenation interferes with bimolecular recombination. This effect will be especially pronounced with hv bituminous coals, which normally generate an abundance of mobile fragments that may either recombine into larger refractory fragments or depolymerize further into tar precursors.

The impact of hydrogenation on bimolecular recombination is expressed by reducing the recombination rate constant in proportion to the fraction of metaplast fragment ends that contain hydrogenated peripheral groups, according to

$$k_R \to k_R\left(1 - p_H^e\right) \tag{9.3}$$

where k_R is an Arrhenius rate constant for recombination and p_H^e is the probability that a metaplast fragment contains a hydrogenated peripheral group. If all the ends of metaplast fragments contained hydrogenated ends, recombination would cease; otherwise, the recombination rate is reduced in proportion to the fraction of ends that carry hydrogenated peripheral groups.

The mechanism also estimates the H_2 pressure within a reacting coal particle from a transport analysis that balances the inward diffusion of H_2 against the outward convective volatiles flux with H_2 consumption in the hydrogenation reactions (Niksa, 2018a). According to this analysis, H_2 fully penetrates the coal particles to achieve the same internal H_2 pressure as the ambient value provided that heating rates are slower than 10^3°C/s with 70 μm particles. Appreciably lower internal H_2 pressures were determined for 10^4°C/s over a broad range of H_2 pressure. But the lower internal H_2 pressures did not affect tar production, because such rapid heating does not provide sufficient contact time for bridge hydrogenation anyway.

9.4.2.2 *Char hydrogasification kinetics*

Char hydrogasification is described with a special version of the Carbon Burnout Kinetics Model called "CBK/G" (Liu and Niksa, 2004). CBK/G predicts the rate of char conversion, the char particle temperature, and the changes in the particle diameter and bulk density as hydrogasification proceeds, given profiles of gas temperature and radiative exchange temperature, and the H_2 partial pressure. Char reactivity is a dynamic function of heat treatment severity, based on a DAEM model of thermal annealing. The theory also tracks the impact of pore diffusion with an effectiveness factor defined by a Thiele analysis, and a random pore model to account for the loss of surface area with conversion. The analysis covers Zone I (reaction control), Zone II (pore diffusion control), and Zone III (film diffusion control) and their transitional regimes. It also includes a submodel of the effect of ash inhibition during the later stages of gasification of high-ash coals.

The H_2 gasification rate is several orders of magnitude slower than those for CO_2 or steam (Harris and Patterson, 1995). Of necessity, the char hydrogasification kinetics in CBK/G are oversimplified because so few rate studies on this chemistry have been reported. This analysis uses a single nth-order rate law in the surface concentration of H_2, as follows:

$$R_{HG} = k_{HG}(p_{H_2,S})^{n_{HG}} \tag{9.4}$$

where k_{HG} is an Arrhenius rate constant; $p_{H2,S}$ is the H_2 partial pressure on the char surface; and n_{HG} is the reaction order for hydrogasification specified to be one-half. In the data validation work, E_{HG} was assigned as 81.5 kJ/mol for all coals and A_{HG} was adjusted for each coal to quantitatively interpret reported extents of char conversion. The assigned values are discussed in connection with Fig. 9.8 below.

In CBK/G, the impact of pore transport is assessed from a conventional Thiele analysis in terms of the ratio of scales for chemical conversion to pore diffusion (Thiele modulus), and the effective diffusivity for pore transport is normally evaluated as the ratio of total porosity to a tortuosity factor. This simplification gives negligible transport resistance for hydrogasification of even the largest char sizes of interest. To interpret the reported size dependence in the hydrogasification of softening coals, the effective diffusivity is evaluated from the random pore model developed by Wakao and Smith (1964), with micropore and transitional pore sizes of 25 and 250 Angstroms, respectively, and micropore voidages that are six times larger than the voidage of transitional pores. These values are suitable for softening coals that lose their original pore systems during devolatilization, and can be used with ranks from subbituminous through mv bituminous. However, the much more open structures in chars from nonsoftening coals are assumed to provide full H_2 accessibility to the internal surface area.

9.4.2.3 Mechanistic interpretations for coal conversion

To accurately interpret the impact of all the major operating conditions on coal conversion during primary hydropyrolysis, the pseudo-frequency factors for bridge hydrogenation (A_{HY}) were specified to simulate the datasets in Figs. 9.1–9.6. The assigned values are compiled in Fig. 9.8. However, most qualified validation tests imposed heating rates of at least 10^3°C/s, which is too fast for appreciable bridge hydrogenation, so these assignments are tentative because the enhancements to total

Table 9.1 Measured (M) and predicted (P) tar yields after heating at 210°C/s to 1000°C with 2 s IRP under various H_2 pressures and the predicted enhancements from hydrogenation (E), in daf wt%.

P	Coal A			Coal B			Coal C			Coal D		
MPa	M	P	E	M	P	E	M	P	E	M	P	E
0.01	9.0	11.1	0.0	16.3	19.0	0.0	24.4	27.5	0.1	25.6	27.5	0.2
0.1	–	10.2	0.2	–	17.1	0.1	–	25.2	0.3	22.2	25.0	0.4
1.0	7.0	7.0	0.3	12.5	11.8	0.4	17.9	18.6	1.3	19.3	18.5	1.1
9.0	4.8	4.8	0.5	8.9	8.3	1.2	15.4	15.0	3.5	16.0	15.0	3.9

Reproduced from Arendt P, van Heek K-H. Comparative investigations of coal pyrolysis under inert gas and H_2 at low and high heating rates and pressures up to 10 MPa. Fuel 1981;60(9):779–787, with permission from Elsevier.

and tar yields were usually small. One notable exception is the evaluation in Table 9.1, which shows the impact of H_2 pressure variations on tar yields for a heating rate slow enough for appreciable hydrogenation of four coals. These WMR tests imposed a moderate, uniform heating rate to a temperature hot enough to achieve asymptotic, ultimate primary hydropyrolysis yields in the imposed reaction time (Arendt and van Heek, 1981). In addition to the measured (M) and predicted (P) tar yields, the predicted enhancements over and above the tar yields under inert atmospheres (E) are shown. The predicted tar yields are within 3 daf wt% of measured values for all pressures and coal samples. This agreement reflects negligible enhancements for H_2 pressures below 1 MPa, with relatively large enhancements at 9 MPa, especially with both hv bituminous coals. Clearly, tar yield enhancements during hydropyrolysis are strongly and independently mediated by both heating rate and H_2 pressure.

The assigned values for the bridge hydrogenation rate, A_{HY}, in Fig. 9.8 fall within a narrow band for ranks from lignite to hv bituminous. For higher ranks, they surge and reach up to four times the mean of the band for lower ranks. This abrupt transition coincides with the near-elimination of oxygen from the labile bridges in coal, which leaves primarily aliphatic, hydroaromatic (naphthenic), and olefinic functional groups. It is a reasonable transition because purely hydrocarbon linking structures are more easily hydrogenated than links that contain carboxylic acids, ethers, and esters as well.

The validation database on char hydrogasification covered substantial extents of char hydrogasification and a broad domain of operating conditions, so the assigned activation energy of 81.5 kJ/mol and the order of one-half are secure. Although these values were fixed for all coals, the pseudo-frequency factor for hydrogasification was tuned-in for each sample. Fig. 9.8 also shows the assigned frequency factors for hydrogasification for all the coals in the validation cases in Sections 9.4 and 9.5. For the bulk of these coals, the assigned hydrogasification reactivity values are slower than 7000 $atm^{0.5}\ s^{-1}$, and show no consistent tendencies with rank. The most remarkable feature is that the values for most coals lie within 50% of a mean value of 4000 $atm^{0.5}\ s^{-1}$. In contrast, the frequency factors for gasification in steam and

CO_2 vary by as much as two orders of magnitude for coals of the same rank, and those for char oxidation vary by up to an order of magnitude. Both gasification rates increase for coals of progressively lower rank. However, most of the values in the lower band in Fig. 9.8 express a variation of only a factor of four, and show no consistent trend with rank. This fundamentally different rank independence suggests that hydrogasification may not share any common adsorbed intermediates or primary reaction sites with the surface chemistry for gasification by steam and CO_2.

However, this interpretation does not cover the assignments for the hv bituminous coals tested by Tang et al. (1999) whose A_{HG} values are 5 to 20 times greater than the rest. In isolation, the assignments for Tang et al.'s dataset appear to indicate that hydrogasification reactivities increase for progressively higher coal rank until they diminish for low volatility coals. But in the context of all the assignments in Fig. 9.8 only Tang et al.'s bituminous coals have markedly faster reactivities. It is worth reiterating that hydrogasification reactivities vary by only a factor of four for the bulk of coals tested across the rank spectrum.

All the curves in the figures in Section 9.4.1 are simulation results from the mechanisms based on FLASHCHAIN® and CBK/G. With few exceptions, the simulations here and elsewhere (Niksa, 2011; Niksa, 2018a,b) depict the tendencies for all the major operating conditions within the measurement uncertainties. This performance demonstrates that remarkably little additional chemistry is needed to quantitatively interpret coal conversion for hydropyrolysis and hydrogasification: Two additional hydrogenation reactions for bridges and peripheral groups and suppression of fragment recombination in FLASHCHAIN®, and a single half-order reaction in the CBK/G framework for char hydrogasification.

Elevated H_2 pressures can affect both primary devolatilization and char conversion, depending on the processing conditions. However, since hydrogenation chemistry is relatively slow, its impact during devolatilization is governed by the time scale for spontaneous primary devolatilization during particle heating. Tar yields are not appreciably enhanced under the H_2 pressures associated with entrained-flow coal gasification technology, whereas tar yields under slow heating conditions are enhanced at elevated H_2 pressures. The proposed mechanism for bridge hydrogenation and its associated impact on fragment recombination in FLASHCHAIN® accurately depicts these tendencies. The predicted yield enhancements due to bridge hydrogenation are modest at even the highest test pressures for rapid heating, but become much more substantial at slower heating rates. Consequently, the proposed reaction mechanisms correctly depict the inversion of the heating rate dependence for the greatest H_2 pressures of interest.

During primary hydropyrolysis, three new independent mechanisms come into play: (1) inward H_2 penetration against the outward flux of volatiles; (2) the direct promotion of bridge scission by bridge hydrogenation; and (3) the suppression of bimolecular recombination by hydrogenated peripheral groups on fragment ends. The first mechanism never affects tar production, because intraparticle H_2 pressures equal the ambient values for all heating rates that provide sufficient time for the hydrogenation of coal components. Consequently, it is not necessary to accurately describe the evolution of char structure to simulate hydropyrolysis, although char morphology can factor into the char hydrogasification rate. This finding might seem to be at odds with the accurate interpretation of the reduced extents of hydrogasification for progressively larger particle sizes in Fig. 9.6, which has been cited as evidence for important transport mediations in the overall hydrogasification rate.

Primary devolatilization under inert gases is unaffected by variations in particle size provided the size is smaller than a few millimeters. In contrast, reported hydrogasification yields diminish for progressively larger sizes, as seen in Fig. 9.6. This behavior cannot be attributed to any aspect of hydropyrolysis, because the heating rates in these tests were too fast for appreciable hydrogenation in the coal phase during devolatilization. Instead, it is due to mediation of the hydrogasification rate by pore transport, as evident from the accurate predictions for the entire size range based on CBK/G's transport submodels. CBK/G includes three explicit size dependences in the char conversion analysis: The film diffusion rate; the transport rate through the internal pore system; and an empirical expression that describes how size and density vary throughout a gasification history. Although the first and third factors are negligible for the slow hydrogasification kinetics, pore transport resistances do explain the apparent size dependence, provided that the effective diffusivity is evaluated for micro- and transitional pores, as explained above (cf. Section 9.4.2.2).

The other two new hydropyrolysis processes were characterized with simulations with hv bituminous under 7 MPa H_2 for heating rates from 1 to 10^3°C/s to 900°C (cf. Fig. 9.1). As seen in Fig. 9.9, ultimate tar yields were achieved before the ends of the heating period for all rates except 10^3°C/s, and that rate achieved 84% of the ultimate tar yield. The tar release dynamics indicate whether hydrogenation occurred during the interval for maximum tar production. Bridge hydrogenation increases the average number of H-atoms per bridge which, in turn, promotes bridge scission and the associated accumulation of tar precursors. The indirect impact is apparent as the reduction in the bimolecular recombination rate compared to the rate during primary devolatilization without H_2 for the same thermal history and pressure. Both the direct and indirect impacts of bridge hydrogenation are strong for 1°C/s, and both are synchronized with tar production. The average H/bridge-number increased from 13 to 25, which increased the fraction of bridge decompositions that break bridges from the initial value for this hv bituminous sample, 0.41, to the maximum value of 0.71 for this coal. In other words, hydrogenation skewed bridge decomposition as far as possible toward scission, and the hydrogenation kinetics did not limit the impact. Similarly, the recombination rate was reduced to 10% of the original value by interference from hydrogenated fragment ends, and this minimum arose immediately after the maximum in the tar production rate. Conversely, both these impacts nearly vanished for 10^3°C/s, because the hydrogenation kinetics are too slow to keep pace with this thermal time scale. The average H/bridge increased to only 13.5 at this heating rate, which hardly perturbed the selectivity toward bridge scission from its coal-based initial value. The recombination rate remained at its original value throughout tar production as well. Both of the intermediate heating rates sustained hydrogenation effects that were between the two extremes, as were the associated tar yields.

Calculations without any effect of hydrogenation on the recombination rate isolate the impact of bridge hydrogenation on the selectivity for bridge scission. Without the

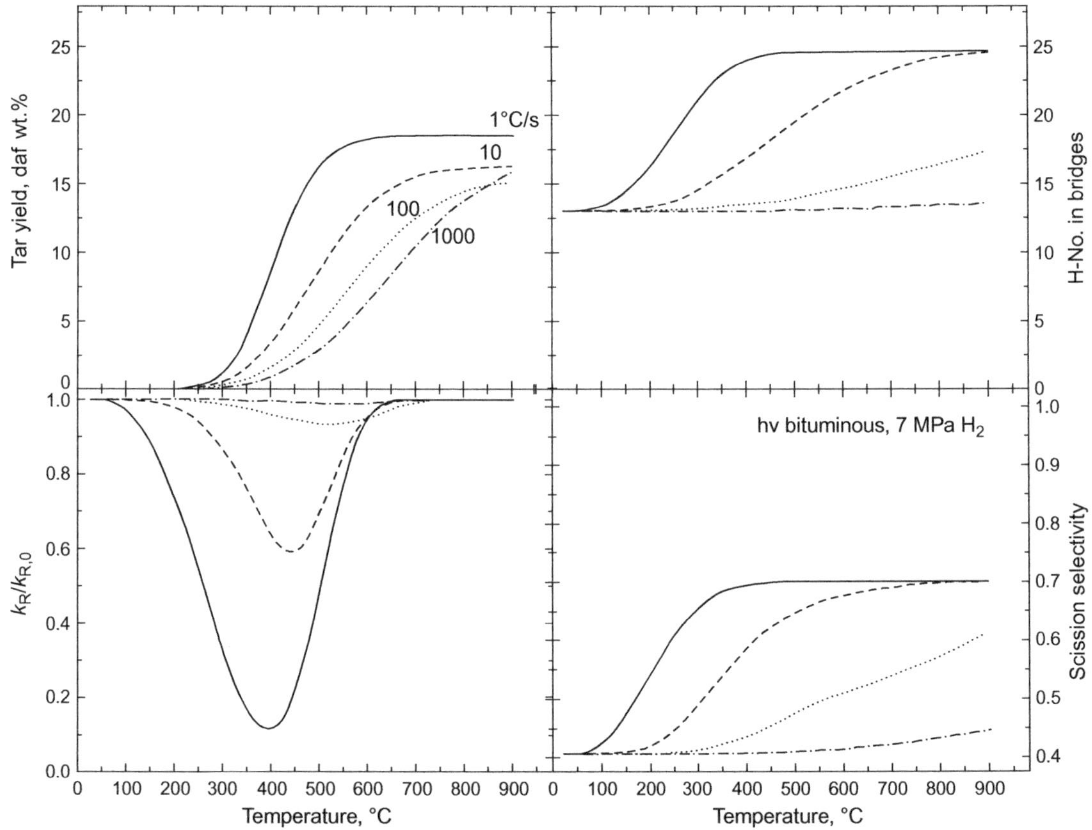

Fig. 9.9 In clockwise order from upper left, tar yields; number of H-atoms per bridge; scission selectivity coefficient; and ratio of modified to original rate constant for recombination for hv bituminous coal heated at 4 heating rates to 900°C under 7 MPa H_2. The x-axes show temperature during heatup. Reproduced from Niksa S. Flashchain theory for rapid coal devolatilization kinetics. 10. Extents of conversion for hydropyrolysis and hydrogasification of any coal. Energy Fuels 2018a;32:384–95, with permission from the American Chemical Society.

diminished recombination rate, the tar yield decreased from 21.4 daf wt% to 18.9%. This change accounts for one-quarter of the total enhancement due to hydrogenation, which increased the tar yield from 11% to 21.4%. Both of the impacts from hydrogenation are substantial under the heating conditions that provide sufficient contact time to appreciably enhance tar yields.

Generally speaking, the theory exposes the term "flash hydropyrolysis" as the oxymoron that it is. Rapid or "flash" heating is incompatible with appreciable enhancements of primary tar yields during hydropyrolysis because the slow hydrogenation chemistry cannot keep pace with the relatively very fast time scale for primary devolatilization. Since nominal rates for primary devolatilization increase in proportion to increases in heating rate, and since primary devolatilization chemistry is spontaneous, hydrogenation can only affect the distribution of primary products under slow heating conditions, because only slow heating enables bridge hydrogenation to occur before the bridges decompose through the normal, spontaneous channels.

9.5 Product distributions for hydropyrolysis and hydrogasification

This section covers the product distributions from hydropyrolysis and hydrogasification, and how they are affected by coal quality and the most important operating conditions. The presentation first covers primary hydropyrolysis in isolation, followed by the impact of tar hydroconversion, and then the ultimate products, which include substantial CH_4 yields from hydrogasification. The resolution of these stages is not nearly as sharp as the resolution of primary devolatilization and tar decomposition under inert atmospheres, primarily because some aspects of tar hydrogenation overlap with the initial release of primary tars even near the onset temperatures for devolatilization. After the product database has been surveyed, a tar hydroconversion mechanism accurately interprets the evolution of the product distributions in time with any coal across the commercial operating domain.

9.5.1 Primary hydropyrolysis in isolation

In principle, the products of primary hydropyrolysis can be recovered before tars are radically transformed by hydroconversion in the same reactor configurations used to resolve primary devolatilization from tar decomposition under inert atmospheres. In practice, the relatively rapid rates of tar hydroconversion mean that even minor aspects of reactor design that inadvertently affect volatiles quench rates can affect product distributions. The most reliable index on whether or not tar hydroconversion has come into play is the oils yield, specifically in comparison with the oils yield for primary devolatilization under an inert gas, all other operating conditions the same as in the hydropyrolysis test. Without this index, it is essentially impossible to estimate an extent of tar hydroconversion.

9.5.1.1 Char compositions and physical characteristics

Char compositions are unaffected by tar hydroconversion, but often reflect conversion by both hydropyrolysis and hydrogasification. Char compositions resolved in time throughout only the hydropyrolysis stage in one test series, and throughout primary devolatilization in another for the same thermal history and pressure can clearly determine if elevated H_2 pressures affect the release of heteroatoms, provided that heating rates are slow enough to enhance the tar yields. Unfortunately, compositions for such test series have not yet been reported, although sufficient data is available to indirectly identify the most important char transformations.

The char H– and O-contents in Fig. 9.10 were obtained in an EFR and FFR over comparable temperature ranges with similar coals. Heating rates in both reactors were too fast for enhanced tar yields during hydropyrolysis. The main differences in these test series are that the FFR was run at 1 MPa H_2, whereas the EFR was operated at 12.9 MPa H_2, and the solids contact time in the EFR was too short for extensive char hydrogasification, whereas hydrogasification during several seconds in the FFR was substantial. As seen in Fig. 9.10, the char-H levels decay over progressively hotter temperatures with comparable sensitivities, which confirms that char-H is released during hydropyrolysis but not hydrogasification. But the temperature sensitivities in the reported char-O levels are markedly different, and it appears that hydrogasification, rather than hydropyrolysis, is responsible for the much greater earlier O-release in the EFR tests. These datasets also showed that char N-contents are unaffected by hydropyrolysis and hydrogasification, and that char C-contents increase for

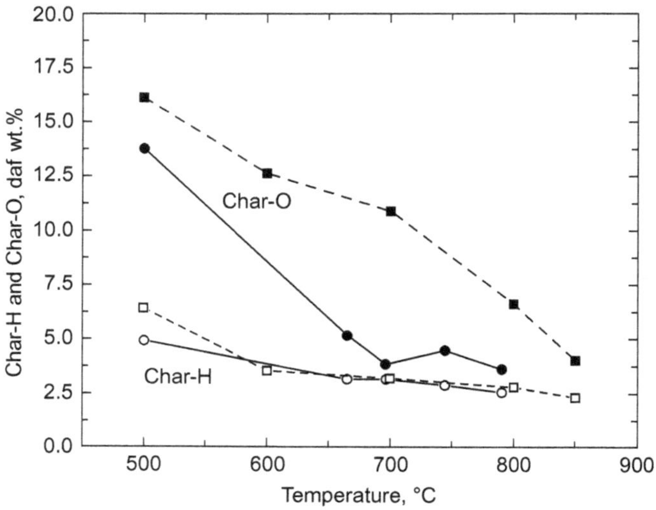

Fig. 9.10 Mass percentages of H and O in chars from *(solid curves)* an EFR at 12.9 MPa H_2 (Ikura and Last, 1988) and *(dashed curves)* a FFR at 1 MPa H_2 (Xu et al., 2003a). Coal values appear at 500°C.

Table 9.2 Char compositions for heating at 5°C/s to 575°C over 10 min under three pressures of H_2 and N_2, in daf wt%.

	P	%C		%H		%O	
Coal	**MPa**	H_2	N_2	H_2	N_2	H_2	N_2
Brown coal	0.1	88.1	85.8	3.2	3.1	8.9	11.4
	5	93.9	87.0	3.3	3.1	4.1	10.2
	30	88.6	84.4	3.3	3.5	8.5	12.2
hv bituminous	0.1	90.0	89.5	3.2	3.3	5.0	5.6
	5	91.8	89.8	3.3	3.0	2.5	4.3
	30	95.6	90.0	3.5	3.4	2.4	5.6
Anthracite	0.1	92.5	93.6	3.1	3.3	1.8	1.8
	5	93.9	94.3	3.2	3.3	1.6	1.8
	30	94.3	93.2	3.3	3.1	1.5	1.8

Reproduced from Fynes G, James RG, Ladner WR, Newman JOH. Structural differences in the tars and char from the pyrolysis of coals of different rank in hydrogen and in nitrogen. Fuel 1984;63:897–903, with permission from Elsevier.

progressively greater weight loss, due to the preferential elimination of heteroatoms. Such changes are also evident in the char compositions reported by Fynes et al. (1984).

The accelerated elimination of char-O under elevated H_2 pressures is seen more clearly in Table 9.2, albeit for very long contact times under the same pressures of H_2 and N_2. The chars were prepared in a HRR heated at 5°C/s to 575°C with a total run time of 10 min, so both enhanced tar yields from hydropyrolysis and substantial hydrogasification factor into some of the reported char yields (Fynes et al., 1984). Neither effect is important at atmospheric pressure and, consequently, the char compositions under 0.1 MPa H_2 and N_2 are similar with all three coals. But for both elevated pressures, the chars prepared under H_2 contained more carbon and much less oxygen, and very similar amounts of hydrogen.

Elevated H_2 pressures promote O-release, but hardly perturb the release of C, H, and N. (Both N and S release are considered further in Section 9.5.4.) O-release is more extensive under H_2 even for temperatures around 600°C, suggesting that hydrogenations in the condensed coal phase or char hydrogasification are responsible. Hydrogasification is probably responsible for the promotion, although tests like those behind Table 9.2 but with time resolution are needed to definitively identify the participating stage.

Char morphologies are also affected by elevated H_2 pressures although, here too, the available data are limited. Stanczyk and Kisielow (1994) monitored surface areas and pore characteristics for chars prepared from a lignite and subbituminous coal in a swept PBR heated at 0.25°C/s to 450–800°C with 30 min IRP under 0.5–5 MPa of H_2 and Ar. Compared to the coals' morphologies, hydropyrolysis and hydrogasification did not radically alter any of the physical characteristics. Both CO_2 and N_2 surface areas were greater under Ar than under H_2 at moderate temperatures, but the order was reversed at 850°C. In fact, both surface areas decreased monotonically for progressively higher H_2 pressures at 550–600°C, despite greater coal conversion across this range. Porosities were insensitive to temperature under both H_2 and Ar, whereas mean pore diameters diminished for hotter temperatures. For progressively higher pressures, porosities remained the same while mean pore diameters expanded. None of these tendencies pertain to bituminous coals whose physical characteristics are radically transformed during the molten plastic stage of hydropyrolysis.

9.5.1.2 Primary tar compositions

Tars are exclusively generated during hydropyrolysis, so contributions from hydropyrolysis and hydrogasification do not need to be resolved to interpret their compositions. But as for primary devolatilization (cf. Section 4.2.7), secondary tar hydroconversion transforms primary hydropyrolysis tars beyond recognition, which makes it difficult to see how hydrogenations in the condensed coal phase affect the primary tar product. Atomic H/C ratios of tars from primary devolatilization clearly expose the impact of secondary tar decomposition in inert gases (cf. Fig. 4.20).

Unfortunately, H/C ratios for hydropyrolysis tars do not gauge tar hydroconversion in the same way. This is apparent in Table 9.3, which shows H/C ratios of ultimate tar samples and their parent coals for hydropyrolysis at moderate temperatures and very high H_2 pressures in two HRRs and an EFR. There are weak influences from the parent coal quality and test temperature, but the most striking feature is that all the tar H/C ratios relax toward unity, despite very significant differences in the enhancements to primary tar yields and in the extents of tar hydroconversion among these tests. The reason is simply that benzene, whose H/C ratio is unity, is the predominant intermediate aromatic product of tar hydroconversion, as demonstrated in Section 9.5.2.

The tars least affected by hydroconversion in Table 9.3 are those for 15 MPa (Bolton et al., 1987; Li et al., 1996), whose H/C ratios are close to unity except for the brown coal tar, whose relatively high H/C ratio affects the tar H/C but only at the cooler test temperature. Moreover, the tar/oil mixture whose hydroconversion was essentially complete in the test at 665°C also gave an H/C ratio of 0.99, because benzene was the predominant component (Ikura and Last, 1988). So tar H/C ratios are unsuitable indices for extents of tar hydroconversion, unless the tar sample is collected and processed to be free of oils.

Table 9.3 Ultimate H/C ratios of tar/oil mixtures and their parent coal ratios for hydropyrolysis at very high H_2 pressures in different reactor configurations.

Coal-C,daf wt%	Coal H/C	P_{H2} (MPa)	T (°C)	Tar H/C
68.7	1.01	15	520	1.22
		15	600	1.08
70.2	0.76	30	575	1.06
76.8	0.77	12.9	665	0.99
80.7	0.79	15	520	1.07
80.8	0.80	30	575	0.97
81.6	0.80	15	650	0.92

Table 9.4 Elemental compositions and proton aromaticities of primary hydropyrolysis tars prepared under 15 MPa H_2, in daf wt%.

Coal type	T (°C)	%C	%H	%O	${}^{H}f_a{}'$
Lignite coal	–	68.1	5.7	14.3	–
Tar	520	79.5	8.1	11.4	0.25
Tar	600	83.5	7.5	8.0	0.35
hv bit 1 coal	–	81.0	5.4	10.3	–
Tar	520	83.0	7.2	6.0	0.33
hv bit 2 coal	–	81.6	5.4	8.6	–
Tar	650	85.4	6.6	6.4	0.44

Reproduced from Bolton C, Snape CE, O'Brien JO, Kandiyoti R. Influence of carrier gas flow and heating rates in fixed bed hydropyrolysis of coal. Fuel 1987;66:1413–17; Li B, Mitchell SC, Snape CE. Effect of heating rate on normal and catalytic fixed-bed hydropyrolysis of coals. Fuel 1996;75:1393–96, with permission from Elsevier.

Extents of sooting are the most reliable index for tar decomposition under inert gases for temperatures hotter than 850°C. But the pathways that convert tar into soot are disrupted by hydroconversion, so this index also cannot pertain to tar hydroconversion. As noted above, oils yields for both hydropyrolysis and primary devolatilization for the same thermal history and pressure are the only suitable index for tar hydroconversion.

Elemental compositions for primary tars prepared by hydropyrolysis at 15 MPa at moderate temperatures are compiled in Table 9.4. These tars are enriched in carbon and hydrogen but depleted of oxygen, compared to their parent coals. Although the carbon enrichments and oxygen depletions become more severe for hotter temperatures (and also with faster heating rates (Bolton et al., 1987)), the hydrogen enrichments diminish. All these tendencies pertain to primary tars from inert atmospheres as well, so tar compositions are not distinguishing characteristics of primary hydropyrolysis tars. Hydropyrolysis tars become more aromatic for progressively hotter temperatures and for parent coals of higher rank, as seen in the reported proton aromaticities. Also, these values increase for progressively higher H_2 pressures (Fynes et al., 1984; Bolton et al., 1987). However, this index is also potentially misleading because the aromaticities must approach unity as secondary tar hydroconversion converts primary tars into mostly benzene.

One of the few available direct comparisons of tars from hydropyrolysis and devolatilization under N_2 appears in Table 9.5 (Fynes et al., 1984). At atmospheric pressure, the tars under both atmospheres are essentially the same, as expected. But for both higher pressures, hydropyrolysis tars contain more carbon, less hydrogen, and much less oxygen. It is unclear whether these differences are due to hydrogenations in the condensed coal phase or secondary tar hydroconversion, because these tests used 10 g samples. Nonetheless, they demonstrate that, under comparable test conditions, hydropyrolysis tars are appreciably more aromatic than primary devolatilization tars, as corroborated by Bolton et al. (1987); and that the hydrogenation chemistry preferentially eliminates oxygen. Primary devolatilization tars become more aromatic whenever peripheral groups are eliminated from fragments in the

Table 9.5 Tar compositions for heating at 5°C/s to 575°C over 10 min under three pressures of H_2 and N_2, in daf wt%.

Coal	P	%C		%H		%O	
	MPa	H_2	N_2	H_2	N_2	H_2	N_2
Brown coal	0.1	80.4	79.6	7.8	8.3	11.0	11.1
	5	84.7	80.2	7.9	8.7	6.5	9.9
	30	85.2	81.3	7.5	8.7	6.0	9.9
Hv bituminous	0.1	81.4	81.8	7.9	8.1	8.6	8.1
	5	83.8	82.5	7.2	8.5	6.2	7.5
	30	86.4	83.0	7.0	8.0	4.5	7.8

Reproduced from Fynes G, James RG, Ladner WR, Newman JOH. Structural differences in the tars and char from the pyrolysis of coals of different rank in hydrogen and in nitrogen. Fuel 1984;63:897–903, with permission from Elsevier.

condensed phase. Since tar fragments of a specified degree of polymerization are lighter without peripheral groups, greater aromaticity is usually associated with lower tar yields. In contrast, hydropyrolysis tars are more aromatic and more abundant (Fynes et al., 1984; Bolton et al., 1987), which necessarily implies that hydrogenations in the condensed phase must skew bridge decomposition toward scission, away from spontaneous condensation, and thereby increase the population of tar precursors.

Primary tars prepared under inert gases become lighter for progressively higher pressures. The M_n-values increased from 240 to 280 g/mol for tars from hydropyrolysis from 2.5 to 15 MPa (Bolton et al., 1987), which overlap the weights expected for primary devolatilization tars at such pressures. The values in Bolton et al.'s hydropyrolysis dataset stay the same across broad ranges of heating rate and pressure, whereas other data show diminishing weights for progressively higher H_2 pressures (Gibbins et al., 1991). Accurate MWDs for hydropyrolysis tars are unavailable, except for inferences from uncalibrated GPC signals (Moliner et al., 1989; Guell et al., 1993).

9.5.1.3 Volatile product distributions

Hydrogenation in the condensed coal phase is precluded for heating rates faster than 10^3°C/s, and tar hydroconversion is minimal if volatile products are quenched as soon as they are released. Under these conditions, the distributions of primary products should be the same for hydropyrolysis and devolatilization at even the highest H_2 pressures. Both conditions were satisfied in the tests with a swept CPP that generated the distributions in Fig. 9.11. At 9000°C/s, heating rates were definitely fast enough to prevent hydrogenation in the coal phase, so the tar yields exhibit the substantial decay from 0.1 to 2 MPa expected for primary devolatilization, then saturate to an asymptotic value. BTX yields are no >2 daf wt% across the entire pressure range, so tar hydroconversion was inconsequential. The only potential complication is that the IRP of 10 s was definitely long enough for appreciable hydrogasification, but this factor is only responsible for the substantial growth in the CH_4 yields for progressively higher H_2 pressures. In all

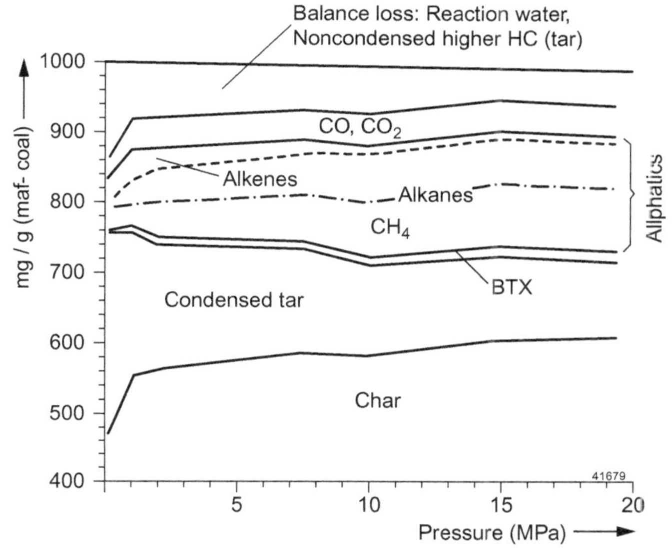

Fig. 9.11 Product distributions from hv bituminous coal for 9000°C/s to 800°C with 10 s IRP under various H_2 pressures.
Reproduced from Kaiser M, Wanzl W. Characterization and comparison of liquid products from coal pyrolysis in laboratory and process development units. Fuel Process Technol. 1988;20:23–32, with permission from Elsevier.

other aspects, these primary hydropyrolysis distributions are indistinguishable from those for primary devolatilization across the same pressure range, as they should be for rapid heating without tar hydroconversion.

Similar comparisons were reported for 7 MPa under H_2 and He in a swept WMR that heated six diverse coals at 10^3°C/s to 1000°C with 2 s IRP (Strugnell and Patrick, 1995). These distributions corroborate the uniformity of the tar yields from hydropyrolysis and devolatilization, albeit subject to larger uncertainties because the reported liquids yields included chemically formed moisture and were evaluated by difference. However, the detailed resolution of all major noncondensable products showed markedly greater yields of CH_4 and C_2H_6 and slightly less CO for hydropyrolysis. Since the changes to the CH_4 yields were comparable to the reductions in the char yields, these enhancements are undoubtedly due to char hydrogasification. The enhanced C_2H_6 yields and diminished CO yields probably reflect reforming of olefins into alkanes with water gas shifting among noncondensable gases, presumably because the gas sweep did not completely counteract the very hot temperature in these tests. Even though tar hydroconversion was inconsequential, noncondensables were affected by secondary reforming chemistry under these relatively severe conditions.

In contrast to the minimal impact of hydrogenation on primary products at fast heating rates, the product distributions for devolatilization at slow heating rates and elevated H_2 pressures are radically altered by hydrogenation in the condensed coal phase.

Table 9.6 Measured hydropyrolysis products compared to predicted primary devolatilization products for 5°C/s to 500°C under 15 MPa H_2.

Products, daf wt%	Primary devolatilization	Hydropyrolysis
Tar	9.4	26
H_2O	5.4	6[a]
$CO + CO_2$	2.7	2
GHCs	5.4	7
Char	76.6	59

[a]Estimated.
Reproduced from Bolton C, Riemer C, Snape CE, Derbyshire FJ, Terrer MT. Effect of low temperature catalytic hydrogenation on pyrolysis and hydropyrolysis of a bituminous coal. Fuel 1988;67:901–5, with permission from Elsevier.

The huge enhancements to tar yields are clearly apparent in Table 9.6, which compares the products of primary devolatilization and hydropyrolysis of an hv bituminous for heating at 5°C/s to 500°C with 10 min IRP under 15 MPa H_2. The hydropyrolysis products were measured by Bolton et al. (1988) in a HRR, which was subsequently operated at 520–600°C to produce tar yields in excess of 50 daf wt% from low rank and bituminous coals with the same heating rate (Klavetter et al., 1993). The primary devolatilization products in Table 9.6 are from FLASHCHAIN® simulations which were tuned-in to reported products for rapid heating at atmospheric pressure in a fluidized bed, then extrapolated to the thermal history and pressure in the hydropyrolysis tests, but without H_2. The large reduction in the char yield for hydropyrolysis cannot be attributed to hydrogasification because it is much greater than the minimal enhancement to the GHC yields; in fact, the temperature in this test was too cool for hydrogasification. Essentially all of the additional weight loss reflects the enhancement to the tar yield. The yields of oxygenated permanent gases, moisture, and even GHCs are hardly perturbed by the elevated H_2 pressure, which implies that tar hydroconversion is also inconsequential at 500°C and short contact times. Hydrogenation in the condensed coal phase during devolatilization at slow heating rates dramatically enhances primary tar yields and supplements GHC yields but hardly perturbs the yields of oxygenated noncondensables, provided that tar hydroconversion is suppressed.

9.5.2 Tar hydroconversion products

9.5.2.1 Hydroconversion dynamics

Tar hydroconversion largely explains why the ultimate product distributions from hydropyrolysis are radically different than those for devolatilization under inert gases with or without tar decomposition. To identify the major transformations, this section reviews product distributions that were either resolved in time at fixed temperature, or resolved in temperature at a uniform gas contact time. These distributions are not the same because extents of char hydrogasification increase for progressively hotter temperatures, but this tendency is easy to factor out because it only affects CH_4 yields as, for example, in the description of Fig. 9.11.

Transformations among the major product lumps are resolved in temperature and across a broad range of H_2 pressures in Fig. 9.12. These tests heated an hv bituminous coal at 2000°C/s to the stated temperatures in roughly 1.5 to 2 s in a FFR, with a nominal gas contact time of 3 s (Xu et al., 2003a). These contact times were sufficient to achieve ultimate hydropyrolysis yields at all but the coolest temperature, but not complete tar hydroconversion, even at the hottest temperature. The hydrocarbon liquids (HCL) are mixtures of BTX, PCX, and naphthalenes, although none of the naphthalene yields exceed 2% of coal-C, so HCL is nearly the same as oils. Even at 625°C (898 K), appreciable portions of primary tar were hydrogenated into HCL and noncondensables in 3 s. Extents of hydroconversion increase for progressively hotter temperatures, although some secondary tar persists even at 850°C (1123*K*). Oils yields increase for progressively hotter temperatures but are insensitive to H_2 pressures higher than about 1 MPa. Increases in gas yields are much greater than changes to the char yields and ultimate HCL yields are considerably smaller than the sum of tar plus HCL at the coolest temperature. Consequently, roughly two-thirds of the carbon in primary tar was converted into oils, and the rest formed noncondensable gases. As in tar decomposition in inert gases, tar hydroconversion releases substantial portions of the heteroatoms and peripheral groups as noncondensable gases, as seen shortly in detail.

In the pressure sweep, the total yield passes through a weak maximum at 1 MPa, which reflects the diminishing tar yields from 0.25 to 2 MPa vs. increasing extents of hydrogasification for progressively higher H_2 pressures. Tar-C gradually diminishes and HCL-C gradually increases for progressively higher H_2 pressures, except for a surge in HCL-C from 0.25 to 1 MPa. The surge suggests that oils are the earliest tar hydroconversion products and that noncondensables are released from tar at a slower rate. Beyond 1 MPa, the reduction in char-C parallels the expansion of gas-C due to hydrogasification. The apparent reaction order for H_2 in tar hydroconversion appears to be small, which probably reflects the high dilution of coal in H_2 at all pressures in these tests, rather than the actual order under commercial coal loadings.

The major C-bearing noncondensable gases are resolved in both gas contact time and temperature in Fig. 9.13. The total gas yield, which excludes HCL, increased only slightly over the range of contact times because the coal conversion was uniform, reflecting uniform solids transit times in these tests. The sum of tar-C plus HCL-C (not shown) was also uniform throughout. At the shortest contact time, the total GHC yields are at least double those for primary devolatilization under inert gases, and remain at this level throughout. The amounts of CO and CO_2 account for about half of coal-O, and the remainder presumably appears as moisture from the release of coal-O and tar-O, or remains in char. The only significant chemical transformation is the reforming of all GHCs into a mixture of CH_4 and C_2H_6. The yields of CO and CO_2 remain the same throughout. HCL yields diminish slightly, which reflects the net change between additions of tar hydroconversion products and loss of aliphatic components from oils as additional GHCs. These same trends are evident in the C-distribution of gases vs. temperature through 800°C (1073 K), except that tar hydroconversion is more apparent as the growth in C_2-C_3 GHCs and BTX for

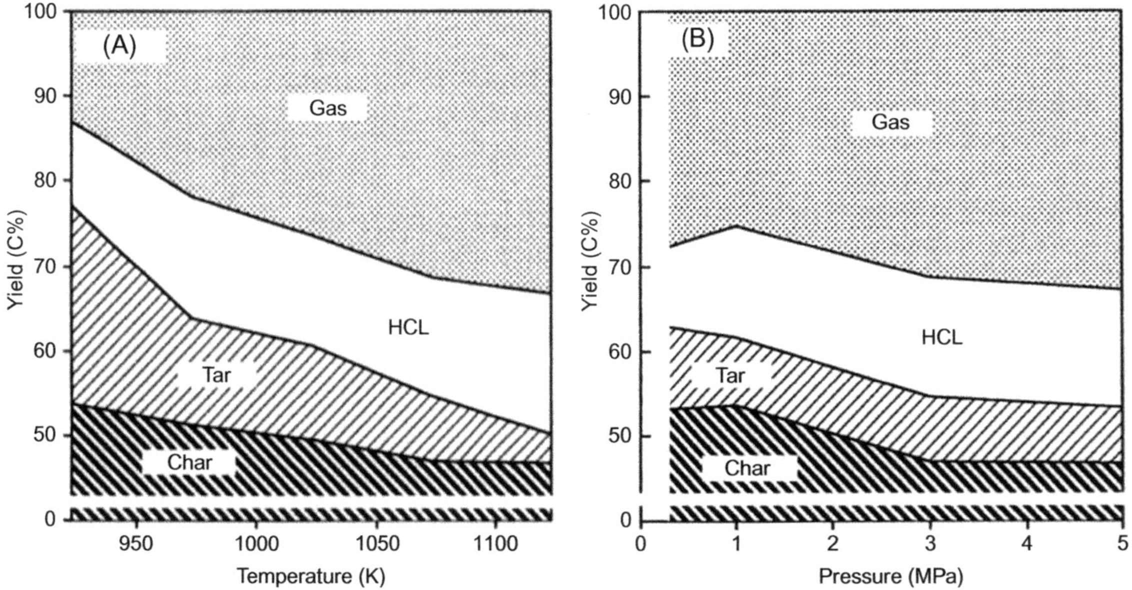

Fig. 9.12 Carbon distributions from hv bituminous coal for 2000°C/s (left) to various temperatures under 3 MPa H_2; and (right) to 800°C under various H_2 pressures. Gas contact times were about 3 s.
Reproduced from Xu W-C, Matsuoka K, Akiho H, Kumagai M, Tomita A. High pressure hydro-pyrolysis of coals by using a continuous free-fall reactor. Fuel 2003a;82:677–85, with permission from Elsevier.

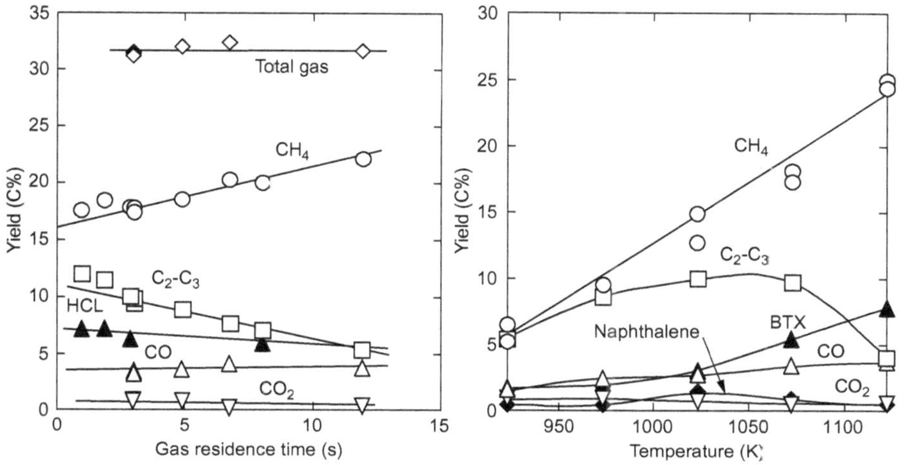

Fig. 9.13 Carbon distributions for noncondensable gases and HCL from hv bituminous coal for heating at 2000°C/s (left) to 800°C under 3 MPa H_2 for various gas contact times; and (right) to various temperatures at 3 MPa H_2 with 3 s contact time.
Reproduced from Xu W-C, Matsuoka K, Akiho H, Kumagai M, Tomita A. High pressure hydro-pyrolysis of coals by using a continuous free-fall reactor. Fuel 2003a;82:677–85, with permission from Elsevier.

progressively hotter temperatures. In comparison to the change in the tar yield over this temperature range in Fig. 9.12, light GHCs are a major tar hydroconversion product, accounting for at least as much of the converted tar as HCL. The C-distribution at the hottest temperature indicates near-complete reforming of C_2-C_3 GHCs into CH_4, and a surge in the BTX component of HCL. Contributions from CO and CO_2 diverge toward more CO.

Across the pressure range in Fig. 9.14, the contributions for CH_4 and BTX nearly double while those for CO and CO_2 are halved, and those from C_2-C_3 GHCs stay the same. Tar yields did decrease slightly across this range, but not by enough to explain the growth in BTX, so other HCL components must have either eliminated aliphatic peripheral groups or ruptured rings to form BTX. Since hydrogasification is responsible for most of the additional CH_4, GHC reforming appears to be very insensitive to H_2 pressure although, again, this probably reflects the very low H_2/coal ratio in these tests. Tests on tar hydroconversion in a two-stage reactor at only 500°C across the same range of H_2 pressure with bituminous coal recorded comparable changes to BTX and CH_4, but GHC and CO_X levels also surged at the highest H_2 pressure (Chareonpanich et al., 1994).

9.5.2.2 *Ultimate tar hydroconversion products*

Given sufficient contact time, the product distributions for tar hydroconversion relax to their ultimate, stable compositions. In reviewing the literature database on maximum oils yields, one gets the impression that oils are stable products whose yields

Fig. 9.14 Carbon distributions of noncondensable gases and BTX from hv bituminous coal for 2000°C/s to 800°C under various H_2 pressures for 3 s gas contact time. Reproduced from Xu W-C, Matsuoka K, Akiho H, Kumagai M, Tomita A. High pressure hydro-pyrolysis of coals by using a continuous free-fall reactor. Fuel 2003a;82:677–85, with permission from Elsevier.

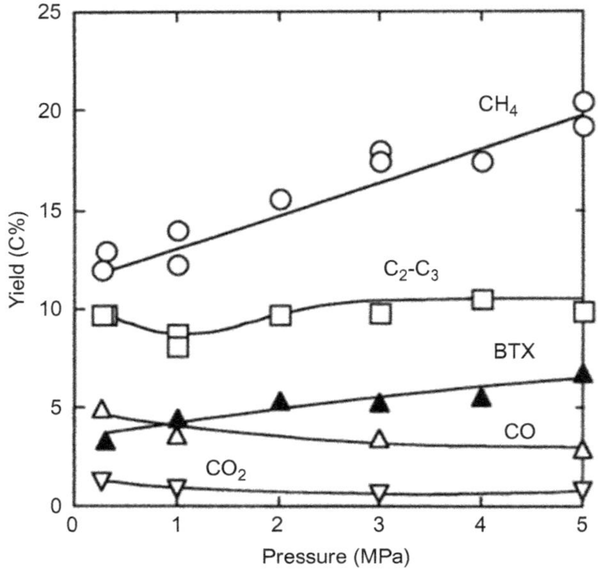

reach an ultimate, asymptotic value for any particular coal sample. But this view is biased by researchers' common focus on the narrow ranges of temperature and pressure that maximized oils yields for virtually all coal types. In actuality, oils yields pass through relatively sharp maxima that depend on temperature and H_2 pressure, as seen in Fig. 9.15. The increasing oils yields for progressively hotter temperatures up to the maximum yield reflect the appreciable activation energies in the kinetics of oils formation. The rapid decay beyond the maximum indicates hydrocracking of single-ring compounds into CH_4 and C_2 GHCs, at an accelerating rate for progressively higher H_2 pressures. Consequently, the maximum oils yields shift toward appreciably cooler temperatures for progressively higher H_2 pressures. Although few datasets in the literature cover the domain of conditions to illustrate this instability of oils, it is important to recognize that oils are only stable, and therefore recoverable, up to about 850°C for H_2 pressures to 15 MPa (2176 psi). Succeeding cases in this section emphasize the maximum oils yields for somewhat less severe hydroconversion conditions.

Ultimate oils yields for coals across the rank spectrum appear in Fig. 9.16 for low (3–4 MPa) and high (13–15 MPa) H_2 pressures. The main tendency in both groups of yields is for fairly uniform oils yields for ranks through hv bituminous, followed by an abrupt reduction through the low volatility ranks. The sample-to-sample variability is substantial across the entire rank spectrum. Oils yields for the higher pressure range are uniformly higher than the low-pressure set for all coal types. However, this ranking cannot be directly attributed to H_2 pressure, because all but one of the coals in the higher pressure range were processed with slow heating under 15 MPa H_2, which markedly increased the primary hydropyrolysis tar yields, and thereby enhanced the oils yields as well. All other tests in Fig. 9.16 reflect rapid heating.

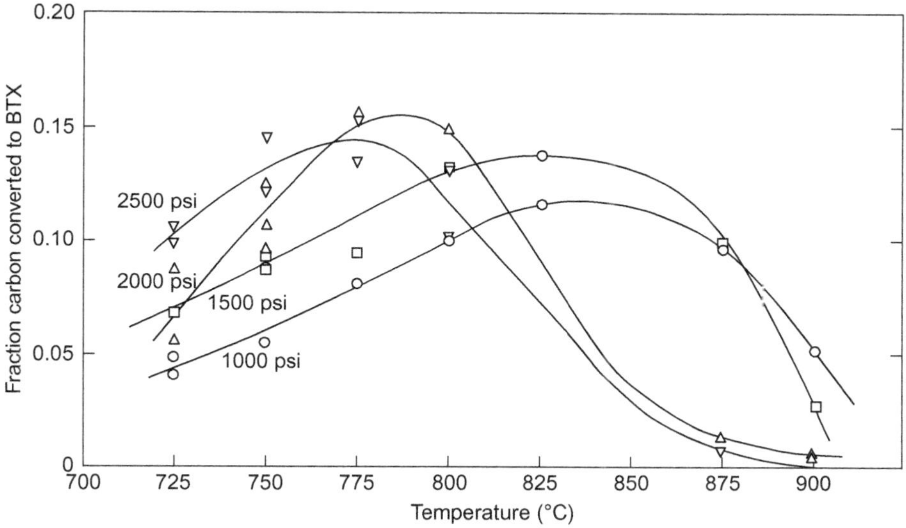

Fig. 9.15 C-fractions converted into BTX from subbituminous coal for rapid heating under various H_2 pressures for contact times of several seconds.
Reproduced from Fallon PT, Bhatt B, Steinberg M. The flash hydropyrolysis of lignite and subbituminous coals to both liquid and gaseous hydrocarbon products. Fuel Frocess Technol. 1980;3:155–68, with permission from Elsevier.

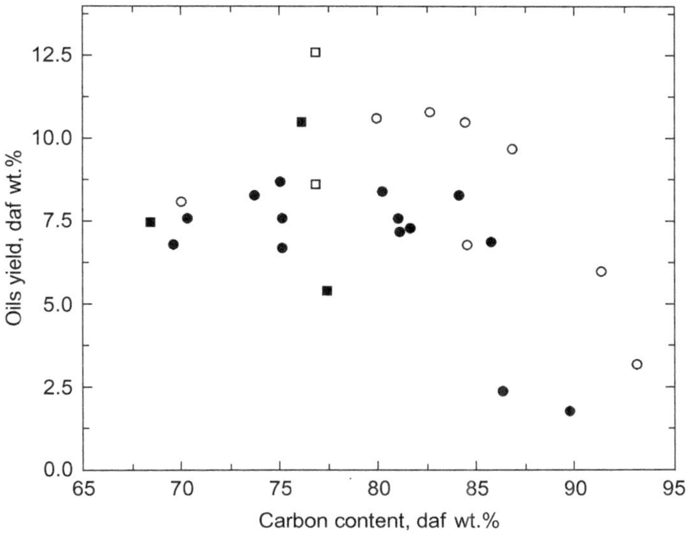

Fig. 9.16 Oils yields after extended contact at 750–875°C under *(filled symbols)* 3–4 and *(open symbols)* 12.9–15 MPa H_2.

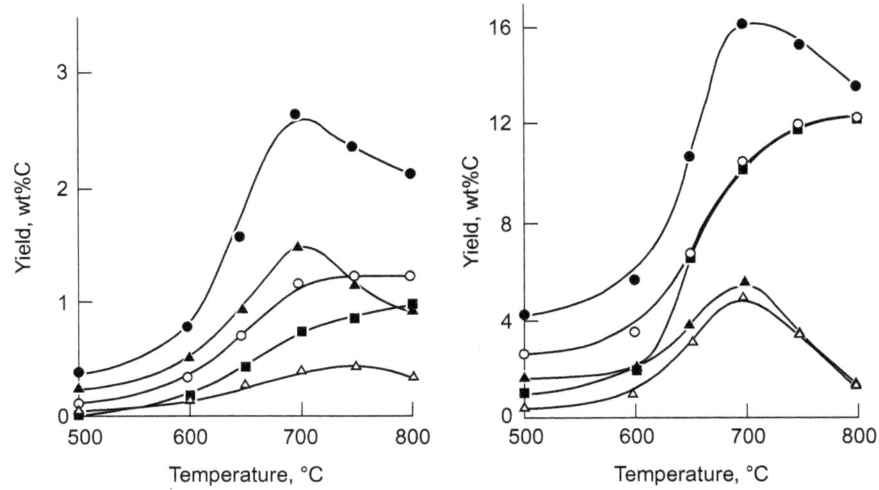

Fig. 9.17 Yields of (●) oils, (○) BTX, (■) benzene, (▲) PCX, and (△) xylenol from a lignite after heating at 7000°C/s with 42 s gas contact time under (left) 0.1 and (right) 6 MPa H_2. Reproduced from Zhu Z, Ma Z, Zhang C, Jin H, Wang X. Flash hydropyrolysis of northern Chinese coal. Fuel 1996;75:1429–33, with permission from Elsevier.

Ultimate oils yields as a function of temperature for two disparate H_2 pressures for a lignite appear in Fig. 9.17. Note the grossly different yield scales to cover the much lower oils yields for 0.1 MPa H_2. As expected, the oils yields pass through a maximum around 700°C for both pressures; this temperature is cooler than those in Fig. 9.15 because the gas contact time for Fig. 9.17 is relatively very long at 42 s. The most interesting feature is the evidence for the elimination of peripheral groups from BTX and PCX at both pressures, and for hydrocracking of PCX into BTX plus GHCs at the higher pressure. At 6 MPa H_2, BTX has been rendered into pure benzene at 650° C and hotter, while PCX is rendered into xylenol at 750°C and hotter. Once the PCX yield passes through its maximum at 700°C, only a portion of the converted single-ring compounds appear as benzene and, consequently, the aggregate oils yields diminish. From 700 to 800°C, the benzene yields saturate to an "ultimate" value, whereas PCX vanishes and the oils yield diminishes.

At 0.1 MPa H_2, peripheral groups are also eliminated from BTX and PCX, but the elimination is only apparent above 700°C and neither benzene nor xylenol become the exclusive species in their respective lumps. Once the PCX yield passes through its maximum at 700°C, the BTX yield has saturated so oils yields diminish for progressively hotter temperatures. Elevated H_2 pressures accelerate the rendering of BTX and PCX into benzene and xylenol, respectively, and also accelerate the destruction of PCX and the associated reduction in the oils yield. Although the benzene yields appear to saturate at the hottest temperature for both pressures in Fig. 9.17, the trends in Fig. 9.15 indicate that benzene destruction simply requires more severe conditions than those behind Fig. 9.17.

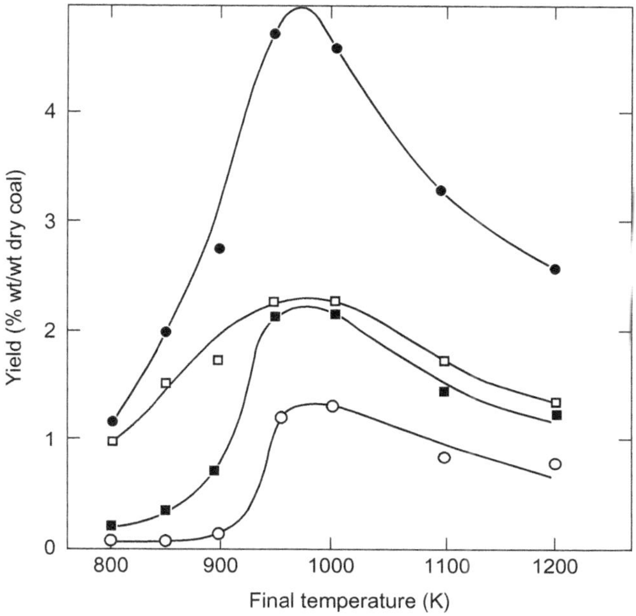

Fig. 9.18 Yields of (●) oils, (□) BTX, (■) benzene, and (○) PCX from a hv bituminous coal after heating at 7°C/s under 15 MPa H_2 with 11 s gas contact time.
Reproduced from Finn MJ, Fynes G, Ladner WR, Newman JOH. Light aromatics from the hydropyrolysis of coal. Fuel 1980;59:397–404, with permission from Elsevier.

Benzene destruction is resolved in the partial product compositions for slow heating of a hv bituminous coal at 15 MPa in Fig. 9.18. The gas contact time was much shorter in these tests than in Fig. 9.17, yet the temperatures for significant changes are very similar. For example, benzene became the predominant product in BTX at 675°C (948 K) in Fig. 9.18. The new and more interesting feature from Fig. 9.18 is that the maxima in oils yields, BTX, and PCX coincide at 700°C (973 K), which indicates that BTX and PCX are destabilized at the same temperature under 15 MPa H_2. Both product lumps are converted into GHCs, since the reduction in the oils yield is only slightly less than the sum of the reductions in the BTX and PCX yields, and secondary tar yields always diminish for progressively hotter temperatures.

The GHC yields associated with maximum oils yields appear in Fig. 9.19. These yields belong to the same dataset as the oils yields in Fig. 9.16 for 4 MPa (Tang et al., 1999). The population of GHCs is rendered into mixtures of CH_4 plus C_2 species, and C_2's never comprise as much as a quarter of the GHC yield with any coal type. The CH_4 yields are three-to-four times greater than those for primary hydropyrolysis, and the C_2 yields are comparable to total GHC yields for primary hydropyrolysis. These particular data show that CH_4 yields increase for progressively higher coal ranks through hv bituminous. However, the hydrogasification reactivities assigned for the cluster of six hv bituminous samples are much faster than for many other hv

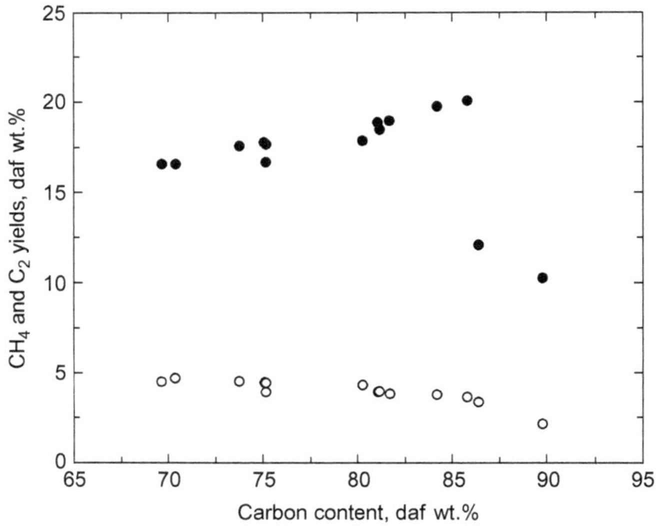

Fig. 9.19 Yields of (●) CH_4 and (○) C_2 GHCs after heating at 7000°C/s to 750°C with 42 s gas contact time under 4 MPa H_2.
Reproduced from Tang L, Zhu Z, Gu H, Zhang C. The effect of coal rank on flash hydropyrolysis of Chinese coal. Fuel Process Technol. 1999;60:195–202, with permission from Elsevier.

bituminous samples from other datasets (as seen in Fig. 9.8). So at face value, these data indicate that CH_4 yields increase for coals of progressively higher rank, then fall off sharply for low volatility coals. But in combination with data from other testing programs, the tendency is for relatively uniform CH_4 yields for ranks through hv bituminous and abrupt reductions for low volatility coals, so that the coal quality impacts on the yields of oils and CH_4 are the same. The C_2 yields in Fig. 9.19 slightly diminish for ranks through mv bituminous, then fall off more sharply through the low volatility ranks.

Having already seen the clear indications that peripheral groups are eliminated from BTX and PCX in Figs. 9.17 and 9.18, and knowing that the methyl groups on toluene, xylene, and xylenol are abundant peripheral groups in oils, one could suppose that oils are a major source of GHCs during tar hydroconversion. In fact, nearly all the enhancements to CH_4 yields in Fig. 9.19 come from char hydrogasification, as evident in quantitative estimates for this particular dataset in Fig. 9.26 below.

Trends in the yields of CO and CO_2 with variations in temperature and H_2 pressure appear in Fig. 9.20. Since the gas contact time was 42 s in these tests, the tendencies reflect an equilibration of the gas compositions, especially at the hottest temperatures and highest pressures. The yields of both CO and CO_2 grow for progressively hotter temperatures; CO yields also increase for higher H_2 pressures, while CO_2 yields diminish. Oxygen is eliminated from carbon oxides for progressively higher pressures, since the change in CO_2 is not nearly compensated by a commensurate increase in CO.

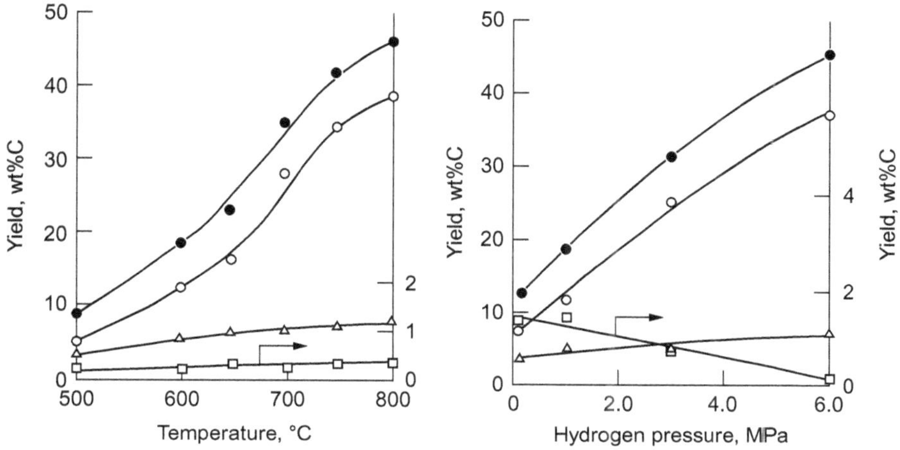

Fig. 9.20 Yields of (●) C-bearing gasses, (○) CH_4, (△) CO, and (□) CO_2 after heating a lignite at 7000°C/s with 42 s gas contact time vs. (left) temperature under 6 MPa H_2 and (right) vs. H_2 pressure at 800°C.
Reproduced from Zhu Z, Ma Z, Zhang C, Jin H, Wang X. Flash hydropyrolysis of northern Chinese coal. Fuel 1996;75:1429–33, with permission from Elsevier.

In other words, moisture yields also grow for progressively higher H_2 pressures with this coal and conditions. The gas yields in Fig. 9.14 for hv bituminous coal have CO and CO_2 yields diminishing in tandem for progressively higher H_2 pressures. These tendencies also reflect the extremely dilute coal loadings and, therefore, extremely high H_2 concentrations in all these tests.

The cumulative C-distributions in Fig. 9.21 concisely summarize the main features in the product distributions from tar hydroconversion plus hydrogasification. The tendencies are illustrated for progressively longer contact times in Fig. 9.21, although they would essentially be the same for temperature sweeps for fixed contact time and H_2 pressure and for H_2 pressure sweeps for fixed contact time and temperature. For progressively more severe hydroconversion conditions, oils first lose their substituents, then hydrocrack into GHCs and moisture. GHCs heavier than C_2's are reformed first, followed by the C_2's, so that CH_4 is the ultimate hydrocarbon product under the most severe hydrogasification conditions. Indeed, for the longest contact time in Fig. 9.21, the only C-bearing products are methane and the carbon oxides. The rendition of this data in Fig. 9.21 shows less C-conversion for 10 s contact time than at 7.5 s. However, an earlier rendition of the data reported at the time the tests were conducted (Fallon et al., 1980) has the conversion saturate to an asymptotic level for contact times longer than 5 s. It is hard to fathom how total conversions can diminish for progressively longer contact times at fixed temperature and pressure, because char hydrogasification always continues to gasify char.

Due to the impact of hydrocracking, oils yields are maximized within a temperature window that shifts toward cooler temperatures for progressively higher H_2 pressures

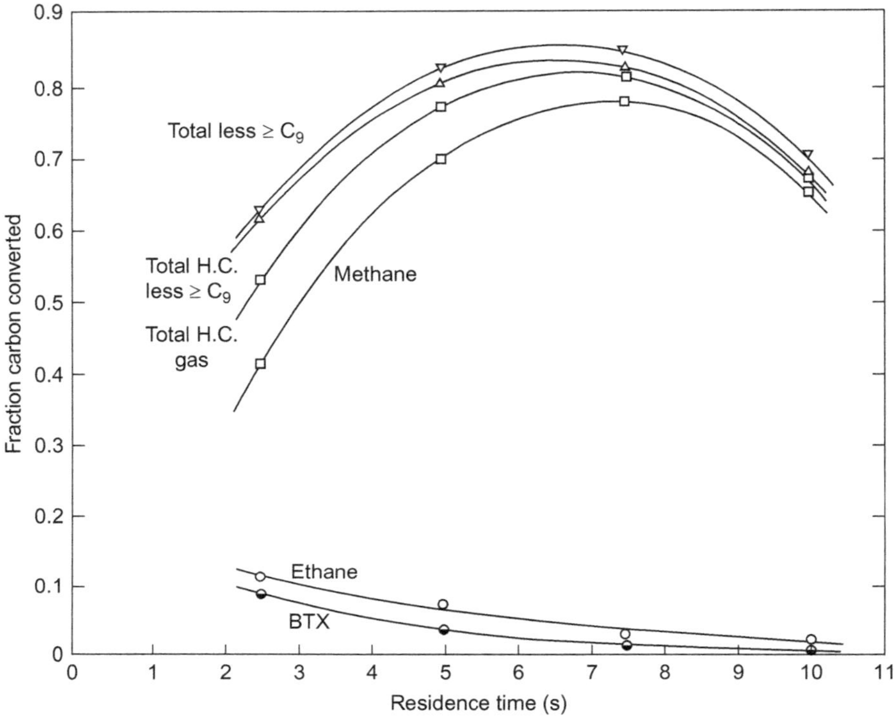

Fig. 9.21 Cumulative distribution of C-conversion vs. contact time during flash hydrogasification of lignite at 825°C and 20.7 MPa H_2. In ascending order, the curves depict absolute conversions for BTX, C_2's, and CH_4; then cumulative conversions for $CH_4 + C_2$'s; GHCs+BTX; GHCs+BTX + CO_X. Secondary tars are excluded but were <0.01C-fraction for these conditions.
Reproduced from Steinberg M. The flash hydropyrolysis and methanolysis of coal with hydrogen and methane. Int J Hydrogen Energy 1987;12:251–66, with permission from Elsevier.

and, to a lesser extent, for progressively longer gas contact times. For example, the window spans 725 to 775°C for pressures of several MPa H_2. The lower temperature must be sufficient to achieve a truly ultimate primary hydropyrolysis tar yield, because oils yields always increase in tandem with the primary tar yield. The upper temperature must be cool enough to minimize oils hydrocracking. Otherwise, oils are hydrocracked into CH_4. Under the most favorable conditions, essentially all secondary tar is hydroconverted to achieve the maximum oils yield, while negligible amounts of oils are hydrocracked into CH_4.

Trends among GHCs are almost completely obscured by the predominant CH_4 contribution from char hydrogasification. C_2's are produced during the hydroconversion and decomposition of heavier tar fragments, and as heavier GHCs are reformed away. Eventually even the C_2's are hydrogenated into CH_4. Methane yields are also supplemented by the elimination of peripheral groups from oils, albeit at much lower levels than by char hydrogasification.

The carbon oxides are relatively minor repositories for tar-O, which mostly ends up in moisture.

9.5.3 Mechanistic interpretations for tar hydroconversion

The only reaction mechanism developed for tar hydroconversion is based on FLASHCHAIN® and was presented only recently (Niksa, 2018b). The following section is based on this extension. Throughout the model development, all chemistry is homogeneous and involves reactants, intermediates, and products that remain vapor species, except for soot. Heterogeneous tar chemistry on surfaces of char, inorganics, and metals is omitted.

9.5.3.1 Phenomenology

The mechanism combines the FLASHCHAIN®-based mechanism for tar decomposition under inert gases with aspects of the coal hydropyrolysis mechanism plus new steps for hydrocracking and oils production. The goal is to accurately predict oils yields from any coal at any hydrogasification conditions. Oils yields from hydropyrolysis alone are small fractions of the ultimate yields. So the most direct evidence for tar hydroconversion during coal hydrogasification is an abundance of oils, light GHCs, and moisture with very little PAH at moderate temperatures, and that the product distributions relax to methane and moisture at hotter temperatures, with minor amounts of ethane, carbon oxides, NH_3, and H_2S. The radical simplification of primary products under elevated H_2 pressures contrasts with the abundance of PAH and, ultimately, soot under inert atmospheres, along with GHCs, oxygenated gases, H_2, HCN, and H_2S. Once all polynuclear tar compounds have been hydroconverted, the distribution of oils and noncondensable gases is determined by reforming chemistry in the gas phase, which explains why CH_4 and moisture are the ultimate products under elevated H_2 pressures given sufficient contact time and thermal severity.

Since tar constitution shares numerous features in common with coal macromolecules, the tar reactant is specified with the same four structural components as coal: Aromatic nuclei (A), labile bridges (B), labile peripheral groups (S), and char links (C), along with new structural components for hydrogenated bridges (B*) and peripheral groups (S*). Open fragment ends may have lost either hydrogenated (I) or labile (E) peripheral groups, which determines whether or not they recombine with other fragment ends.

The tar hydroconversion mechanism describes the disintegration of a tar MWD into oils and additional noncondensables. It does not describe the elimination of substituents from oils; the decomposition of oils into alkanes and olefins; the conversion of GHCs into CH_4 and C_2's; water-gas shifting; and other prominent stages of homogeneous reforming chemistry. Validated elementary reaction mechanisms are already available to simulate the chemistry of mixtures of single-ring compounds and lighter organic species, including pollutant formation. Consequently, the mechanism describes the homogeneous hydroconversion of C/H/O/N/S in primary hydropyrolysis tar into oils, GHCs, CO, CO_2, H_2O, H_2S, NH_3, and H_2, but omits the succeeding conversion of oils into additional GHCs and the reforming of noncondensables into CH_4 and moisture.

The tar hydroconversion mechanism incorporates all the steps in the FLASHCHAIN®-based mechanism for tar decomposition under inert gases (cf. Chapter 7), including those for soot production. Even though sooting would be completely disrupted under commercial hydrogasification conditions, its inclusion enables the analysis to span the range of low-to-moderate H_2 pressures in conventional gasification applications where hydroconversion can interfere with, but not completely eliminate soot production under some conditions. Unfortunately, the available database does not identify the threshold H_2 pressure that completely suppresses soot production. In addition, three new roles for H_2 factor into tar hydroconversion, the first of which is analogous to primary hydropyrolysis in the condensed coal phase: (1) hydrogenation of labile bridges and peripheral groups in tar fragments without any transformations of aromatic nuclei or bridge scissions, accompanied by suppression of recombinations among the hydrogenated ends of tar fragments; (2) hydrocracking of the labile and hydrogenated bridges and char links in tar to produce smaller tar fragments without any transformations of aromatic nuclei; and (3) hydrogenation of isolated tar monomers into oils, moisture, H_2S, and NH_3. All three processes utilize ambient H_2 at rates that accelerate for progressively higher H_2 pressures.

Hydrogen may react with the components of labile bridges in various ways, including conversion of carboxylic acids into alcohols, which destroys crosslinking agents that skew bridge conversions toward the spontaneous generation of refractory char links away from scission of longer fragments into smaller ones. Once a tar bridge has been hydrogenated, its only decomposition pathway is scission because its potential for spontaneous conversion into a char link has been eliminated. Accordingly, bridge hydrogenation shifts the tar MWD toward lighter fragments, away from the size range that may spontaneously nucleate into soot, and also enhances oils yields by promoting the production of tar monomers.

Once a bridge has been hydrogenated, it may also interfere with another charring process, bimolecular recombination of tar fragments. Whenever a hydrogenated bridge breaks, its remnants remain attached to the newly formed ends of the two new fragments. A fragment end that had been hydrogenated, either with or without the remnants of its parent hydrogenated bridge, is regarded as capped by stable functionalities and therefore unable to produce a char link in combination with another fragment end. According to the tar decomposition mechanism for inert atmospheres, without hydroconversion the ends of tar fragments can recombine regardless of whether or not the end contains the remnants of broken bridges. Since hydrogenation eliminates unsaturated components from bridges, and since such components readily coalesce further into aromatics, and since aromatics are essential to the refractory character of char links, hydrogenation interferes with bimolecular recombination.

Hydrocracking destroys labile and hydrogenated bridges and char links in a tar fragment. Each of these channels produces two smaller product fragments which have labile, hydrogenated, or capped ends, depending on the original linkage. These reactions incorporate ambient H_2, although the stoichiometric coefficients are deemed to be negligible because the explicit purpose of hydrocracking is to disintegrate tar fragments into smaller ones and, ultimately, to promote monomer production, and thereby enhance oils production. None of these channels produces noncondensable gases, so hydrocracking is solely a scission process.

In the tar decomposition mechanism, oils form from a tar monomer, provided that at least one of the two sites for end groups contains either a labile or hydrogenated peripheral group to act as a hydrogen donor. All the carbon in the nucleus is incorporated into the oils, whereas all carbon in the fragment ends becomes noncondensable gases. In contrast, the proposed monomer hydrogenation process generates oils from monomers with two end groups of any form attached to an aromatic nucleus, including capped and open ends. The types of the end groups determines how much oils form, because all the carbon in a monomer – both aromatic and aliphatic – is converted into oils, and also whether H_2 and H_2O act as reactants or products, because oils contain oxygen and hydrogen in fixed proportions. For example, with two hydrogenated peripheral groups, H_2 could conceivably be released if their H-contents exceeds the H-requirement of oils; and with two open or capped ends, the oxygen requirement in oils may need to be supplied by moisture as a reactant. Regardless of the forms of the end groups, all the carbon in the monomer is incorporated into oils so that the stoichiometry for oils production will be a weighted sum over all possible end group configurations.

9.5.3.2 Reaction scheme

The reactant in the mechanism is a stream of primary tars from rapid primary hydropyrolysis. Primary tars' macromolecular structure is rendered in terms of the structural components of FLASHCHAIN® as a mixture of chains ranging in size from a monomer to a specified degree of polymerization, J*. In principle, tar weights extend to 2000 to 4000 g/mol, depending on the parent coal constitution, but for the elevated H_2 pressures in the target applications, extents of the MWDs of primary tar are only half as large.

Fifteen chemical processes, diagrammed in Table 9.7, represent the spontaneous hydroconversion of tar fragments: Hydrogenation of labile bridges and peripheral groups; hydrocracking; monomer hydrogenation into oils; scission of labile and hydrogenated bridges; spontaneous charring of labile bridges; bimolecular recombination; elimination of labile and hydrogenated peripheral groups; release of oxygen from nascent char links; release of nitrogen from nuclei; release of S_{th} from nuclei; oil production from tar monomers; nucleation of the largest tar molecules into soot; addition of smaller tar molecules onto soot; and addition of oils onto soot. Rates of these reactions only partially determine rates of product evolution, because of the independent influence of fragment statistics. The tar hydroconversion mechanism comprises the 11-step reaction mechanism in the FLASHCHAIN®-based tar decomposition mechanism plus four new reactions. The only changes to the tar decomposition mechanism are that scission now opens both labile and hydrogenated

Table 9.7 Fifteen reactions for tar hydroconversion.

Reaction	Diagram
Hydroconversion	
Labile bridge hydrogenation	$t_{j\text{-}k}–B–t_k + \nu_{HY}H_2 \rightarrow t_{j\text{-}k}–B^*–t_k$
Labile peripheral group hydrogenation	$t_j{\cdot}S + (\nu_{HY}/2)H_2 \rightarrow t_j{\cdot}S^*$
Hydrocracking	$t_{j\text{-}k}\text{-}(B,B^* \text{ or } C)–t_k \rightarrow (S,S^* \text{ or } E){\cdot}t_{j\text{-}k} + (S,S^* \text{ or } E){\cdot}t_k$
Monomer hydrogenation	$(S,S^*,E \text{ or } I)\text{-}t_1\text{-}(S,S^*, E \text{ or } I) + \nu_{H2,MH}H_2 \rightarrow \nu_{O,MH}O + \nu_{H2O,MH}H_2O$
Tar decomposition	
Scission of labile and hydrogenated bridges	$t_{j\text{-}k}–(B \text{ or } B^*)–t_k \rightarrow (S \text{ or } S^*){\cdot}t_{j\text{-}k} + (S \text{ or } S^*){\cdot}t_k$
Spontaneous charring	$t_{j\text{-}k}–B–t_k \rightarrow t_{j\text{-}k} = t_k + \nu_C G$
Bimolecular recombination	$t_j{\cdot}S + t_k \rightarrow t_j = t_k + (\nu_C/2)G$
Elimination of labile and hydrogenated peripheral groups	$t_j{\cdot}(S \text{ or } S^*) \rightarrow t_j + (\nu_{G^*}G^* \text{ or } (\nu_C/2)G)$
Release of char-O	$C\text{-}O\text{-}C \rightarrow CO + C'$
Release of N in nuclei	$A\text{-}\eta N \rightarrow \eta HCN + A'$
Release of Sth in nuclei	$A\text{-}\xi_{th}\, S_{th} \rightarrow \xi_{th}H_2S + A'$
Oil production	$t_1{\cdot}s \rightarrow b_O O + \Sigma b_i G_i$
Soot nucleation	$t_j \rightarrow \nu_{Rj}\, R + \Sigma c_{ij} G_i;\ J^* + 1 < j < 2\,J^*$
Soot addition	$t_j + R \rightarrow (\nu_{Rj} + 1)R + \Sigma c_{ij} G_i;\ 1 < j < J^*$
Oil addition	$O + R \rightarrow (\nu_{R,O} + 1)R + \Sigma c_{i,O} G_i$

Reproduced from Niksa S. Flashchain theory for rapid coal devolatilization kinetics. 11. Tar hydroconversion during hydrogasification of any coal. Energy Fuels 2018b;32:7569–84, with permission from the American Chemical Society.

bridges; gases are produced via elimination of both labile and hydrogenated peripheral groups; and oils form from monomers with either labile or hydrogenated peripheral groups. Otherwise their characteristics and rate expressions remain as described in detail in Chapter 7.

The labile bridge hydrogenation reaction is based on

$$B + \nu_{HY}H_2 \rightarrow B^*$$

where B and B^* are the molar concentrations in tar of labile and hydrogenated bridges, respectively; and ν_{HY} is the stoichiometric coefficient for complete hydrogenation. The global bridge hydrogenation rate is given by

$$R_{HY} = k_{HY}BC_{H_2}^{n_{HY}} \tag{9.5a}$$

where k_{HY} is a distributed-energy rate constant, C_{H2} is the H_2 concentration in the gas phase, and n_{HY} is an empirical reaction order.

The stoichiometric requirement for hydrogenation is based on complete hydrogenation of all carbon, oxygen, and sulfur in the labile bridge. The carbon is hydrogenated into a mixture of CH_4 and C_2H_4, in proportions that maintain an H/C ratio of three. All oxygen becomes H_2O and all sulfur becomes H_2S. Hence, hydrogenation leaves only short methylene hydrocarbon chains capped by methyl groups and alcohol functional groups in the hydrogenated bridge, which is the theoretical maximum extent possible. This limiting structure specifies the stoichiometric requirement for hydrogenation as the difference between the H-numbers of hydrogenated and labile bridges, according to

$$\nu_{HY} = \frac{3\,{}^{C}N\mathrm{B} + 2\left({}^{O}N\mathrm{B} + {}^{S}N\mathrm{B}\right) - {}^{H}N\mathrm{B}}{2} \tag{9.5b}$$

where ${}^{E}N_{B}$ with $E = $ C, H, O, S are the numbers of E-atoms per labile bridge.

As always, labile bridges decompose in independent pathways for scission and spontaneous condensation, and the selectivity for bridge conversion is fixed throughout. But hydrogenated bridges never spontaneously condense into refractory char links plus noncondensables, because of insufficient aromaticity. Hydrogenated bridges break into two hydrogenated peripheral groups. Noncondensables are not released during the scission, although these peripheral groups may subsequently decompose into noncondensables at a distinctive rate. Bridge hydrogenation also suppresses bimolecular recombination reactions if either one of the participating fragment ends holds the remnants of a broken hydrogenated bridge, or is bare but had been hydrogenated previously.

Consequently, hydrogenation shifts the bridge decomposition process toward scission by eliminating the decomposition pathway that produces char links, and further suppresses char link production by eliminating bimolecular recombinations for selected fragment end configurations. It thereby generates smaller fragments for progressively greater extents of bridge hydrogenation. Since more monomers will also be generated, hydrogenation enhances oils yields. And since all bridge elements may ultimately be converted into noncondensables because no char links form from a hydrogenated bridge, hydrogenation also enhances noncondensables yields. Bridge scissions also inhibit the production of the largest fragments which are the only ones that can nucleate soot. So bridge hydrogenations suppress soot production when tars are converted under H_2, albeit as only one factor among others.

Hydrogenation of labile peripheral groups and their subsequent decomposition into noncondensables are analogous to their counterparts for labile bridges because labile peripheral groups have the same compositions as labile bridges.

The hydrocracking process is uniformly applied to labile and hydrogenated bridges and char links with a common set of kinetic parameters. In actuality, the hydrocracking kinetics should be different among these linkages, but neither the theoretical foundations nor the empirical validation database are sufficient to specify distinctive kinetic parameters for each linkage type. For the same reasons, the reaction order with respect to H_2 pressure is the same for each linkage. None of the hydrocracking channels produce noncondensable gases, so hydrocracking is solely a scission process. Each channel produces two smaller product fragments but the ends may be labile, hydrogenated, or capped, depending on the reactant linkage.

Similarly, the rate of monomer hydrogenation into oils is deemed to be independent of the types of fragment ends attached to the nucleus, although the stoichiometry differs among the various end configurations because all carbon in the monomer is incorporated into oils. Oils are mixtures of BTX and PCX. To simplify the accounting, the stoichiometric coefficients are specified for a fixed composition of oils from all coals, based on the reported yields of BTX and PCX from the rapid pyrolysis of 17 diverse coals (Xu and Tomita, 1987). On average, oils contain 13.7, 17.5, and 15.5 mole% BTX, respectively, and 28.5% and 24.8% phenol and cresol, which gives 6.73C-atoms/mol oil; 7.47H; and 0.53 O, for a mean molecular weight of 96.7 g/mol. These assignments determine the stoichiometric coefficients for monomer hydrogenation (Niksa, 2018b).

The rate of oils production from monomer hydrogenation is given by

$$R_{O,MH} = \nu_{O,MH} k_{MH} t_1 C_{H_2}^{n_{MH}} \tag{9.5c}$$

where k_{MH} is an Arrhenius rate constant and n_{MH} is an empirical reaction order. Analogous expressions describe the rates of H_2 consumption and H_2O production with their respective coefficients.

Rate equations for the proposed reaction set were developed with the analytical approach used for FLASHCHAIN® and have been reported (Niksa, 2018b).

9.5.3.3 Tar constitution submodel and rate parameter assignments

Tar composition is markedly different than a parent coal's because primary tars are always enriched in hydrogen, and because they contain different proportions of the structural components. So all aspects of tar constitution and all the rate parameters are evaluated from the elemental composition of the whole primary hydropyrolysis tar sample predicted by FLASHCHAIN®, rather than the standard coal properties. In addition, both carbon and proton aromaticities are evaluated from a database for petroleum asphaltenes, which covers the range of H/C ratio for primary coal tars.

All stoichiometric coefficients are evaluated from tar constitution, and variations in these coefficients are primarily responsible for interpreting the large variations in the product distributions reported for diverse coal types. All the parameters in the 11-reaction tar decomposition mechanism were fixed for all coals at the values from the independent validation cases in Chapter 7 and elsewhere. Similarly, the kinetic parameters for hydrogenation of bridges and tar monomers were fixed for all coals; those for hydrocracking were fixed as well, except that the frequency factor had to be adjusted by a factor of four to interpret one of the validation datasets (Niksa, 2018b).

For char hydrogasification, the assigned activation energy of 81.5 kJ/mol and the order of one-half are more secure, since these are the same values used to previously interpret hydrogasification conversions for 21 coals in Section 9.4. Although these parameters were fixed for all coals, the pseudo-frequency factor for hydrogasification was tuned-in to match the reported total weight loss for each sample, and presented in Fig. 9.8.

9.5.3.4 Predicted behavior

The first case in Fig. 9.22 covers variations in H_2 pressure with a subbituminous coal for rapid heating and 3 s gas contact time. The measured oils yields surge from 0.3 to 1 MPa, then remain essentially constant for all higher H_2 pressures, while the measured secondary tar yields decay very gradually across the entire pressure range (Xu et al., 2003a). The predicted oils yields are within measurement uncertainties throughout, whereas secondary tar yields diminish slightly faster than the measured values, although the discrepancy is significant only at 0.3 MPa. This case is the only one in this chapter that included soot nucleation and growth in the simulations, so that the 12.7 daf wt% tar yield for 0.3 MPa in Fig. 9.22 is actually a sum of 5.2% tar plus 7.5% soot. The decay in the predicted tar yields corresponds with the decay in the predicted primary tar yields from 32% at 0.3 MPa to 21% at 5 MPa. Despite this reduction in the precursors to oils, the tar hydroconversion mechanism accurately depicts the saturation in the oils yields above 1 MPa. Higher H_2 pressures accelerate hydrocracking and monomer hydrogenation to compensate for the diminishing primary tar yields.

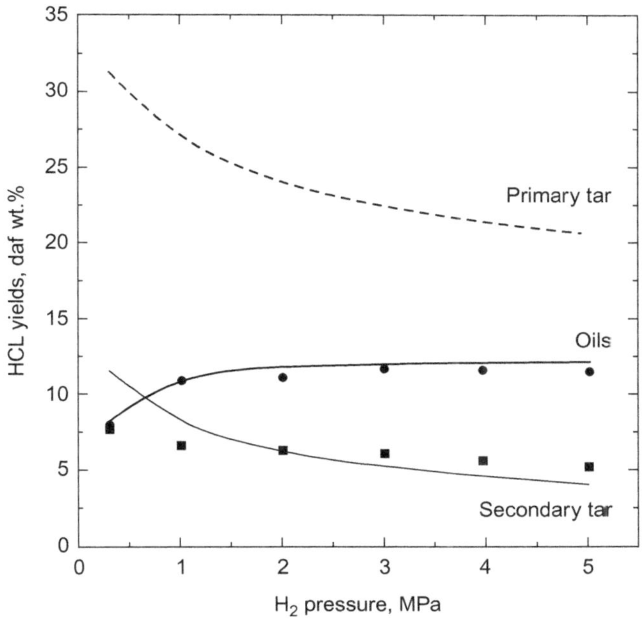

Fig. 9.22 Evaluation of yields of (●) oils and (■) secondary tar from a subbituminous coal for heating at 2000°C/s to 800°C under various H_2 pressures. Predicted primary tar yields appear as the *dashed curve*. Tests imposed 1.8 s coal contact time and 3 s gas contact time in a FFR. Reproduced from Niksa S. Flashchain theory for rapid coal devolatilization kinetics. 11. Tar hydroconversion during hydrogasification of any coal. Energy Fuels 2018b;32:7569–84, with permission from the American Chemical Society.

The EFR tests in Fig. 9.23 covered a temperature sweep from 665 to 790°C with gas contact times from 14 to 23 s (Ikura and Last, 1988). The measured tar yields diminish from 3% to 1% while oils yields increase from 6.5% to roughly 10%, so that the sums gradually increase across the temperature range, in tandem with the predicted primary tar yields in Fig. 9.23. The predicted secondary tar yields are within measurement uncertainties throughout, and the oils yields accurately depict the trend, albeit within the considerable scatter in the measured yields. The notable exception is the very large over-prediction for the oils yield at 790°C, at 11% vs. 1.5%. This discrepancy reflects the omission of oil hydrogenation into CH_4 in the hydroconversion mechanism, which is relatively slow and therefore comes into play only for very long contact times and very high H_2 pressures (Finn et al., 1980; Zhu et al., 1996; Steinberg, 1987), like the 20 s contact time and 12.9 MPa H_2 in this particular test. This effect can be depicted by combining the hydroconversion mechanism with an elementary reforming mechanism.

The contribution from char hydrogasification is also resolved in Fig. 9.23 as the difference between the measured or predicted weight loss and the predicted hydropyrolysis yield for each test condition. In these simulations, the same hydrogasification rate was used for each set of test conditions. Predicted weight loss is within measurement uncertainties for all tests, and exhibits the step-change at 745°C when the coal contact time was increased from 3.7 to 5.5 s. For this particular coal and these operating conditions, hydrogasification was responsible for about 45% of the total coal conversion, which is so high because of the extreme H_2 pressure in this test series. Hydrogasification generated 35–40 daf wt% CH_4, and minor amounts of H_2O, NH_3, and H_2S.

The previous two evaluations covered tests with rapid heating rates, for which tar yield enhancements during hydropyrolysis would be minimal. The evaluation in Fig. 9.24 covers heating at 5°C/s to 600°C in a first reactor stage, with tar hydroconversion in a second unpacked flow reactor at 850°C with 4 s transit time (Finn et al., 1980). As seen in Fig. 9.24, the predicted primary tar yields are as large as those for heating rates approaching 10^4°C/s in Fig. 9.23, which definitely reflects tar yield enhancement due to hydrogenation in the condensed coal phase under these slow heating conditions. Consequently, the oils yields are also as large as those for much faster heating rates. The predicted oils yields depict the observed coal quality impact, whereby oils yields pass through a weak maximum for hv bituminous coals before falling off sharply for low volatility coals. They are within measurement uncertainties through 85 daf wt% C, except that they do not depict the distinctive low oils yield for one of the coals with 84% C because its primary tar yield is also not much smaller than those for the other hv bituminous coals. The accuracy deteriorates for the low volatility coals; this must be due to under-predicted primary tar yields for the two coals of highest rank rather than problems in the hydroconversion mechanism, because the predicted primary tar yields are much lower than the measured oils yields. The measured secondary tar yields are approximately 2% until they diminish for both anthracites, whereas the predicted secondary tar yields are insignificant for all coals.

The sample-to-sample variability among oils yields from coals of the same nominal rank is apparent in Fig. 9.25, based on Tang et al.'s dataset (1999, 2003) for rapid

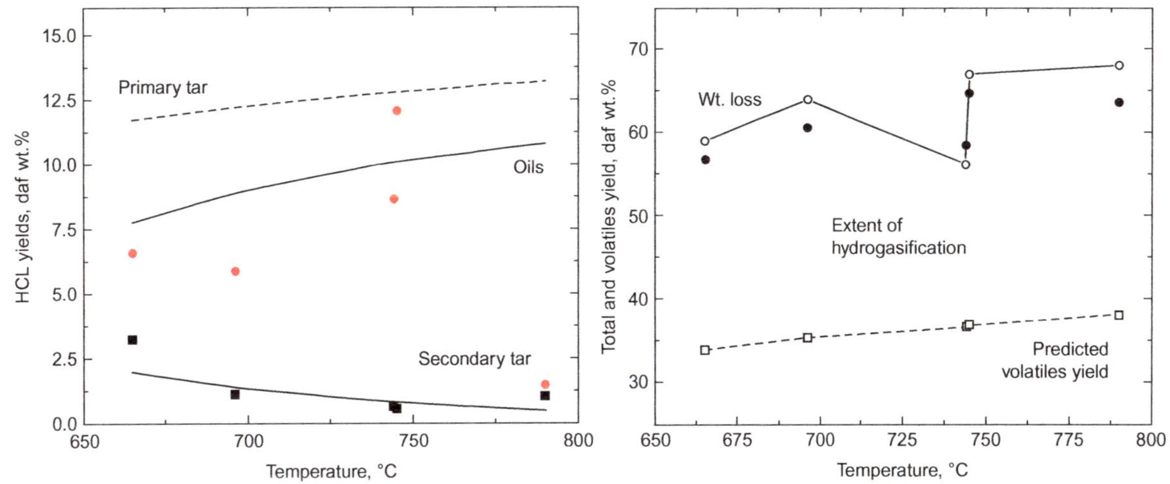

Fig. 9.23 (Left) Evaluation of yields of (●) oils and (■) secondary tar from a hv bituminous coal blend for heating at 8000°C/s to different temperatures under 12.9 MPa H_2 with coal contact times from 3.7 to 6.5 s and gas contact times from 14 to 23 s. Predicted primary tar yields appear as the *dashed curve*. (Right) (●) Measured and (○ connected by line segments) predicted total weight loss and (□ connected by *dashed line* segments) predicted volatiles yields for the same test conditions.
Reproduced from Niksa S. Flashchain theory for rapid coal devolatilization kinetics. 11. Tar hydroconversion during hydrogasification of any coal. Energy Fuels 2018b;32:7569–84, with permission from the American Chemical Society.

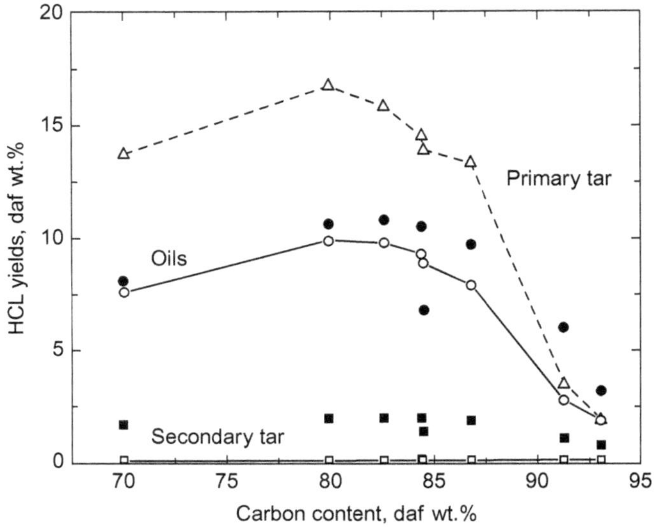

Fig. 9.24 Evaluation of yields of (●) oils and (■) secondary tar from diverse coals for heating at 5°C/s to 600°C for 900 s under 15 MPa H_2 in a first reactor stage with hydroconversion is a second reactor stage at 850°C with 4 s gas contact time. Predicted products appear as corresponding *open symbols* (○, □) connected by *solid line* segments. Predicted primary tar yields appear as (△) connected by *dashed* segments.
Reproduced from Niksa S. Flashchain theory for rapid coal devolatilization kinetics. 11. Tar hydroconversion during hydrogasification of any coal. Energy Fuels 2018b;32:7569–84, with permission from the American Chemical Society.

heating of 14 coals to 750°C under 4 MPa H_2. Although the coal contact time of 2 s is comparable to the contact times in Xu et al.'s FFR and Ikura et al.'s EFR, the 42 s gas contact time in Tang et al.'s tests is the longest, by far, in the validation database. This distinctive condition factors into the explanation for an apparent discrepancy in the predicted oils yields. The hydroconversion mechanism regards oils as mixtures of BTX and PCX in fixed proportions for all coals. These predicted values in Fig. 9.25 are consistently greater than the measured oils yields by 3 to 5 daf wt% across the rank spectrum. However, the literature database on tar hydroconversion firmly establishes that oils mixtures are not stable under hydrogasification conditions; rather, they eliminate their substituents as CH_4 and H_2O, which eventually renders oils mixtures into pure benzene (Ikura and Last, 1988; Finn et al., 1980; Zhu et al., 1996). Given sufficient time at a sufficiently high H_2 pressure, benzene is reformed into CH_4, so that reported oils yields often pass through a maximum for progressively hotter temperatures (Finn et al., 1980; Fallon et al., 1980; Steinberg, 1987), as seen in Fig. 9.21.

The discrepancies for the oils mixtures in Fig. 9.25 were almost completely eliminated by arithmetically converting the predicted BTX + PCX mixtures into benzene, by applying a ratio of the molecular weights of benzene to oils or 78/96.7 to the

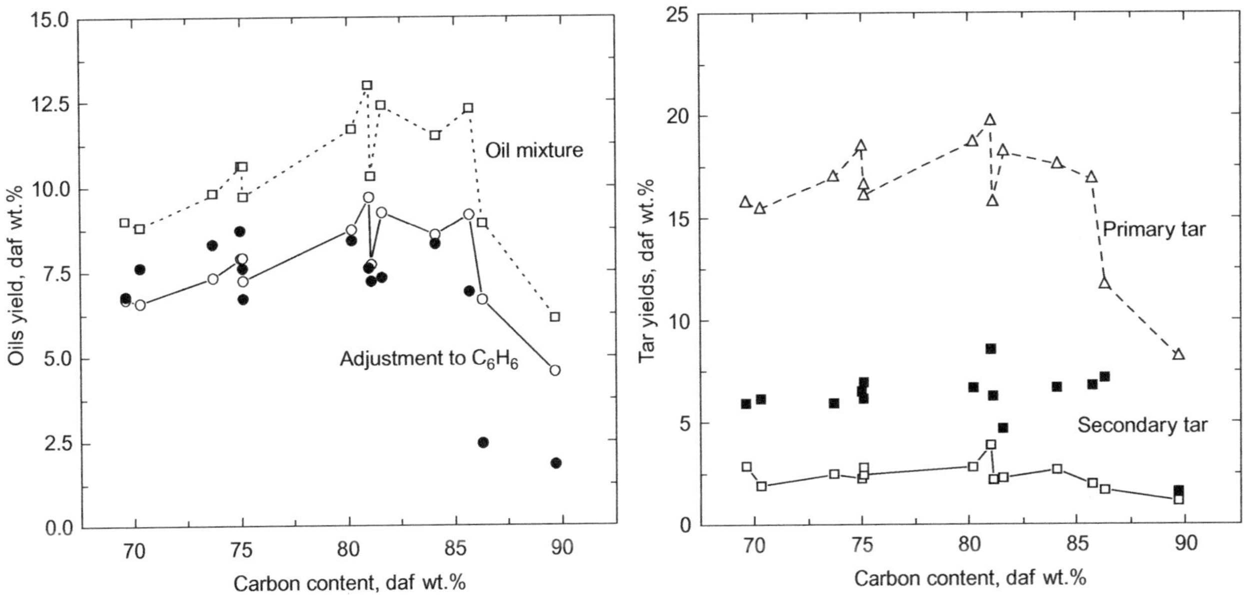

Fig. 9.25 Evaluation of yields of (●) oils and (■) secondary tar from diverse coals for heating at 7000°C/s to 750°C for 2 s under 4 MPa H_2 in a FFR with 42 s gas contact time. Predicted products appear as corresponding *open symbols* connected by *solid line* segments. The left panel also shows the predicted oils mixtures of substituted BTX + PCX as (□) connected by *dotted line* segments. Predicted primary tar yields appear in the right panel as (△) connected by *dashed* segments.

Reproduced from Niksa S. Flashchain theory for rapid coal devolatilization kinetics. 11. Tar hydroconversion during hydrogasification of any coal. Energy Fuels 2018b;32:7569–84, with permission from the American Chemical Society.

predicted oils mixtures. The predicted benzene yields are within the measurement uncertainties for all coals through 85 daf wt% C, except for two coals with about 81% C. Although the predicted benzene yields diminish for the three low volatility coals, the discrepancies are larger; however, the discrepancies are over-predictions for these tests, in contrast to the under-predictions for the two anthracites tested by Finn et al. (cf. Fig. 9.24). Although Tang et al. did not resolve oils as molecular species, they reported independent yields for the BTX and PCX lumps. Their oils contained, on average, <15 wt% PCX, which is four times smaller than the 60% PCX in the predicted oils mixture. In addition, Zhu et al. (1996) imposed the same conditions as Tang et al. on two lignites and reported that BTX and PCX were entirely reduced to benzene and xylenol, respectively, in tests at 750°C with 42 s gas contact time, albeit at 6 MPa H_2. So the adjusted oils yields in Fig. 9.25 are corroborated by direct measurements for these conditions.

This explanation raises the question as to why the oils were not completely eliminated for the very long gas contact times in these tests, when complete elimination was invoked to rationalize the very low oils yield reported for 790°C in only 20 s in Fig. 9.23. The hotter temperature is one factor to counteract the shorter gas contact time behind Fig. 9.23, but the predominant consideration is the much greater H_2 pressure, at 12.9 vs. 4 MPa. A quantitative resolution to this issue remains to be demonstrated with an elementary reforming mechanism.

The predicted secondary tar yields in Fig. 9.25 vary from 2 to 4 daf wt% vs. measured tar yields from 5% to 8%. Although the quantitative discrepancies are comparable to those in Fig. 9.24, note that the hydroconversion mechanism correctly predicts residual secondary tar for 42 s under 4 MPa H_2, but none for 4 s under 15 MPa. Another important feature in Fig. 9.25 is that the predicted primary tar yields clearly illustrate that the sample-to-sample variability in the tar yields is responsible for the variability in the yields of oils and secondary tar, rather than any coal quality impacts on any aspects of the tar hydroconversion mechanism. In other words, it is impossible to depict the sample-to-sample variability of hydroconversion products without first predicting the distinctive tar yields for particular coal samples from hydropyrolysis.

Predicted extents of char hydrogasification for these tests (Tang et al., 1999) appear in Fig. 9.26 as the difference between the predicted ultimate hydropyrolysis yields and the measured total weight loss. The predicted weight loss is not shown because this dataset represents only one set of operating conditions, and A_{HG} was adjusted to match the predicted weight loss to the measurement for each coal. The hydrogasification contributions are approximately 20% of the reported coal conversion for all coals except for the bituminous coal with 85.5% C, which has a much greater contribution. The relatively low contributions for nonbituminous coals reflect the coal contact time of only 2 s and the moderate H_2 pressure, whereas the enhanced conversions of the bituminous coals reflect accelerated reactivities, as shown in Fig. 9.8. Hydrogasification generated from 11 to 18 daf wt% CH_4, except for half as much from the coals of lowest rank and for 27% from the coal with 85.5% C.

According to this tar hydroconversion mechanism, oils form in two stages controlled by monomer hydrogenation and hydrocracking, respectively. The hydrogenation of labile bridges and peripheral groups is relatively unimportant. In the first stage,

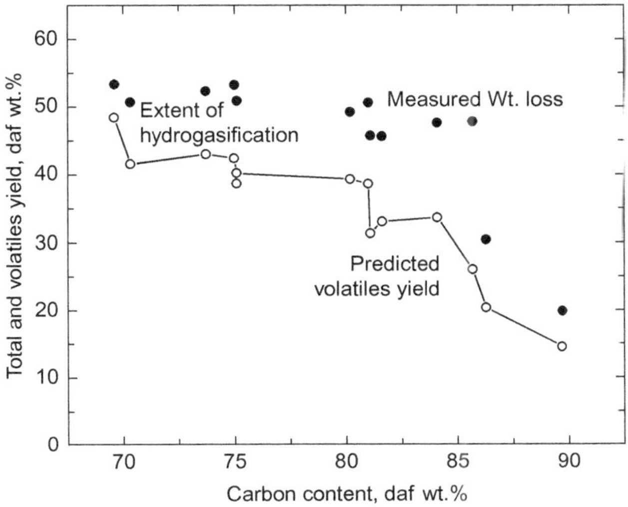

Fig. 9.26 (●) Measured total weight loss and (○ connected by *line* segments) predicted volatiles yields for the test conditions in Fig. 9.25.
Reproduced from Niksa S. Flashchain theory for rapid coal devolatilization kinetics. 11. Tar hydroconversion during hydrogasification of any coal. Energy Fuels 2018b;32: 7569–84, with permission from the American Chemical Society.

the tar monomers released as primary tars are rapidly hydrogenated into oils at the monomer hydrogenation rate. Since elevated pressures always shift primary tar MWDs toward lighter species, monomers constitute substantial fractions of primary tar, approaching one-half under some operating conditions. Consequently, as much as half the ultimate oils yield is produced soon after the onset of tar hydroconversion. In the second stage, additional tar monomers are gradually released by hydrocracking of larger tar molecules and then hydrogenated into oils, while control of the oils production rate shifts from monomer hydrogenation to hydrocracking. Tests with fine time resolution are needed to establish that monomer hydrogenation occurs in two distinct stages, as proposed here, and that the apportioning of the ultimate oils yield is accurate.

The mechanism has been validated across a broad range of coal and gas contact times, and correctly predicts that secondary tar persists for all but the highest H_2 pressures. But an ambiguity persists in the predicted dynamics: The predicted conversion of secondary tar may be too fast, since measured ultimate tar yields usually exceeded the predicted values. This could signal a need for more robust kinetics although, alternatively, the small amounts of secondary tar that persist at even the longest gas contact times could also reflect linkage-specific hydrocracking rates. Currently, the same hydrocracking kinetics are used for labile and hydrogenated bridges and char links for lack of supporting information to resolve any differences. But if char links were more resistant to hydrocracking than bridges, then tar fragments bound by only char links would persist longer than the rest.

Primary tar compositions and, especially, their structural components, determine oils yields from different coals. The main trend is that oils yields pass through a weak maximum for hv bituminous coals, then decay sharply for low volatility coals. The most interesting aspect of the coal quality impacts is the exceedingly large percentage of the material in aromatic nuclei incorporated into oils under the most favorable operating conditions with any coal type. Fig. 9.27 presents the ratio of the predicted oils yield to the mass of all aromatic nuclei in primary tar, A_{TAR}, vs. the C-content of the parent coal. Since tar nuclei can only ultimately appear in secondary tar and oils in the proposed mechanism, values less than unity indicate that some primary nuclei remain in secondary tar. Conversely, values greater than unity indicate that aliphatic tar components have also been incorporated into oils. The ratio values in Fig. 9.27 segregate into two groups as a function of H_2 pressure. At 3–4 MPa, aliphatics make up one-third of the oils with low rank coals and slightly less with bituminous and low volatility coals, whereas at 12.9–15 MPa, they constitute half or more of the oils mass. Consequently, the notions that aromatic nuclei in primary tar, which have two to four rings with bituminous coals and fewer or more for coals of lower and higher rank, respectively, are either winnowed down or completely converted into single-ring compounds are incompatible with reported ultimate oils yields. Indeed, a trial rendition of the monomer hydrogenation process in which only the mass in nuclei became oils could

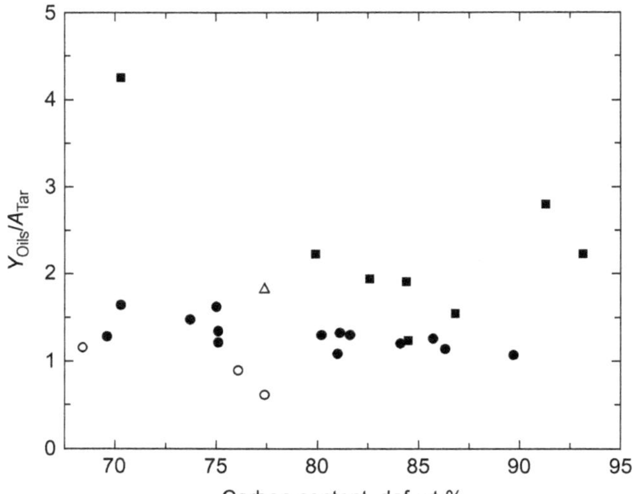

Fig. 9.27 Ratio of predicted oils yields to the amounts of aromatic nuclei in primary tar vs. coal-C content for (●) 14 coals from Tang et al. (1999) where Y_{OILS} are the predicted BTX/PCX mixtures rather than the benzene-only values; (○) 3 coals from Xu et al. (2003b); (△) 1 coal blend from Ikura and Last (1988) and (■) 8 coals from Finn et al. (1980).
Reproduced from Niksa S. Flashchain theory for rapid coal devolatilization kinetics. 11. Tar hydroconversion during hydrogasification of any coal. Energy Fuels 2018b;32:7569–84, with permission from the American Chemical Society.

not predict sufficient oils yields. However, the ultimate rendition demonstrates that monomer hydrogenation must also incorporate aliphatic peripheral groups into oils as well.

9.5.4 Nitrogen and sulfur species

This section covers the release of N- and S-species during hydrogasification. Specific contributions to the release of these species have not yet been resolved for hydropyrolysis, tar hydroconversion, and char hydrogasification. But the most important gaseous pollutant precursors have been identified.

9.5.4.1 N-species transformations

The impacts of variations in temperature and H_2 pressure on N-species distributions from three hv bituminous coals appear in Figs. 9.28 and 9.29, respectively. As seen in Fig. 9.28, the conversion of both char-N and tar-N into gaseous N-species increases in direct proportion to increases in temperature, as also found by Tang et al. (2003). Tar-N vanishes by 850°C, which indirectly indicates that oils are nitrogen free. Ammonia is the predominant gaseous N-species through 800°C, while the contribution for N_2 grows to about 10% at hotter temperatures. Cyanide makes inconsequential contributions throughout. Presumably, the appearance of N_2 at the hotter temperatures reflects chemistry in the gas phase, rather than a shift in the release mechanisms from char and tar.

For progressively higher pressures, the N-release increases, especially for pressures from 0.1 to 3 MPa, as seen in Fig. 9.29. Tang et al. (2003) also reported greater N-release for progressively higher pressures, but with greater sensitivity for H_2

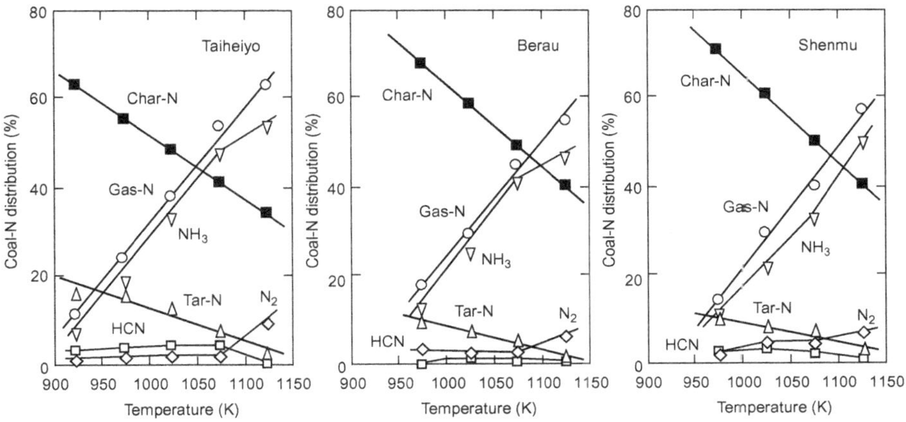

Fig. 9.28 Coal-N distributions vs. temperature for rapid heating under 3 MPa H_2 with coal and gas contact times of 2 and 3 s, respectively.
Reproduced from Xu W-C, Kumagai M. Nitrogen evolution during rapid hydropyrolysis of coal. Fuel 2002;81:2325–34, with permission from Elsevier.

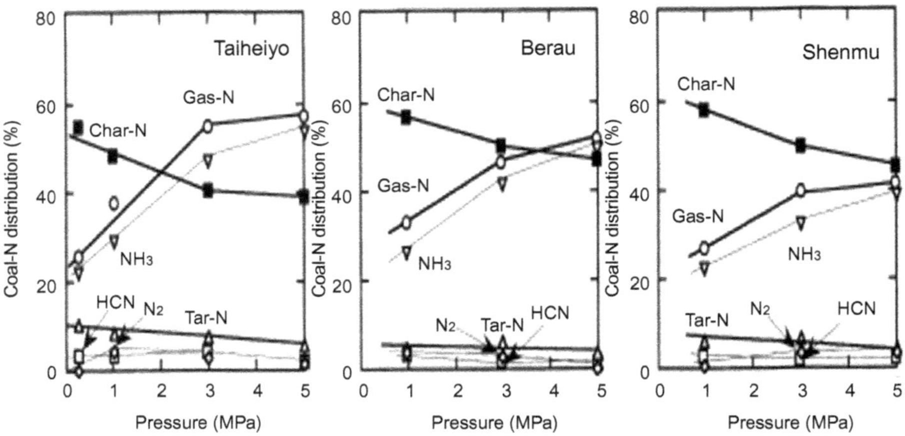

Fig. 9.29 Coal-N distributions vs. H_2 pressure for rapid heating to 800°C with coal and gas contact times of 2 and 3 s, respectively.
Reproduced from Xu W-C, Kumagai M. Nitrogen evolution during rapid hydropyrolysis of coal. Fuel 2002;81:2325–34, with permission from Elsevier.

pressures higher than 3 MPa. Tar-N diminishes for progressively higher H_2 pressures. Ammonia is the predominant gaseous N-species at all pressures, and contributions from N_2 and HCN remain negligible throughout.

The coal quality impacts on N-release are apparent in Fig. 9.30. The expected tendency for uniform N-release through the hv bituminous rank with an abrupt reduction for low volatility coals is hardly discernable amidst the excessive sample-to-sample variability. The nominal N-release of 35%–45% for most of these coals is somewhat lower than the 40%–50% expected for devolatilization at atmospheric pressure, but this small difference probably reflects the moderate temperature in the tests behind Fig. 9.30 (Tang et al., 2003). Under strictly comparable operating conditions under the same pressure of H_2 and an inert gas, one expects greater N-release from hydrogasification, primarily because the N-release during char conversion supplements that during hydropyrolysis. Such a comparison was staged by Xu and Kumagai (2002) to find that retentions of coal-N in char diminished by about 20% during hydrogasification of three hv bituminous coals, compared to the N-retention during devolatilization at the same pressure.

In most respects, coal-N release abides by the same reaction processes during hydrogasification as in devolatilization, with two important differences. First, NH_3 is the predominant gaseous N-species under all commercial hydrogasification conditions. It has not yet been determined whether this reflects different reaction mechanisms in the condensed coal phase or secondary chemistry under the excessively reducing atmosphere of hydrogasification. But this distinction is inconsequential because NH_3 will inevitably predominate under such highly reducing conditions, regardless of the primary volatile-N species. The second difference is that extents of N-release during hydropyrolysis are supplemented by the contribution for char

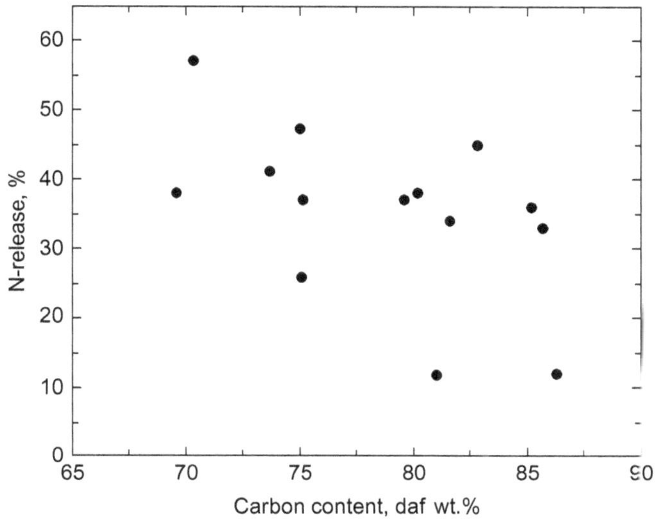

Fig. 9.30 Extents of N-release from diverse coals for rapid heating to 750°C under 6 MPa H_2 with 42 s gas contact time (Tang et al., 2003).

hydrogasification. Presumably, this supplement is directly proportional to extents of char conversion, as reported for one of the few test series that directly monitored it (Xu and Kumagai, 2002).

9.5.4.2 S-species transformations

The partitioning of coal-S among char-S, tar-S, and gas-S appears in Fig. 9.31 for three bituminous coals rapidly heated in a FFR under H_2 pressures to 5 MPa. As expected, S-release increases both for progressively hotter temperatures and for higher H_2 pressures as char-S is converted into gas-S, which was exclusively H_2S in all these tests. Features in these contributions closely mimic each other with changes of equal magnitude in opposite directions. Tar-S levels rarely exceed 10% and show no consistent tendencies for variations in both temperature and H_2 pressure. Only coal BR exhibits the expected tendency for tar-S to vanish at the hottest temperatures and highest pressures. It is most sensible to regard the tar-S levels across the pressure sweep as uniform within the measurement uncertainties, because the levels of secondary tar and oils would have been fairly uniform over this pressure range (cf. Fig. 9.22). At such a moderate temperature and short gas contact time, thiophene-S in secondary tar could not be released at even the highest H_2 pressures in the series. But the tar-S levels are similarly uniform across the temperature sweep with samples TH and SM, which is difficult to explain because more secondary tar is converted into oils for progressively hotter temperatures (cf. Fig. 9.23). Based on the conditions used to recover tar in these tests, the tar-S levels included sulfur in both secondary tar and oils. So uniform tar-S levels

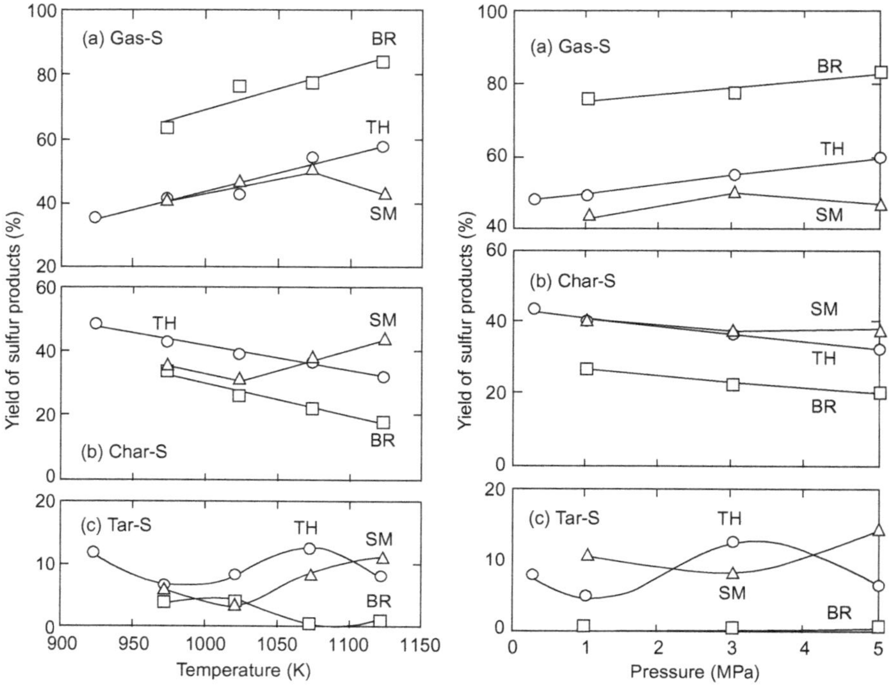

Fig. 9.31 S-partitioning from three hv bituminous coals labeled as BR, TH, and SM (left) vs. temperature for rapid heating under 3 MPa H_2 with 2 s gas contact time; and (right) vs. H_2 pressure for rapid heating to 800°C with coal and gas contact times of 2 and 3 s, respectively. Reproduced from Xu W-C, Matsuoka K, Akiho H, Kumagai M, Tomita A. Sulfur transformation during rapid hydropyrolysis of coal under high pressure by using a continuous free-fall pyrolyzer. Fuel 2003b;82:245–54, with permission from Elsevier.

across the temperature sweep with some coals could be an indirect indication that thiophene-S may be incorporated into oils, although this important implication remains to be verified.

The transformations of sulfur in char are resolved in the S-distributions in Fig. 9.32. Organic-S was released exclusively as H_2S for this particular coal since the tar-S levels were negligible across the sweeps in temperature and H_2 pressure. Pyritic-S was directly converted into FeS, although the level of pyrite in the coal was too low to resolve other products. Evidently, the complete decomposition of pyrite into metallic Fe and H_2S expected under elevated H_2 pressures did not materialize at the moderate temperatures and brief contact times in these tests. Sulfate-S was mostly converted into sulfite-S, and it appears that hotter temperatures are required for complete conversion. Tests that imposed the same operating conditions under 3 MPa H_2 and He found that the S-transformation in char were the same with three

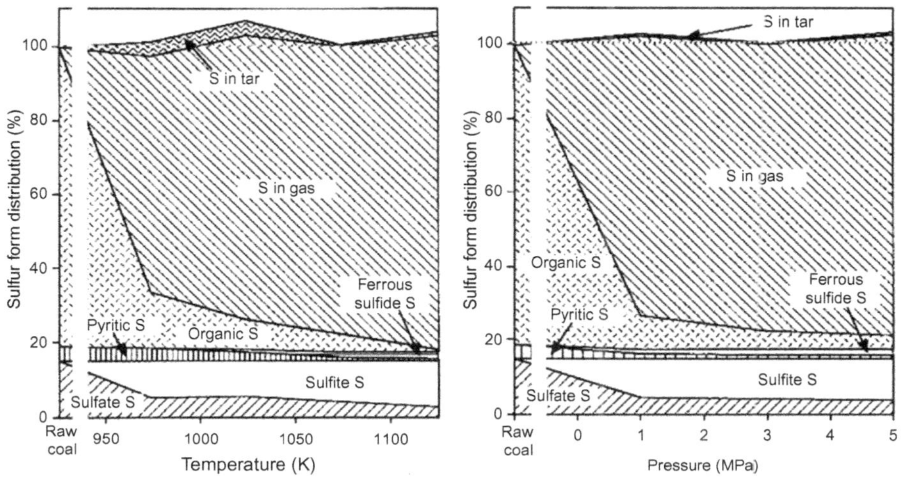

Fig. 9.32 S-species distributions from a hv bituminous coal (left) vs. temperature for rapid heating under 3 MPa H_2 with 2 s gas contact time; and (right) vs. H_2 pressure for rapid heating to 800°C with coal and gas contact times of 2 and 3 s, respectively.
Reproduced from Xu W-C, Matsuoka K, Akiho H, Kumagai M, Tomita A. Sulfur transformation during rapid hydropyrolysis of coal under high pressure by using a continuous free-fall pyrolyzer. Fuel 2003b;82:245–54, with permission from Elsevier.

hv bituminous coals (Xu et al., 2003b). Organic-S levels in char were 15%–20% lower under H_2, as expected for the supplement to S-release from char-S hydrogasification.

These same tendencies are apparent in the complete S-distributions reported for three diverse samples during slow heating to 500°C with hour-long soaking under 2.9 MPa H_2 (Sugawara et al., 1988). Even with such long coal contact times, FeS was the primary Fe-product of pyrite decomposition, presumably because of the very low temperature.

As with coal-N transformations during hydrogasification, organic-S release abides by the same reaction processes during hydrogasification as in devolatilization, albeit with one important difference because H_2S is the predominant gas-S species under both atmospheres. The main difference is that extents of S-release during hydropyrolysis are supplemented by the contribution for char hydrogasification. Presumably, this supplement is directly proportional to extents of char conversion, as reported for one of the few test series that directly monitored it (Xu et al., 2003b). In addition, pyrite decomposes into FeS with short contact times at moderate temperatures. However, S-transformations during the extended coal contact times for batch hydrogasification processing have not yet been characterized at hydrogasification temperatures, so pyrite conversion into metallic-Fe and H_2S have not yet been reported. Sulfate-S is completely converted into sulfite-S during hydrogasification.

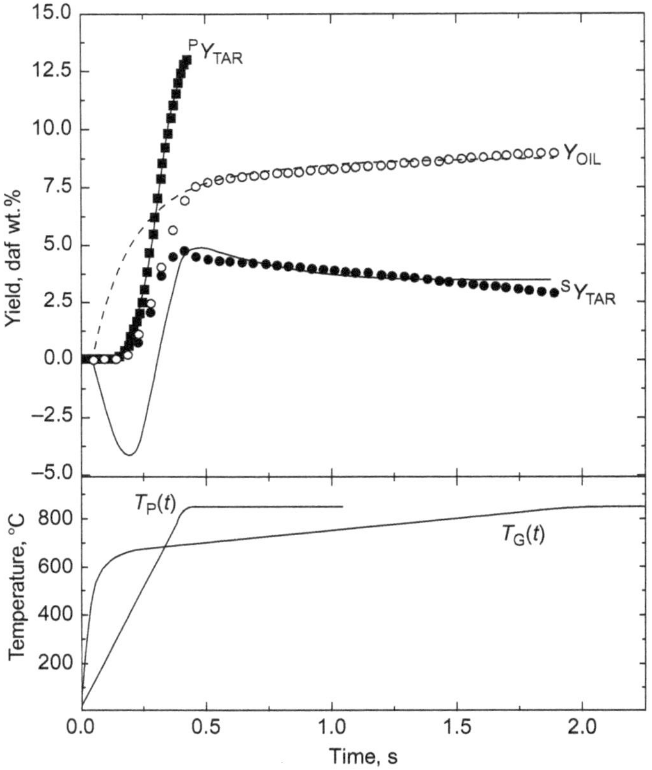

Fig. 9.33 Predicted yields of primary and secondary tar and oils from *(points)* the full tar hydroconversion mechanism and *(curves)* the 3-step global scheme in Eqs. (9.6a)–(9.6c) for hydrogasification with a bituminous coal heated at 200°C/s to 850°C under 15 MPa H_2. The lower panel shows the thermal histories for coal and gas in the simulation.

9.6 Global rates for tar hydroconversion and oils production at moderate temperatures

Like the nominal rates of tar decomposition developed in Chapter 7, simple first-order reactions and competitive reactions in parallel are fundamentally inconsistent with the dynamics of tar hydroconversion. Our aim is to predict the yields of secondary tar and oils throughout tar hydroconversion. To illustrate the complexity, the predicted transient yields of primary and secondary tar and oils for typical hydrogasification conditions appear in Fig. 9.33. The coal was heated at 2000°C/s to 850°C, to represent the rapid heating rates imposed on coal suspensions near the injectors in hydrogasification reactors. The gas temperature history rapidly increased to 650°C at the injection point, then gradually increased to 850°C at the reactor outlet. The H_2 pressure is 15 MPa. As seen in the simulation results from the full reaction mechanism, tar hydroconversion resembles a sequential hydrocracking process in which primary tar is converted into secondary tar, a portion of which is hydrogenated into lighter oils. But the form of the secondary tar yield in time is clearly not amenable to representation with a SFOR, because conventional first-order rate expressions necessarily relax to some ultimate yield parameter, V^{∞}, without any overshoot. In fact, it is impossible to depict the secondary tar history in Fig. 9.32 with any of the rate laws applied to primary devolatilization. The scheme developed in the next section accurately describes levels of secondary tar and oils throughout tar hydroconversion at moderate temperatures.

To identify a suitable rate law for tar hydroconversion at moderate temperatures, first notice that the difference between the yields of primary and secondary tar in Fig. 9.33 exhibits the characteristic dynamic behavior of a SFOR, initially growing at an accelerating rate before it relaxes to a saturation value. Hence, the release of primary tar into the gas phase during hydropyrolysis, secondary tar conversion, and oils production are well described by the following three reactions:

$$\frac{d^1Y_T}{dt} = {}^1R_T = {}^1k_T\left({}^1Y_T^{\infty} - {}^1Y_T(t)\right) \tag{9.6a}$$

$$\frac{d\Delta Y_T}{dt} = {}^2RT = k_T\left(\Delta Y_T^{\infty} - \Delta Y_T(t)\right) \tag{9.6b}$$

$$\frac{dY_{OILS}}{dt} = R_{OILS} = k_{OILS}\left(Y_{OILS}^{\infty} - Y_{OILS}(t)\right) \tag{9.6c}$$

The first rate in Eq. (9.6a) describes the release of primary tar with a conventional SFOR; and the second describes the difference between the instantaneous primary and secondary tar yields; i.e., $\Delta Y_T(t) = {}^1Y_T(t)\,{}^2Y_T(t)$. Once ${}^1Y_T$ and ΔY_T are evaluated, ${}^2Y_T(t)$ is determined as ${}^1Y_T - \Delta Y_T$. The third rate describes oils production in terms of the ultimate oils yield, Y_{OILS}^{∞}.

These global reactions contain three rate constants, ${}^1k_T$, k_T, and k_{OILS}, each in Arrhenius form, so there are three activation energies, ${}^1E_T$, E_T, and E_{OILS}; three pseudo-frequency factors, ${}^1A_T$, A_T, and A_{OILS}, and three hypothetical ultimate yield parameters, ${}^1Y_T^{\infty}$, ΔY_T^{∞}, and Y_{OILS}^{∞} to assign. The parameters for the devolatilization rate of primary tar are specified with the method of Niksa and Lau (1993) from FLASHCHAIN® predictions for hydropyrolysis. Then the predicted values of ΔY_T(t) and Y_{OILS} from the tar hydroconversion mechanism are analyzed in the same way for the second and third sets of parameters. The ultimate oils yield is specified as the final predicted value in the tar hydroconversion simulation. The trends in these parameter values for different heating rates are the same as those in the global rates for tar decomposition (cf. Fig. 7.22). Ultimate yields for primary tar and, therefore, ultimate oils yields change for different operating conditions (and also for different coals). This is not a limitation because sets of rate parameters for primary and secondary tars and ultimate tar and oils yields can be evaluated from FLASHCHAIN® for each set of conditions in seconds on ordinary personal computers, given only the proximate and ultimate analyses of the parent coals. In addition to the rate parameters, the detailed mechanism also specifies the mass-based stoichiometric coefficients for all secondary tar, oils, and noncondensable products of tar hydroconversion, in g/g primary tar, like those in Table 7.5. These stoichiometric coefficients simply multiply the secondary tar hydroconversion rate, $d\Delta Y_{TAR}/dt$, to evaluate the product formation rates.

9.7 Summary

Hydrogasification converts coal under elevated H_2 pressures at moderate temperatures into oils, CH_4 and other light GHCs, moisture, and the carbon oxides along with residual char and minor amounts of H_2S and NH_3. Tars can be eliminated almost completely because the extremely reducing atmosphere thoroughly disrupts the spontaneous conversion of tar into soot. Elevated H_2 pressures may also promote more extensive devolatilization by enhancing tar yields, and gasify char into CH_4 when coal contact times are extended. Consequently, hydrogasification product distributions reflect how all the major stages of coal conversion – primary hydropyrolysis, tar hydroconversion, secondary volatiles reforming, and char hydrogasification – operate under elevated H_2 pressures. Not surprisingly, the complex interplay among these four mechanisms confounds facile interpretations, and accurate interpretations of hydrogasification product distributions require robust mechanisms for all four stages.

Across a broad range of H_2 pressure, weight loss often passes through a minimum near 1 MPa then increases in direct proportion to the H_2 pressure. The minimum reflects the acute sensitivity to pressure in the primary tar yields for the lowest test pressures and the positive reaction order for H_2 pressure in the char hydrogasification kinetics. Higher H_2 pressures give greater extents of hydrogasification, which counteracts the lower tar yields for progressively greater pressures with rapid heating. Clearly, hydropyrolysis and hydrogasification must be resolved as distinct conversion mechanisms because the time scale for hydrogasification is approximately two orders of magnitude longer than times for hydropyrolysis. Hydrogasification rates become appreciable around 600°C for contact times of several seconds, then extents of hydrogasification continuously increase for progressively hotter temperatures. They also increase for progressively higher H_2 pressures and longer contact times, and diminish for progressively larger particle sizes. Hydrogasification does not become slower for coals of progressively higher rank, like the rates of gasification by steam and CO_2.

Hydropyrolysis enhances tar yields, but only if heating rates are much slower than 10^3°C/s and H_2 pressures are >1 MPa. Weight loss grows for progressively faster heating rates at the lowest pressures; is nearly neutral at intermediate pressures; and then diminishes for progressively faster heating rates at the highest H_2 pressures. Tar yields are enhanced by faster heating at lower pressures; insensitive to heating rate at intermediate pressures; and diminish for progressively faster heating at the highest H_2 pressures. So for progressively greater H_2 pressures, the variations in tar yields due to faster heating change from yield enhancements to negligible changes to yield suppression. Chapter 4 presented the shift from tar yield enhancements due to faster heating to negligible changes at elevated pressures for primary devolatilization under inert atmospheres (due to hindered flash distillation of tar precursors). Elevated H_2 pressures invert the heating rate dependence.

The key to understanding the crucial role for heating rate is that hydrogenation chemistry in the condensed coal phase is relatively slow compared to the rate of spontaneous primary devolatilization chemistry. The inverted heating rate dependence at the highest H_2 pressures indicates that the hydrogenation of coal components can be more forceful than the suppression of tar release by hindered flash distillation, and thereby enhance the tar yields. Most importantly, hydrogenation wins this competition only at the slowest heating rates, because hydrogenation is slow chemistry. Only the tests with the slowest heating rates provide enough time for hydrogenation to appreciably enhance the tar yields before the coal components are converted by spontaneous devolatilization channels. At the fastest heating rates, there is too little time for hydrogenation to enhance tars yields before the spontaneous channels take over, so tar yields diminish monotonically for progressively greater pressures, as they would under inert atmospheres.

Three independent mechanisms govern primary hydropyrolysis, (1) Inward H_2 penetration against the outward flux of volatiles; (2) The direct promotion of bridge scission by bridge hydrogenation; and (3) The suppression of bimolecular recombination by hydrogenated peripheral groups on fragment ends. The first mechanism never affects tar production, because intraparticle H_2 pressures equal the ambient values for all heating rates that provide sufficient time for the hydrogenation of coal components. Conversely, the impacts from both hydrogenation and suppression of recombination are substantial under the heating conditions that provide sufficient contact time to enhance tar yields. Bridge hydrogenation increases the average number of H-atoms per bridge which, in turn, promotes bridge scission and the associated accumulation of tar precursors. Suppressing bimolecular recombination inhibits the growth of tar fragments out of the size range for tar precursors, and thereby promotes tar production.

Generally speaking, FLASHCHAIN® for hydropyrolysis exposes the term "flash hydropyrolysis" as the oxymoron that it is. Rapid or "flash" heating is incompatible with appreciable enhancements of primary tar yields during hydropyrolysis because the slow hydrogenation chemistry cannot keep pace with the relatively very fast time scale for primary devolatilization. Since nominal rates for primary devolatilization increase in proportion to increases in heating rate, and since primary devolatilization chemistry is spontaneous, hydrogenation can only affect the distribution of primary products under slow heating conditions, because only slow heating enables bridge hydrogenation to occur before the bridges decompose through the spontaneous channels.

Tar hydroconversion largely explains why the ultimate product distributions from hydropyrolysis are radically different than those for devolatilization under inert gases with or without tar decomposition. Oils yields are maximized within a temperature window that shifts toward cooler temperatures for progressively higher H_2 pressures and, to a lesser extent, for progressively longer gas contact times. For example, the window spans 725 to 775°C for pressures of several MPa H_2. The lower temperature must be sufficient to achieve a truly ultimate primary hydropyrolysis tar yield, because oils yields always increase in tandem with the primary tar yield. The upper temperature must be cool enough to minimize oils hydrocracking. Otherwise, oils are hydrocracked into CH_4. Under the most favorable conditions, essentially all secondary tar is hydroconverted to achieve the maximum oils yield, while negligible amounts of oils are hydrocracked into CH_4. For progressively more severe hydroconversion conditions, oils first lose their substituents, then hydrocrack into GHCs and moisture. GHCs heavier than C_2's are reformed first, followed by the C_2's, so that CH_4 is the ultimate noncondensable hydrocarbon product under the most severe hydrogasification conditions. The carbon oxides are relatively minor repositories for tar-O, most of which ends up in moisture.

According to the FLASHCHAIN®-based tar hydroconversion mechanism, oils form in two stages controlled by monomer hydrogenation and hydrocracking, respectively. The hydrogenation of labile bridges and peripheral groups is relatively unimportant. In the first stage, the tar monomers released as primary tars are rapidly hydrogenated into oils at the monomer hydrogenation rate. Since elevated pressures always shift primary tar MWDs toward lighter species, monomers constitute substantial fractions of primary tar, approaching one-half under some operating conditions. Consequently, as much as half the ultimate oils yield is produced soon after the onset of tar hydroconversion. In the second stage, additional tar monomers are gradually released by hydrocracking of larger tar molecules and then hydrogenated into oils, while control of the oils production rate shifts from monomer hydrogenation to hydrocracking. The mechanism has been validated across a broad range of coal and gas contact times, and correctly predicts that secondary tar persists for all but the highest H_2 pressures.

Primary tar compositions and, especially, their structural components, determine oils yields from different coals. The main trend is that oils yields pass through a weak maximum for hv bituminous coals, then decay sharply for low volatility coals. The most interesting aspect of the coal quality impacts is the exceedingly large percentage of the material in aromatic nuclei in tar incorporated into oils under the most favorable operating conditions with any coal type. The notions that aromatic nuclei in primary tar, which have two to four rings with bituminous coals and fewer or more for coals of lower and higher rank, respectively, are either winnowed down or completely converted into single-ring compounds are incompatible with reported ultimate oils yields. Monomer hydrogenation must also incorporate aliphatic peripheral groups into oils as well.

In most respects, release of coal-N and coal-S abide by the same reaction processes during hydrogasification as in devolatilization. Ammonia and H_2S are the predominant gaseous species under all commercial hydrogasification conditions. Extents of both N- and S-release during hydropyrolysis are supplemented by the contributions from char hydrogasification, which are probably directly proportional to extents of char conversion. In addition, pyrite decomposes into FeS with short contact times at moderate temperatures. However, the extended coal contact times for batch hydrogasification processing have not yet been characterized at hydrogasification temperatures, so pyrite conversion into metallic-Fe and H_2S has not yet been reported. Sulfate-S is completely converted into sulfite-S during hydrogasification.

Char hydrogasification kinetics are accurately depicted by a single half-order surface reaction within the CBK/G framework. The assigned activation energy of 81.5 kJ/mol and the order of one-half are secure but, as with all CBK kinetics, the pseudo-frequency factor for hydrogasification must be tuned-in to calibration data for each sample. For the bulk of the validation coals, the assigned hydrogasification reactivity values show no consistent tendencies with rank. The most remarkable feature is that the values for all but coals of highest rank lie within 50% of a mean value. In contrast, the frequency factors for gasification in steam and CO_2 vary by as much as two orders of magnitude for coals of the same rank, and those for char oxidation vary by up to an order of magnitude (Niksa et al., 2003; Liu and Niksa, 2004). Both gasification rates accelerate for coals of progressively lower rank. However, most hydrogasification reactivities vary by only a factor of four, and show no consistent trend with rank. The fundamentally different rank independence suggests that hydrogasification may not share any common adsorbed intermediates or primary reaction sites with the surface chemistry for gasification by steam and CO_2.

However, this interpretation does not cover the assignments for the hv bituminous coals tested by Tang et al. (1999) which are 5–20 times greater than the rest (cf. Fig. 9.8). In isolation, the assignments for Tang et al.'s dataset appear to indicate that hydrogasification reactivities increase for progressively higher coal rank until they diminish for low volatility coals. But in the context of all the assignments with CBK/G only Tang et al.'s bituminous coals have markedly faster reactivities.

References

Anthony DB, Howard JB, Hottel HC, Meissner HP. Rapid devolatilization of pulverized coal. Proc Combust Inst 1975;15:1303.

Arendt P, van Heek K-H. Comparative investigations of coal pyrolysis under inert gas and H_2 at low and high heating rates and pressures up to 10 MPa. Fuel 1981;60(9):779–87.

Bolton C, Snape CE, O'Brien JO, Kandiyoti R. Influence of carrier gas flow and heating rates in fixed bed hydropyrolysis of coal. Fuel 1987;66:1413–7.

Bolton C, Riemer C, Snape CE, Derbyshire FJ, Terrer MT. Effect of low temperature catalytic hydrogenation on pyrolysis and hydropyrolysis of a bituminous coal. Fuel 1988;67:901–5.

Chareonpanich M, Takeda T, Yamashita H, Tomita A. Catalytic hydrocracking reaction of nascent coal volatile matter under high pressure. Fuel 1994;73:666–70.

Eklund H, Wanzl W. Pyrolysis and hydropyrolysis of solid fuels at high heating rates using Curie point technique, In: Int Conf Coal Sci. IEA; 1981. p. 701–7.

Fallon PT, Bhatt B, Steinberg M. The flash hydropyrolysis of lignite and subbituminous coals to both liquid and gaseous hydrocarbon products. Fuel Process Technol 1980;3:155–68.

Finn MJ, Fynes G, Ladner WR, Newman JOH. Light aromatics from the hydropyrolysis of coal. Fuel 1980;59:397–404.

Fynes G, James RG, Ladner WR, Newman JOH. Structural differences in the tars and char from the pyrolysis of coals of different rank in hydrogen and in nitrogen. Fuel 1984;63:897–903.

Gibbins JR, Gonenc ZS, Kandiyoti R. Pyrolysis and hydropyrolysis of coal: Comparison of product distributions from a wire-mesh and hot-rod reactor. Fuel 1991;70:621–6.

Guan Q, Dapeng B, Xuan W, Zhang J. Kinetic model of hydropyrolysis based on the CPD model. Fuel 2015;152:74–9.

Guell AJ, Kandiyoti R. Development of a gas sweep facility for the direct capture of pyrolysis tars in a variable heating rate high pressure wire mesh reactor. Energy Fuel 1993;7:943–52.

Guell AJ, Wu F, Li C-Z, Madrali ES, Dugwell DR, Kandiyoti R. Effect of H_2-pressure on yields and structures of liquids from the hydropyrolysis of maceral concentrates. Fuel Process Technol 1993;36:327–32.

Harris DJ, Patterson JH. Use of Australian bituminous coals in IGCC power generation technologies. Aust Inst Energy J 1995;13:22.

Heyd LE. Weight loss behavior of coal during rapid pyrolysis and hydropyrolysis. M.S. thesis, Princeton, NJ: Princeton University; 1982.

Howard JB. In: Elliot MA, editor. Chemistry of coal utilization. New York: Wiley-Interscience; 1981. 2nd Suppl. Vol. [chapter 12].

Ikura M, Last AJ. Flash hydropyrolysis of bituminous caking coal. Fuel Process Technol 1988;20:257–68.

Klavetter E, Mitchell SC, Snape CE. High tar yields from low-rank coals in non-catalyzed hydropyrolysis, In: International conference on coal science, Vol. 1. IEA; 1993.

Li B, Mitchell SC, Snape CE. Effect of heating rate on normal and catalytic fixed-bed hydropyrolysis of coals. Fuel 1996;75:1393–6.

Liu G-S, Niksa S. Coal conversion submodels for design applications at elevated pressures. Part II. Char gasification. Prog Energy Combust Sci 2004;30(6):697–717.

Moliner R, Ibarra JV, Lagarma MD. Size exclusion chromatography of pyrolysis and hydropyrolysis tars from Spanish low rank coal. Fuel 1989;68:1487–8.

Niksa S. Flashchain theory for rapid coal devolatilization kinetics. 4. Predicting ultimate yields from ultimate analyses alone. Energy Fuel 1994;8:659–70.

Niksa S, Lau C-W. Global rates of devolatilization for various coal types. Combust Flame 1993;94:294–307.

Niksa S. Interpreting coal conversion under elevated H_2 pressures with FLASHCHAIN® and CBK, In: Proc. 2011 int. conf. on coal science and technolOviedo, Spain: IEA; 2011.

Niksa S. Flashchain theory for rapid coal devolatilization kinetics. 10. Extents of conversion for hydropyrolysis and hydrogasification of any coal. Energy Fuel 2018a;32:384–95.

Niksa S. Flashchain theory for rapid coal devolatilization kinetics. 11. Tar hydroconversion during hydrogasification of any coal. Energy Fuel 2018b;32:7569–84.

Niksa S, Liu G-S, Hurt RH. Coal conversion submodels for design applications at elevated pressures. Part I. Devolatilization and char oxidation. Prog Energy Combust Sci 2003;29 (5):425–77.

Stanczyk K, Kisielow W. Influence of hydropyrolysis parameters on char yield, properties, and combustion reactivity. Fuel 1994;73:660–7.

Steinberg M. The flash hydropyrolysis and methanolysis of coal with hydrogen and methane. Int J Hydrog Energy 1987;12:251–66.

Strugnell B, Patrick JW. Hydropyrolysis yields in relation to coal properties. Fuel 1995;74:481–6.

Strugnell B, Patrick JW. Rapid hydropyrolysis studies on coal and maceral concentrates. Fuel 1996;75:300–6.

Sugawara T, Sugawara K, Ohashi H. Dynamic behavior of sulfur forms in hydropyrolysis of coals. Fuel 1988;67:1263–8.

Suuberg EM. In: Schlosberg RH, editor. Chemistry of coal conversion. New York: Plenum; 1985 [chapter 4].

Tang L, Zhu Z, Gu H, Zhang C. The effect of coal rank on flash hydropyrolysis of Chinese coal. Fuel Process Technol 1999;60:195–202.

Tang L, Zhu Z, Zhu H, Zhang C. Study of coal flash hydropyrolysis denitrogentation. Fuel Process Technol 2003;81:103–8.

Wakao N, Smith JM. Diffusion and reaction in porous catalysts. Ind Eng Chem Fundam 1964;3:123.

Xu W-C, Kumagai M. Nitrogen evolution during rapid hydropyrolysis of coal. Fuel 2002;81:2325–34.

Xu WC, Tomita A. Effect of coal type on the flash pyrolysis of various coals. Fuel 1987;66 (5):627.

Xu W-C, Matsuoka K, Akiho H, Kumagai M, Tomita A. High pressure hydro-pyrolysis of coals by using a continuous free-fall reactor. Fuel 2003a;82:677–85.

Xu W-C, Matsuoka K, Akiho H, Kumagai M, Tomita A. Sulfur transformation during rapid hydropyrolysis of coal under high pressure by using a continuous free-fall pyrolyzer. Fuel 2003b;82:245–54.

Zhu Z, Ma Z, Zhang C, Jin H, Wang X. Flash hydropyrolysis of northern Chinese coal. Fuel 1996;75:1429–33.

Further reading

Kaiser M, Wanzl W. Characterization and comparison of liquid products from coal pyrolysis in laboratory and process development units. Fuel Process Technol 1988;20:23–32.

Suuberg EM, Peters WA, Howard JB. Product composition and kinetics of lignite pyrolysis. Ind Eng Chem Process Des Develop 1978;17:37.

Index

Note: Page numbers followed by *f* indicate figures, *t* indicate tables, and *np* indicate footnotes.

D

R

S

T

U

V

www.ingramcontent.com/pod-product-compliance
Lightning Source LLC
Chambersburg PA
CBHW050748080126
37798CB00016B/2962
9780128187135